POLARIS Personal Watercraft

1992-97 REPAIR MANUAL

SELOC®

Managing Partners Dean F. Morgantini
Barry L. Beck

Executive Editor Kevin M. G. Maher, A.S.E.

Production Managers Melinda Possinger
Ronald Webb

Manufactured in USA

104 Willowbrook Lane
West Chester, PA 19382
ISBN-13: 978-0893300456
ISBN-10: 0893300454
4567890123 9876543210

www.selocmarine.com
1-866-SELOC55

SELOC PUBLISHING'S FULL-LINE MASTER LIST

	ISBN	PART NO	TITLE/DESCRIPTION	YEARS
OUTBOARDS				
089330018-7	018-7	1000	Chrysler Outboards, All Engines	1962-84
089330055-1	055-1	1100	Force Outboards, All Engines	1984-99
089330048-9	048-9	1200	Honda Outboards, All Engines	1978-99
089330007-1	007-1	1300	Johnson/Evinrude Outboards, 1-2 Cyl	1956-70
089330008-X	008-X	1302	Johnson/Evinrude Outboards, 1-2 Cyl	1971-89
089330026-8	026-8	1304	Johnson/Evinrude Outboards, 1-2 Cyl	1990-95
089330009-8	009-8	1306	Johnson/Evinrude Outboards, 3-4 Cyl	1958-72
089330010-1	010-1	1308	Johnson/Evinrude Outboards, 3, 4 & 6 Cyl	1973-91
089330040-3	040-3	1310	Johnson/Evinrude Outboards - All V Engines	1992-96
089330063-2		1310	Johnson/Evinrude Outboards - All V Engines	1992-01
089330052-7		1312	Johnson/Evinrude Outboards, All In-line engines/2 & 4 Stroke	1996-01
089330015-2	015-2	1400	Mariner Outboards, 1-2 Cyl	1977-89
089330016-0	016-0	1402	Mariner Outboards, 3, 4 & 6 Cyl	1977-89
089330012-8	012-8	1404	Mercury Outboards, 1-2 Cyl	1965-91
089330013-6	013-6	1406	Mercury Outboards, 3-4 Cyl	1965-89
089330014-4	014-4	1408	Mercury Outboards, 6 Cyl	1965-89
089330051-9	051-9	1416	Mercury/Mariner Outboards, All Engines	1990-00
089330050-0	050-0	1600	Suzuki Outboards, All Engines	1988-99
089330021-7	021-7	1700	Yamaha Outboards, 1-2 Cyl	1984-91
089330022-5	022-5	1702	Yamaha Outboards, 3 Cyl	1984-91
089330023-3	023-3	1704	Yamaha Outboards, 4 & 6 Cyl	1984-91
089330047-0	047-0	1706	Yamaha Outboards, All Engines	1992-98
STERN DRIVES				
089330029-2	029-2	3000	Marine Jet Drive	1961-96
089330005-5	005-5	3200	Mercruiser Stern Drive	1964-91
089330053-5		3206	Mercruiser Stern Drive - All	1992-01
089330004-7	004-8	3400	OMC Stern Drive	1964-86
089330025-X	025-X	3402	OMC Cobra Stern Drive	1985-95
089330056-X	025-X	3402	OMC Cobra Stern Drive	1985-99
089330011-X	011-X	3600	Volvo/Penta Stern Drives	1968-91
089330038-1	038-1	3602	Volvo/Penta Stern Drives	1992-93
089330041-1	041-1	3604	Volvo/Penta Stern Drives	1992-95
089330057-8	041-1	3604	Volvo/Penta Stern Drives	1992-02
INBOARDS				
089330049-7	049-7	7400	Yanmar Inboards	1975-98
PERSONAL WATERCRAFT				
089330032-2	032-2	9200	Kawasaki	1973-91
089330042-X	042-X	9202	Kawasaki	1992-97
089330045-4	045-4	9400	Polaris	1992-97
089330033-0	033-0	9000	Sea-Doo/Bombardier	1988-91
089330043-8	043-8	9002	Sea-Doo/Bombardier	1992-97
089330034-9	034-9	9600	Yamaha	1987-91
089330044-6	044-6	9602	Yamaha	1992-97
Seloc-On-Line (Internet Access)				
089330075-6	5000		One Mfg/Model - Subscription per user 3 years	1990-01
	5002		Master Pack -Case of 6 CD's including POP Display	
PROFESSIONAL TECHNICIANS MANUALS				
089330060-8		4500	Labor Guide - Johnson and Evinrude	1980-00
089330061-6		4550	Labor Guide - Yamaha	1980-00
089330062-4		4600	Labor Guide - Mercury	1980-00
BRIGGS AND STRATTON				
096376610-4	610-4	610	Briggs & Stratton 4 Stroke Horizontal Crankshaft	1950+
096376611-2	611-2	611	Briggs & Stratton 4 Stroke Vertical Crankshaft	1950+
096376612-0	612-0	612	Briggs & Stratton 4 Stroke Overhead Crankshaft	1950+

SelocOnLine

SelocOnLine is a maintenance and repair database accessed via the Internet.
Always up-to-date, skill level and special tool icons, quick access buttons to wiring diagrams, specification charts, maintenance charts, and a parts database.
Contact your local marine dealer or see our demo at www.seloconline.com

Evinrude / Force / Honda / Johnson / Mariner / Mercruiser
Mercury / OMC / Suzuki / Volvo / Yamaha / Yanmar

Purchase price provides access to one single year / model engine and drive system. The purchase of this product allows the user unlimited access for three years.

SelocOnLine

Step 1

Select your manufacturer

Year / Model

Engine

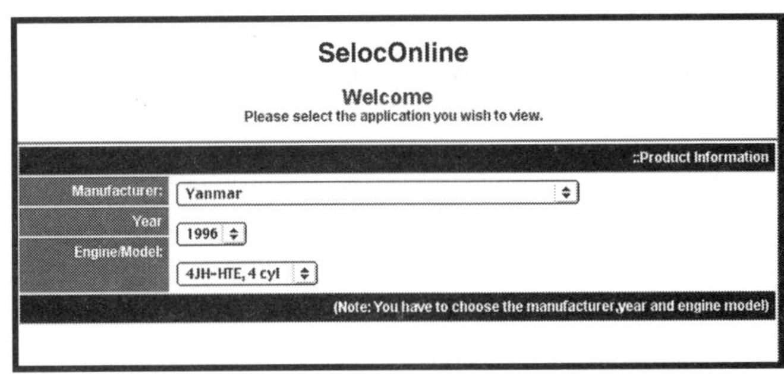

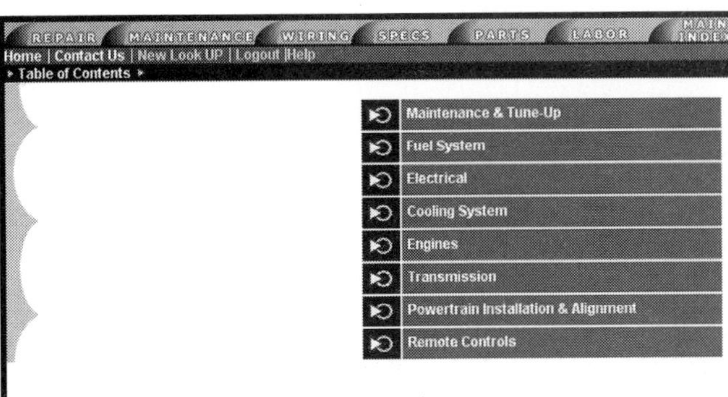

Step 2

Select an Engine System

Step 3

Select the repair

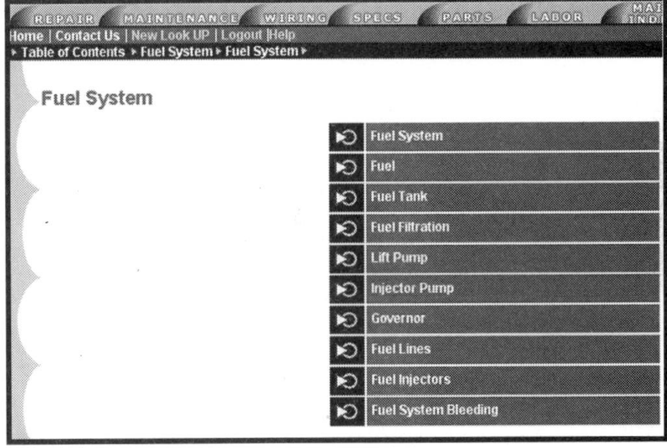

REPAIR | MAINTENANCE | WIRING | SPECS | PARTS | LABOR | MAIN INDEX

Home | Contact Us | New Look UP | Logout |Help

▸ Table of Contents ▸ Fuel System ▸ Fuel System ▸ Lift Pump ▸ Removal & Installation ▸

JH Series Engines

Fig. 12 On JH engines, the fuel lift pump mounts to the injector pump ▸ see image

1. Turn the fuel tank petcock to the **OFF** position.
2. Place a small receptacle under the lift pump to contain any excess fuel.
3. Remove the fuel line fittings attached to the lift pump. Be careful not to lose the copper washers.

Be sure to mark the fuel lines before they are removed from the pump. If the lines are reversed, the lift pump will not supply fuel to the injector pump.

4. Remove the two bolts that attach the lift pump to the engine block.
5. Draw the lift pump from the engine block.

To install:

6. Carefully scrape any gasket residue from the mating surfaces of the lift pump and injector pump.
7. Lightly grease the end of the lift pump piston where it rides on the cam.
8. Using a new gasket, install the lift pump to the injector pump.
9. Install the fuel lines to the lift pump in the proper direction.
10. Once the lines are attached, turn the fuel petcock to the **ON** position.
11. Loosen the air bleed screw on the secondary (between the lift pump and injector pump) fuel filter.
12. Operate the priming lever on the lift pump until fuel runs out of the air bleed screw opening, then tighten the screw.

Step 4

Print the repair procedure

Seloc OnLine Solutions (SOS)

FEATURES AND BENEFITS

Seloc Publishing has been an innovator in helping boaters and technicians maintain and repair outboard and inboard engines and/or drive systems for over 30 years. From performing actual teardown procedures to digital photography, Seloc has been there all the way. Now Seloc wants to assist you in solving those difficult problems. We are as close as your fingertips or your phone. Not 100% sure about that procedure? Email or phone us, we can help you with your engine and boat related questions.

REAL WORLD SOLUTIONS IN REAL TIME!!

- Technical questions answered by factory trained technicians.
- Response to questions in less that 24 hours.
- Priority Response System (ability to have technician call you in 30 minutes or less).
- Technical documentation and diagrams
- Monthly newsletter containing helpful hints, and "How to" features.
- Future access to on-line repository of SelocSOS inquiries

One visit to SelocSOS could potentially save you hundreds of dollars. Above all, our priority is to get you back in the water. You have invested thousands of dollars on your boating investment. Now it is time to get a return on your investment. Visit us at www.selocsos.com or selocmarine.com and see what Seloc can do for you.

TABLE OF CONTENTS

1
DESCRIPTION & OPERATION

1-1 BRIEF HISTORY

The jet drive system for propelling a craft through the water arrived on the scene in the mid 1960's with the jet drive boat. In those early days, the jet drive system was mated only with high performance powerplants -- engines in the 454 cu. in. class and larger. For this reason, during the "gas crunch" in the 1970's the jet drives were labeled as inefficient and as "gas hogs".

In addition to these two negative terms, they earned the reputation as "bad boy" boats due to their noisy "straight" exhaust, high rpm operation, and their almost unbelievable maneuverability. These combined factors did little to enhance their image and certainly restricted their popularity.

With new and improved technology, personal watercraft arrived on the scene about the mid 1970's. Personal watercraft, as we know them today, were developed using the same principles as the jet boats, and originally powered with a single cylinder two-stroke engine.

In order to meet the demand for more speed and the ability to carry more than just

A typical inner harbor summer weekend with scores of personal watercraft preparing to leave or just returning from a "fun day" on the water in the "outer harbor" or at sea close to shore. Just a reasonable amount of "TLC", will reward the owner and his/her friends with hours of trouble free enjoyment.

one person, watercraft manufacturers were quick to respond. Today, most craft are powered with a two or three-cylinder two-stroke powerplant, coupled to a single stage pump.

Aftermarket shops and manufacturers have come into existence all across the United States and Canada. Their output of specialty products and services permit the owner to gain more speed and win over the competition at racing events wherever enough water is available.

As mentioned in the "Foreword", this book has been designed and written to cover stock factory engines and jet drives. Modifications for higher than manufacturer's rated performance are so extensive and varied, no attempt has been made to include them in this volume. In such cases the publisher's recommendation is to follow the after market instruction with the particular product or service.

Series Covered

The following Polaris series produced from 1992 thru 1997 are covered in this manual.

Model	Approx. Yr. of Production
SL650 Series	1992 - 1995
SL750 Series	1993 - 1995
SLT750 Series	1994 - 1995
SLX780 Series	1995 Only
SL700 Series	1996 - 1997
SLT700 Series	1996 - 1997
Hurricane Series	1996 - 1997
SL780 Series	1996 - 1997
SLT780 Series	1996 - 1997
SL900 Series	1996 - 1997
SLTX Series	1996 - 1997
SL700 Deluxe	1997
SL1050 Series	1997

Other Seloc Personal Watercraft Manuals

This manual is only one in a series of Personal Watercraft manuals produced by Seloc Publications, as follows:

Volume I - Kawasaki -- early days thru 1991
Volume II - Bombardier SeaDoo -- early days
Volume IIA - Bombardier SeaDoo -- 1992 thru 1997
Volume III - Yamaha - early days thru 1991
Volume IIIA - Yamaha -- 1992 thru 1997
Volume IA - Kawasaki -- 1992 thru 1998

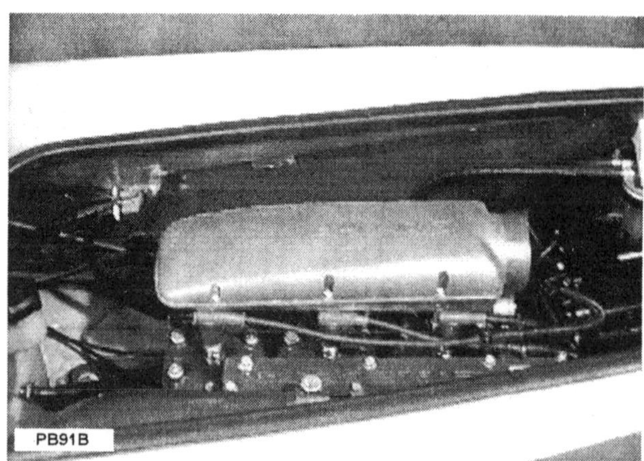

View of a "squared away" SLT750 Model with a three-cylinder installation, typical of all Models covered in this manual.

1-2 PRINCIPLES OF OPERATION

One of the first lessons to be learned in any elementary physics class is Newton's basic law: "For every force, there is an opposite and equal force". This statement is the basic principle of the jet pump. Water is "sucked" and "scooped" in from under the craft by a powerful pump rotating at incredible speed and then discharged, "blown" out, sternward in the opposite direction. In this manner the watercraft is propelled forward.

The personal watercraft covered in this manual are all equipped with a twin or 3-cylinder water cooled two-stroke engine, matched with a single stage (one impeller) jet pump.

On certain models, a reverse gate is swung down over the pump outlet nozzle forcing the exhausted water back in a forward direction thus moving the craft sternward.

Personal watercraft jet pumps may be classified as "axial flow" or "mixed flow". Today, all Polaris Personal Watercraft are equipped with an axial flow jet pump.

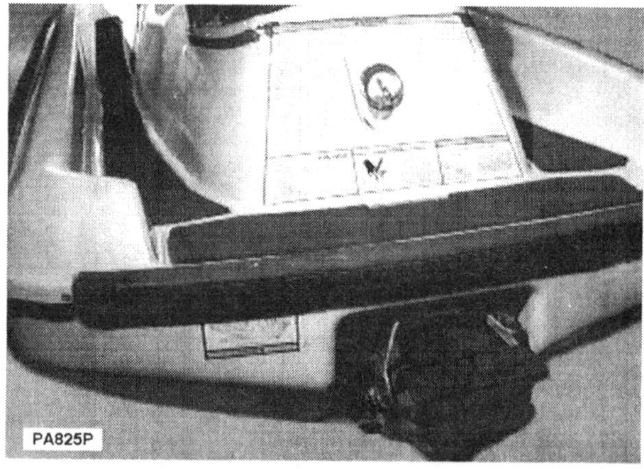

This watercraft is equipped with a reverse gate, shown in the full reverse position.

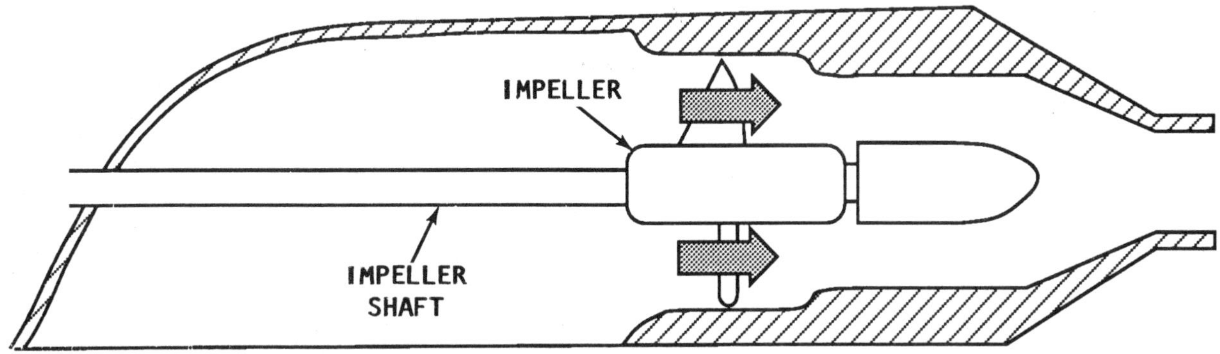

Cross-section line drawing to depict water flow through an axial flow jet pump. The water passes through parallel to the axis of impeller rotation.

Axial Flow

Water in an axial pump moves in a single axis, as depicted in the adjacent illustration -- thus the word "axial" is used. In simple layman's terms -- water is ingested and discharged parallel to the axis of impeller rotation, as shown.

1-3 ENGINE AND JET DRIVE

The basic principle and arrangement of engine and pump are almost identical for all manufacturers.

The engine crankshaft is coupled either directly to the pump shaft through a coupler containing a rubber "shock absorber" or through a short driveshaft without any gear reduction. An impeller mounted on the pump shaft draws water into an opening in the hull through a suction and intake casting. Once the craft has attained forward motion the intake also serves as a "scoop" adding to the volume of water moved through the pump.

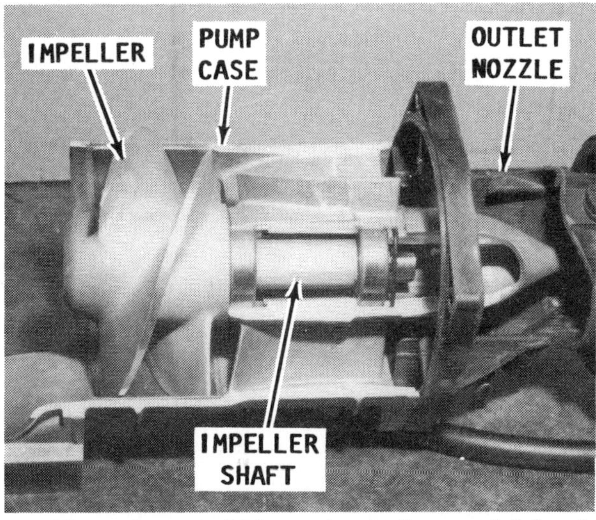

Photo of a typical "demonstration" type axial flow jet pump. A few major parts are identified.

Volume and Velocity

The outlet nozzle is slightly funnel shaped. This design coupled with the capacity of the pump (impeller) causes the volume of water entering the pump to exit the nozzle with increased velocity. This principle is similar to air passing through the venturi of a carburetor.

The amount of water ejected sternward from the nozzle at high velocity is the force propelling the craft through the water. This force can actually be measured and calculated in foot pounds or Newton meters. The greater the velocity of water mass moving through the nozzle, the greater the thrust to move the craft.

The vanes in the pump case straighten water flow and the conical shape of the outlet nozzle increase flow velocity in much the same way as air passing through a carburetor venturi. The pump shown was used extensively in salt water, and was not cleaned regularly. Corrosion decreases performance and may lead to expensive repair or replacement.

Cooling System

A fitting on the jet pump, aft of the impeller, siphons off cooling water. This water is delivered to the engine and exhaust system via a tee fitting.

Good Words

On some early SL650 Models, a Cooling Kit is available, using a water manifold separator, to distribute cooling water. This cooling system was carefully engineered to distribute water properly, via carefully sized passages, fittings and hoses. If this kit has been installed and cooling hoses must be replaced, **NO** deviation in hose size or fitting size is allowable. Severe overheating in the exhaust system or engine may occur.

Water passes through the exhaust manifold first, then enters the crankcase, is channeled around the cylinder walls, cylinder head then directed to the outlet manifold, atop the cylinder caps. An outlet hose delivers the water to the exhaust thru-hull fitting, where the water exits the craft.

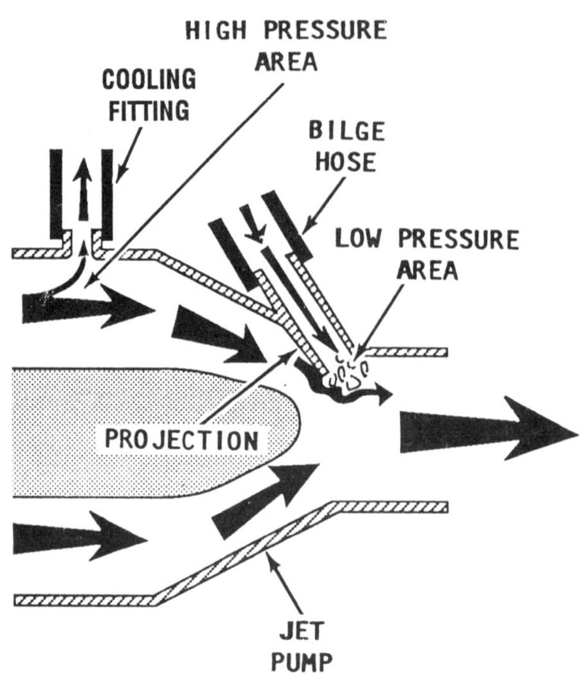

The direction of water flow through the cooling hose is opposite to the water flow moving through the bilge hose. The cooling hose routes water from the pump to the engine. Water is actually "vacuumed" from the bilge through the bilge hose to the pump where it is forced out through the nozzle.

Cooling Water and Bilge Hoses

On most pumps, two hoses -- or a hose and one tube outlet fitting -- are attached to the pump case and outlet nozzle. The tube outlet fitting or hose (varies by Model) routes water flowing through the pump to the exhaust manifold -- the hottest part of the engine -- to cool the block during operation. The other hose fitting siphons water out of the bottom of the hull, and is referred to as the bilge system. Some models may have two bilge siphon fittings on the nozzle.

Water flows **FROM** the jet pump to the engine for cooling, the bilge system siphons water **TO** the jet pump.

What determines the direction of water flow? The answer to this question is in a simple explanation of high and low pressure areas along the inside surface of the jet pump.

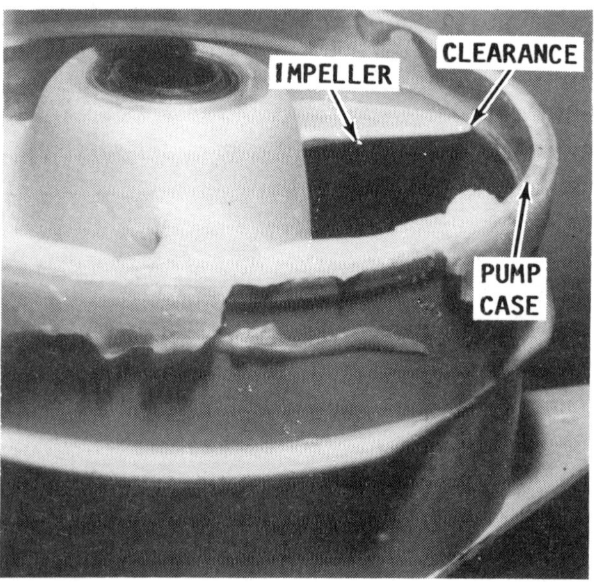

A feeler gauge may be used to measure the clearance between the impeller blades and the pump case.

Close view of the nozzle with the cooling water tube and the bilge hose fittings clearly visible.

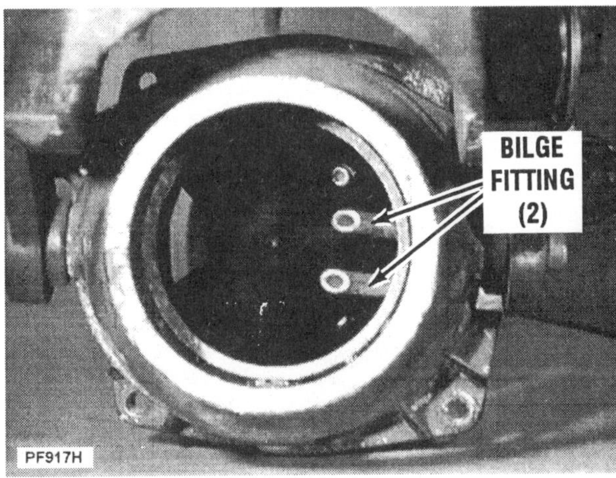

The bilge fitting protrusion extends into the water flow area. Water passing through causes a low pressure area a "vacuuming" -- siphoning -- effect, sucking water out of the bilge.

Close inspection of the area inside the pump at the cooling water hose fitting reveals a smooth rounded shoulder with no obstruction or obstacle to impede the flow of water. A high pressure area develops here and draws the water down the hose attached to the fitting.

Further inspection of the area inside the hosing at the bilge hose fitting reveals a small protrusion around the opening. This protrusion causes disturbance to the water flow and creates an area of low pressure. This low pressure area will have the effect of emptying air and water from the hose aft with the impeller water flow -- similar to the action of a vacuum cleaner, as depicted in the accompanying illustration. If the other end of the hose is submerged in water inside the hull, the water will be "vacuumed" out. In this manner the bilge system drains the bilge.

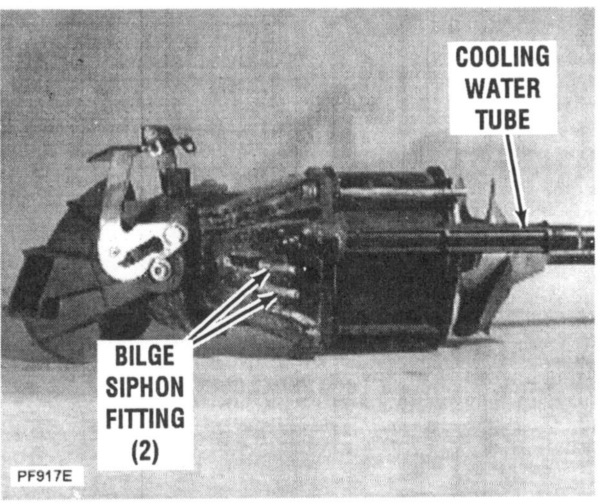

Location of the bilge siphon fittings attached to the outlet nozzle. Other pumps in the Series have only one siphon fitting. The cooling water tube is also shown.

Many an owner -- with good intentions -- thinking his action would smooth water flow and increase pump performance -- has filed the protrusion described in the previous paragraph smooth with the surrounding area of the pump. During operation of the water craft, the bilge system worked in **REVERSE.** Water was pumped from the impeller housing into the bilge area, quickly filling the bilge and possibly sinking the craft before the rider had time to realize what was happening.

Bilge Breather Fitting

A bilge breather fitting is installed at the highest point of the bilge line. This fitting has a small breather hole to prevent siphoning water from the pump into the bilge. If the engine was shut down with the bilge hose filled with water and the jet pump fitting was positioned higher then the end of the bilge line, water would flow back and fill the bilge area. The small breather hole in the fitting prevents this back-siphoning.

*A bilge breather fitting may be installed in different locations, depending on the engine model. However, it is **ALWAYS** installed at the highest point in the bilge hose. The fitting has a small "breather" hole which **MUST** be kept clean and unobstructed.*

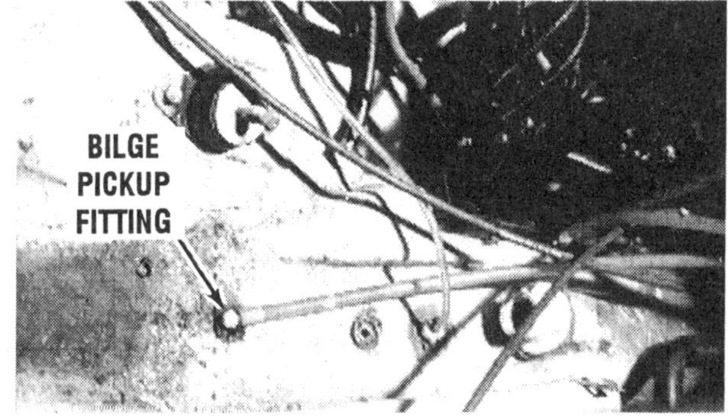

Location of the bilge pickup in the lowest part of the engine compartment. (Photograph taken with the engine removed for clarity.)

1-4 IMPELLERS

Impellers provide thrust using a combination of water flow and water pressure in a closed environment.

The blades of an impeller overlap, thus water is trapped and forced through the impeller into the pump case and nozzle. The water moving past the impeller resists cavitation because it is under pressure. Blades in the pump case are angled exactly opposite to those of the impeller. These blades redirect the water to establish a concentrated flow through the nozzle.

The manufacturer recommends a radius of about 1/64" (0.3-0.5mm) for the leading edge of the impeller blade. A sharper radius will result in cavitation and a greater radius will reduce pump efficiency.

Cavitation Burns

Cavitation burns are the worst enemy of an impeller. These burns literally "eat away" material from the impeller, leaving holes and weakening the impeller structure. In extreme cases entire blades have been known to "depart" from the impeller hub because the cavitation burns at the base of the blade were so severe.

Cavitation burns are the result of imperfections (damage) on the impeller blades or air mixed with the water flow.

Cavitation burns may be caused by:

Wave jumping - air is sucked into the pump as the craft leaves the surface of the water.

Bad seal between the pump case and outlet nozzle -- allowing air to enter.

Leading edge of the impeller becoming damaged.

To explain exactly what causes cavitation burns, first, let us examine the last cause. If the impeller leading edge is damaged and "mushrooms" over the blade, a low pressure area will form under the mushroomed lip.

At atmospheric pressure, water boils at 212°F. Anytime there is an imperfection on the surface of an impeller blade and water passes over that imperfection, the water pressure is lowered. If the pressure is lowered, water will boil at a much lower temperature. Therefore, air bubbles will form in the water boiling under the mushroomed lip. The bubbles will creep down the surface of the blade and accumulate at the impeller hub. A high pressure area is formed at the base of the blade. Here, the air bubbles will collapse and reform back into water with a release of energy. This energy is absorbed by the impeller and results in material being eaten away.

The impeller should be dressed to a slightly rounded edge. It is impossible to achieve a perfectly straight sharp edge, as one side will always be convex and the other side concave. The concave side will form an area of low pressure and encourage cavitation as described above.

The other two causes of cavitation -- defective seals and wave jumping -- occurs when air is sucked in with the water flow. The same principles apply, as just described, in the previous paragraphs. Air bubbles will creep down the surface of the blade and accumulate at the impeller hub. A high pressure area is formed at the base of the blade. Here, the air bubbles will collapse and reform back into water with a release of energy. This energy is absorbed by the impeller and results in material being eaten away.

1-5 REPLACEABLE PARTS

Any time the engine or jet pump is disassembled, all manufacturers and the authors heartily recommend all parts included in an overhaul kit be installed. An engine kit usually includes all gaskets, seals, O-rings, and other parts required to rebuild the engine. Pistons, rings, bearings, crankshafts and even blocks should be inspected to determine if they are fit for further service. The pump kits usually include all gaskets, seals, O-rings, and other parts required to restore full efficiency and "like new" condition to the jet pump unit.

1-6 DEBRIS REMOVAL AND ENGINE OVERHEATING

As may be expected, a clogged impeller will slow the craft. The volume of water passing through the nozzle and water velocity will be reduced. A clogged impeller may cause restricted cooling water through the engine.

Many times the cooling water supply hose, if not properly secured with a hose clamp, may "blow off". The engine loses its supply of cooling water and internal damage quickly occurs.

Some units are equipped with an overheating warning horn. If the horn sounds, **SHUT DOWN** the engine immediately and remove the restriction or check the cooling hose connection.

Caution must be exercised when moving the craft off a beach to avoid sucking in mud, sand, pebbles, or rocks through the intake. Such ingestion could cause damage to the impeller, the vanes in the impeller duct, or clog engine cooling passages.

The engine cooling passages must be flushed with clear water on a regular basis. This is accomplished by simply connecting a garden hose to the cooling water hose or fitting on the manifold, if so equipped. Such an easy maintenance practice should become a habit and will prevent accumulation of debris which could restrict engine coolant water flow and cause overheating.

1-7 REVERSE CAPABILITY

Certain Models of Polaris Watercraft are equipped with a reverse "gate". This gate is installed in such a manner to swing down over the steering nozzle and forces the water from the jet pump nozzle to be directed forward -- thus moving the craft sternward. The gate is controlled by the operator through a cable.

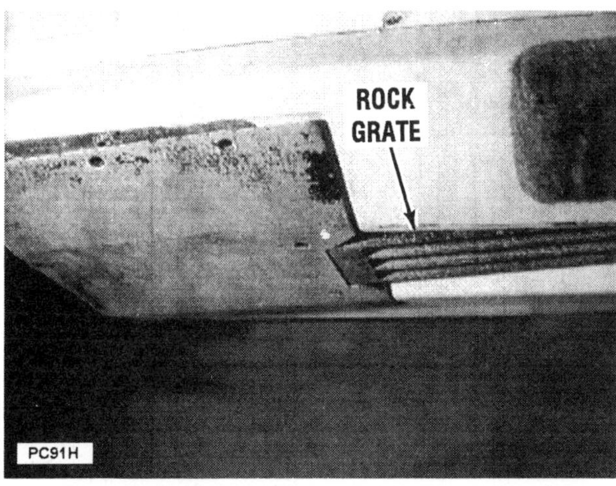

Any obstruction, such as a plastic bag, sucked up against the rock grate will dramatically restrict water volume through the pump. A reduction in water flow affects craft performance and cooling water to the engine. Such a condition may quickly lead to a serious and expensive overheating problem.

With the gate in the fully raised position, all water is directed to the rear of the craft and the craft moves forward.

If the gate is lowered partially, some of the water is directed forward and some is allowed to move sternward -- the craft is "static" in the water and the gate could be considered in a "neutral" position.

When the gate is allowed to move to the fully lowered position, almost all of the water is directed forward and the craft moves sternward -- the gate could be considered in a "reverse" position.

Adjustment of the reverse cable is covered in Chapter 11, Control Adjustments.

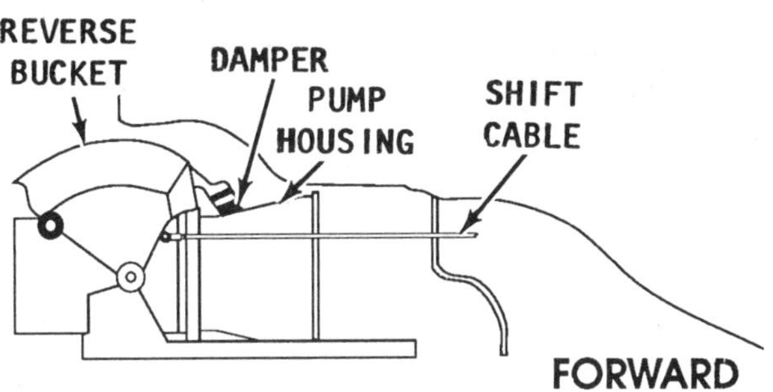

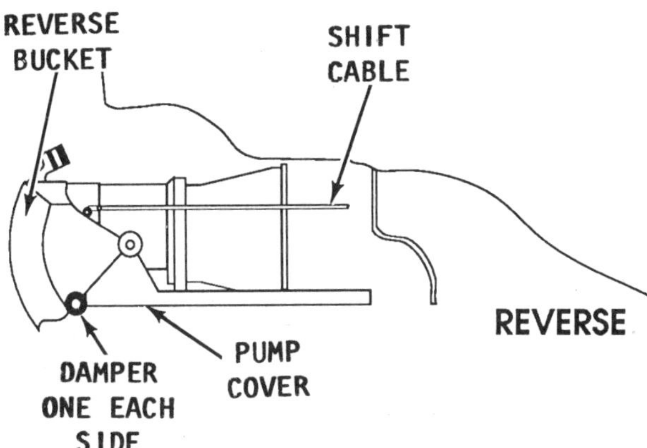

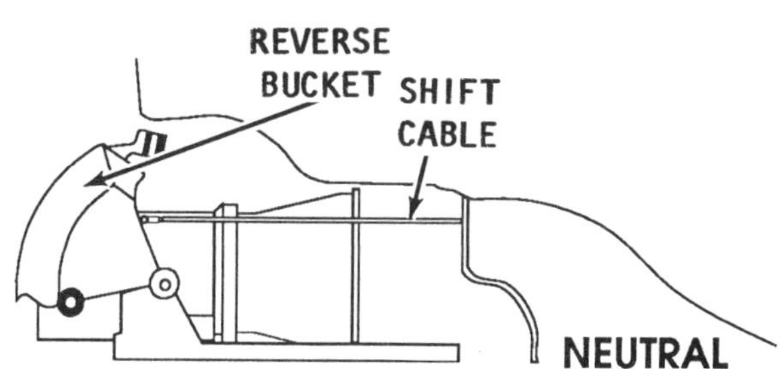

*Line drawing of a typical jet pump with reverse capability, depicting the reverse gate in full **FORWARD** (top), **REVERSE** (center), and in the **NEUTRAL** position (bottom).*

1-8 SPECIAL FEATURES

The Polaris Series covered in this manual have a few unique features worth mentioning at this time. The following warning systems serve to protect the engine -- and operator -- against damage to the craft due to overheating of the engine, or a low fuel or oil state.

Depending on Model, the warnings used vary from an audible buzzer, to a visual warning light located on the MFD, to the engine rpm limiter based on overheat or low fuel. Some Models have a combination of these various warning systems.

RPM Limiter

A few of the Models covered in this manual are equipped with electronic circuitry designed to limit engine rpm in the event of engine overheat. This condition is detected by a temperature sensor located on the water outlet manifold or the exhaust pipe.

As the operating temperature reaches the activation threshold, the sensor sends a signal to the limiting circuit, inducing a reduction in maximum power available. The rpm limit is set at 4,200 or 4,500 rpm, depending on the Model. An accompanying HEAT indication is shown on the MFD, identifying the source of the reduced maximum power condition. In such an occurrence, immediately reduce the throttle to idle -- for maximum cooling benefit, maneuver to a safe location and shut down the engine as soon as possible.

Other Warning Systems

Most watercraft have a warning light indicator for an overheat condition, some have the light indicator and/or an electrical warning buzzer, alerting the operator of an impending overheat condition.

Many Models are equipped with a sensor in both the fuel and oil tanks. If either fuel or oil level should drop below specific requirements, the audible warning buzzer activates, or the warning light illuminates to alert the operator of the condition. A few Models limit rpm only in the event of an overheat or low fuel condition -- on these Models, the low oil condition activates a buzzer or warning light only.

Operators of these craft have been known to report poor engine operating conditions, only to discover the fuel state was low! Check to be sure the fuel and oil levels are sufficient before determining the problem must exist in the engine.

On the 1996 SL900 and SLTX, the rpm limiter affects the ignition by retarding the timing to 8° BTDC during the event of a low fuel or overheat condition. The craft will exhibit reduced performance during acceleration and reduced top speed. The light will illuminate also, indicating the probable cause.

When the rpm limiter circuitry activates to retard maximum available rpm, the operator **MUST** satisfy the following two conditions before the limiter deactivates:

1. The condition causing the problem -- low fuel or allow the engine to cool -- must be remedied.

2. The throttle must be released long enough for the engine rpm to fall below the limiting threshold -- 4,200 or 4,500rpm.

Procedures for testing these warning systems are presented in detail, refer to Chapter **7** -- Ignition or Chapter **9** -- Electrical for further information.

2
SAFETY

2-1 INTRODUCTION

Personal watercrafting is one of the fastest growing leisure activites in the world. Riding a personal watercraft can be an exciting, exhilarating, and safe experience provided the operator exercises prudent behavior and remains in constant control of the vehicle.

Personal watercraft may be likened to a firearm. Weapons do not injure people, but the individual handling them is the true culprit. In a similar manner, if the personal watercraft is operated in an unresponsible manner the driver, passengers, and others using the same waterway may be in danger.

The greatest hazard in operating a personal watercraft stems not from the craft itself, but from the behavior of the operator. The Personal Watercraft Industry Association (PWIA) and the National Marine Manufacturers Association (NMMA) recognize there are a number of problems due to irresponsible craft operators contributing to tarnishment of the sport. Specifically, these problems range from alcohol and substance abuse to excessive noise, speed, and reckless maneuvering. Insensitive operators can disrupt an otherwise enjoyable day for other craft users through misconduct on the water. These individuals often endanger their own lives as well as the lives of others by losing full control of their craft.

Organized events, races, and exhibitions, properly planned and controlled have proven to be a spectacular event appealing to a wide range and number of participants and spectators.

Craft Classification

Any craft equipped with propulsion machinery is classified as a motorboat.

Personal watercraft are classified as "Class A inboard boats" by the Coast Guard and therefore are subject to most of the same laws and requirements as more conventional larger craft. A decal, usually affixed somewhere in the stern area, lists exceptions the Coast Guard has granted for this type of water vehicle.

Information

In 1989, several states recognized the need to impose specific regulations governing operation of personal watercraft. Rules and regulations differ from state to state and are constantly being revised. Therefore, it would be an impossible task to present current legislation on personal watercraft activites. Before the printing ink for this publication was dry, additions, deletions, and revisions would already be in effect.

For current information, the reader may call 1-800-336-BOAT for the name and ad-

*PWC owners and their friends prepare for a **SAFE** fun day on the water. By paying attention to U.S. Coast Guard safety requirements and using good judgement, their enjoyment will actually be increased many fold.*

dress of the local Boating Law Administrator. The reader may also contact the PWIA Government Relations Representative, Washington, D.C.

Regulation Enforcement

The U.S. Coast Guard and state enforcement officers have the authority to stop any craft to check for compliance with federal or state law. They can order a craft to return to the closest dock and remain out of the water until the hazardous condition of the craft is corrected.

Examples are:

Inadequate number of Personal Flotation Devices.

Missing/Inoperative fire extinguisher.

Overloading.

Operation at night without proper lights.

Fuel leakage.

Fuel transportation (other than approved fuel tank).

Failure to meet ventilation requirements.

Failure to meet carburetor backfire flame arrestor requirements.

Excessive leakage or accumulation of water in the bilge.

The U.S. Coast Guard and state enforcement officers have the authority to stop any craft if they determine the craft is being operated in a negligent manner. Negligent operation is defined as "the failure to excercise the degree of care necessary to prevent the endangering of life, limb, or property of any person".

Examples are:

Operating while under the influence of drugs or alcohol.

Excessive speed in a congested area.

Excessive speed in stormy or foggy conditions.

Coming too close to another vessel.

Operating in a swimming area where bathers are present.

Towing water skiers where obstructions exist or where a fall might cause skier to be struck by another vessel.

Operating in the vicinity of a dam.

Cutting through a marine parade or regatta.

2-2 MINIMUM LEGAL REQUIREMENTS FOR EQUIPMENT ONBOARD CRAFT

All personal watercraft owners **MUST** provide the following equipment:

One approved Type I, II, III or IV PFD (personal flotation device) for each person on board or being towed on water skis, etc.

An efficient sound producing device.

A B-I approved hand portable fire extinquisher.

Visual distress signals for nighttime use, if operated on coastal waters.

Personal Flotation Devices (PFDs)

The Coast Guard requires an approved life-saving device be worn by each person on board. Devices approved are identified by a tag indicating Coast Guard approval. Such devices may be life preservers, or buoyant vests. Ring buoys, or buoyant cushions are not acceptable.

Life preservers have been classified by the Coast Guard into five type categories. All PFDs presently acceptable on recreational craft fall into one of these five designations. Only four of the five catagories are stipulated by legal requirements. Type V is not acceptable for use on a personal watercraft.

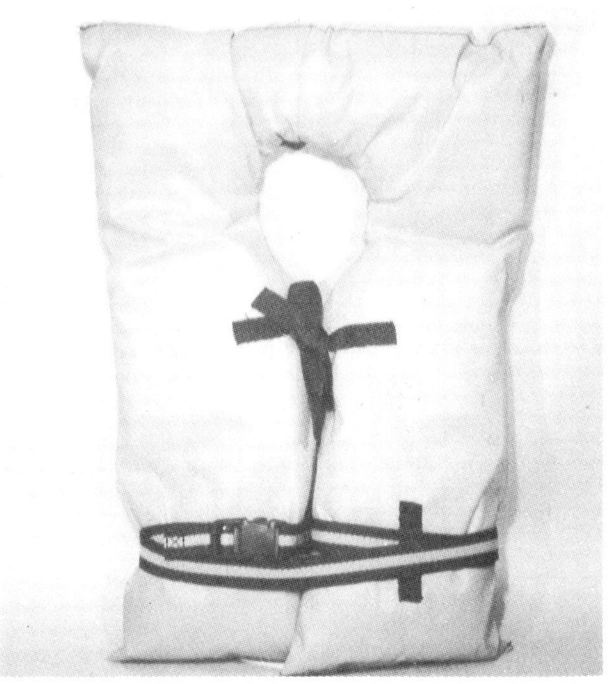

Type I PFD Coast Guard approved life jacket. This type flotation device provides the greatest amount of buoyancy.

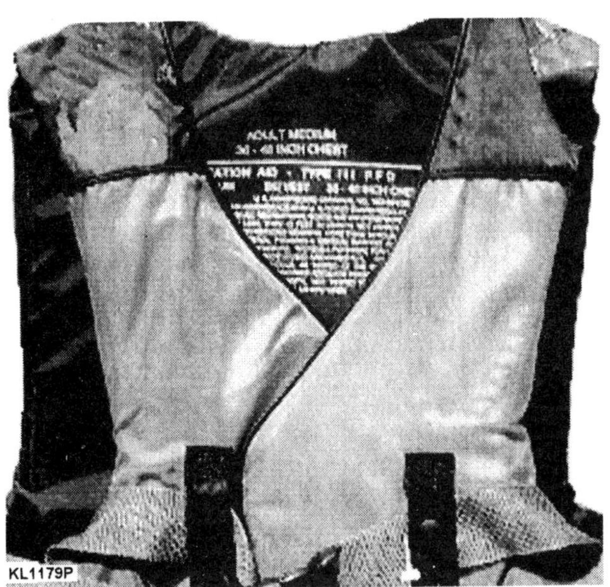

A Type III PFD requires the wearer to be active to remain upright in the water. This type device is comfortable and allows the wearer to actually swim while in the water.

A PFD **MUST** be U.S. Coast Guard approved, in good and serviceable condition, and of an appropriate size for the person intending to wear it.

It is an accepted fact that most boating people own life preservers, but too few actually wear them. There is little or no excuse for not wearing one because the modern comfortable designs available today do not subtract from an individual's boating pleasure.

PFDs may have a long serviceable life, if proper care is exercised and the device is not abused. Wipe them clean and dry before storing in a well ventilated area during the off-season. **NEVER** use a PFD as a fender. A crushed PFD loses its buoyancy and becomes useless.

Type I PFD has the greatest required buoyancy and is designed to turn most **UNCONSCIOUS** persons in the water from a face down position to a vertical or slightly backward position. The adult size device provides a minimum buoyancy of 22 pounds and the child size provides a minimum buoyancy of 11 pounds. The Type I PFD provides the greatest protection to its wearer and is most effective for all waters and conditions.

Type II PFD is designed to turn its wearer in a vertical or slightly backward position in the water. The turning action is not as pronounced as with a Type I. The device will not turn as many different type persons under the same conditions as the Type I. An adult size device provides a minimum buoyancy of $15\frac{1}{2}$ pounds, the medium child size provides a minimum of 11 pounds, and the infant and small child sizes provide a minimum buoyancy of 7 pounds.

Type III PFD is designed to permit the wearer to place himself (herself) in a vertical or slightly backward position. The Type III device has the same buoyancy as the Type II PFD but it has little or no turning ability. Many of the Type III PFD are designed to be particularly useful when water skiing, sailing, hunting, fishing, or engaging in other water sports. Several of this type will also provide increased hypothermia protection.

Type IV PFD is designed to be thrown to a person in the water and grasped and held by the user until rescued. It is **NOT** designed to be worn. The most common Type IV PFD is a ring buoy or a buoyant cushion.

ONE LAST WORD

Common sense dictates the "cheap and cheerful" approach does not pay when applied to a life preserving device. Select and buy the very best. The life you save may be your own.

NOW, THESE WORDS

Due to the nature and maneuvering characteristics of personal watercraft, and from a "practical" standpoint, the following paragraph on bells, horns, and whistles hardly apply. The information is included to make the manual "complete" and because according to the U.S. Coast Guard they are required. Also, the authors feel the reader may "skipper" or "crew" on a larger craft and therefore, might be required to use the information presented here.

Bells, Horns and Whistles

All craft must carry some means of producing an efficient sound signal and the sound must be audible for a specific distance and duration, depending on the size of the craft. For personal watercraft, an audible sound must carry for one half mile and be of four to six second duration.

Law enforcement craft are the **ONLY** boats allowed to use sirens.

Personal watercraft operators are required to sound fog and maneuvering signals when underway in fog or rain as a means of communication between craft which are not visible. This system was devised to alert nearby craft of the presence of others in the water and as a means of communicating maneuvering intentions. Signals indicating maneuvering tactics are short one second blasts.

One blast indicates the craft is turning to starboard.
Two blasts indicate the craft is turning to port.

Three blasts indicate the craft is in reverse.
Five or more blasts indicate - stay away - danger.

A prolonged blast lasting 4 to 6 seconds duration every two minutes indicates a straight line course in poor visibility.

These signals should be memorised by every personal watercraft operator, not necessarily because the operator would give them, but in the event of a head on collision with say a large powerboat, the skipper may signal his intention to turn either to port or starboard using this system of communication.

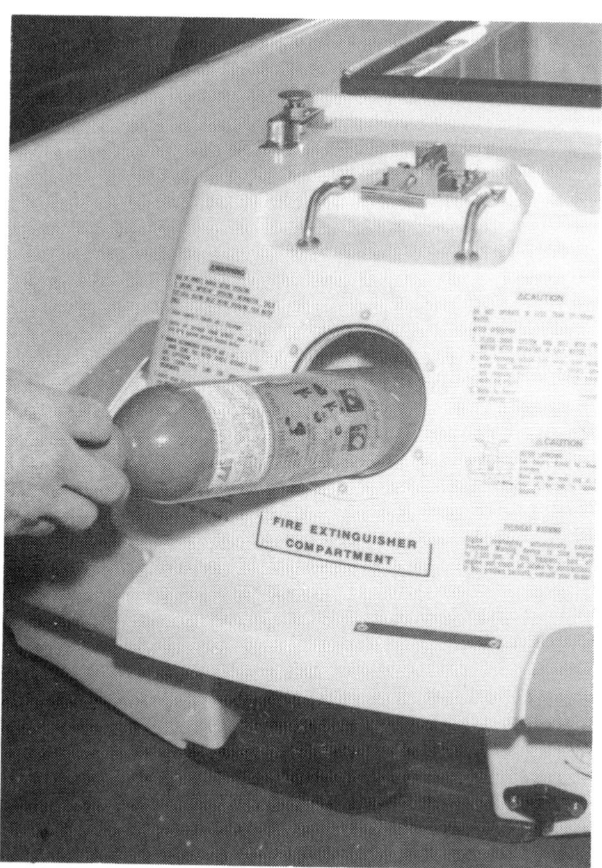

Personal watercraft are required to carry a Type B-1 fire extinguisher aboard. On Polaris models, the extinguisher is located in a forward hatch.

Fire Extinguishers

Personal watercraft are classified as "Class A inboard boats" by the Coast Guard and therefore are required to carry a B-I fire extinguisher.

All fire extinguishers must bear Underwriters Laboratory (UL) "Marine Type" approved labels. With the UL certification, the extinguisher does not have to have a Coast Guard approval number.

B-I contains 1-1/4 gallons foam, or 4 pounds carbon dioxide, or 2 pounds dry chemical agent, or 2-1/2 pounds Halon.

READ labels on fire extinguishers. If the extinguisher is U.L. listed, it is approved for marine use.

Any personal watercraft can be rolled over and held under water for a few seconds to extinguish the fire. Turning the craft over in the water will prevent oxygen from entering the engine compartment. Without oxygen, the fire will soon exhaust the remaining oxygen in the compartment and then go out. **DO NOT** open the engine compartment until the smoke has completely stopped, then open with great caution and have a fire extinguisher at hand.

Visual Distress Signals

Coast Guard Regulations require personal watercraft to carry night visual distress signals when used on coastal waters - even though it is illegal to operate a personal watercraft at night.

Coastal waters include: the Great Lakes, the territorial seas and those waters directly connected to the Great Lakes and the territorial seas, up to a point where the waters are less than two miles wide.

These signals assist rescuers in locating and aiding people in distress. This information is useful to every personal watercraft owner, to know how to respond to others in distress.

For night use: Pyrotechnic visual distress signaling devices (red flares, hand held or aerial) **MUST** be Coast Guard Approved, in serviceable condition and stowed to be readily accessible. If they are marked with a date showing the serviceable life, this date must not have passed.

Coast Guard Approved pyrotechnic devices carry an expiration date. This date can **NOT** exceed 42 months from the date of manufacture and at such time the device can no longer be counted toward the minimum requirements.

The only non-pyrotechnic visual distress signaling device approved by the coast guard for use at night is an electric distress light -- not a flashlight but an approved electric distress light which **MUST** automatically flash the international **SOS** distress signal (. . . - - - . . .)four to six times each minute.

2-3 MINIMUM LEGAL REQUIREMENTS FOR REGISTRATION OF CRAFT

Federal regulations require all personal watercraft to be issued a registration number and have the number displayed in a specific manner on the craft.

The "Certificate of Number" is issued by the state and must be carried on board at all times. The Hull Identification Number (HIN) is necessary to register the craft with the state.

The registration number must be displayed as follows:

The figures must be read from left to right.

The figures must be displayed on the forward half of each side of the bow of the craft.

The figures must be in bold, block letters of good proportion and must not be less than 3 inches high.

The figures must be of contrasting color to the craft hull, or background.

The figures must be as high above the waterline as practical.

No figures other than those assigned to the craft can be displayed on the forward half of the craft.

Letters must be separated from numbers by spaces or hyphens.

Validation decals (if required by state) must be displayed within 6 inches of the registration number.

Elaborate color schemes and custom detailing are very much a part of the fun of owning a personal watercraft. **HOWEVER,** the paint design must not obstruct the registration numbers or the validation decal. Like an automobile without a license plate, sooner or later a craft with no visible registration numbers will attract the attention of the law.

Hull Identification Number

Federal regulations require all personal watercraft to have a Hull Identification Number (HIN) permanently attached to the craft in two separate locations: on the starboard side of the transom, above the waterline. The HIN must also be displayed in an unexposed location, which is usually left to the discretion of the manufacturer.

2-4 SAFETY PRACTICES

MOST IMPORTANT WORDS IN THIS MANUAL

Under normal vehicle operation -- in an automobile, truck, motorcycle, tractor -- if an emergency situation arises in front of the vehicle, the immediate response is foot off the throttle **-- reducing SPEED --** and hit the brake.

WRONG, WRONG, WRONG, in a personal watercraft. Water rushing through the jet pump and out the steering nozzle is the **ONLY** means of control. Once the throttle is fully backed off, direction of the craft can **NOT** be changed regardless of how drastic the operator may move the handle bars. The craft will continue on its course.

THEREFORE, if an emergency arises dead ahead, **DO NOT** back off the throttle completely. Reduce speed, but keep control.

Full Throttle Operation

Prolonged operation at full throttle will raise the cylinder head temperature by

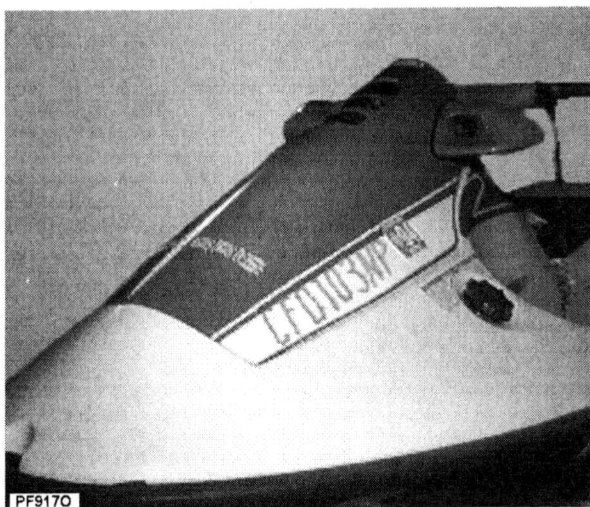

PF9170

The validation decal, required by most states, and the registration number, required by all states, must be displayed port and starboard on the forward portion of the hull.

about 100° above normal. Some units are equipped with a thermo-switch engine temperature sensor. This sensor is connected directly to the CDI unit and grounded to the engine kill switch. If the engine temperature exceeds a maximum value specifed by the manufacturer, the ignition will be grounded and the engine shut down to prevent damage through overheating.

Jumping Waves

In the very early days of personal watercraft, jumping could cause extensive internal engine damage. When the craft left the water, unchecked engine rpm rapidly increased to as high as 12,000 rpm. As soon as the craft re-entered the water, engine rpm instantly decreased to about 6,000 rpm. This sudden increase and decrease in engine speed dramatically shortened engine life. In those days, repeated jumps were known to cause bent crankshafts, sheared flywheel keys, and broken connecting rods.

Polaris engineers have developed an rpm limiter system as an integral part of the CDI unit. This limiter prevents high engine rpm and an almost "runaway" condition, during wave or ramp jumping. On "stock", out of the door factory units, this limiter restricts engine speed to a preset maximum rpm value. Thus, wave jumping, using a ramp, and a host of other almost unbelievable maneuvers can be performed without undue stress on the engine. Today, with an rpm limiter system installed, a bent crankshaft or broken connecting rod is almost non-existent.

Polaris engineers developed the limiter system to make it very difficult to bypass, as was done many times in the past. No!, we shall not reveal operation or possibilities for bypass.

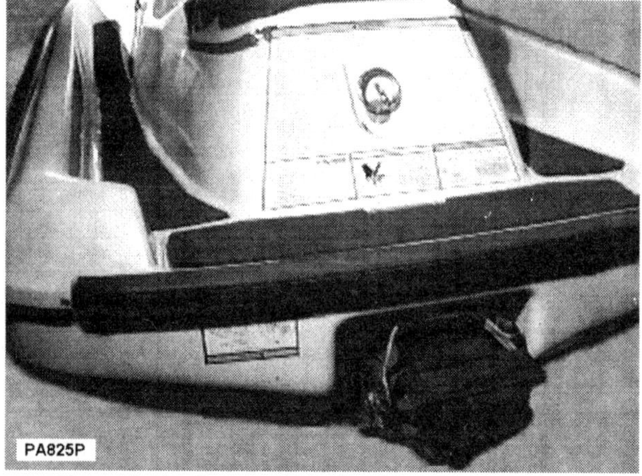

Several decals may be affixed to the stern of the hull to indicate compliance with or exceptions granted to normal Coast Guard regulations.

Alcohol and Substance Use

Operating a personal watercraft while under the influence of either alcohol or drugs is both illegal and life threatening. This fact cannot be stated too strongly. Such substances affect vision and slow reaction times. The danger is quadrupled in a craft on the water compared to on the shore.

Age Restrictions

Children under 14 years of age should not operate a watercraft unless accompanied on a two passenger craft by an adult experienced operator. Specific minimum age limits have been imposed by many states.

Speed Restrictions

Where no speed limits are posted, commonsense dictates the craft should be operated so it will pose no danger to other water users or the operator. Many states now have enforced speed limits for certain areas and conditions.

When entering "no wake" or 5 mph zones, such as fishing areas, swimming areas or marinas, the operator must slow to headway speed, the slowest speed at which appropriate steering capability can be maintained.

Individuals responsible for law enforcement on the water are keeping an ever vigilant eye on personal watercraft in many areas. Horseplay such as "wetting down" or "buzzing" other water users may result in a "ticket" for negligent operation.

Operation at Night

Personal watercraft are not equipped with lights and therefore, it is illegal to operate such a craft on the water after sunset.

Shallow Water Operation

Under normal conditions, personal watercraft may be operated in very shallow waters, and it is possible to maneuver the craft directly onto a sloping beach.

Caution must be exercised when moving the craft off a beach to avoid sucking in mud, sand, pebbles, or rocks through the intake. Such ingestion could cause damage to the impeller, the vanes in the pump bowl, or clog engine cooling passages.

WARNING
Expelled sand and small pebbles may cause INJURY to persons in the area.

A grated section set into the hull bottom helps prevent the ingestion of small rocks and pebbles.

Carry about a foot of stiff wire aboard for use in probing into the impeller area to dislodge debris where fingers cannot reach.

Engine Compartment Ventilation

All motorboats (personal watercraft included) built after April 25, 1940 and before August 1, 1980, powered by a gasoline engine or by fuels having a flashpoint of 110° F or less **MUST** have the following, which is quoted from a recent Coast Guard publication:

> At least two ventilation ducts fitted with cowls or their equivalent for the purpose of properly and efficiently ventilating the bilges of every engine and fuel tank compartment. Each duct must be at least two inches in internal diameter.
>
> There shall be at least one exhaust duct installed so as to extend from the lower portion of the bilge to cowls in the open air, and at least one intake duct installed so as to extend to a point at least midway to the bilge or at least below the level of the carburetor air intake.

Flame Arrestors

A flame arrestor, as the name suggests, safeguards against flame caused by engine backfire.

A gasoline engine installed in a motorboat or motor vessel after April 25, 1940, except outboard motors, must have a Coast Guard Approved flame arrestor fitted to the carburetor. This requirement applies to personal watercraft because the engine is enclosed.

Fuel System

All parts of the fuel system should be selected and installed to provide maximum service and protection against leakage.

The capacity of the fuel filter must be large enough to handle the demands of the engine as specified by the engine manufacturer.

All fittings and outlets must come out the top of the tank.

In order to obtain maximum circulation of air around fuel tanks, the tank should not come in contact with the hull except through the necessary supports. The supporting surfaces and hold-downs must fasten the tank firmly and they should be insulated from the tank surfaces. This insulation material should be non-abrasive and non-absorbent material.

Taking On Fuel

The fuel tank should be kept almost full to allow for expansion, but still prevent water from entering the system through condensation caused by temperature changes. Water droplets forming are one of the greatest enemies of the fuel system. By keeping the tank almost full, the air space in the tank is kept to an absolute minimum and there is less room for moisture to form. It is a good practice not to store fuel in the tank over an extended period, say for six months. Today, fuels contain ingredients that change into gums when stored for any length of time. These gums and varnish products will cause carburetor problems and poor spark plug performance. An additive (Sta-Bil) is available and can be used to prevent gums and varnish from forming.

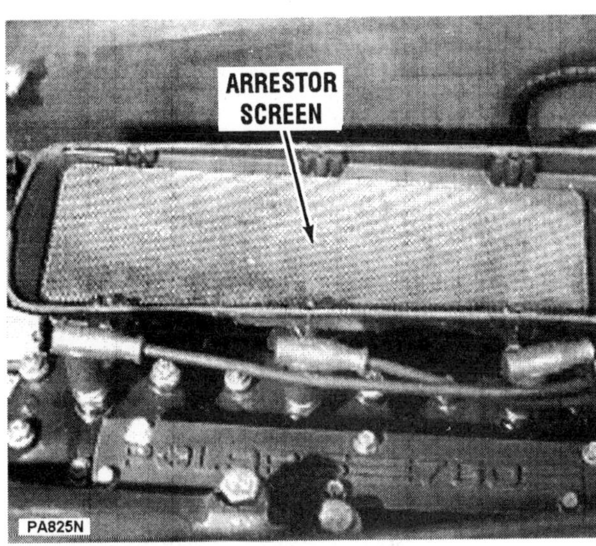

Carburetors on personal watercraft are required to be fitted with a flame arrestor because the engine compartment is an enclosed area.

Excessive Noise

Every manufacturer designs a stock exhaust system to keep noise to a minimum. A poorly maintained or modified exhaust system will produce excessive and disturbing noise. This is a form of air polution and is illegal in most states.

Automotive Replacement Parts

When replacing fuel, electrical, and other parts, check to be sure they are marine type and Coast Guard Approved. Automotive parts are not made to the high standards of marine parts. The carburetors must **NOT** leak fuel; electrical equipment **MUST** be able to operate in a gas fume enclosed area without exploding; etc. Automotive parts could cause a fire endangering the operator, passengers, and the craft. The part may look the same and even have a similar number, but if it is not **MARINE** it is not safe to use and will not pass Coast Guard inspection.

Overloading

Each personal watercraft is designed to carry a specific number of persons. Attempting to board more passangers than the craft is designed to carry, will upset weight distribution causing a marked decrease in performance and possible loss of control.

2-5 BOATING ACCIDENT REPORTS

New federal and state regulations require an accident report to be filed with the nearest state boating authority within 48 hours if a person is lost, disappears, or is injured to the degree of needing medical treatment beyond first aid. The time limit is reduced to 24 hours if the accident results in a loss of life.

Accidents involving only property or equipment damage **MUST** be reported within 10 days, if the damage is in excess of $500.00. Some states require reporting of accidents with propery damage less then $500.00 or a total boat loss. A $1,000.00 **PENALTY** may be assessed for failure to submit the report.

WORDS OF ADVICE

Take time to make a copy of the report to keep for your records or for the insurance company. Once the report is filed, the Coast Guard will not give out a copy, even to the person who filed the report.

The report must give details of the accident and include:

1- The date, time, and exact location of the occurrence.

2- The name of each person who died, was lost, or injured.

3- The number and name of the vessel.

4- The names and addresses of the owner and operator.

If the operator cannot file the report for any reason, each person on board **MUST** notify the authorities, or determine that the report has been filed.

In nautical terms, the front of the craft is the **bow**; the rear is the **stern**; the right side, when facing forward, is the **starboard** side; and the left side is the **port** side. One easy way to remember this basic fundamental is to consider the words "port" and "left" both have four letters and go together.

All directional references in this manual use this terminology. Therefore, the direction from which an item is viewed is of no consequence, because **starboard** and **port** **NEVER** change no matter where the individual is located, or standing, even on his or her head.

2-6 SECURITY

As mentioned in the opening paragraphs of this chapter, personal watercrafting is one of the fastest growing leisure activites in the world. Like most personal possessions, personal watercraft have value and valuables are subject to be lost to unscrupulous characters -- thieves. The best advice for a personal watercraft owner is "lock it or lose it".

Do not leave the craft unattended on the beach or dock. **ALWAYS** remove the kill switch tether from the switch and of course, the ignition key, if the unit is equipped with these features. Certain aftermarket security devices are available for installation to prevent theft of personal watercraft. These include: A "secret" ignition grounding switch, which grounds one side of the ignition coil, and a hidden fuel shut-off valve.

Most owners transport their craft on trailers or in the bed of a pickup truck. If loaded on a trailer, chain and lock the craft to the trailer. Invest in a hitch lock so the trailer cannot be detached from the towing vehicle. If the craft is loaded on a pickup truck, chain and lock it to the bed, or to a large and heavy object.

3
TUNING

3-1 INTRODUCTION

The efficiency, reliability, fuel economy and enjoyment available from engine performance are all directly dependent on having the engine tuned properly. The importance of performing service work in the sequence detailed in this chapter cannot be overemphasized. Before making any adjustments, check the specifications in the Appendix. **NEVER** rely on memory when making critical adjustments.

Before beginning to tune any engine, check to be sure the engine has satisfactory compression. An engine with worn or broken piston rings, burned pistons, or scored cylinder walls, cannot be made to perform properly no matter how much time and expense is spent on the tune-up. Poor compression must be corrected or the tune-up will not give the desired results.

A practical maintenance program that is followed throughout the year, is one of the best methods of ensuring the engine will give satisfactory performance at any time.

The extent of the engine tune-up is usually dependent on the time lapse since the last service. A complete tune-up of the entire engine would entail almost all of the work outlined in this manual. A logical sequence of steps will be presented in general terms. If additional information or detailed service work is required, the chapter containing the instructions will be referenced.

Each year higher compression ratios are built into modern marine engines and the electrical systems become more complex, especially with electronic (capacitor discharge) units. Therefore, the need for reliable, authoritative, and detailed instructions becomes more critical. The information in this chapter and the referenced chapters fulfill that requirement.

3-2 TUNE-UP SEQUENCE

During a major tune-up, a definite sequence of service work should be followed to return the engine to the maximum performance desired. This type of work should not be confused with attempting to locate problem areas of "why" the engine is not performing satisfactorily. This work is classified as "troubleshooting". In many cases, these two areas will overlap, because many times a minor or major tune-up will correct the malfunction and return the system to normal operation.

The following list is a suggested sequence of tasks to perform during the tune-up service work. The tasks are merely listed here. In most cases, procedures are

A clean, properly tuned, and adequately maintained power unit can multiply by many fold the enjoyment derived from owning and operating a personal watercraft.

given in subsequent sections of this chapter. For more detailed instructions, see the referenced chapter, for the unit being seviced.

1- Perform a compression check of each cylinder. See Ignition Chapter.
2- Inspect the spark plugs to determine their condition. Test for adequate spark at the plug. See Ignition Chapter.
3- Start the engine in a body of water or connect a "Flushing Kit" with a garden hose and check the water flow through the engine, See Engine Chapter.
4- Check the carburetor adjustments and the need for an overhaul. See Fuel Chapter.
5- Check the fuel pump for adequate performance and delivery. See Fuel Chapter.
6- Make a general inspection of the ignition system. See Ignition Chapter.
7- Test the cranking motor and the solenoid. See Electrical Chapter.
8- Check the internal wiring.

3-3 COMPRESSION CHECK

A compression check is extremely important, because an engine with low or uneven compression between cylinders **CANNOT** be tuned to operate satisfactorily. Therefore, it is essential that any compression problem be corrected before proceeding with the tune-up procedure.

If the engine shows any indication of overheating, such as discolored or scorched paint, inspect the piston skirts visually thru the transfer ports, using a mirror and flashlight, for possible scoring. It is possible for a cylinder with satisfactory compression to be scored slightly.

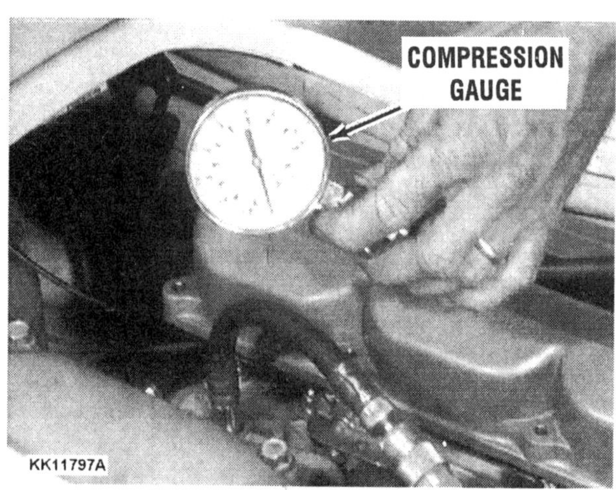

Performing a compression test on a typical three-cylinder engine. The spark plug high tension leads have been grounded to the engine -- to prevent placing an unnecessary extra load on the ignition coil.

Checking Compression

Remove the spark plug high tension leads. **ALWAYS** grasp the molded cap and pull it loose with a twisting motion to prevent damage to the connection. Remove all spark plugs and remember from which cylinder they came for evaluation later. Ground all spark plug high tension leads to the engine to render the ignition system inoperative while performing the compression check.

Insert a compression gauge into the forward spark plug opening. Crank the engine with the cranking motor, through at least four complete revolutions of the crankshaft, with the throttle at the wide-open position, to obtain the highest possible reading. Repeat the test and record the compression for the other cylinders.

The manufacturer specifies the compression pressure for all engines covered in this manual, should indicate between 100psi - 150psi. (690kPa - 1,034kPa).

GOOD WORDS

On all 1997 Model 1050 units, the No. 2 -- **CENTER** cylinder will have an approximate 15psi lower reading than the No. 1 -- foward and No. 3 -- aft cylinders.

A variation between cylinders is far more important than the actual readings. A variation of more than 5% between the cylinders (service limit) indicates the lower compression cylinder is defective. The problem may be worn, broken, or sticking piston rings, scored pistons or worn cylinders.

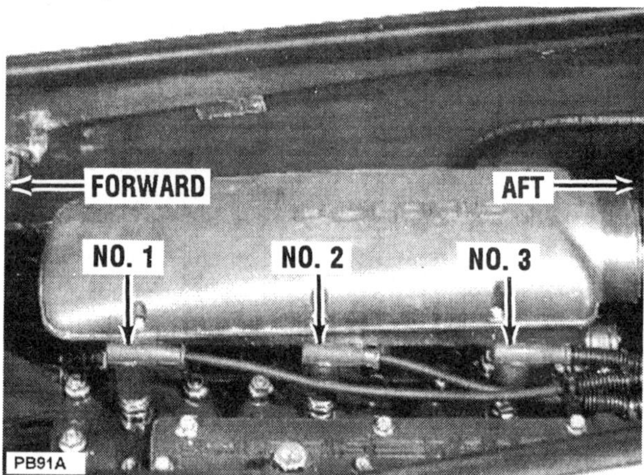

The cylinders are identified as No. 1, (Forward); No. 2, (Center) and No. 3, (Aft).

Use of an engine cleaner will help to free stuck rings and to dissolve accumulated carbon. Several different brand names are available from the local marine or automotive parts store. Follow the directions on the can closely.

3-4 SPARK PLUG INSPECTION

Mark each spark plug, indicating which cylinder it was removed from. Inspect all plugs for badly worn electrodes, glazed, broken, blistered, or lead fouled insulators.

Make an evaluation of the cylinder performance by comparing the spark or "firing" condition with those shown in Chapter 7 -- Ignition. Because different engines may require different configurations, an increased number of spark plugs are manufactured to meet the specifications. Check each spark plug to be sure they are both of the same manufacturer and have the same heat range rating. If the new spark plugs are not of the same heat range, misfiring of the engine may occur.

When purchasing new spark plugs, **ALWAYS** ask the dealer if there has been a spark plug change for the engine being serviced. The spark plug gap standard is 0.028" (0.7mm) for all series engines covered in this manual.

Crank the engine through several revolutions to blow out any material which might have become dislodged during cleaning.

Install the spark plugs and tighten them to a torque value of 18 ft lbs (24Nm) -- **NEW**; 11 ft lbs (15Nm) -- **USED**. **ALWAYS** use a new gasket and wipe the seats in the head clean. The gasket must be fully compressed on clean seats to complete the heat transfer process and to provide a gas tight seal in the cylinder. If the torque value is too high, the heat will dissipate too rapidly. Conversely, if the torque value is too low, heat will not dissipate fast enough.

After a compression check or inspection of the spark plugs, the high-tension leads should be firmly connected to each plug, and then covered with the rubber boot.

Check to ensure the spark plug caps are screwed completely onto the end of the plug wire.

Apply a small amount of dielectric grease to the inside of the plug caps and install each cap to the corresponding plug.

Broken Reed

A broken reed is usually caused by metal fatigue over a long period of time. The failure may also be due to the reed flexing too far because the reed stop has not been adjusted properly or the stop has become distorted.

If the reed is broken, the loose piece **MUST** be located and removed, before the engine is returned to service. The piece of reed may have found its way into the crankcase, behind the by-pass cover. If the broken piece cannot be located, the engine must be completely disassembled until it is located and removed.

An excellent check for a broken reed on an operating engine is to hold an ordinary business card in front of the carburetor. Under normal operating conditions, a very small amount of fine mist will be noticeable, but if fuel begins to appear rapidly on the card from the carburetor, one of the reeds is broken and causing the backflow through the carburetor onto the card.

A broken reed will cause the engine to operate roughly and with a "pop" back through the carburetor.

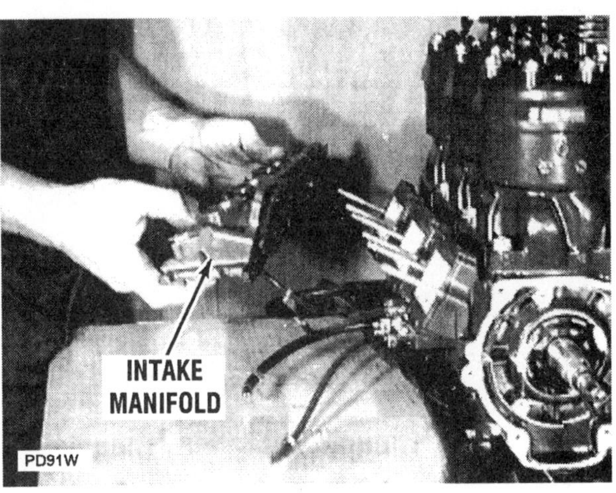

Removing/installing the intake manifold on a Model SLT750 Series engine prior to or following reed block service. Where differences occur, the differences will be clearly identified.

The reeds must **NEVER** be turned over in an attempt to correct a problem. Such action would cause the reed to flex in the opposite direction and the reed would break in a very short time.

3-5 ELECTRICAL POWER SUPPLY

Remove the battery from the engine compartment for service. Inspect and service the battery, cables and connections. Check for signs of corrosion. Inspect the battery case for cracks or bulges, dirt, acid, and electrolyte leakage.

Make sure the breather pipe is attached to the battery and not pinched by any part of the engine compartment.

Clean the top of the battery. The top of a 12-volt battery should be kept especially clean of acid film and dirt, because of the high voltage between the battery terminals and prevent contact between the acid and the two rubber hold down bands. For best results, first wash the battery with diluted ammonia or baking soda solution to neutralize any acid present. Flush the solution off the battery with clean water. Keep the vent plugs tight to prevent the neutralizing solution or water from entering the cells.

"Maintenance Free Batteries"

Many batteries installed in personal watercraft are considered "Maintenance Free" batteries and do not require regular maintenance.

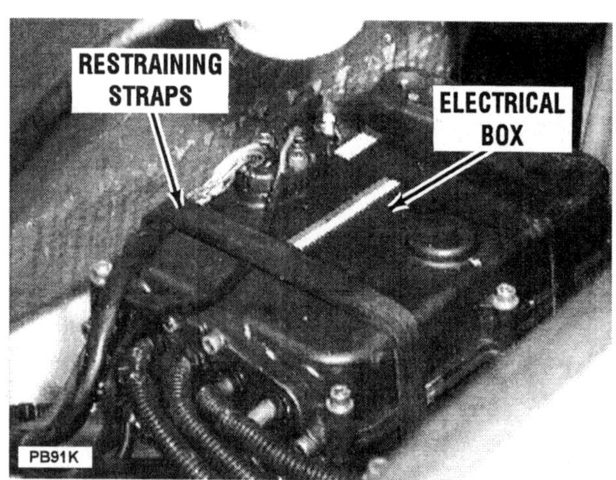

*Due to the violent maneuvers experienced by a personal watercraft -- more so than any other type craft -- the battery **MUST** be extremely well secured with **NO** movement possible. The battery is secured in place with the electrical box placed above it and strapped onto the battery.*

Standard Batteries

If the battery installed in the craft being serviced is not a "Maintenance Free" battery proceed with the following tasks.

Remove each filler cap using a pair of pliers. Fill each cell to the proper level with distilled water. The level of electrolyte should be between the upper and lower level marks during operation. When filling the battery, fill to the upper line and allow the battery to stand for twenty minutes. Check the level again and replenish as necessary.

Use a temperature corrected hydrometer to test the specific gravity of the electrolyte. At 20°C (68°F) the hydrometer reading should be 1.26 on the scale. If it is necessary to charge the battery, leave the filler caps lightly resting on the cell openings to allow the gases to escape. The charging current should not exceed 1.9 amps for 10 hours.

After charging, push each filler cap into place and wipe the top of the battery clean before installation.

Clean the battery posts and battery cable ends with a wire brush to ensure good, clean connections. Clean the top surface of the battery and identify the "POS" or "+" and "NEG" or "-" embossed symbols. Correctly connect the battery cables to the battery observing polarity. If the cables are connected backwards, the ignition system **WILL** be destroyed the first time the engine is started.

Check to be sure the battery is fastened securely in position. The hold-down rubber straps should be tight enough to prevent any movement of the battery in the holder under the most violent maneuvers of the watercraft.

If the battery posts or cable terminals are corroded, the cables should be cleaned separately with a baking soda solution and a wire brush. Apply a thin coating of Multipurpose Lubricant to the posts and cable clamps before making the connections. The lubricant will help to prevent corrosion.

Jumper Cables

If booster batteries are used for starting an engine the jumper cables must be connected correctly and in the proper sequence to prevent damage to either battery, or the rectifier diodes.

ALWAYS connect a cable from the positive terminals of the dead battery to the positive terminal of the good battery **FIRST. NEXT,** connect one end of the other cable to the negative terminal of the good battery. Finally, connect the other end of the cable to the far side of the **ENGINE** for a good ground. By making the ground connection on the engine, any spark created will not be near the battery. An arc near the battery could cause an explosion, destroying the battery and causing serious personal injury.

If a trickle charger is used on a dead battery installed in the craft, one battery lead **MUST** be disconnected prior to connecting the charger, to prevent destroying the diodes in the rectifier.

NEVER use a trickle charger as a booster to start the engine because the diodes in the rectifier will be **DAMAGED.**

3-6 CARBURETOR ADJUSTMENT

Fuel and Fuel Tanks

Take time to check the fuel tank and all of the fuel lines, fittings, couplings, valves, flexible tank fill and vent. Turn on the fuel supply valve. If gas was not drained at the end of the previous season, make a careful inspection for gum formation. When gasoline is allowed to stand for long periods of time, particularly in the presence of copper, gummy deposits form. This gum can clog the filters, lines, and passageways in the carburetor.

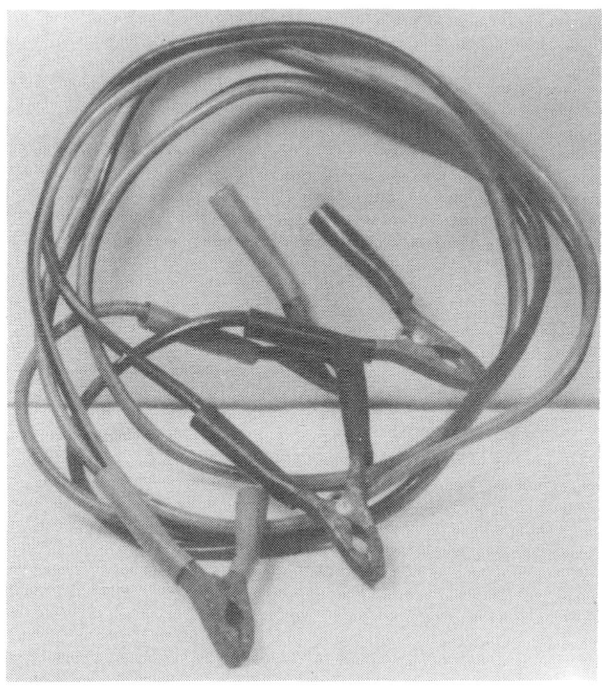

*Common set of jumper cables for use with a second battery to crank and start the engine. **EXTREME** care should be exercised when using a second battery, as explained in the text.*

Draining Fuel Tank

If the condition of the fuel is in doubt, drain, clean, and fill the tank with fresh fuel. On most models, this is "conveniently" achieved by removing the fuel tank from the front hatch. Other manufacturers may require the engine to be removed first, before the fuel tank can be lifted from the craft. However, using a siphon method, as mentioned in the picture caption, is a much more practical method.

Add a 50:1 oil mixture to the first fuel at the beginning of a new season in addition to the oil supplied by the oil injection system.

Low Speed and Idle Mixture Adjustment

The idle mixture and idle speed are set at the factory. Due to local conditions, it may be necessary to adjust the carburetor while the craft is operating in a test tank or secured in a body of water. For maximum performance, the idle mixture and the idle rpm should be adjusted under actual operating conditions.

Set the low speed adjustment screw at the specified number of turns open from a lightly seated position. Refer to the Appendix for the model being serviced.

Start the engine and allow it to warm to operating temperature.

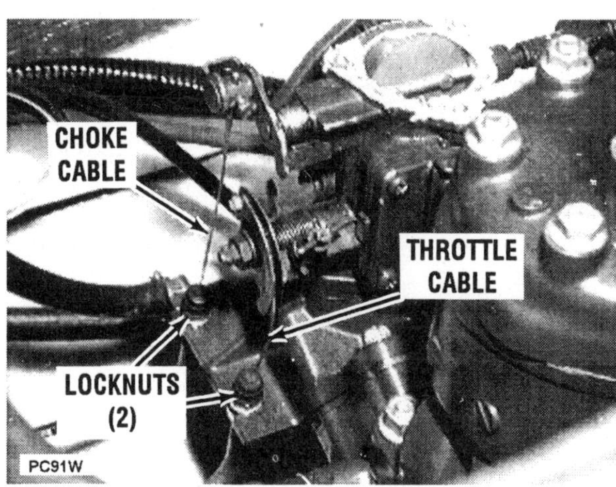

Typical setup for the choke cable and throttle cable adjustment through the locknut at each bracket. Chapter 6 -- Fuel and Oil covers adjustment of both cables. The bracket and routing of the cables will differ depending on the Model Series being serviced.

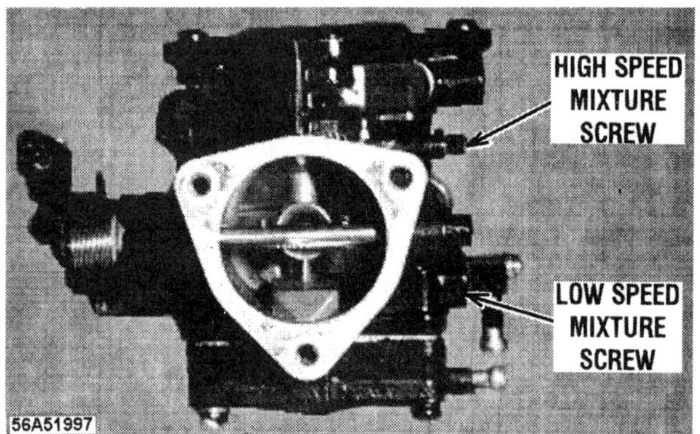

The Mikuni carburetor, with a remote fuel pump is used the Model SL650, SL750, SLT750, SLT780, and SLX780 Series engines.

CAUTION

Water must circulate from the jet pump to and from the engine anytime the engine is operating. Just a few minutes without cooling circulating water may cause extensive damage or seizure of the engine.

NEVER, AGAIN NEVER, operate the engine at high speed with a flush device attached. An engine operating at high speed with such a device attached, would **RUNAWAY** from lack of a load on the impeller shaft, causing extensive damage.

Connect a tachometer to the engine.

SPECIAL WORDS ON TACHOMETERS AND CONNECTIONS

A tachometer connected to the engine must be used to accurately determine engine speed during idle and high-speed adjustments.

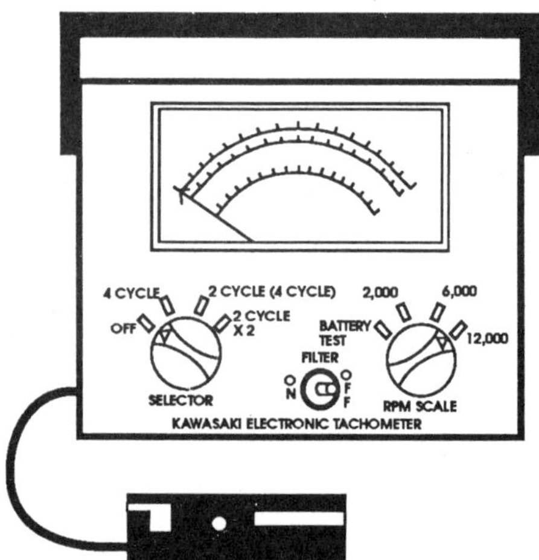

Maximum engine performance can only be obtained through proper tuning using a tachometer.

View of a Keihin carburetor with location of the low-speed and high-speed adjustment screws and the throttle stop screw identified.

Theoretically, it should be possible to connect a tachometer to any engine and receive an accurate count of crankshaft revolutions -- **RIGHT? WRONG!**

Unfortunately, this is not the case. Tachometer manufacturers admit, on certain two-stroke applications where points, and an ignition coil are **NOT** used, an attempt to connect a tachometer to the unit could "go up in smoke".

The general rules for connecting a tachometer to a marine two-stroke engine are:

1- The tachometer **MUST** be calibrated at the factory for two-stroke applications. Some tachometers have a dial indicator which may simply be switched from two-stroke to four-stroke and back again.

2- A tachometer may have as many as **FOUR** leads. The meter will always have a minimum of an input lead and a ground lead. The colors of these leads may vary between manufacturers, but the instructions provided with the meter will (or should), identify these leads. The input lead is connected to the primary **NEGATIVE** side of the ignition coil and the ground lead is connected to a suitable engine ground.

3- Some tachometers have a third wire which is considered the "hot" lead. The "hot" lead is attached to the **POSITIVE** battery terminal. On other tachometers, this "hot" lead is only used for an internal light bulb to illuminate the meter face. On still other tachometers this "hot" lead is necessary to the operation of the meter.

Once again, the instructions with the tachometer should identify the "hot" lead.

4- A fourth lead from the tachometer is used as the ground lead to the light bulb.

FIRST, last, and **ALWAYS**, check the tachometer manufacturer's instructions for use with a two-stroke unit.

Procedure

With the craft in the water or test tank, start the engine and allow it to idle. With the engine running, slowly turn the idle speed screw in (clockwise) to increase; or out (counterclockwise) to decrease idle speed. until the engine idles at 1,250 ± 50rpm -- All models except Hurricane Series, which should idle at 1,350 ±50rpm. For further information on specific rpm settings per model, refer to the Appendix.

High Speed Screw Adjustment

Set the high speed adjustment screw at the specified number of turns open from a slightly seated position. Refer to the Appendix for the model being serviced. This is the best setting for sea level. For operation at higher elevations, a leaner mixture is obtained by rotating the screw **CLOCKWISE** slightly, 1/8 turn at a time. **FURTHER HIGH ELEVATION INFORMATION IS AVAILABLE IN CHAPTER 6 -- FUEL.**

3-7 FUEL PUMPS

Many times, a defective fuel pump diaphragm is mistakenly diagnosed as a probe in the ignition system. The most common problem is a tiny pin-hole in the diaphragm. Such a small hole will permit gas to enter the crankcase and foul the spark plugs at idle-speed. During high-speed operation, gas quantity is limited, the plug is not fouled and will therefore fire in a satisfactory manner.

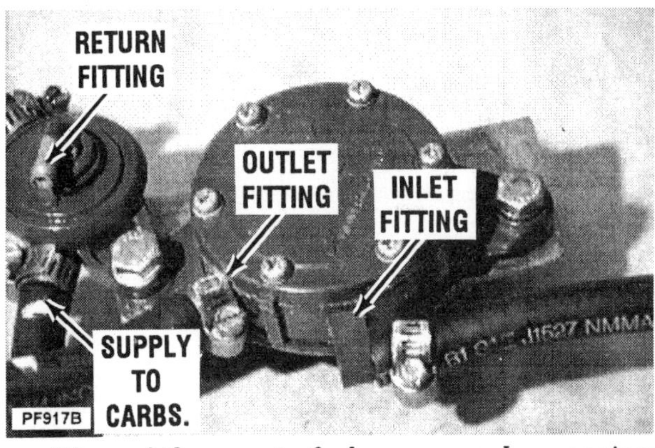

View of the remote fuel pump used on engines equipped with Mikuni carburetors. On the Model SLT750 Series, this pump is installed on the aft end of the flame arrestor box.

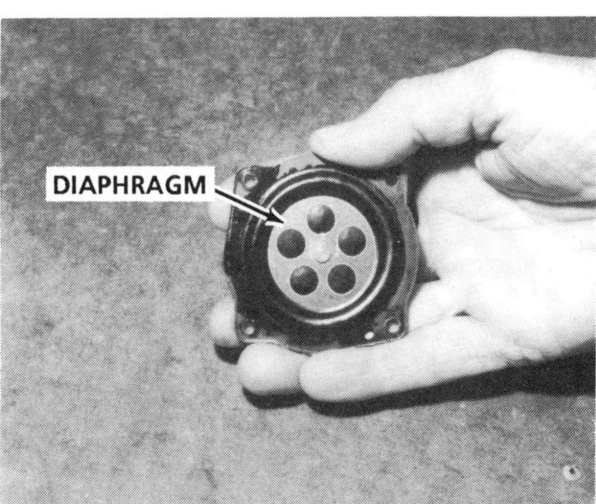

Close view of the fuel pump diaphragm from a Keihin carburetor with integral fuel pump. The diaphragm must be carefully inspected during a carburetor overhaul for the tiniest hole or tear. Such damage will "wet foul" the spark plug at idle speed.

Remote Fuel Pump

All models in the 650, 750, and 780 Series covered in this manual are equipped with Mikuni Super BN carburetors -- one per cylinder. These carburetors receive fuel from a remotely located fuel pump, operating on a vacuum created by the engine.

Detailed instructions for removal, disassembling, servicing and cleaning the remote fuel pump are given in Chapter 6 -- Fuel.

Integral Fuel Pump

All other models -- SL/T700, Hurricane, SL900, SLTX1050, and SLTX Series -- are equipped with Keihin CDKII carburetors, one carburetor per cylinder. The Keihin units have an integral fuel pump, with a "slight" exception. The SL700 and SL700 Deluxe Series are equipped with an accelerator fuel pump.

Triple-carburetor systems are equipped with two fuel pumps -- the No. 2 (Center) carburetor does **NOT** have an integral fuel pump; it receives fuel from the pump within the No. 1 and No. 3 carburetor.

If the fuel pump fails to perform properly, an insufficient fuel supply will be delivered to the carburetor. This lack of fuel will cause the engine to run lean, lose rpm or cause piston scoring.

NEVER use liquid Neoprene on fuel line fittings. Always use Permatex when making fuel line connections. Permatex is available at almost all marine and hardware stores.

To service the fuel pump, see Chapter 6 -- Fuel.

3-8 CRANKING MOTOR AND SOLENOID

Cranking Motor Test

Check to be sure the battery is fully charged. Would you believe, many cranking motors are needlessly disassembled, when the battery is actually the culprit.

Lubricate the pinion gear and screw shaft with No. 10 oil.

Connect one lead of a voltmeter to the positive terminal of the cranking motor. Connect the other meter lead to a good ground on the engine. Check the battery voltage under load by depressing the ignition switch and observing the voltmeter reading.

If the reading is 9-1/2 volts or greater, and the cranking motor fails to operate, repair or replace the unit. See the Electrical Chapter.

Solenoid Test

An ohmmeter is required for this test. Select the Rx1000 ohm scale. Calibrate the meter by making contact with the meter leads to each other, and then rotating the calibration knob on the meter until the needle indicates zero ohms. Separate the meter leads.

Make contact with the Red meter lead to one of the large terminals on the solenoid. Make contact with the Black meter lead to the other large solenoid terminal. **NEVER** connect the battery leads to the large terminals of the solenoid, or the meter will be damaged.

Using battery jumper leads, connect the positive battery lead to the small terminal of the solenoid. Connect the negative battery lead to the small terminal of the solenoid. If the meter needle indicates continuity, the solenoid is serviceable. If the meter fails to indicate continuity, the solenoid is defective and **MUST** be replaced.

3-9 JET PUMP

Reverse capability is becoming a popular feature indeed. Seldom used in the past, this feature is now common. If the unit being serviced has a reverse gate, the gate **MUST** be properly adjusted to obtain maximum performance from the craft. When properly adjusted, the gate will permit the pump to deliver its full potential of thrust with no drag.

Gate Position

The shift rod lever adjustment should be checked from time to time to ensure the gate is firmly against the pump housing, when the unit is in the **FORWARD** position (wide open).

With the gate against the housing, any rattle noises will be avoided as the craft moves through the water. Proper positioning of the gate in forward gear will prevent wave action from accidently shifting the gate into reverse as the craft is operated through violent maneuvers.

A simple "tuning" task with the jet pump is to make a thorough check of the linkage and gate movement. Such a check will ensure maximum thrust and use of the horsepower developed by the engine.

Impeller

Excessively rounded jet impeller edges will reduce the efficiency of the jet drive.

The term cavitation "burn" is a common expression used throughout the world among people working with pumps, impeller blades, and forceful water movement.

"Burns" on the impeller blades are caused by cavitation air bubbles exploding with considerable force against the impeller blades. The edges of the blades may develop small "dime size" areas resembling a porus sponge, as the aluminum material is actually "eaten" by the condition just described.

4
MAINTENANCE

4-1 INTRODUCTION

The watercraft being prepared for service probably represents a sizeable investment In order to protect this investment and for the owner to receive the maximum amount of enjoyment from the craft it must be cared for properly while being used and when it is out of the water. Always store the watercraft with the bow higher than the stern and be sure to remove the inner hull drain plug/s. If you use any type of cover to protect craft, plastic, canvas, whatever, be sure to allow for some movement of air through the hull. Proper ventilation will assure evaporation of any condensation that may form due to changes in temperature and humidity.

INITIAL TASKS

The authors estimate 75% of engine repair work can be directly or indirectly attributed to lack of proper care for the engine. This is especially true of care during the off-season period. There is no way on this green earth for a mechanical engine, particularly a marine engine, to be left sitting idle for an extended period of time, say for six months and then be ready for instant satisfactory service.

Imagine, if you will, leaving your automobile for six months, and then expecting to turn the key, have it roar to life, and be able to drive off in the same manner as a daily occurrence.

It is critical for a marine engine to be run at least once a month, preferably, in the water, but if this is not possible, then a flush attachment **MUST** be connected either directly to the engine block or in the hose line from the jet pump to the engine.

Move the craft to a body of water or connect a garden hose to the engine cooling water supply fitting or flush fitting on the cylinder head.

Start the engine and allow the rpm's to stabilize at idle speed **FOR JUST A FEW SECONDS**, and then turn the water on.

Adjust the water flow until a small trickle is discharged from the bypass outlet on the port side of the hull.

When the engine is to be shut down turn the water off **FIRST** -- raise the aft portion of the hull -- **WHILE THE ENGINE IS OPERATING AT IDLE** -- "rev" the engine just a **COUPLE** times to clear water from the exhaust system -- and then shut it down.

NEVER allow the engine to operate without cooling water for more than 15 seconds.

Operating the engine for more than 15 seconds without circulating water, will quickly lead to an overheat condition. If the engine is not shut down almost immediately, serious and expensive damage may be caused to internal engine components.

It is quite possible for the engine to "freeze" -- lock-up -- because of excessive heat. In such cases it becomes impossible to rotate the crankshaft and a complete removal, tear down, and overhaul is necessary.

A "squared away" SLT750 Series engine compartment. There is no substitute and few valid excuses for not developing good habits leading to regular maintenance practices. The results of a good program are always evident in top performance, maximum enjoyment, and "pride of ownership", all at minimum cost.

NEVER, AGAIN NEVER, operate the engine at high speed with a flush device attached. The engine, operating at high speed with such a device attached, would **RUNAWAY** from lack of load on the impeller, causing extensive damage.

At the same time, the shift mechanism, if the unit is so equipped, should be operated through the full range several times. The steering should also be operated from hard-over to hard-over.

Only through a regular maintenance program can the owner expect to receive long life and satisfactory performance at minimum cost.

The material presented in this chapter is divided into four general areas.

1- General information every owner should know.

2- Maintenance tasks that should be performed periodically to keep the craft operating at minimum cost.

3- Care necessary to maintain the appearance of the craft and to give the owner that "Pride of Ownership" look.

4- Winter storage practices to minimize deterioration during the off-season when the craft is not in use.

In nautical terms, the front of the craft is the **bow**; the rear is the **stern**; the right side, when facing forward, is the **starboard** side; and the left side is the **port** side. All directional references in this manual use this terminology. Therefore, the direction from which an item is viewed is of no consequence, because **starboard** and **port** **NEVER** change no matter where the individual is located or his position -- even standing on his/her head.

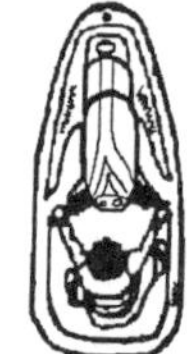

*Common terminology used throughout the world, and of course extensively in this manual, for reference designation on watercraft of **ALL** sizes. "**Port**", "**Starboard**", "**Forward**", and "**Aft**", never change even if an individual is standing on his/her head.*

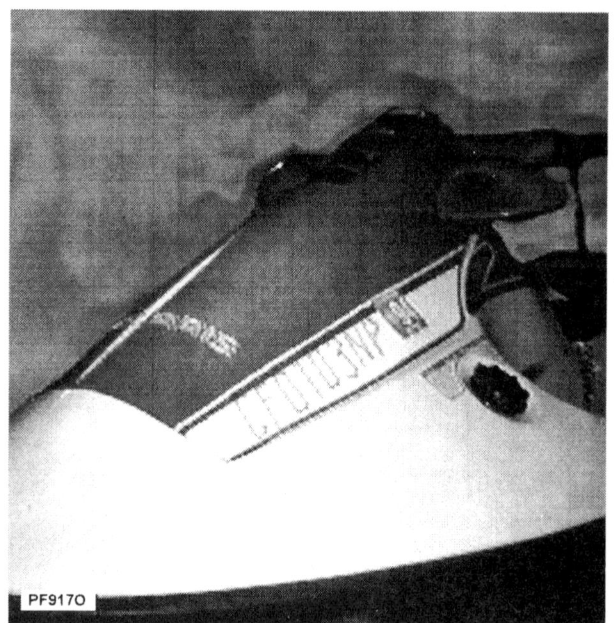

*After a fun day on the water and before moving the craft out of the water, the operator of the craft should **SLOWLY** open the fuel tank cap to release built-up pressure in the tank.*

After Use Tasks

At the end of a day's use, make a habit of opening the engine compartment to allow fumes to escape and sponging out all water in the bottom of the hull before trailering the craft. Leave the engine cover off when the craft is stored overnight to allow any moisture in the compartment to evaporate. Slowly open the fuel tank cap to release built up pressure. Pressure in the tank will increase considerably during hot weather. Releasing the pressure reduces the chance of an explosion and ensures smooth fuel flow when the engine is next started.

If the craft will remain unused for a week, spray storage oil inside the engine and a light film of lubricant, such as WD-40, Super Lube or Justice Bros. 80, on the engine exterior. Keeping the engine exterior lightly oiled at all times will result in water sliding off the engine and draining away rather than accumulating in casting webs -- resulting in corrosion.

Cooling

The engine cooling passages must be flushed with clear water on a regular basis. This is accomplished by simply connecting a "Flushing Kit" per the manufacturer's instructions **OR** by disconnecting the cooling hose from the pump to the engine and connecting a garden hose to the disconnected hose. Such an easy maintenance practice should become a habit and will prevent accumulation of debris which could restrict engine coolant water flow and cause overheating.

Connect a garden hose to the engine cooling water supply fitting or flush fitting on the cylinder head.

Start the engine and allow the rpm's to stabilize at idle speed **FOR JUST A FEW SECONDS,** and then turn the water on.

Adjust the water flow until a small trickle is discharged from the bypass outlet on the port side of the hull.

When the engine is to be shut down, turn the water off **FIRST** -- raise the aft portion of the hull -- **WHILE THE ENGINE IS OPERATING AT IDLE** -- "rev" the engine just a **COUPLE** times to clear water from the exhaust system -- and then shut it down.

NEVER allow the engine to operate without cooling water for more than 15 seconds.

Cooling System Flushing

Connect the garden hose to the engine cooling water supply fitting or flush fitting on the cylinder head.

The manufacturer recommends the engine be started and engine rpm allowed to stabilize at idle speed **FOR A FEW SECONDS,** before turning on the water. If the water is turned on before the engine is started, the water may flow back through the exhaust pipe into the engine. Should this occur the cylinder would fill with water and if the engine was cranking at the time -- trying to start -- the piston would be attempting to compress water and the connecting rod would buckle under the stress.

As soon as the engine stabilizes at idle rpm, turn on the cooling water and adjust the water flow until a small trickle is discharged from the bypass outlet on the portside of the hull.

Perform whatever adjustments or flushing is necessary and then turn off the water **WHILE THE ENGINE IS OPERATING AT IDLE.**

Work **QUICKLY** -- raise the aft section of the craft and "rev" the engine a few times to clear water from the exhaust system.

DO NOT allow the engine to operate without cooling water for more than 15 seconds.

Shut down the engine.

Controlling Corrosion

Since man first started out on the water, corrosion has been his enemy. The first form was merely rot in the wood and then it was rust, followed by other forms of destructive corrosion in the more modern materials.

The higher the concentration of salt in the water, the worse the corrosion which will take place.

One defense against corrosion is to use similar metals in the manufacture of the parts exposed to the water.

A second defense against corrosion is to insulate dissimilar metals. This can be done by using an exterior coating of Sea Skin or by insulating them with plastic or rubber gaskets.

Pump Impeller

The pump impeller is a precisely machined and dynamically balanced aluminum spiral. Observe the drilled recesses at exact locations to achieve the delicate balancing.

Excessive vibration of the jet pump may be attributed to an out-of-balance condition caused by the impeller being struck excessively by rocks, gravel or cavitation "burn".

The term cavitation "burn" is a common expression used throughout the world among people working with pumps, impeller blades, and forceful water movement.

"Burns" on the impeller blades are caused by cavitation air bubbles exploding with considerable force against the impeller

The location of the flush fitting on most Models. First, the cap is removed, then a garden hose attached to the fitting, the engine started, and operated for 10 to 15 seconds before the water is turned on to flush the engine. This is a very important task, especially if the craft has been used in salt water. The water is to be turned off for a few seconds before the engine is shut down.

blades. The edges of the blades may develop small "dime size" areas resembling a porous sponge, as the aluminum is actually "eaten" by the condition just described.

Excessive rounding of the impeller edges will reduce efficiency and performance. Therefore, the impeller should be inspected at regular intervals. If rounding is detected, the impeller should be removed from the housing, placed on a work bench and the edges restored to as sharp a condition as possible using a file. Refer to the Jet Pump chapter for detailed procedures on removal and installation of the impeller.

4-2 SERIAL NUMBERS

Most manufacturers use three different identification numbers stamped, embossed or decaled somewhere on the craft.

The engine serial number is the manufacturer's key to engine changes. This number identifies the year of manufacture, the qualified horsepower rating, and the parts book identification. If any correspondence or parts are required, the engine serial number **MUST** be used or proper identification is not possible.

Federal regulations require all personal watercraft to have a Hull Identification Number (HIN) permanently attached to the craft in two separate locations: one on the transom on the starboard side, above the waterline and the other in an unexposed location, which is usually left to the discretion of the manufacturer This "HIN" number can be most helpful when attempting to retrieve a stolen craft. The "HIN"

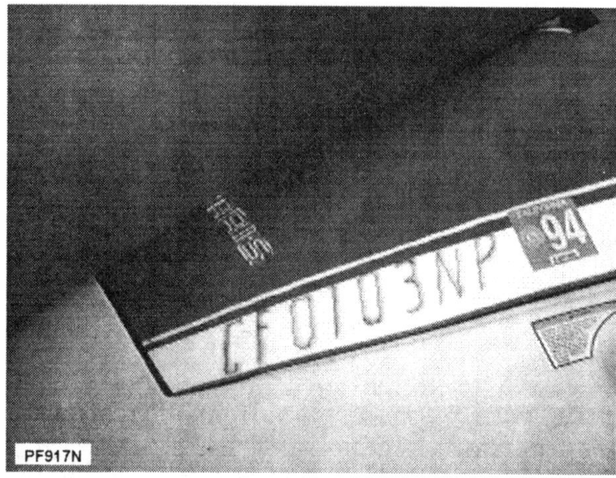

All states in the U.S. and provinces in Canada require a license identification number to be prominently displayed on the forward portion of the hull -- port and starboard.

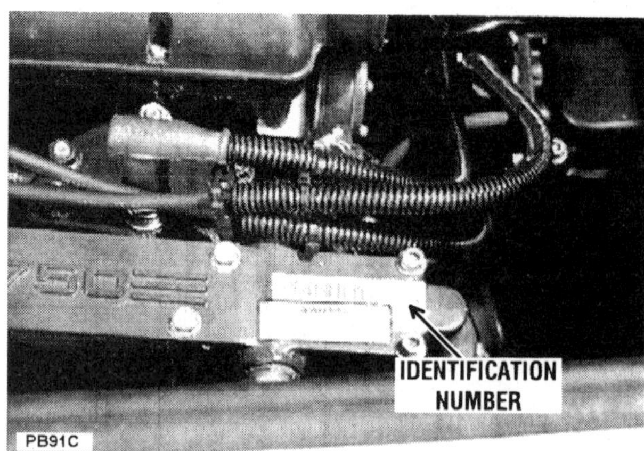

The engine ID number is typically located on the water outlet manifold of most engines covered in this manual.

number is required in most states in order to properly register the craft.

Some manufacturers also use a Primary I.D. Number stamped on a plate attached to the inside of the engine compartment.

ONE MORE WORD

As a theft prevention measure, all metal plates with any type of identification number are especially made wafer thin. Any attempt to remove these plates will result in cracks across the serial number.

4-3 LUBRICATION - COMPLETE UNIT

As with every type mechanical invention with moving parts, lubrication plays a prominent role in operation, enjoyment, and longevity of the unit.

If a personal watercraft is operated in salt water the frequency of applying lubricant to fittings is usually cut in half as compared with the unit being used in fresh water. The few minutes involved in moving around the craft applying lubricant and at the same time making a visual inspection of its general condition will pay in rich rewards with years of continued service.

The first personal watercraft arrived on the scene in the United States in the early 1970s. It is not uncommon to see outboard units well over 20 years of age moving a boat through the water as if the unit had recently been purchased from the current line of models. An inquiry with the proud owner will undoubtedly reveal his main credit for its performance to be regular periodic maintenance. There is no reason this same longevity of performance cannot be true for personal watercraft.

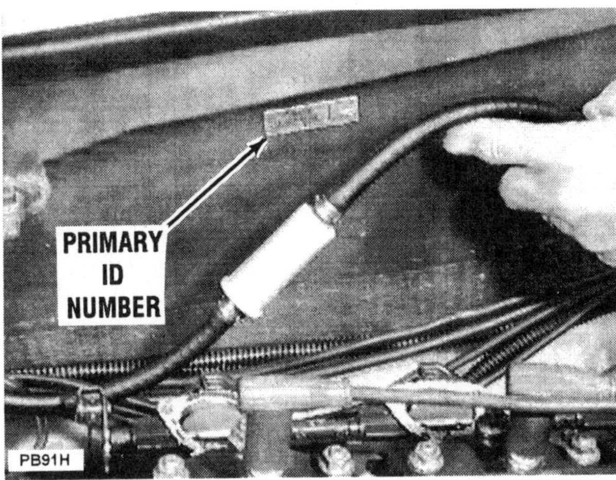

Typical location of the primary ID number. The flame arrestor cover has been removed for photographic clarity.

The accompanying chart can be used as a guide to the frequency of maintenance while the craft is being used during the season. Unless otherwise stated use a good brand name multi-purpose water resistant lubricant. Detailed lubrication instructions are given below for locations requiring special attention.

In addition to the normal lubrication listed in the lubrication chart, the prudent owner will inspect and make checks on a regular basis as listed in the accompanying chart.

Throttle and Choke Cable Lubrication

Apply a small amount of lubrication -- Polaris Cable Lube P/N 2870510 -- to the inner throttle cable and choke cable at the carburetor. Check for smooth throttle opening and closing

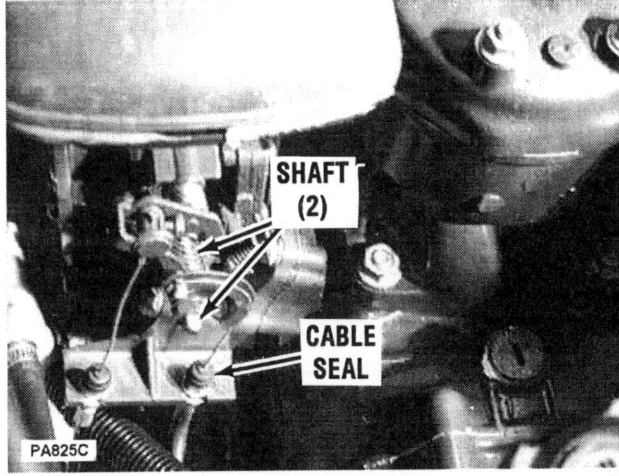

Periodically, apply water-resistant grease to the throttle and choke shafts, springs and cables. This is important if the watercraft is used in a salt water environment, as evidenced by the accompanying illustration on this page of a corroded steering arm.

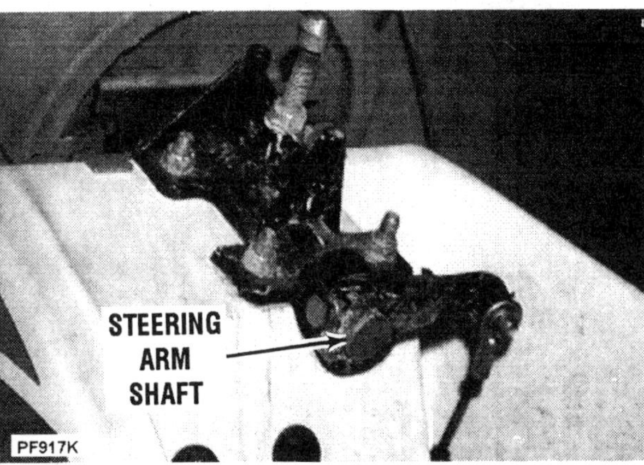

The steering arm shaft requires periodic lubrication. Other model joints are different, but the principle and requirement for lubrication is the same.

in all handlebar positions. Operation of the throttle should be smooth and the lever should return without binding.

Hold the throttle lever in the full open position against the grip. Spray rust inhibitor along the exposed length of cable, or use the Polaris Cable Lube. The cable seals can be moved, to allow lubrication of the cable. Apply an amount of oil to the cable, then push and release the throttle several times to work the lubricant down the throttle cable.

The manufacturer recommends applying water resistant grease to the throttle and choke shafts and springs, and idle stop screw. This should be performed often, if the craft is used in salt water.

A small amount of lubricant may be applied to the choke knob shaft by pulling out the choke knob until it stops.

Steering Cable Lubrication

Spray a small amount of rust inhibitor on the steering cable. Lubricate the handle pivot shaft and bushing using a water resistant grease. Move the handlebars left-to-right several times to allow the lubrication to work its way between the two surfaces.

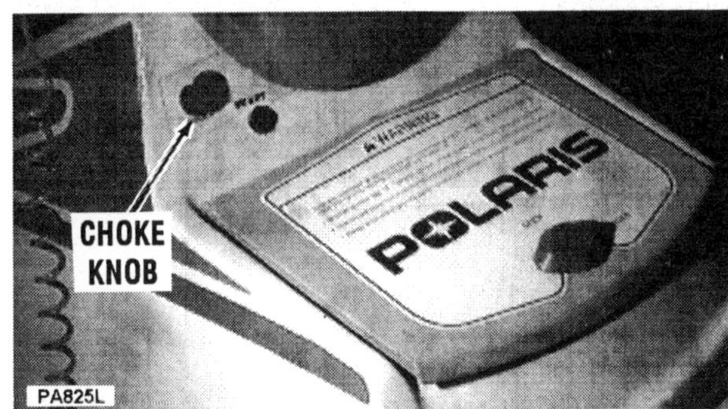

Location of the choke knob on an SLT750 Series watercraft. Lubricate the shaft, then manipulate the knob to "work" the grease into the cable sleeve.

Lubrication Points:	Initially every 10hrs	Initially every 50hrs	Initially every 100hrs	Then every 100hrs	Then every 200hrs
Carburetor throttle shaft			❖	❖	
Throttle cable - engine end			❖	❖	
Throttle cable - handlebar stop			❖	❖	
Choke cable - engine end			❖	❖	
Choke knob shaft			❖	❖	
Steering nozzle pivot shaft			❖	❖	
Steering post arm ball joint			❖	❖	
Reverse gate pivot pins			❖	❖	
Jet pump			❖	❖	
Seat latch			❖	❖	
Maintenance Tasks:					
Check engine alignment					❖
Clean and gap spark plugs	❖	❖	❖	❖	
Adjust carburetor idle	❖		❖	❖	
Adjust carburetor mixture	❖		❖	❖	
Drain and inspect fuel tank					❖
Clean and inspect fuel filter	❖	❖	❖	❖	
Inspect fuel line			❖	❖	
Inspect check valve			❖	❖	
Adjust throttle control			❖	❖	
Adjust choke system			❖	❖	
Adjust steering system			❖	❖	
Inspect impeller		❖	❖	❖	
Check impeller to pump case clearance			❖	❖	
Inspect rubber coupling					❖
Inspect electrical connections	❖	❖	❖	❖	
Check battery electrolyte level	❖	❖	❖	❖	
Clean bilge strainer			❖	❖	
Inspect rubber water seal	❖		❖	❖	

Fuel/Oil Mixture

All engines covered in this manual are equipped with an oil injection system. Such a system replaces the age-old method of mixing oil with gasoline prior to filling the fuel tank. However, additional oil should be added to the fuel during a "break-in" period following an engine overhaul or following a long storage period.

"Break-in" Lubrication

In order to obtain extra lubrication during the first 10 hours of "break-in", the manufacturer recommends a premix of 25:1 mixture be used in the fuel tank. Therefore, any existing unused fuel in the tank should be removed before adding the premixed solution. This ratio will **ENSURE** adequate lubrication of moving parts which have been drained of oil during the storage period.

Professional mechanics who own oil injected units add an extra ounce of oil per each gallon of fuel. The addition of this extra oil may prevent engine seizure in the event of an oil pump failure or clog in an oil delivery line.

Use only outboard marine oil in the mixture, never automotive oils. Four-stroke automotive engine oil is not formulated to burn completely -- only to lubricate. Therefore, automotive oil, if used in a two-stroke engine will leave an undesirable residue.

Jet Pump Lubrication

The forward end of the pump shaft is supported by the bearing housing at the bulkhead. The aft end of the driveshaft is supported by the bearing and seal housing. Both of these bearing and supports are sealed units with "lifetime" lubrication ability. Therefore, no maintenance lubrication is necessary.

Location of the oil injection pump on some Series engines -- starboard side. Additional oil should be added to the fuel during the "break in" period, or following an engine overhaul.

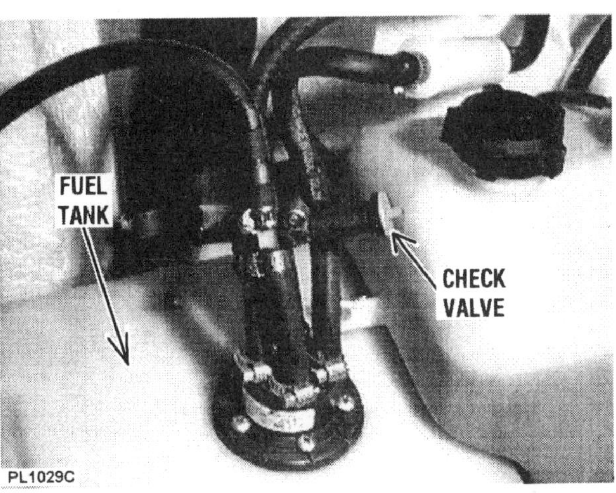

Typical location of the one-way check valve installed in the fuel tank vent line.

4-4 INSPECTION AND SERVICE

Fuel Tank Check Valve
SL650, SL/T750 Only

A check valve is installed at the end of the fuel tank pickup (main) line. If the valve is believed to be defective, it may only be removed with kit Part No. 2200620 installed. The valve will permit air to pass into the tank to replace the fuel drawn out, but will not allow fuel to flow in the opposite direction.

Air Inlet Check Valve
Most Models

To test the air inlet check valve, obtain a portable vacuum device, such as the Mity Vac. Disconnect the hose from the valve and lift the valve free of the compartment. Apply approximately 5 psi pressure to the valve end marked "VAC". The valve should hold steady pressure. Connect the tester to the other end and apply pressure. Air should pass through the valve

Typical installation of a fuel/water separator bowl, installed on all watercraft covered in this manual. Actual location on the starboard hull may vary by model.

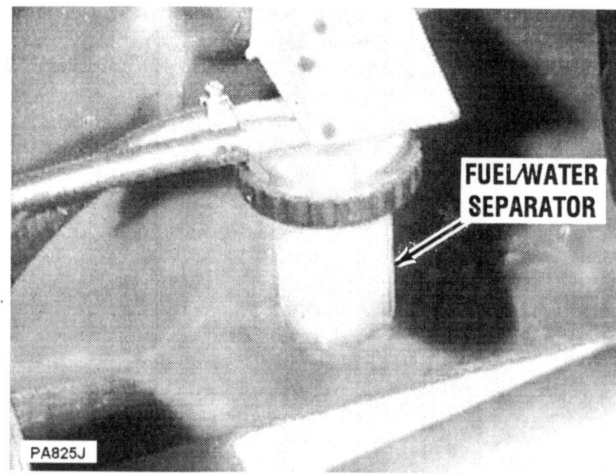

After 1994, the manufacturer equipped the water/fuel separator with an integral screen to aid in visual detection of water or other contaminants in the fuel supply.

with no pressure build up. If the valve holds pressure both ways, or fails to hold pressure one way, the valve must be replaced.

If a portable vacuum device is not available, the basic principle applies. Attempt to blow air through the valve end. Air should pass through one way, but not from the other end. If air will pass through the valve in either direction or if no air will pass through -- the valve is defective and must be replaced.

Fuel/Water Separator

A fuel/water separation bowl is installed on most models covered in this manual. From 1995 on, these devices have an integral screen -- some having a float, making water detection easier.

To service the unit, turn the fuel valve selector to the OFF position. Grasp the bowl with a shop rag or towel to catch any fluid which might spill during removal. Unscrew the ring nut, remove the bowl and inspect it for water and sediment. Discard the contents and replace the bowl -- tightening the ring nut just hand tight.

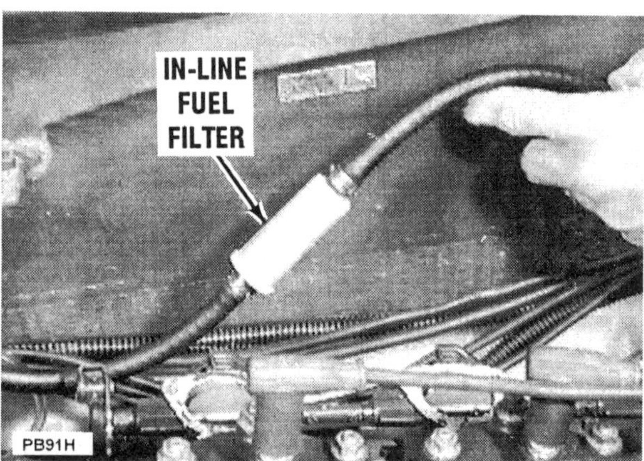

Typical in-line fuel filter, located on the starboard hull. Some Models are equipped with two fuel filters. The flame arrestor has been removed for photographic clarity.

Fuel Filter

All units covered in this manual have an in-line fuel filter, typically "free standing" -- not mounted on the inner hull, as shown in the accompanying illustration. Some models will have two in-line filters.

Inspect the filter/s for evidence of sediment. If contaminants are visible, the filter should be replaced -- it cannot be serviced. It is strongly recommended to obtain the correct in-line filter, do not use a substitute.

If replacing the filter, "pinch off" both hoses near the ends, with clamps. Now remove the filter. Install the new unit, ensuring it is placed in the proper direction of fuel flow. Install the fuel hose clamps. Remove the other clamps once the filter is secured in place.

Oil Filter

Most models are equipped with an oil filter, similar to the in-line fuel filter. This filter must be replaced if necessary, it is not serviceable. Replace the oil filter in the same fashion the in-line fuel filter was replaced.

Drain Plug

A drain plug installed in the transom, permits water in the bilges to be easily removed. Inspect the condition of the **O**-ring - threaded plugs only. Replace the plug as required.

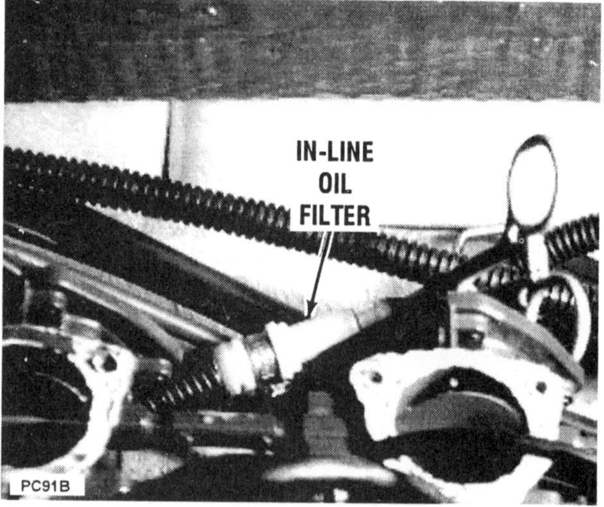

Location of the in-line oil filter on the starboard hull. Actual location of the filter will vary by Model.

Pressure seal type plugs may be replaced with threaded type plugs -- more convenient to use -- by installing kit P/N 2871751 and following the included instructions.

If the craft is to remain out of the water -- in storage but outside -- the plug may be left out to permit any rain water or melted snow which has found its way into the bilge, to drain free of the craft. **REMEMBER** to replace the plug before moving the craft into the water during the next outing or season!

FLUSHING -- COOLING AND BILGE SYSTEMS

Cooling System

A regular maintenance program should include flushing the cooling system to prevent sand or salt deposits from accumulating in the system. It is certainly worth the time and effort to flush after each outing and especially if the craft is used in salt water.

Remove the fitting cap, as shown in the accompanying illustration, and then connect a garden hose to the fitting.

Start the engine and allow it to idle for just a short time **BEFORE** turning the water on. The engine **MUST** be idling before the water is turned on to prevent water from back flowing through the exhaust pipe into the engine. Such action could cause severe and expensive damage to engine internal parts.

After the water is turned on, adjust the amount of flow until a "tattle-tale" stream of water is observed exiting from the bypass outlet on the starboard side of the craft.

Now, allow the engine to operate at idle speed for a few minutes with the water running, and then turn the water off, but allow the engine to continue idling. Increase engine rpm a couple of times for just a few seconds -- not more than 15 seconds -- to clear water from the exhaust system. Operation of the engine without water for a longer period could cause severe and expensive damage.

Shut down the engine, disconnect the garden hose and cover the fitting with the cap.

Bilge System

The bilge system must be kept in an operable condition at all times in order to perform properly in an emergency situation.

To **ENSURE** a "ready" condition of the bilge system and to prevent any clogging, the system should be inspected and cleaned, whenever debris is observed.

Location of the flush fitting at the aft end of the water outlet manifold on most Models.

Most models have two bilge siphon fittings and hoses with pick-up screens. A few models will only have one siphon hose and an electric bilge pump, while some models may have only the electric bilge pump.

The siphon and breather hoses connect together via elbow fittings, attached to the upper hull, inside the engine compartment. Check to be sure the small .040" breather hole in the side of the elbow fittings are clear to prevent siphoning water into the bilge.

4-5 IMPELLER-TO-PUMP CASE CLEARANCE

SPECIAL WORDS

The following procedure may be performed with the watercraft elevated on saw horses enabling a person to work underneath, or with the water craft on its side allowing access to the rock grate and impeller.

If the impeller-to-pump case clearance is to be determined with the water craft raised on

The bilge breather assembly is secured to the aft bulkhead in the engine compartment. Check to be sure the breather hole is free of obstructions.

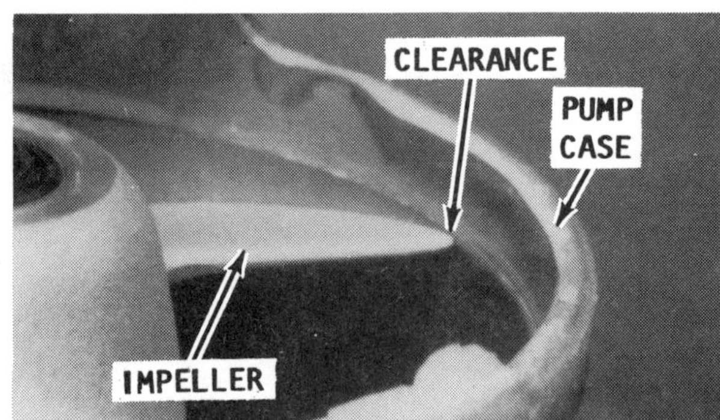

Typical location to show clearance area between the impeller blades and the pump case, as explained in the text. Clearance limits are also given in the text.

sawhorses, the battery may be left secured in the craft. If the watercraft is to be positioned on its side, first disconnect the electrical leads at both battery terminals, and then remove the battery from the craft.

Remove the bolts and washer securing the rock grate to the jet pump. If the seal between the hull and the pump housing is broken, or the foam packing is damaged, reseal the housing and intake area with silicone sealant to prevent impeller cavitation.

Insert a feeler gauge between each impeller blade and the pump case and determine the clearance. Make a note of the clearance, for each blade.

Determine the average of the clearances, by first adding them for a total, and then dividing by the number of impeller blades.

The clearance between the impeller blade and pump case on a new jet pump is referred to as "standard" clearance. On all models, the standard clearance is 0.002 - 0.008" (0.05 - 0.20mm). The maximum allowable clearance, or "service limit", is 0.020" (0.5mm).

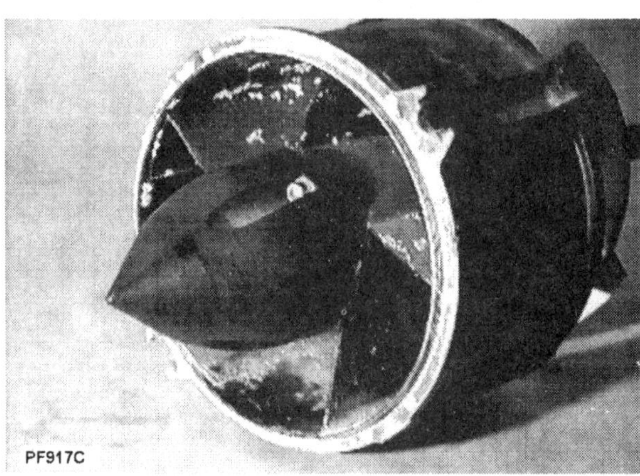

The jet pump shown was not thoroughly rinsed after use in salt water. Various subsystems of the watercraft will suffer the same fate as the jet pump. Not only will performance suffer, due to corrosion, it will probably be necessary to replace expensive parts.

Clearance Results

If the clearance is less than the maximum value listed, no action is necessary.

If the clearance is more than the service limit, inspect the condition of the pump case. If the pump case has scratches deeper than 0.04" (1mm), replace the pump case. If the pump case is satisfactory, the problem must be with the impeller.

Visually inspect the impeller for nicks, scratches, pitting or a "mushroomed" edge. If the cause of the excessive clearance cannot be determined visually, the pump **MUST** be disassembled and the pump case measured and compared to specifications.

The jet pump must be removed from the craft to perform this work. Detailed illustrated procedures are presented in Chapter 10 -- Jet Pump.

If it is determined the impeller must be replaced, it is **IMPERATIVE** that the correct impeller is obtained. Any other impeller will most definitely not be matched up to the specific engine, therefore -- serious damage to the engine may result due to the mismatch (unbalancing) of equipment.

4-6 PRE-SEASON PREPARATION

Satisfactory performance and maximum enjoyment can be realized if a little time is spent in preparing the unit for service at the beginning of the season. Assuming the unit has been properly stored, as outlined in Section 4-10, a minimum amount of work is required to prepare the unit for use.

The following steps outline an adequate and logical sequence of tasks to be performed before using the craft the first time in a new season.

1- Lubricate the craft according to the manufacturer's recommendations. Refer to the lubrication table on Page 4-6. Remove, clean, inspect, adjust, and install the spark plugs with a new gasket (if they require a gasket). Make a thorough check of the ignition system. This check should include: the wiring and the battery electrolyte level and charge. Many modern craft use a maintenance free battery, therefore this check is not necessary. If the exhaust outlet was plugged at the end of the last season, remove the plug.

2- Take time to check the gasoline tank and all fuel lines, fittings, couplings, valves, including the flexible tank fill and vent. If the fuel

Opening the electrical box exposes the rectifier/regulator, cranking motor relay, ignition coils and CDI unit, and many ignition and electrical leads which are all easily accessible.

was not drained at the end of the previous season, it may have "soured" and give off an order of rotten eggs.

Check carefully for any gum deposits which may have formed. Such gum can clog the filters, lines, and passageways in the carburetor. To prevent gum from forming, a fuel additive such as "Sta-Bil" can be added to the fuel at the end of the season. Such an additive will prevent fuel from "souring" for up to twelve full months.

GOOD WORDS

All manufacturers recommend the fuel filters be replaced -- or cleaned if applicable -- at the start of each season or at least once a year. The in-line filter/s are easily accessible. Substitutions should not be made when replacing these units. Ensure the arrow embossed on the filter housing points in the direction of the flow of fuel.

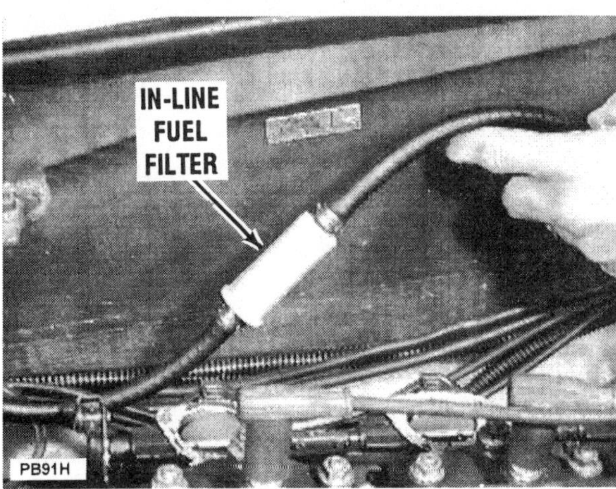

The in-line fuel filter is non-serviceable and should be replaced at the beginning of each new season. Check to be sure the embossed arrow is aligned with the fuel flow during installation.

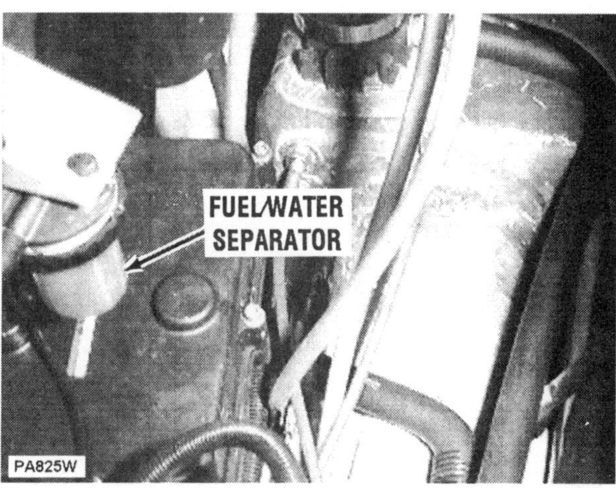

The fuel/water separator bowl should always be removed and drained as a preseason preparation task.

Most models have a fuel/water separator unit, which is also easily accessible. The separator bowl should be drained as a preseason preparation task.

3- Inspect all water drains, making sure they are fit for service. Close all water drains. Check and replace any defective water hoses. Check to be sure the connections do not leak. Replace any spring-type hose clamps, if they have lost their tension -- or if they have distorted the water hose -- with band-type clamps.

4- The engine can be run with the jet pump in water to flush it. If this is not practical, a flush attachment may be used. This device is attached to the engine water supply hose, between the engine and the jet pump.

Connect a garden hose to the engine cooling water supply fitting, or flush fitting connected to

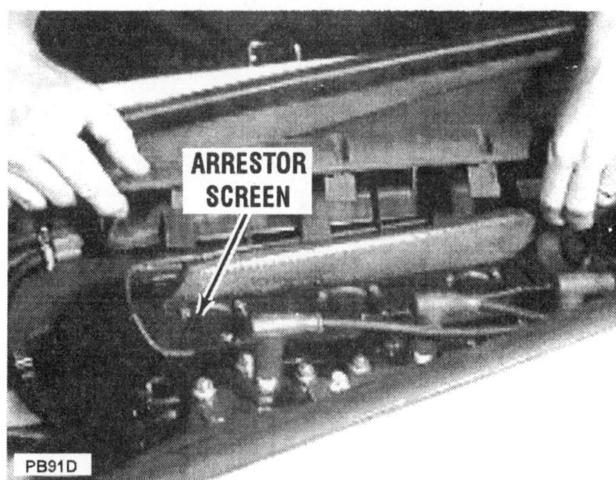

The flame arrestor screen should be removed and cleaned at the beginning of each season and on a regular basis as a maintenance task. Some Models are equipped with separate arrestor screens -- one per carburetor.

Check all hose clamps to be sure they are in good condition and tightened properly.

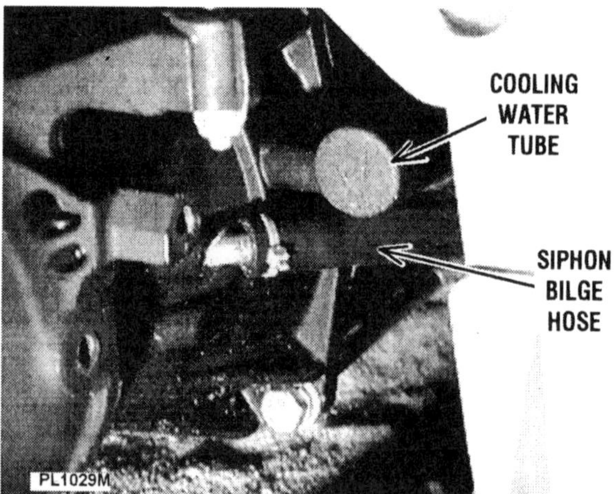

Close view of the cooling water connection and the bilge water system fitting at the aft end of the jet pump.

the water outlet manifold located atop the cylinder heads.

Start the engine and allow the rpm's to stabilize at idle speed **FOR JUST A FEW SECONDS,** and then turn the water on.

Adjust the water flow until a small trickle is discharged from the bypass outlet fitting on the hull of the watercraft.

When the engine is to be shut down, turn the water off **FIRST** -- raise the aft portion of the hull -- **WHILE THE ENGINE IS OPERATING AT IDLE** -- "rev" the engine just a **COUPLE** of times to clear water from the exhaust system -- and then shut it down.

NEVER allow the engine to operate without cooling water for more than 15 seconds.

Check the exhaust outlet for water discharge. Check for leaks.

5- Check the electrolyte level in the battery -- if applicable, due to the growing popularity of "maintenance-free" or sealed batteries -- and check the voltage for a full charge. Clean and inspect the battery terminals and cable connections. **TAKE TIME** to check the polarity, if a new battery is being installed. Cover the cable connections with lubricant or special protective compound as a prevention to corrosion formation. Check all electrical wiring and grounding circuits.

6- Check all electrical parts on the engine and lower portions of the hull to be sure they are of a type to prevent ignition of an explosive atmosphere. Rubber caps help keep spark insulators clean and reduce the possibility of arcing. Electric cranking motors and high-tension wiring harnesses should be of a marine type to prevent an explosive mixture from igniting.

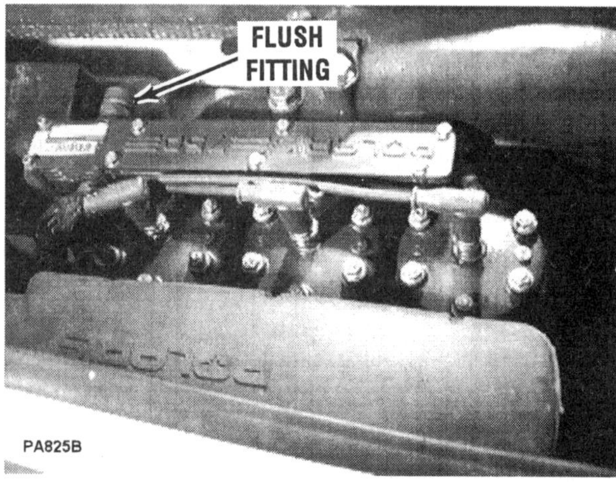

Location of the flush fitting on the water outlet manifold on an SLT750 Series engine.

Many late model units are equipped with what is termed a "Maintenance Free" battery. As the name implies, regular maintenance regarding electrolyte level and hydrometer reading is not necessary. The battery must be secured to prevent even the slightest movement during the most violent maneuvers of the craft.

4-7 SEALANTS, ADHESIVES, LUBRICANTS, AND FUEL STABILIZERS

It is common practice for the larger manufacturers of personal watercraft to market their own line of products for use on their craft. Polaris chemical engineers have developed such a line available for use with Polaris personal watercraft. Throughout this manual, the authors recommend the application of the manufacturer's products as a first choice. All products listed are alternatives of equal value, and may be used with confidence if the manufacturer's line is not available.

Sealants and Adhesives

Four sealants are recommended and they are **NOT** interchangeable. Each is designed to perform under a different set of conditions. Follow the directions on the package for cleaning and preparing surfaces. Sealants and adhesives **MUST** be applied **ONLY** to clean and dry parts. Apply sparingly -- excessive amounts may block oil passageways and cause serious damage.

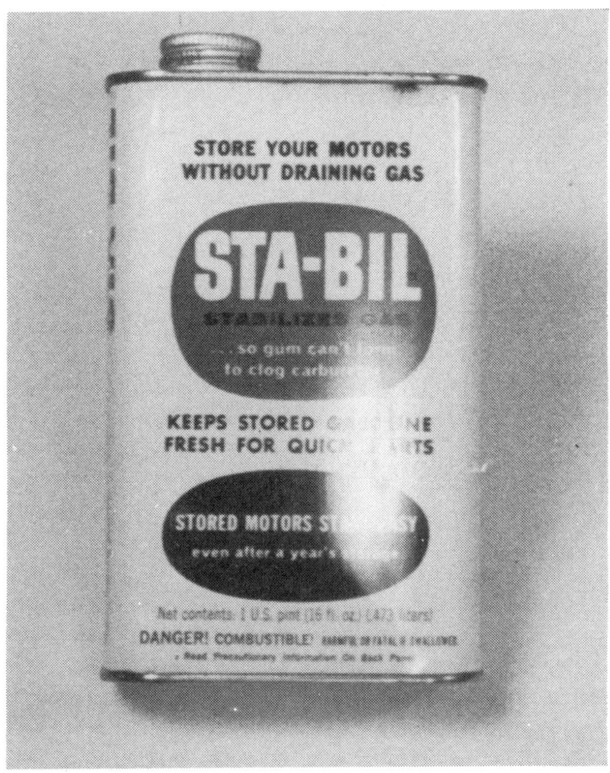

*A gasoline stabilizer and conditioner may be used to prevent fuel from "souring" for up to twelve full months. However, draining fuel and adding new at the beginning of each season is still the best practice to **ENSURE** top performance.*

Loctite Lock N' Seal is a non-hardening, non-permanent locking compound. This material is recommended for application to the threads of load bearing fasteners. Loctite helps prevent loosening of the bolt due to vibration, thread wear, and corrosion.

Loctite Stud N' Bearing Mount compound is also a non-hardening, non-permanent locking compound. This material is recommended for application to the threads of load bearing fasteners submerged under water.

Loctite Superflex is a water resistant silicone sealer which provides a very effective flexible seal and is able to withstand high temperatures. This sealer is recommended for use where gaskets are required next to a metal surface at high temperature, for example, on the exhaust manifold cover gaskets. Loctite Superflex is blue in color.

3-Bond 1215 - Recommended by Polaris (Polaris P/N 2871557), this sealant is a non-hardening material and is recommended for metal-to-metal joints such as crankcase halves. This substance is resistant to oil and gasoline.

Lubricants

Different lubricants are recommended in the lubrication procedures presented in this manual. Generally, these lubricants are **NOT** interchangeable, each is designed to perform under varying conditions. Following are descriptions of several lubricants recommended.

Polaris All Season grease is a general marine lubricant, recommended for application to bearings, bushings and oil seals.

Shell Alvania EP1 is a general marine lubricant, chemically formulated to resist salt water. This lubricant is also recommended for application to bearings, bushings, and oil seals.

Polaris Premium 2 Cycle lubricant is a two-stroke engine oil. This oil reduces carbon deposits and ensures maximum protection against engine wear. It contains an ashless detergent to keep piston rings "free". Oil additives are usually not recommended by the manufacturer, and in some cases the use of such a substance may invalidate the warranty.

Fuel Stabilizer

"Sta-Bil" Fuel Conditioner and Stabilizer is recommended during engine operation and during the storage period. This fluid absorbs water in the fuel system and protects against corrosion.

If used during operation, this fuel additive will prevent the formation of gum and varnish deposits and greatly extend the period between required carburetor overhauls.

When added to the fuel during storage, the additive will prevent the fuel from "souring" for up to twelve full months.

4-8 FIBERGLASS HULLS

Fiberglass reinforced plastic hulls are tough, durable, and highly resistant to impact. However, like any other material they can be damaged. One of the advantages of this type of construction is the relative ease with which it may be repaired. Because of its break characteristics, and the simple techniques used in restoration, these hulls have gained popularity throughout the world.

A fiberglass hull has almost no internal stresses. Therefore, when the hull is broken or stove-in, it retains its true form. It will not dent to take an out-of-shape set. When the hull sustains a severe blow, the impact will be either absorbed by deflection of the laminated panel or the blow will result in a definite, localized break. Repairs are usually confined to the general area of the rupture.

Most modern marinas, especially those in or with salt water access, provide the boater with fresh water to rinse their craft before heading for home. Such a practice certainly helps to preserve the wax and paint job for that "Pride of Ownerhip" appearance.

Cleaning, Waxing, and Polishing

Any craft should be washed with clear water after each use to remove surface dirt and any salt deposits from use in salt water. Regular rinsing will extend the time between waxing and polishing. It will also give "pride of ownership", by having a sharp looking piece of equipment. Elbow grease, a mild detergent, and a brush will be required to remove stubborn dirt, oil, and other unsightly deposits.

Avoid harsh abrasives or strong chemical cleaners. A white buffing compound can be used to restore the original gloss to a scratched, dull, or faded area. The finish should be thoroughly cleaned, buffed, and polished at least once each season. Take care when buffing or polishing with a marine cleaner not to overheat the surface being worked, because the area will become "burned".

4-9 SUBMERGED ENGINE SERVICE

A submerged engine is always the result of an unforeseen accident. Once the craft is recovered, special care and service procedures **MUST** be closely followed in order to return the unit to satisfactory performance.

NEVER, again we say **NEVER** allow an engine that has been submerged to stand more than a couple hours before following the procedures outlined in this section and making every effort to get it running. Such delay will result in serious internal damage. If all efforts fail and the engine cannot be started after the following procedures have been performed, the engine should be disassembled, cleaned, assembled, using new gaskets, seals, and O-rings, and then started as soon as possible.

Submerged engine treatment is divided into three unique problem areas: submersion in salt water; submerged while powerhead was running; and a submerged unit in fresh water.

The most critical of these three circumstances is the engine submerged in salt water, with submersion while running, a close second.

Salt Water Submersion

NEVER attempt to start the engine after it has been recovered. This action will only result in additional parts being damaged and the cost of restoring the engine increased considerably. If the engine was submerged in salt water the complete unit **MUST** be disassembled, cleaned, and assembled with new gaskets, O-rings, and seals. The corrosive effect of salt water can only be eliminated by the complete job being properly performed.

Submerged While Running
Special Instructions

If the engine was running when it was submerged, the chances of internal engine damage is greatly increased. Remove the spark plugs to prevent compression in the cylinders. Use a socket wrench on the flywheel nut to rotate the crankshaft. If the attempt fails, the chances of serious internal damage, such as: bent connecting rod, bent crankshaft, or damaged cylinder, is greatly increased. If the crankshaft cannot be rotated, the engine must be completely disassembled.

CRITICAL WORDS

Never attempt to start an engine that has been submerged. If there is water in the cylinder, the piston will not be able to compress the liquid. The result will most likely be a bent connecting rod.

Submerged Engine — Fresh Water
SPECIAL WORDS

As an aid to performing the restoration work, the following steps are numbered and should be followed in sequence. However, illustrations are not included with the procedural steps because the work involved is general in nature.

In the United States, Coast Guard regulations require PWC's to contain sufficient built-in floatation -- usually styrofoam -- to prevent the craft from sinking when filled with water and the authorized number of persons aboard.

1- Recover the craft as quickly as possible.

2- Remove the lanyard cord, seat and spark plugs.

3- Remove the drain plugs to empty water out the bilge.

4- Flush the outside of the engine with fresh water to remove silt, mud, sand, weeds, and other debris. **DO NOT** attempt to start the engine if sand has entered any part of the engine. Such action will only result in serious damage to engine components. Sand in the engine means the unit must be disassembled.

Tip the watercraft to the **PORT** side until the spark plug holes are slightly below the horizontal position.

CRITICAL WORDS

Never attempt to start an engine that has been submerged. If there is water in the cylinder, the piston will not be able to compress the liquid. The result will most likely be a bent connecting rod.

5- Remove as much water as possible from the engine. Most of the water can be eliminated by first holding the engine in a horizontal position with the spark plug holes **DOWN**, and then rotate the flywheel with a socket wrench on the flywheel nut. Rotate the crankshaft through at least 10 complete revolutions. If you are satisfied there is no water in the cylinders, proceed with Step 6 to remove moisture.

6- Alcohol will absorb moisture. Therefore, pour alcohol into the carburetor throat and again rotate the crankshaft.

7- Pour alcohol into the spark plug openings and again rotate the crankshaft.

8- With assistance, rotate the craft into the horizontal position until the spark plug openings are facing **DOWN.** Pour engine oil into the carburetor throat and, at the same time, rotate the crankshaft to distribute oil throughout the crankcase.

9- With the craft upright, pour approximately one teaspoon of engine oil into each spark plug opening. Rotate the crankshaft to distribute the oil in the cylinders.

10- Install and connect the spark plugs.

11- Install the drain plugs and lower the seat into position.

Move the craft to a body of water or connect a garden hose to the engine cooling water supply fitting or flush fitting on the cylinder head.

12- Obtain **FRESH** fuel and attempt to start the engine.

If using a garden hose, after the engine starts, allow the rpm's to stabilize at idle speed **FOR JUST A FEW SECONDS,** and then turn the water on.

Adjust the water flow until a small trickle is discharged from the bypass outlet on the port side of the hull.

Allow the engine to run for approximately an hour to eliminate any unwanted moisture remaining in the engine.

When the engine is to be shut down, turn the water off **FIRST** -- raise the aft portion of the hull -- **WHILE THE ENGINE IS OPERATING AT IDLE** -- "rev" the engine just a **COUPLE** times to clear water from the exhaust system -- and then shut it down.

NEVER allow the engine to operate without cooling water for more than 15 seconds.

CAUTION

Water must circulate through the jet pump -- to and from the engine, anytime the engine is operating. Circulating water will prevent overheating -- which could cause damage to moving engine parts and possible engine seizure.

13- If the engine fails to start, determine the cause, electrical or fuel, correct the problem, and again attempt to get it running. **NEVER** allow an engine to remain unstarted for more than a couple hours without following the procedures in this

The flush fitting is located on the water outlet manifold for all models covered in this manual. Specific location varies slightly from model to model.

section and attempting to start it. If attempts to start the engine fail, the unit should be disassembled, cleaned, assembled, using new gaskets, seals, and O-rings, as **SOON** as possible.

4-10 WINTER STORAGE

Taking extra time to store the craft properly at the end of each season, will increase the chances of satisfactory service at the next season. **REMEMBER,** idleness is the greatest enemy of a marine engine. The unit should be run on a monthly basis. The steering and shifting mechanism should also be worked through complete cycles several times each month. The owner who spends a small amount of time involved in such maintenance will be rewarded by satisfactory performance, and greatly reduced maintenance expense for parts and labor.

Proper storage involves adequate protection of the unit from physical damage, rust, corrosion, and dirt.

The following steps provide an adequate maintenance program for storing the unit at the end of a season.

1- Empty all fuel from the carburetor.

For many years there has been the widespread belief simply shutting off the fuel at the tank and then running the engine until it stops is the proper procedure before storing the engine for any length of time. Right? **WRONG!**

First, it is **NOT** possible to remove all fuel in the carburetor by operating the engine until it stops. Considerable fuel is trapped in the float chamber and other passages and in the line leading to the carburetor. The **ONLY** guaranteed method of removing **ALL** fuel is to take the time to remove the carburetor, and drain the fuel.

Secondly, if the engine is operated with the fuel supply shut off until it stops, the fuel and oil mixture inside the block is removed, leaving bearings, pistons, rings, and other parts without any protective lubricant.

Remove the spark plugs and pour about one ounce of two-cycle engine oil into each spark plug hole. Wait a minute or two and then slowly rotate the flywheel/crankshaft through two complete revolutions to evenly distribute the oil. Install the spark plugs.

2- Drain the fuel tank and the fuel lines. Pour approximately one quart (0.96 liters) of benzol (benzine) into the fuel tank, and then rinse the tank and pickup filter with the benzol. Drain the tank. Store the craft with the engine compartment door open to allow air to circulate around the tank and with the fuel vent **OPEN** to allow air to circulate through the tank.

3- Clean the in-line fuel filter with benzol, see the Fuel chapter.

4- Lubricate the throttle and shift linkage. Lubricate the steering pivot shaft with water resistant multi-purpose lubricant.

5- Plug the exhaust outlet with a shop towel or similar item to prevent any type of contaminant from entering the exhaust system.

Battery Storage

Remove the battery from the craft and keep it charged during the storage period. Clean the battery thoroughly of any dirt or corrosion, and then charge the battery to the full specific gravity reading. After charging, store the battery in a clean cool dry place where it will not be damaged or knocked over.

NEVER store the battery with anything on top of it or cover the battery in such a manner as to prevent air from circulating around the filler caps. All batteries, both new and old, will discharge during periods of storage, more so if they are hot than if they remain cool. Therefore, the electrolyte level and the specific gravity should be checked at regular intervals. A drop in the specific gravity reading is cause to charge the battery back to a full reading.

One of the largest personal watercraft battery manufacturers recommends that a battery in storage should be checked every two weeks and charged if necessary. The electrolyte level can also be replenished at this time to prevent sulfation occurring.

In cold climates, **EXERCISE CARE** in selecting the battery storage area. A fully-charged battery will freeze at about 60° below zero. A discharged battery, almost dead, will have ice forming at about 19° above zero.

4-11 PRE-SEASON CHECK

Before attempting to start the craft after winter storage, check the steering and throttle action for smooth operation, without binding. If the battery was removed for storage, check to make sure it has a full charge.

Filling the Battery

If the battery is a "Maintenance Free" type battery no maintenance is necessary.

However, if the battery is not a "Maintenance Free" type, proceed as follows.

Remove each filler cap using a pair of pliers. Fill each cell to the proper level with distilled water. The level of electrolyte should be between the upper and lower level marks during operation. When filling the battery, fill to the upper line and allow the battery to stand for a while -- say twenty minutes. Check the level again and replenish, as necessary.

Use a temperature corrected hydrometer to test the specific gravity of the electrolyte. At 20°C (68°F) the hydrometer reading should be 1.26 on the scale. If it is necessary to charge the battery, leave the filler caps lightly resting on the cell open-

The electrical box is secured with a "bungee" type strap atop the battery on several Series watercraft. The box is mounted to an aft bulkhead on other Model configurations.

*The battery **MUST** be so firmly in place, to prevent even the slightest movement during violent maneuvers of the craft. Battery life may be extended considerably if the battery is removed from the craft and stored in a warm dry area with proper air circulation during the off-season period.*

ings to allow the gases to escape. The charging current should not exceed 1.9 amps for 10 hours.

After charging, push each filler cap into place and wipe the top of the battery clean before installation.

Clean the battery posts and battery cable ends with a wire brush to ensure good, clean connections. Clean the top surface of the battery and identify the "POS" or "+" and "NEG" or "-" embossed symbols. Correctly connect the battery cables to the battery observing polarity. If the cables are connected backwards, the ignition system **WILL** be destroyed the first time the engine is started.

When the engine is to be shut down, turn the water off **FIRST** -- raise the aft portion of the hull -- **WHILE THE ENGINE IS OPERATING AT IDLE** -- "rev" the engine just a **COUPLE** times to clear water from the exhaust system -- and then shut it down.

NEVER allow the engine to operate without cooling water for more than 15 seconds.

CAUTION

Water must circulate through the jet pump -- to and from the engine, anytime the engine is operating. Circulating water will prevent overheating -- which could cause damage to moving engine parts and possible engine seizure.

The correct procedure for starting is to obtain an **AEROSOL** can of WD-40. It must be an aerosol canister because the liquid must be atomized, a squirt can will not do the job. **DO NOT** use ether or a commercial starting fluid because they will evaporate almost instantly. These products work well with four-stroke engines equipped with an oil pump, but they do not contain a lubricant essential for a two-stroke engine. WD-40 is an extremely flammable oil.

Make sure both spark plugs are tight, so the mist from the canister will not ignite. Spray the WD-40 through the flame arrestor and continue spraying with **NO** throttle until the engine starts. As soon as the engine is operating, cease spraying the WD-40. The engine may emit some smoke for a few seconds after it starts. Watch the sediment bowl (if so equipped). As soon as it fills, the engine should operate on the regular oil/fuel mixture. Work the throttle to keep the engine at idle speed until it reaches operating temperature, then close the choke. Advance the throttle a few times and the engine should then idle smoothly.

5
TROUBLESHOOTING

5-1 INTRODUCTION

This chapter is divided into six main sections as follows:

5-2 Mechanical engine problems.
5-3 Fuel system problems.
5-4 Ignition system faults.
5-5 Cranking system faults.
5-6 Charging system malfunctions.
5-7 Troubleshooting Charts.

If a change in engine rpm is experienced at normal operating speed, such a change is usually a symptom of an engine problem or trouble in the jet pump. An increase or a decrease in rpm may be an indication of the area to be checked. The following generalizations -- or conditions -- may be helpful in isolating the problem.

GENERAL PROBLEMS

A noticeable change in watercraft performance may be attributed to one or more of the areas listed.

Lower than Normal Engine Rpm

Mechanical problems in the engine:
- Plugged flame arrestor.
- Fouled spark plug/s, incorrect heat range or gap.
- Shorted or weak ignition coil.
- Contaminated fuel.
- Restricted exhaust system.
- Inadequate fuel pump pressure.
- Loss of compression due to blown head gasket.

Mechanical problems in the jet drive:
- Worn or damaged impeller.
- Main bearing failure.
- Rubber coupler failure.

Higher than Normal Engine Rpm

Mechanical problem in the engine:
- None.

Mechanical problem in the jet drive:
- Clogged intake grate -- possibly plastic bag or other debris.
- Line, sea weed, etc. entangled in impeller.
- Ingested sand.

If a reduction in speed is experienced, accompanied with no loss or gain of engine rpm, the loss of performance can fairly accurately be attributed to some type of drag on the craft or restriction of the water ejected from the jet pump.

Drag on the hull is usually caused by damage to the craft.

Many times, drag from the jet pump is caused by a partially lowered reverse gate. If so equipped, or from dragging debris caught in the intake grate.

One of the most hazardous conditions for a personal watercraft is restriction of water through the rack grate caused by a plastic bag, plastic wrap, or by some type of sea weed. Even the smallest reduction in water volume will affect jet pump performance and could quickly lead to engine overheating.

A partially lowered reverse gate may be caused by incorrect cable adjustment or by binding around the gate pivot pins.

Engine Troubleshooting

Troubleshooting must be a well thought-out procedure. Always attempt to proceed with the troubleshooting in an orderly manner. The "shot in the dark" approach will only result in wasted time, incorrect diagnosis, replacement of unnecessary parts, and frustration.

Obviously, if the instructions are to be of maximum benefit as a guide, they must be fully understood and followed in the proper sequence.

When an engine fails to start, the trouble must be localized to one of four general areas: cranking system; ignition system; fuel system; or compression. Each of these areas must be systematically inspected until the trouble is isolated. At that time, detailed tests of the system or area may be made to determine the part causing the problem. The last section of this chapter describes charging system troubleshooting.

Troubleshooting Check

When using this Troubleshooting Check, proceed sequentially through each test until the defect is uncovered. Then skip to the detailed testing procedure and check for that system. For example, if, when using the Troubleshooting Check procedure, the first two systems, cranking and ignition tests OK, but the third test shows there is trouble in the fuel system, then proceed to the detailed test under the Fuel System Troubleshooting Check in Section 5-3.

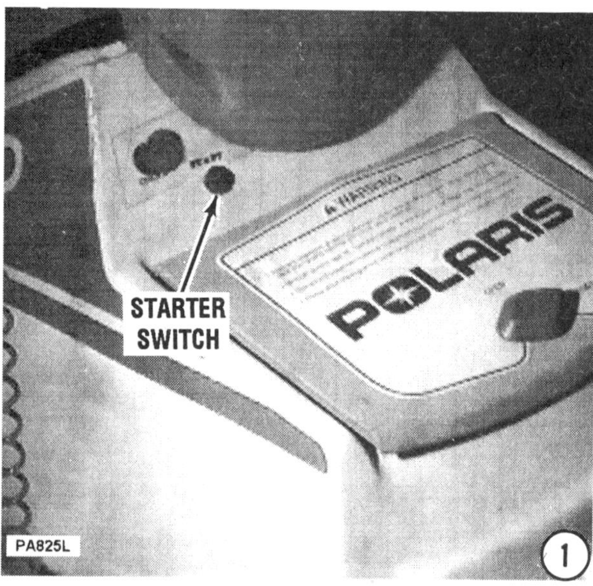

1

Cranking System Test

1- Turn the key switch to the **START** position or press the ignition button. The cranking motor should crank the engine at a normal rate of speed.

If the cranking motor cranks the engine slowly or fails to crank it at all, the trouble is either in the battery or in the cranking system. Test the battery first, then if the battery checks out, proceed directly to Troubleshooting the Cranking System in Section 5-5 for detailed testing procedures to isolate the problem.

Ignition System Test

2- Connect a standard spark tester (available at almost any automotive parts store at a very nominal cost), between the spark plug and the spark plug lead. Hold the spark plug lead with an insulated holder (to avoid receiving a "shock" when the current passes through). Crank the engine with the cranking motor. If there is **NO** spark, or if the spark is very weak, the trouble is in the ignition system. Proceed directly to the Ignition Troubleshooting in Section 5-4 for detailed testing procedures to uncover the problem.

3- Remove the flame arrestor box cover and screen. Look down into the throat of each

2

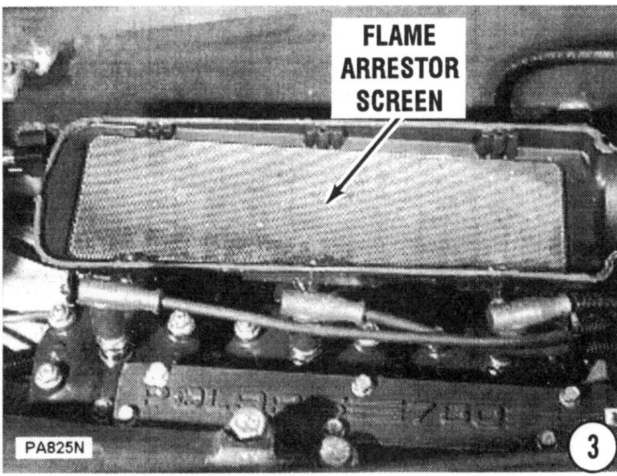

carburetor and at the same time open and close the throttle valve assembly several times. Observe the throttle valves to be sure they move from the fully closed to the fully open positions. Perform the same check with the choke valve assembly.

Attempt to determine the age of the fuel in the fuel tank. Many fuels tend to "sour" (take on the odor of rotten eggs), over a three to four month duration, especially if an additive such as Sta-Bil has not been used. Other fuels take longer to "sour".

In no case, should an attempt be made to start the engine if the fuel in the tank is more than 12 months old. If the tank is drained (siphoned) of fuel, loosen the filler cap to prevent condensation.

Compression Test

4- Good compression is the key to proper engine performance. An engine with worn piston rings or a blown gasket cannot be made to perform satisfactorily until the mechanical defects contributing to low compression are corrected. Generally, a compression gauge is used to determine the cranking pressure within each cylinder.

To make a compression test, remove all spark high tension leads and connect securely to ground on the engine block. Remove the spark plugs and identify from which cylinder they were taken. Evaluation of the spark plug firing end can be most useful in determining how the cylinder is functioning. After the spark plugs have been removed, install a compression gauge into one cylinder and crank the engine.

The throttle valve and choke **MUST** be in the **WIDE OPEN** position in order to obtain maximum readings. Using the starter motor, crank the engine only 3-5 seconds to obtain a maximum reading on the compression gauge. Record the reading. Repeat the process for each of the other cylinders. All readings must be within 5% of each other.

GOOD WORDS

1997 1050cc models only -- The Center cylinder (No. 2) should have approximately 15psi less than the forward (No. 1) cylinder and aft (No. 3) cylinder.

The Standard Compression Value -- open throttle at sea level -- is **120-150lbs/sq**. in. (8.4-10.5kg/sq. cm).

The Standard Service Limit is a **GREATER THAN 5% VARIANCE** between cylinders.

The significance in a compression test is the variation in pressure reading between cylinders.

If a greater variation exists, then the lower reading cylinder should be checked by making a cylinder leak test. A simple and fairly effective leak test can be made by first inserting a teaspoonful of oil into the spark plug opening of the lower reading cylinder, and then cranking the engine a few times to distribute the oil.

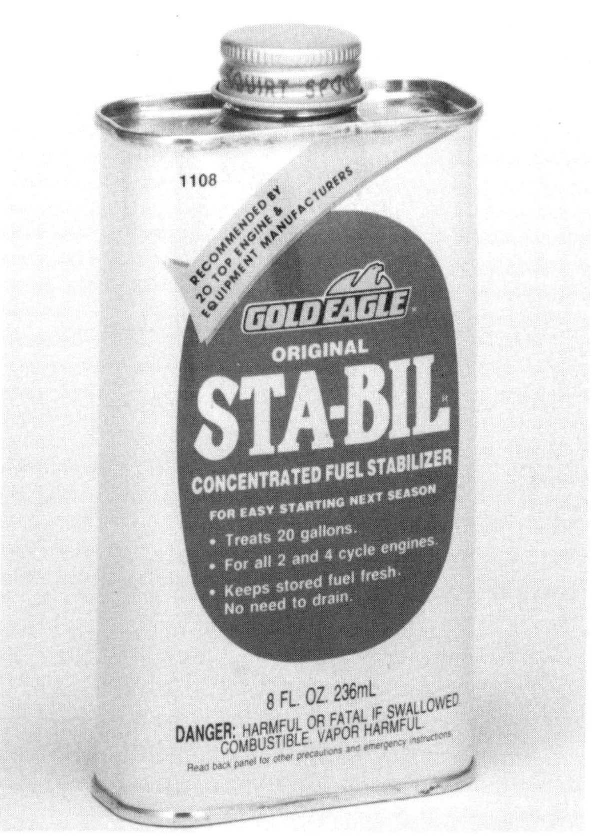

A gasoline stabilizer and conditioner may be used to prevent fuel from "souring" for up to twelve full months.

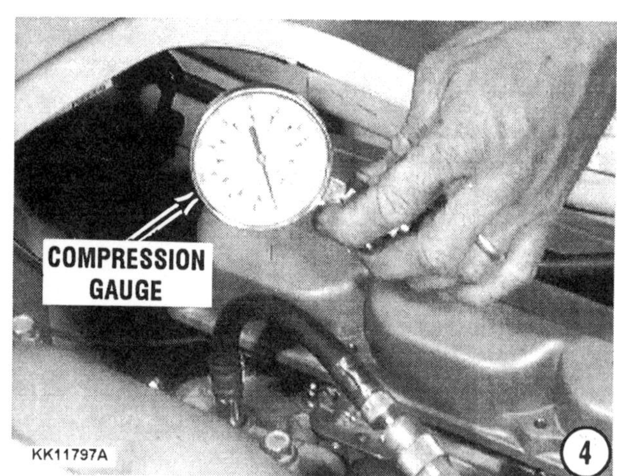

Check the compression again to see if inserting the oil caused a change. The oil helps make a temporary seal around the rings and increases the compression. If the reading increased, the compression loss was probably due to worn rings. If the reading did not change appreciably, the loss may be due to a blown head gasket.

5-2 MECHANICAL ENGINE PROBLEMS

Jumping Waves

Even though "jumping waves" with a personal watercraft may be considered **GREAT** sport, consider the extensive internal engine damage which will surely occur if this type action is consistently practiced.

During wave jumping or using a wet ramp, when the craft clears the water, the engine RPM approaches the limit set by the RPM limiter in an unloaded condition.

Once the craft returns to the water surface, the sudden load on the impeller is transferred along the driveline. Speed of the rotating pump shaft and the crankshaft is suddenly reduced while the flywheel continues to spin rapidly due to its mass. This condition, especially when repeated over a period of time, often shears the Woodruff key and the flywheel becomes repositioned on the crankshaft.

A very small misalignment will result in a change in ignition timing. The craft operator will notice a drastic loss of power. A worse condition would be the flywheel shifting so much to actually prevent engine start.

This sudden increase and decrease in engine speed dramatically shortens engine life. Bent crankshafts and broken connecting rods can usually be directly attributed to this type of "fun play".

Easy Check for Twisted Crankshaft

A twisted crankshaft may be detected by first removing the spark plugs and setting up a dial indicator in one of the spark plug openings. If a dial indicator is not available, a pencil, straw, or other small diameter straight object may be used.

With the indicator, or other device, in place, determine the TDC (top dead center) of the No. 1 piston. Now, make two matching marks -- one on the flywheel and the other adjacent on the stator. Next, rotate the flywheel **EXACTLY** 180° -- preferably using a degree wheel on the flywheel. Make a second mark on the stator adjacent to the mark on the flywheel.

NOW, check for TDC of the other piston/s. The dial indicator should indicate the same reading as for the first piston **AND** the two marks on the stator should be **EXACTLY** opposite each other through the center of the crankshaft. If the indicator is off or if the marks on the stator are not aligned, as described -- the crankshaft is twisted and **MUST** be replaced.

Engine Noises

Engine noises can be generally classified as knocks, slaps, clicks, or squeaks. These noises are usually caused by loose bearings, sloppy pistons, or other moving parts of the engine.

A main bearing knock is usually identified by a dull thud which is noticeable under engine load. Attempting to move the craft under power with it tied to the dock will bring out a main bearing knock. If you pull the high tension lead from one plug at a time and the noise disappears when one of the leads is removed, then the noise is probably coming from that cylinder.

The cause may be either the rod bearing, piston pin, or the piston, which has quieted

Removal of the crankshaft is not necessary to check for a "twisted" condition, as explained in the text.

down because the load has been removed from the part. If a rod bearing has failed, the noise will be loudest when the engine is decelerating. Piston pin noise and piston slap are generally louder when a cold engine is first started.

Carbon build up in the combustion chamber can cause interference with a piston.

Many times the source of an unusual sound may be isolated by using a stethoscope or other listening device. One such device is a long shank screwdriver. Allow your hand to extend over the handle end of the screwdriver and then make contact with the other end against the block at each cylinder or noise area. Take care and use good judgement when using this method of attempting to detect the cause of a problem, because the noise will travel through other metallic parts of the engine and could lead to a false interpretation of what is being heard.

Leak Down Procedure

On a two-stroke engine, the crankcase has to be completely sealed to prevent pressure from escaping during primary compression. This same pressure sealing is also necessary to prevent air/fuel mixture -- resulting in a lean mixture being delivered to the cylinder.

First take measures to seal off the intake and exhaust manifolds Next, obtain a pressure/vacuum gauge and a "MiniVac" device.

Now, plug all vacuum lines at the crankcase. Remove the spark plugs. Using a spark plug adapter threaded into the spark plug opening, pressurize the crankcase through the spark plug opening with the MiniVac device to 6 psi. After six minutes, the pressure loss should not exceed one inch per minute for six minutes.

If the crankcase passes both tests, the unit is in satisfactory condition. If the crankcase failed either test, prepare a solution of soapy water in a "squirt" bottle. Pressurize the crankcase as described above, and then spray the soapy water solution on suspected areas of pressurization loss, such as the head gasket, cylinder base gasket, intake manifold, crankcase mating surface, rear crankshaft seal and the forward crankcase seal. If air bubbles form in the soapy water, even the smallest amount of pressure is being lost at the location.

The forward seal would require the removal of some items in order to use the soapy water, but it is an important spot for possible pressure leak.

After any head has been thoroughly cleaned and checked, according to the procedures outlined in Chapter 8, it is ready for installation, and long service.

5-3 FUEL SYSTEM PROBLEMS

The following paragraphs provide an orderly sequence of tests to pinpoint problems in the system. It is very rare for the carburetor by itself to cause failure of the engine to start.

Many times fuel system troubles are caused by a plugged fuel filter, a defective fuel pump, or by a leak in the fuel line from the fuel tank to the fuel pump or by a vacuum leak in the line between the pump and the crankcase.

Fuel will begin to sour in three to four months and will cause engine starting problems. A fuel additive such as Sta-Bil may be used to prevent gum from forming during storage or prolonged idle periods.

When the engine is hot, the fuel system can cause starting problems. After a hot engine is shut down, the temperature inside the fuel chamber may rise to 200°F and cause the fuel to actually boil. All carburetors are vented to allow this pressure to escape to the atmosphere. However, some of the fuel may percolate over the high-speed nozzle and overflow into the intake manifold.

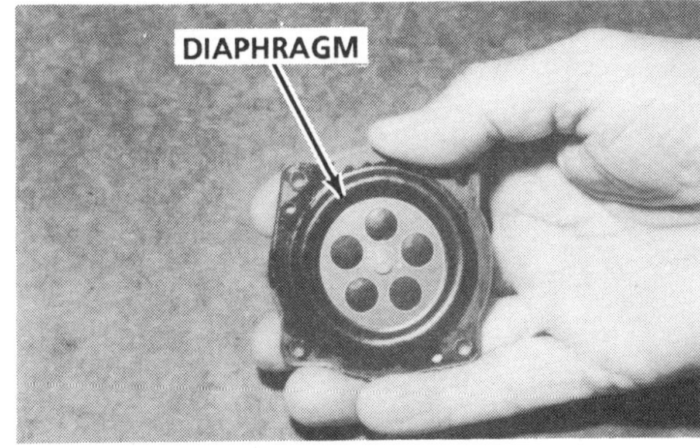

The tiniest "pin" hole in the carburetor diaphragm will seriously affect carburetor performance.

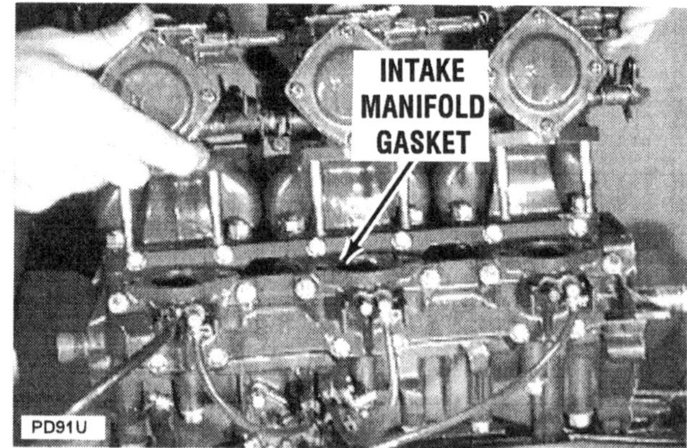

A small leak between the carburetor and the intake manifold, caused by a defective mounting gasket, will result in rough engine performance at idle speed.

In order for this raw fuel to vaporize enough to burn, considerable air must be added to lean out the mixture. Therefore, the only remedy is to open the throttle as wide as possible and to crank the engine until enough air is drawn in to provide the proper mixture for the engine to start. **NEVER** move the throttle lever back-and-forth in an attempt to start a hot engine. This action will only compound the problem by adding more fuel to an already too-rich mixture.

A leak between the fuel tank and the pump many times will not appear when the engine is operating because the suction created by the pump sucking fuel will not allow the fuel to leak. Once the engine is turned off and the suction no longer exists, fuel may begin to leak.

Engine Surge

If the engine operates as if the load on the craft was being constantly increased and decreased, even though the operator attempts to maintain a constant engine speed, the problem can most likely be attributed to the fuel pump.

Check to be sure the clamps securing the fuel hoses to the in-line fuel filter are properly attached. Check other fuel connections -- carburetors and remote fuel pump -- if so equipped.

Rough Engine Idle

Problems preventing an engine from running smoothly include: a fouled spark plug; an air leak in the intake manifold; and uneven compression between the cylinders.

Of course any problem in the carburetor affecting the air/fuel mixture will also prevent the engine from operating smoothly at idle speed. These problems usually include: leaking needle valve and seat; a dirty flame arrestor; defective choke; and improper adjustments for idle mixture or idle speed.

5-4 IGNITION SYSTEM FAULTS

WARNING

Check to be sure the engine compartment is well ventilated and free of any gasoline vapors before starting any of the following tests. A spark will be generated creating a potential fire hazard, if fuel vapors are present.

1- Check to be sure the battery has a full charge. If not, correct the condition by charging the battery or making a substitution. Many times the battery is found to be the culprit.

2- Check all terminals connections inside the electrical box. Check all primary and secondary ignition wiring, the engine stop or "kill" switch and the ignition switch.

3- Use a spark tester and check for spark. If a spark tester is not available, hold the plug wire about 1/4" (6.4mm) from the engine. Crank the engine through a few revolutions using the cranking motor and check for spark. A strong spark over a wide gap must be observed when testing in this manner, because under compression a strong spark is necessary in order to ignite the air/fuel mixture in the cylinder. This means it is possible to think a strong spark is present, when in reality the spark will be too weak when the plug is installed. If there is no spark, or if the spark is weak, the trouble is most likely under the flywheel in the stator assembly.

4- Perform the resistance test procedures given in the Ignition chapter to identify the faulty component.

Intermittent or Multiple Problems

Many ignition problems occur only during engine operation, when the component is subject to vibration and/or an increase in temperature. Many of these defective components return to normal once the engine is shut down and allowed to cool.

There is also the possibility of more than one defective component in the ignition system. One component failure may have caused a surge of power and "fried" another component.

Make every attempt to avoid the **BUY** and **TRY** method of troubleshooting. In **MOST** cases electrical components are not returnable, once the item leaves the store.

SPARK PLUG EVALUATION

Removal: Remove the spark plug wires by pulling and twisting on only the molded cap. **NEVER** pull on the wire or the connection inside the cap may become separated or the boot damaged. Remove the spark plugs and identify from which cylinder they were taken. **TAKE CARE** not to tilt the socket as you remove the plug or the insulator may be cracked.

Examine: Carefully examine the firing end to determine the firing condition in each cylinder.

Correct Color: A proper firing plug should be dry and powdery. Hard deposits inside the shell indicate the engine is starting to use some oil, but not enough to cause concern. The most important evidence is the light gray color of the porcelain, which

Example of overheating, heavy load. Use a spark plug with a lower heat rating.

Spark plug fouled by oil. The engine may be in need of an overhaul.

Red, Brown, or Yellow deposits, are by-products of combustion from fuel and lubricating oil.

is an indication this plug has been running at the correct temperature. This means the plug is one with the correct heat range and also that the air/fuel mixture is correct.

Overheating: A dead white or gray insulator, which is generally blistered, is an indication of overheating and pre-ignition. The electrode gap wear rate will be more than normal and in the case of pre-ignition, will actually cause the electrodes to melt. Overheating and pre-ignition are usually caused by overadvanced timing, detonation from using too-low an octane rating fuel, an excessively lean air/fuel mixture, or problems in the cooling system.

Rich Mixture: A black, sooty condition on both the spark plug shell and the porcelain is caused by an excessively rich air/fuel mixture, both at low and high speeds. The rich mixture lowers the combustion temperature so the spark plug does not run hot enough to burn off the deposits.

Deposits formed only on the shell is an indication the low-speed air/fuel mixture is too rich. At high speeds with the correct mixture, the temperature in the combustion chamber is high enough to burn off the deposits on the insulator.

Powdery deposits have melted and shorted-out this spark plug.

Too Cool: A dark insulator, with very few deposits, indicates the plug is running too cool. This condition can be caused by low compression or by using a spark plug of an incorrect heat range. If this condition shows on only one plug it is most usually caused by low compression in that cylinder. If both plugs have this appearance, then it is probably due to the plugs having a too-low heat range.

Fouled: A fouled spark plug may be caused by the wet oily deposits on the insulator shorting the high-tension current to ground inside the shell. The condition may also be caused by ignition problems which prevent a high-tension pulse to be delivered to the spark plug.

Carbon Deposits: Heavy carbon-like deposits are an indication of excessive oil consumption. This condition may be the result of worn piston rings.

Electrode Wear: Electrode wear results in a wide gap and if the electrode becomes carbonized it will form a high-resistance path for the spark to jump across. Such a condition will cause the engine to misfire during acceleration. If both plugs are in this condition, it can cause an increase in fuel consumption and very poor performance at high-speed operation. The solution is to replace the spark plugs with a rating in the proper heat range and gapped to specification.

Red rust-colored deposits on the entire firing end of a spark plug can be caused by water in the cylinder combustion chamber. This can be the first evidence of water entering the cylinders through the exhaust manifold. This condition **MUST** be corrected at the first opportunity.

5-5 CRANKING SYSTEM FAILURES

Regardless of how or where the solenoid is mounted, the basic circuits of the starting system on all makes of cranking motors are the same and similar tests apply. In the

WARNING

ALWAYS TAKE TIME TO VENT THE BILGE WHEN CONDUCTING ANY OF THE TESTS AS A PREVENTION AGAINST IGNITING ANY FUMES ACCUMULATED IN THAT AREA.

Faulty Symptoms

If the cranking motor spins, but fails to crank the engine, the cause is usually a corroded or gummy Bendix drive. The drive should be removed, cleaned, and inspected for signs of of corrosion or wear.

If the cranking motor cranks the engine too slowly, the following conditions are possible causes. Possible corrective actions are also given.

- **a-** Battery charge is low. Charge the battery to full capacity.
- **b-** High resistance connections at the battery, solenoid, or motor. Clean and tighten all connections.
- **c-** Undersize battery cables. Replace cables with sufficient size.

If the cranking motor must be removed for inspection and/or servicing, most engines covered in this manual require the exhaust elbow, manifold and chamber to be removed first.

Close look at the cranking motor with the two mounting bolts and the electrical connection clearly visible.

All engine cranking problems fall into one of three problem areas:

1. The cranking motor fails to turn.
2. The cranking motor spins rapidly, but does not crank the engine.
3. The cranking motor cranks the engine, but too slowly.

The following paragraphs provide a logical sequence of tests designed to isolate a problem in the cranking system.

Before wasting too much time trouble-shooting the cranking motor circuit, the following checks should be made. Many times, the problem will be corrected.

- **a-** Battery fully charged.
- **b-** Main fuse is "good" (not blown).
- **c-** All electrical connections clean and tight.
- **d-** Wiring in good condition -- insulation not worn or frayed.

The following troubleshooting procedures are presented in a logical sequence.

Do not operate the cranking motor for more than 15 seconds. Prolonged cranking motor operation will cause overheating and damage the motor.

After each test, allow the cranking motor to cool for a minute or so.

Never depress the **START** button to activate the cranking motor while the engine is operating. Such action will damage the pinion and/or flywheel gears.

Cranking Circuit Tests

The following procedures require the cranking motor, solenoid, and battery to be accessible. To gain access to the cranking motor, it may be necessary to remove the exhaust manifold and water box muffler. The cranking motor solenoid is located inside the electrical box. The battery of course, is underneath the electrical box. The cranking motor and solenoid need not be removed. This series of tests will involve dynamic testing of the cranking motor system -- a digital multimeter must be used.

During the dynamic tests, the leads must make contact **WHILE** the engine is being cranked. **ALSO**, both meter leads must be removed from the solenoid terminals **WHILE** the engine is being cranked and **BEFORE** the engine has stopped cranking. If these precautions are not followed, the voltmeter may be damaged. Ask an assistant to press the start switch.

a- Remove all spark plug leads from the plugs and ground the ends to the engine to prevent accidental engine start.

b- Obtain a voltmeter and select the VDC scale. Observe the starter solenoid. Both large terminals have Red leads connected. One terminal has the battery cable connected and the other terminal has the lead from the cranking motor connected.

Disconnect the Red engagement coil wire from the starter solenoid. Connect the Black tester wire to an appropriate ground and the Red lead to the Red harness wire at the solenoid. Press the start button. The meter should read battery voltage. If so, proceed directly to step **"c"**. Otherwise, place the Black tester lead on ground and check for voltage at the large relay terminal, the circuit breaker in and out terminals, and across both sides of the start switch, while the start button is in the **START** position. Repair or replace any defective parts.

c- Reconnect the starter solenoid. Connect the Black tester lead to the positive battery terminal and the Red tester lead to the solenoid end of the battery-to-solenoid cable. Press the start button. If the reading is less than 0.1V DC, proceed with step **"d"**. Otherwise, clean the battery-to-solenoid cable ends or replace the cable.

d- Connect the Black tester lead to the solenoid end of the battery-to-solenoid cable. Connect the Red tester lead to the solenoid end of the solenoid-to-starter cable. Press the start button. If the reading is less than 0.1V DC, proceed to step **"e"**. Otherwise, the starter solenoid must be replaced.

e- Connect the Black tester lead to the solenoid end of the solenoid-to-starter cable. Connect the red tester lead to the starter end of this cable as well. Press the start button. If the reading is less than 0.1V DC, proceed to step **"f"**. Otherwise, clean the ends of the solenoid-to-starter cable or if necessary, replace the cable.

f- Connect the Black tester lead to the cranking motor frame. Connect the Red tester lead to the negative battery terminal. Press the start button. If the reading is less than 0.1V DC, continue to step **"g"**. Otherwise, clean the ends of the negative battery cable at the battery and engine block (or engine mounting plate). Check to be sure the bolts are secure, not loose. The ground path from the cranking motor to the negative battery terminal must be complete.

g- If all these tests indicate a good condition, yet the cranking motor still fails to adequately turn, the motor must be removed for static testing and inspection, see Ch. 9 -- Electrical.

Once the work is complete, close up the electrical box, after checking to be sure all connections are fastened properly. Place the battery cover/electrical tray over the battery. Set the electrical box on top of the tray and secure the assembly with the restraining straps. Insert and tighten the spark plugs to a torque value of 18ft lbs (25Nm). Install the spark plug high tension leads.

5-6 CHARGING SYSTEM MALFUNCTIONS

The lighting coil, rectifier/regulator -- identified as "LR21" or "LR23" on schematics -- battery, and the necessary wiring to connect these all together comprise the charging system.

Before the charging system is blamed for battery problems, consider other areas which may be the cause:

1- Operating at low speed for prolonged periods.

2- Voltage losses due to high resistance.

3- Corroded battery cables, connectors, and terminals

4- Low electrolyte level in the battery cells.

5- Prolonged disuse of the battery causing a self-discharged condition.

Detailed procedures to service the battery is presented in Section 5-3. If, after charging, the battery still fails to give the required specific gravity reading, then the battery will never hold a charge and must be replaced. This problem may have originated with the battery, a shorted cell perhaps, or may be attributed to the failure of the rectifier/regulator or charging coil. Refer to Chapter 7 -- Ignition -- for resistance tests on these two components.

Charging System Testing (Unregulated Voltage)

Obtain a digital multitester. Remove the electrical box cover to gain access of internal components. Set the multitester dial selector to "VAC".

Disconnect the alternator-to-main (stator) harness connector at the electrical box. Connect one of the tester leads to the yellow alternator wire and the other lead to the red/purple alternator wire.

Start the engine.

WARNING

Do not operate the engine for more than 15 seconds with the craft out of the water, as severe damage to the engine will surely occur.

While checking the voltage reading, increase engine rpm to approximately 3,000rpm. Multitester readings should be above 20 VAC. If the output is below this, the stator coils should be tested with an ohmmeter. More information is available in Chapter 7 -- Ignition.

Charging System Testing (Regulated Voltage)

Select VDC with the multitester dial. Connect the leads across the battery terminals. Start the engine. Again, the engine should **NOT** be run for more than 15 seconds while the craft is out of the water, to prevent damage from overheating. **BRIEFLY** run the engine at 3,000rpm. The measured voltage should be around 14.5 VDC. A higher reading may indicate a problem with the voltage regulator or a poor ground at the regulator heat sink. A lower reading may indicate excessive system load, faulty regulator, or alternator problem.

5-7 TROUBLESHOOTING CHARTS

The following six charts are presented as an aid to troubleshooting the engine and jet pump. The uppermost box identifies a specific problem. Start with the problem, then work downward. The probable causes are listed in columns under a specific system, for example: Fuel system, or Electrical system. Once a possible cause is identified, refer to the specific chapter for procedures on testing and service.

When using the charts -- proceed step-by-step through each of the tests or checks until the defect or cause of the problem is uncovered.

FIRST CHART -- Page 5-17 -- Engine fails to start.

SECOND CHART -- Page 5-18 -- Engine starts, then shuts down.

THIRD CHART -- Page 5-19 -- Engine misfires.

FOURTH CHART -- Page 5-20 -- Engine lacks proper power.

FIFTH CHART -- Page 5-21 -- Abnormal engine noises.

SIXTH CHART -- Page 5-22 -- Jet pump problems.

ONE FINAL WORD

Before going out on the water, **TAKE TIME** to verify the drain plug is installed. Countless number of excursions have had a very sad beginning because the craft was eased into the water only to have the hull begin filling with water.

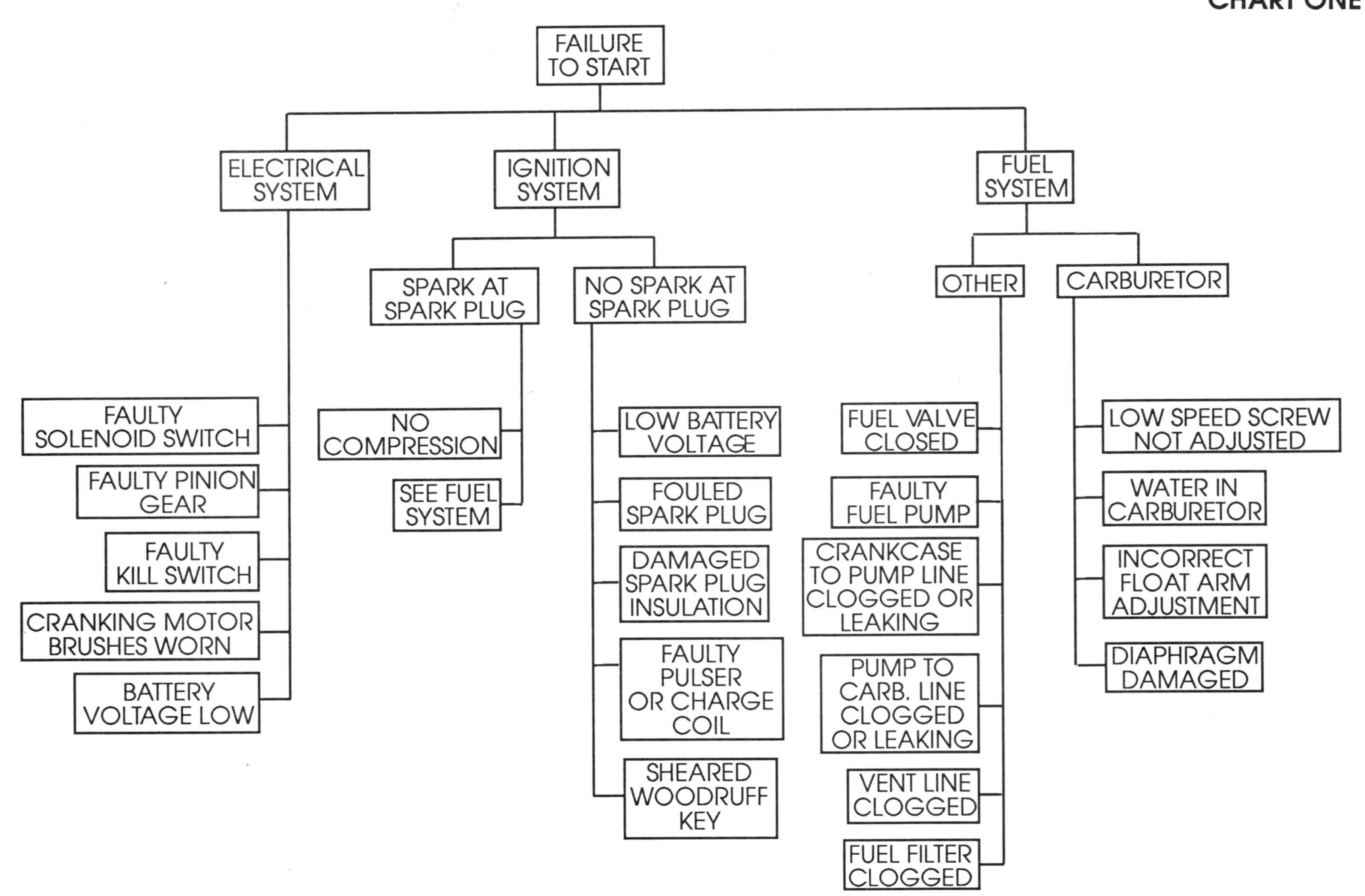
CHART ONE
FAILURE TO START
ELECTRICAL SYSTEM
FAULTY SOLENOID SWITCH
FAULTY PINION GEAR
FAULTY KILL SWITCH
CRANKING MOTOR BRUSHES WORN
BATTERY VOLTAGE LOW
IGNITION SYSTEM
SPARK AT SPARK PLUG
NO COMPRESSION
SEE FUEL SYSTEM
NO SPARK AT SPARK PLUG
LOW BATTERY VOLTAGE
FOULED SPARK PLUG
DAMAGED SPARK PLUG INSULATION
FAULTY PULSER OR CHARGE COIL
SHEARED WOODRUFF KEY
FUEL SYSTEM
OTHER
FUEL VALVE CLOSED
FAULTY FUEL PUMP
CRANKCASE TO PUMP LINE CLOGGED OR LEAKING
PUMP TO CARB. LINE CLOGGED OR LEAKING
VENT LINE CLOGGED
FUEL FILTER CLOGGED
CARBURETOR
LOW SPEED SCREW NOT ADJUSTED
WATER IN CARBURETOR
INCORRECT FLOAT ARM ADJUSTMENT
DIAPHRAGM DAMAGED

CHART TWO

ENGINE STARTS THEN STOPS
IGNITION SYSTEM
FAULTY PULSER OR CHARGE COIL
FOULED SPARK PLUG
POOR WIRING CONNECTION
FUEL SYSTEM
OTHER
CHOKE MISUSE
FAULTY FUEL PUMP
CRANKCASE TO PUMP LINE CLOGGED OR LEAKING
PUMP TO CARB. LINE CLOGGED OR LEAKING
VENT LINE CLOGGED
FUEL FILTER CLOGGED
CARBURETOR
LOW SPEED SCREW NOT ADJUSTED
WATER IN CARBURETOR
INCORRECT FLOAT ARM ADJUSTMENT
DIAPHRAGM DAMAGED
ENGINE COMPRESSION
CRANKCASE JOINT LEAK
HEAD GASKET LEAK
DAMAGED CRANKSHAFT OIL SEAL
SPARK PLUG LEAK
WORN PISTON AND RINGS

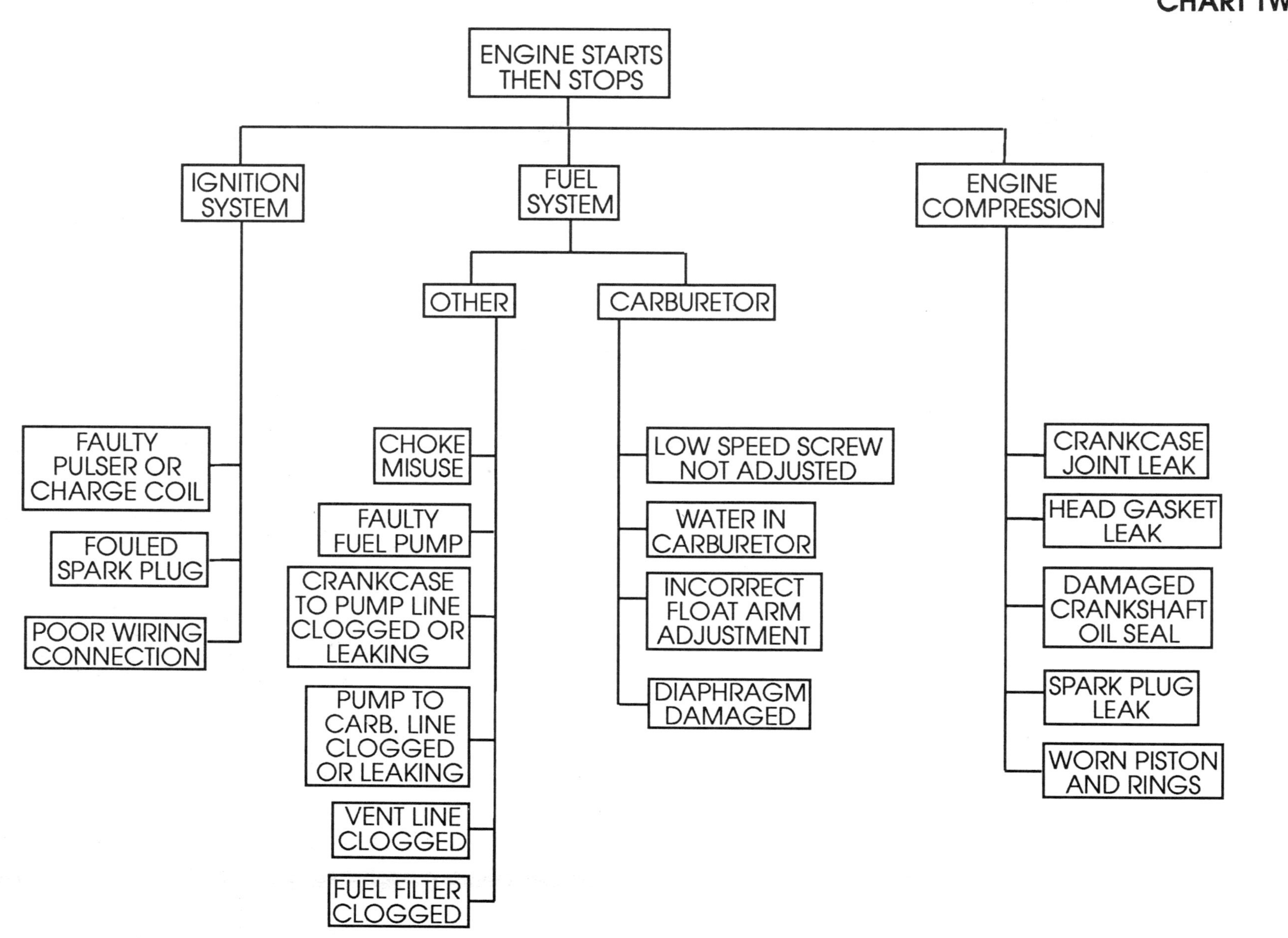

CHART THREE

ENGINE MISFIRES
IGNITION SYSTEM
WEAK SPARK
INSUFFICIENT IGNITION COIL OUTPUT
HIGH TENSION INSULATION BREAKDOWN
FOULED SPARK PLUG
INCORRECT SPARK PLUG GAP
POOR WIRING CONNECTION
FAULTY PULSER OR CHARGE COIL
LOW BATTERY VOLTAGE
INCORRECT IGNITION TIMING DUE TO FAULTY PULSER COIL
FUEL SYSTEM
FUEL MIXTURE TOO RICH
CARBURETOR
CHOKE NOT ADJUSTED
DIAPHRAGM NEEDLE DIRTY OR DAMAGED
LOW SPEED SCREW NOT ADJUSTED
CLOGGED FLAME ARRESTER
FUEL MIXTURE TOO LEAN
OTHER
VENT LINE CLOGGED
FUEL FILTER CLOGGED
FAULTY FUEL PUMP
CARBURETOR LOOSELY MOUNTED
CARBURETOR
NEEDLE VALVE DAMAGED
FUEL PASSAGE BLOCKED
DIAPHRAGM DAMAGED
DAMAGED LOW SPEED SCREW
ENGINE OIL INJECTION SYSTEM ALL MODELS
OIL LINE CLOGGED
OIL PUMP DAMAGED

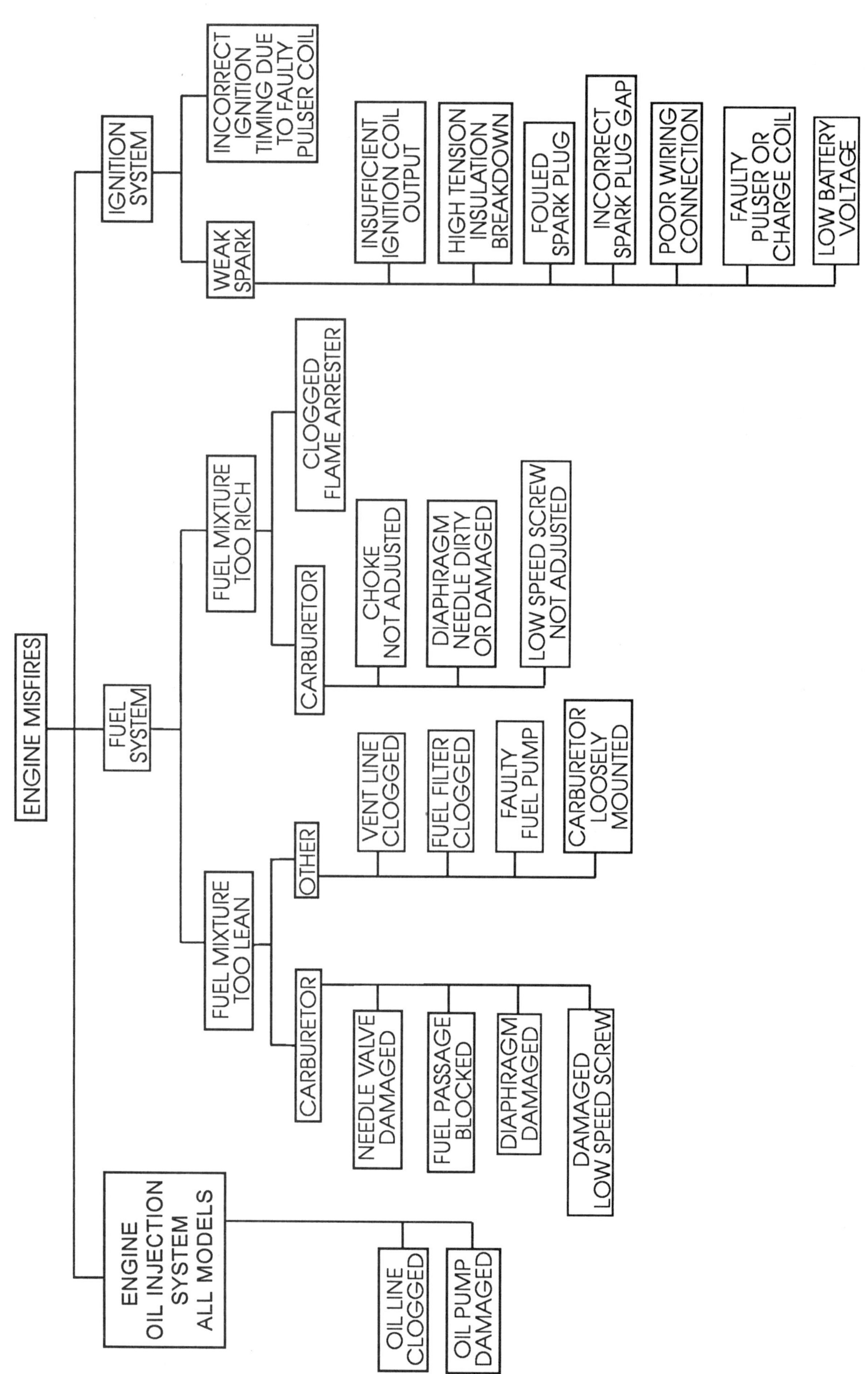

CHART FOUR

LOW ENGINE POWER
FUEL SYSTEM
INSUFFICIENT FUEL SUPPLY TO CARBURETOR
CARBURETOR DIAPHRAGM DAMAGED
PULSE LINE LEAKING OR CLOGGED
THROTTLE VALVE NOT ABLE TO FULLY OPEN
FUEL FILTER CLOGGED
FAULTY FUEL PUMP
DAMAGED LOW SPEED SCREW
POWERHEAD OVERHEATING CONDITION
CARBURETOR MIXTURE INCORRECTLY ADJUSTED
COOLING WATER LINE LEAKING OR CLOGGED
CARBON ACCUMULATION IN COMBUSTION CHAMBER
INCORRECT TYPE OF FUEL (OCTANE) USED IN FUEL MIXTURE
INCORRECT GRADE OF OIL USED IN FUEL MIXTURE
EXHAUST SYSTEM CLOGGED
INCORRECT IGNITION TIMING DUE TO FAULTY PULSER OR CHARGE COIL
FLAME ARRESTOR CLOGGED
IGNITION SYSTEM
SPARK PLUG GAP OR HEAT RANGE INCORRECT
IGNITION SYSTEM CONNECTION LOOSE
REDUCED IGNITION COIL OUTPUT
INCORRECT IGNITION TIMING DUE TO FAULTY PULSER OR CHARGE COIL
SHEARED WOODRUFF KEY
HEAVY FUEL CONSUMPTION
CARBURETOR MIXTURE INCORRECTLY ADJUSTED
FLAME ARRESTOR CLOGGED
EXHAUST SYSTEM CLOGGED
WORN CYLINDER, PISTON, OR RINGS
FUEL FEED LINE LEAKING
CARBURETOR NEEDLE DIRTY OR DAMAGED
OTHER PROBLEMS
FLAME ARRESTOR CLOGGED
CONTAMINATED GASOLINE USED IN FUEL MIXTURE
CONTAMINATED OIL USED IN FUEL MIXTURE
EXHAUST GAS LEAK IN ENGINE COMPARTMENT

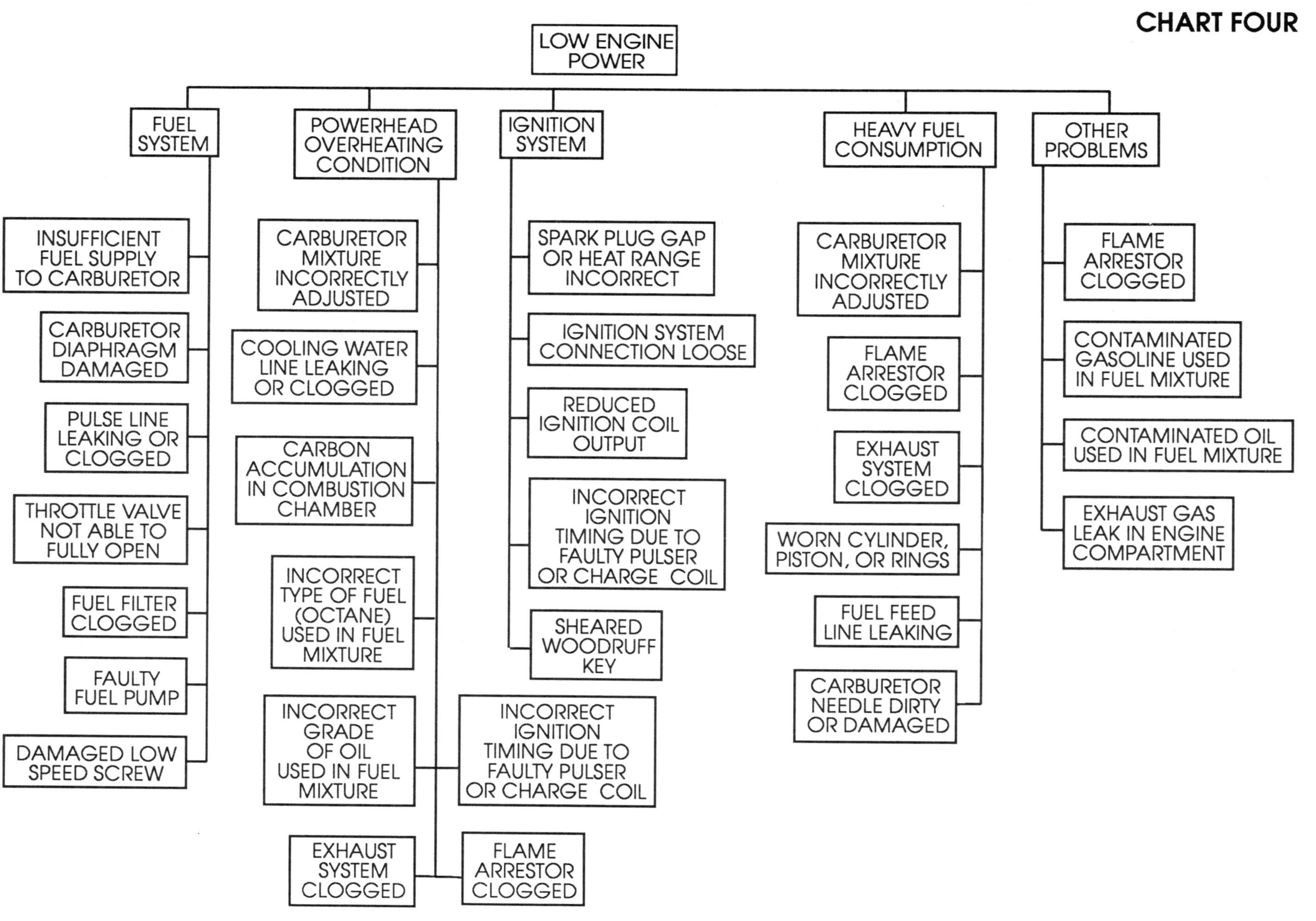

CHART FIVE

ABNORMAL ENGINE NOISE
PINGING
POOR QUALITY FUEL
CARBON ACCUMULATION IN CYLINDER HEAD
SPARK PLUG WRONG HEAT RANGE
IGNITION TIMING TOO ADVANCED DUE TO FAULTY PULSER OR CHARGE COIL
NORMAL CRUISING
PERIODIC PISTON SEIZURE
PISTON PIN STICKING OR BROKEN
MAIN BEARING DAMAGED OR WORN
SUDDEN ACCELERATION
EXCESSIVE CONNECTING ROD BIG END CLEARANCE
EXCESSIVE CONNECTING ROD TO PISTON PIN CLEARANCE
EXCESSIVE PISTON PIN TO PISTON CLEARANCE
AT COLD IDLE
PISTON RINGS WORN
PISTON WORN
CONNECTING ROD TWISTED OR BENT
EXCESSIVE PISTON TO CYLINDER CLEARANCE

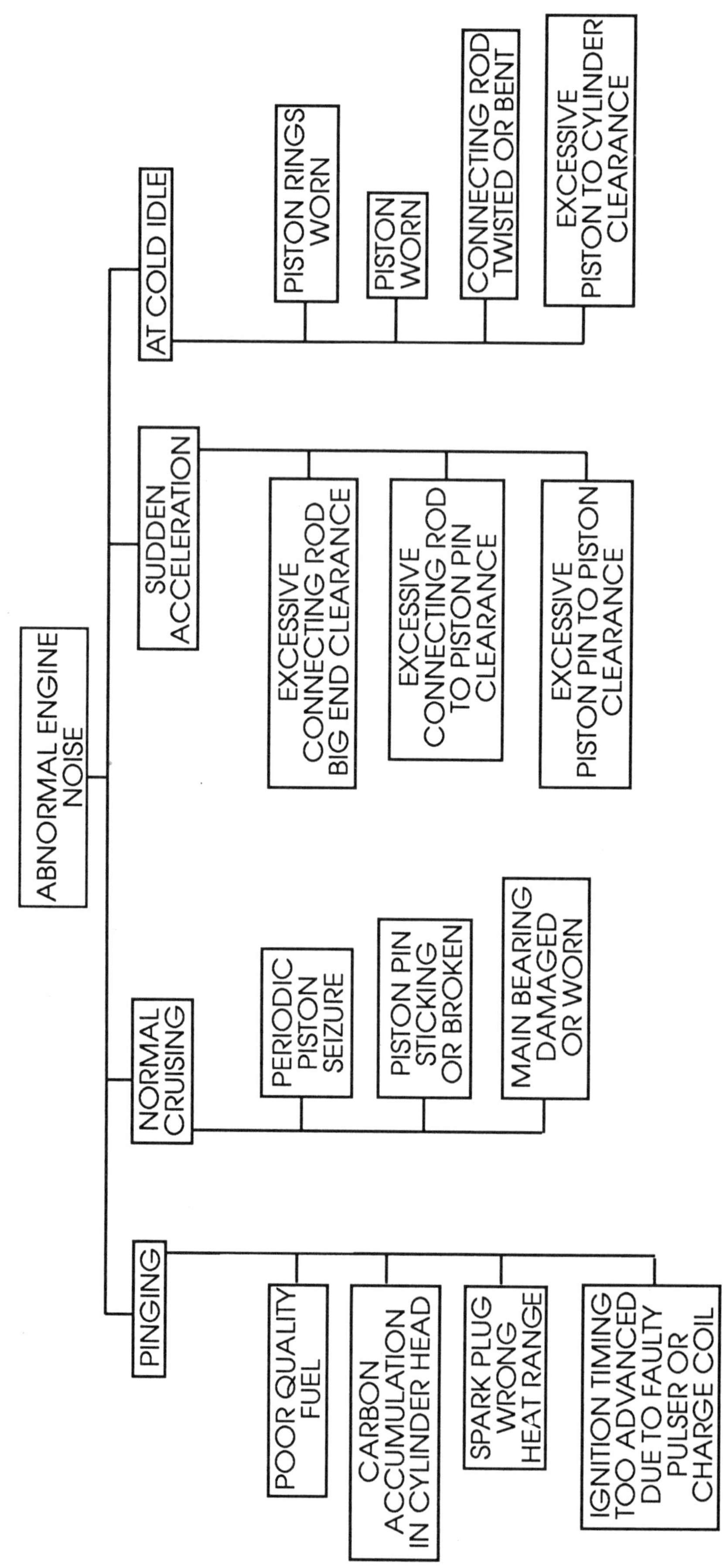

CHART SIX

JET PUMP PROBLEMS
POOR PERFORMANCE
HARD STEERING
ROCK GRATE OBSTRUCTED
IMPELLER DAMAGED
PUMP CASE DAMAGED
EXCESSIVE CLEARANCE BETWEEN IMPELLER AND PUMP CASE
REVERSE GATE INCORRECTLY ADJUSTED
STEERING INCORRECTLY ADJUSTED
BUSHINGS CRACKED OR DAMAGED
STEERING SHAFT BENT
STEERING NOZZLE PIVOT INADEQUATELY LUBRICATED
STEERING CABLE DAMAGED OR IMPROPERLY ROUTED

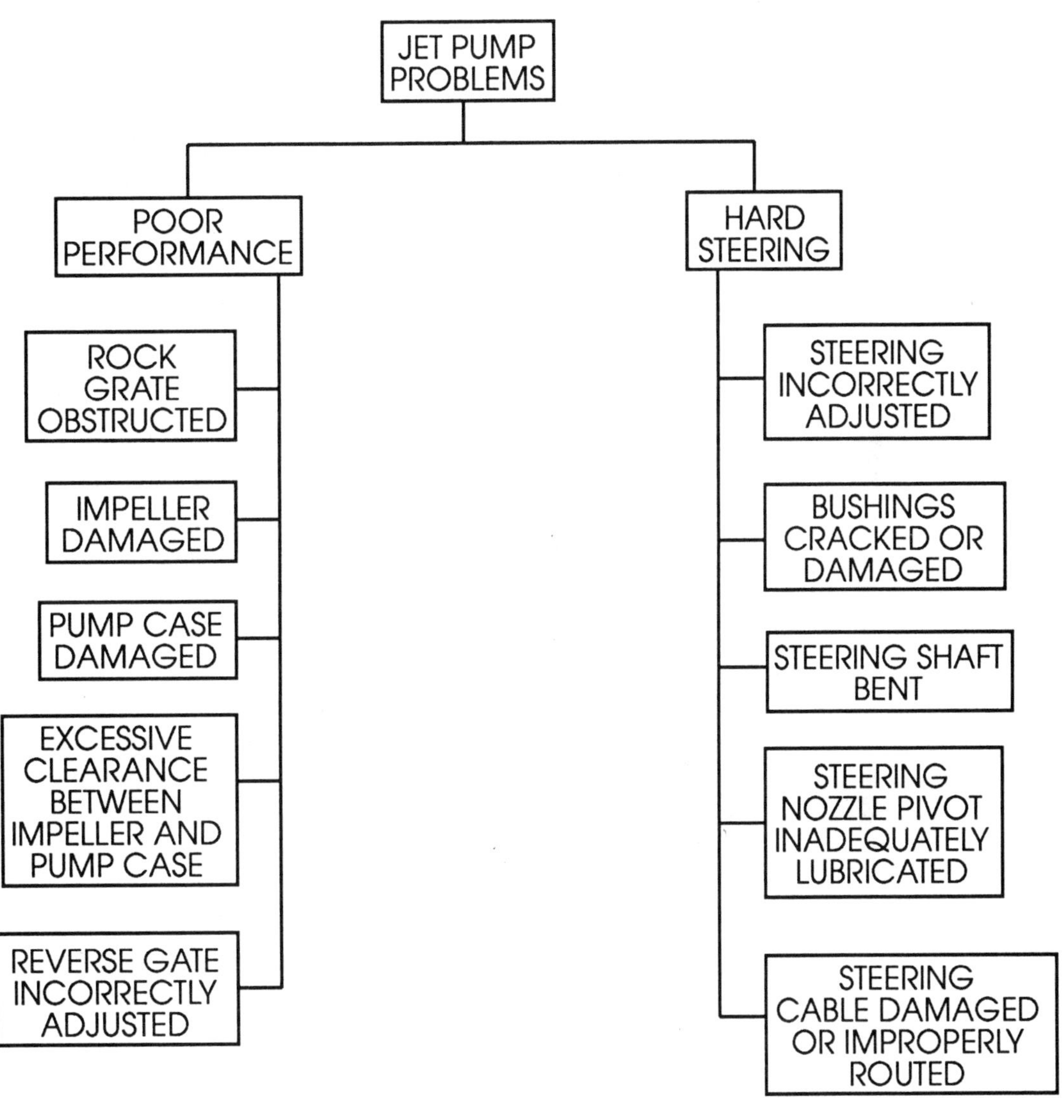

CARBURETOR INSTALLATION AND BENCH ADJUSTMENTS

Year	Model	No. Of Carbs.	Carburetor Identification	High Speed Screw Turns Out	Low Speed Screw Turns Out
1992	SL 650	3	Mikuni Super BN	Fwd 3/8, Ctr 3/8, Aft 1/4	All Three 1-3/8 Note 1
1993	SL650	3	Mikuni Super BN	Fwd 7/8, Ctr 3/8 Aft 5/8	All Three 1/4
	SL 750	3	Mikuni Super BN	Fwd 7/8, Ctr 1/2, Aft 5/8	All Three 1/2
1994	SL 650 Eng. S/N 94-0001 - 94-02010	3	Mikuni Super BN	Fwd 3/4, Ctr 1/4, Aft 1/2	All Three 1-1/4
	SL 650 Eng. S/N 94-02011 & On	3	Mikuni Super BN	Fwd 7/8, Ctr 1/2 Sft 3/4	All Three 1 full turn
	SL 750	3	Mikuni Super BN	Fwd 1-1/4, Ctr 3/8, Aft 7/8	All Three 1/2
	SLT 750	3	Mikuni Super BN	Fwd 1-1/4, Ctr 3/8, Aft 7/8	All Three 1/2
1995	SL 650	3	Mikuni Super BN	Fwd 1-1/8, Ctr 1/4, Aft 7/8	All three 1 full turn
	SL650 STD	3	Mikuni Super BN	Fwd 1-1/8, Ctr 1/4, Aft 7/8	All three 1 full turn
	SL 750	3	Mikuni Super BN	Fwd 1, Ctr 1/2, Aft 3/4	All Three 1/2
	SLT 750	3	MikuniSuper BN	Fwd 1, Ctr 1/2, Aft 3/4	All Three 1/2
	SLX 780	3	Mikuni Super BN	Fwd 7/8, Ctr 3/4, Aft 1-1/8	All Three 1/2
1996	SL 700	2	Keihin CDK II	Both 1-1/2	Both 5/8
	SLT 700	2	Keihin CDK II	Both 1-1/2	Both 5/8
	Hurricane	2	Keihin CDK II	No H/P Screw	Both 5/8

Year	Model	No. Of Carbs.	Carburetor Identification	High Speed Screw Turns Out	Low Speed Screw Turns Out
1996 Cont.	SL780 EC78PWE-02/05	3	Mikuni Super BN	Fwd 3/4, Ctr 3/4 Aft 1 full turn	All Three 5/8
	SLT 780 EC78PWE-03/04	3	Mikuni Super BN	Fwd 1-1/8, Ctr 7/8, Aft 1-1/4	All three 1-3/8
	SLX 780 EC78PWE-04	3	Mikuni Super BN	Fwd 1-1/8, Ctr 7/8, Aft 1-1/4	All three 1-3/8
	SL 900	3	Keihin CDKII	No H/P Screw	All Three 5/8
	SLTX	3	Keihin CDKII	No H/P Screw	1 Full Turn
1997	SL 700	1	Keihin CDK II	No H/P Screw	Single 7/8
	SL 700 Deluxe	2	Keihin CDK II w/accel. Pump	Both 1-5/8 to 1-3/4	Both 5/8
	SLT 700	2	Keihin CDK II w/Accel. Pump	Both 1-5/8 to 1-3/4	Both 5/8
	Hurricane	2	Keihin CDK II	No H/P Screw	Both 5/8
1997	SL 780	3	Mikuni Super BN	All Three 1/8 ± 1/8	All Three 1-1/4
	SLT 780	3	Mikuni Super BN	All Three 1/8 ± 1/8	All Three 1-1/4
	SL 900	3	Keihin CDKII w/Accel. Pump	No H/P Screw	All Three 5/8
	SL 1050	3	Keihin CDKII	No H/P Screw	All Three 7/8
	SLTX	3	Keihin CDKII w/Accel. Pump	No H/P Screw	1 Full Turn

Mekuni carburetors have separate fuel pump.
Keihin carburetors have integral fuel pump.

6
FUEL AND OIL

6-1 INTRODUCTION

The carburetion and ignition principles of two-cycle engine operation **MUST** be understood in order to perform a proper tune-up on a marine engine.

If you have any doubts concerning your understanding of two-cycle engine operation, it would be best to study the Introduction section in the first portion of Chapter 8, before tackling any work on the fuel system.

The fuel system includes the fuel tank, fuel pump, fuel filters, carburetor, and the associated parts to connect it all together. Regular maintenance of the fuel system to obtain maximum performance, is limited to changing the fuel filters at regular intervals and using fresh fuel.

If a sudden increase in gas consumption is noticed, or if the engine does not perform properly, a carburetor overhaul, including boil-out, or replacement of the fuel pump may be required.

6-2 GENERAL CARBURETION INFORMATION

The carburetor is merely a metering device for mixing fuel and air in the proper proportions for efficient engine operation. At idle speed, a marine engine requires a mixture of about 8 parts air to 1 part fuel. At high speed or under heavy duty service, the mixture may change to as much as 12 parts air to 1 part fuel.

DIAPHRAGM CARBURETORS

Description

The two carburetors, Mikuni and Keihin, used on the Polaris PWC's covered in this manual are both diaphragm type carburetors. The Keihin carburetor installed since 1996 has an integral fuel pump. All other models with a Mikuni carburetor, use a separate fuel pump to deliver fuel to the carburetors.

On 1992, '93, '94 and some models in 1995, the fuel pump is mounted on the aft end of the air intake/flame arrestor pan. The separate fuel pump on the Model SLX 780 in 1995 and subsequent models with a Mikuni carburetor have a separate fuel pump. The pump is mounted on the aft outboard starboard bulkhead in the engine compartment. This fuel pump is covered in Section 6-7 of this chapter.

The table on the next page lists the engine model, year of production, number of carburetors installed and carburetor identification.

A diaphragm carburetor does **NOT** have a float bowl or a float. Instead, a diaphragm takes the place of the float.

A hinged arm, still called a float arm, rests against the diaphragm and opens and closes the needle and seat assembly.

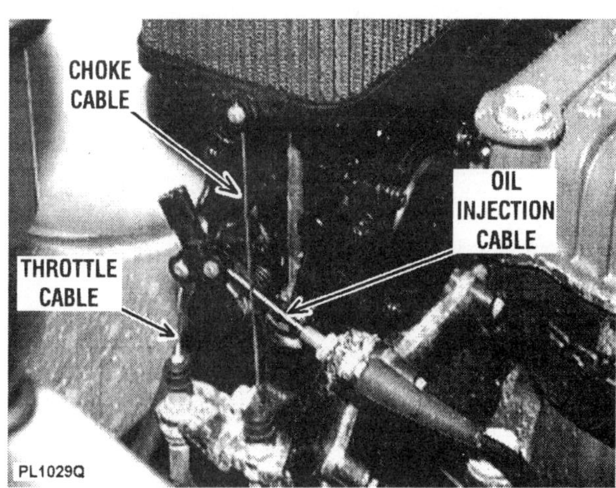

A properly adjusted choke cable, throttle cable and oil injection cable, along with clean flame arrestor screens and the use of fresh fuel, will do much toward maximum performance.

Carburetor Installation

Year	Model	No. Of Carbs.	Carburetor Identification
1992	SL 650	3	Mikuni Super BN
1993	SL650	3	Mikuni Super BN
	SL 750	3	Mikuni Super BN
1994	SL 650	3	Mikuni Super BN
	SL 750	3	Mikuni Super BN
	SLT 750	3	Mikuni Super BN
1995	SL 650	3	Mikuni Super BN
	SL650 STD	3	Mikuni Super BN
	SL 750	3	Mikuni Super BN
	SLT 750	3	MikuniSuper BN
	SLX 780	3	Mikuni Super BN
1996	SL 700	2	Keihin CDK II
	SLT 700	2	Keihin CDK II
	Hurricane	2	Keihin CDK II
	SL 780	3	Mikuni Super BN
	SLX 780	3	Mikuni Super BN
	SLT 780	3	Mikuni Super BN
	SL 900	3	Keihin CDKII
	SLTX	3	Keihin CDKII
1997	SL 700	1	Keihin CDK II w/Accel. Pump
	SL 700 Deluxe	2	Keihin CDK II w/accel. Pump
	SLT 700	2	Keihin CDK II
	Hurricane	2	Keihin CDK II
1997	SL 780	3	Mikuni Super BN
	SLT 780	3	Mikuni Super BN
	SL 900	3	Keihin CDKII w/Accel. Pump
	SL 1050	3	Keihin CDKII
	SLTX	3	Keihin CDKII w/Accel. Pump
Mekuni carburetors have separate fuel pump. Keihin carburetors have integral fuel pump.			

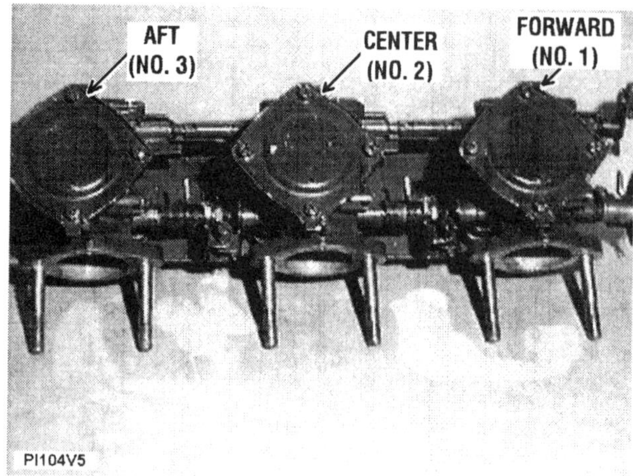

Triple setup of Mikuni carburetors, serviced with replacement parts from carburetor overhaul kits and ready for installation and to give the owner long service.

FUEL FLOW THROUGH THE SYSTEM

Following the fuel through its course, from the fuel tank to the combustion chamber of the cylinder, will provide an appreciation of exactly what is taking place.

Beginning in the fuel tank, an inlet filter screen is attached to the end of primary intake line and on the end of the reserve intake line. Also in the tank, a "roll over" one-way check valve is installed on the end of the vent line. This "roll over" check valve prevents water from entering the fuel tank through the vent line in the event the craft rolls over or is submerged temporarily.

Once the engine starts, the pump draws fuel up through the primary line. Fuel then passes

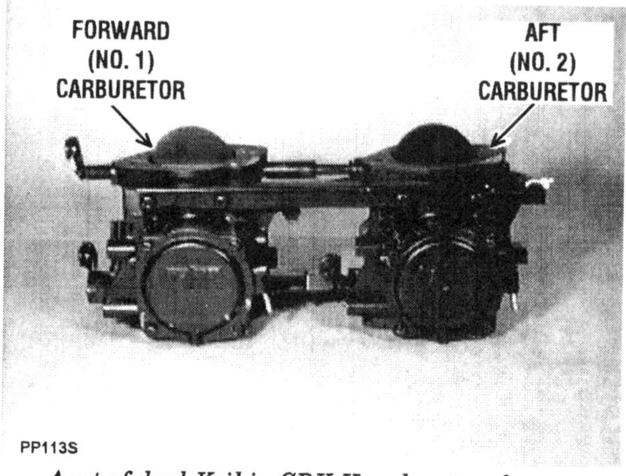

A set of dual Keihin CDK II carburetors from a two-cylinder engine. These carburetors have an integral fuel pump. On a dual installation, only the aft carburetor has the integral pump. On a triple setup, the center carburetor is not equipped with the integral fuel pump.

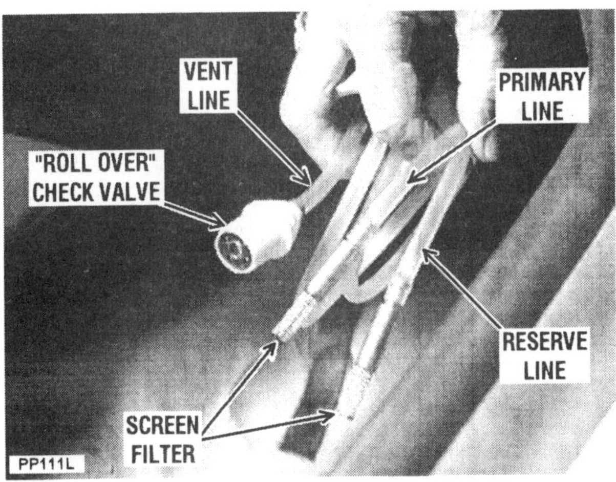

Close view of the fuel lines, inlet screens, and the "roll over" check valve in the fuel tank. The inlet and reserve lines have been coiled for the screens to show in the photo.

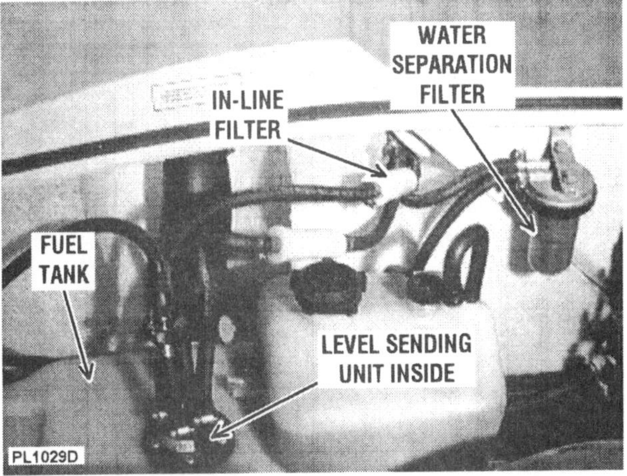

Overall view to show the fuel tank, oil tank, an in-line filter and the water separator filter installed in the engine compartment.

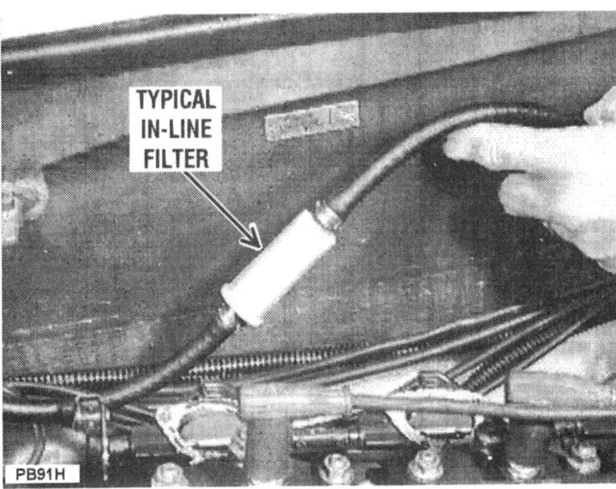

Typical in-line fuel filter which may be attached in any one of convenient locations in the engine compartment.

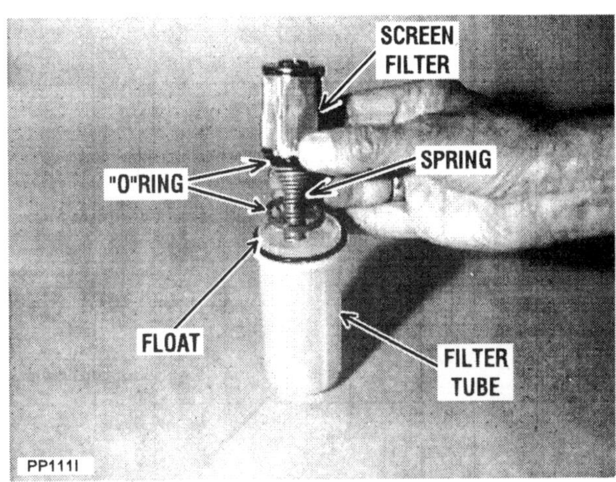

Water separator fuel filter removed from the engine compartment bulkhead and opened to show internal parts.

through an in-line filter, as shown in the two accompanying illustrations, and into a water separator filter, then on to the fuel pump. The water separator contains a screen-type filter, an **O**-ring, a spring, another **O**-ring, and a float which rides on top of any water trapped in the filter tube.

The in-line filter and the water separator filter may be secured to the outboard bulkhead in various locations in the engine compartment. From 1992 into 1995, the fuel pump was mounted on the aft end of the air intake/flame arrestor pan. Since that time, the remote fuel pump has been mounted in various convenient locations on a bulkhead in the engine compartment.

From the pump, fuel is delivered to an inlet valve in the carburetor, This valve is under pressure from the integral or remote fuel

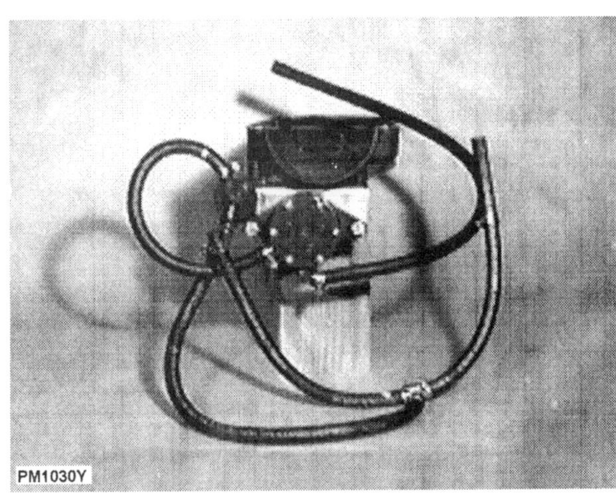

The remote fuel pump secured to the aft end of the air intake/flame arrestor pan. The pump may remain in place with the fuel hoses connected anytime it is necessary to remove the pan. Since sometime in 1995, the remote fuel pump has been attached to the outboard bulkhead in the engine compartment, just aft of the engine.

Close view of the remote fuel pump and the "auto cock" fuel valve.

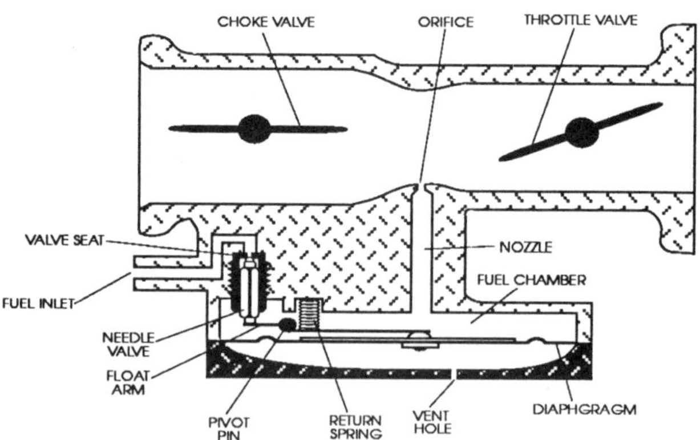

Simple cross-section line drawing of a diaphragm type carburetor used on all series and model years covered in this manual.

pump. When the choke valve is closed and the engine is cranked, or when the piston is on its upward stroke, the pressure in the carburetor throat orifice and the crankcase is less than atmospheric pressure.

The fuel chamber above the diaphragm is connected to the crankcase through a vacuum tube. A nozzle channel connecting the carburetor throat with the fuel chamber also allows the fuel chamber to be at less than atmospheric pressure.

The chamber on the other side of the diaphragm is vented to atmosphere, and therefore, is always at atmospheric pressure.

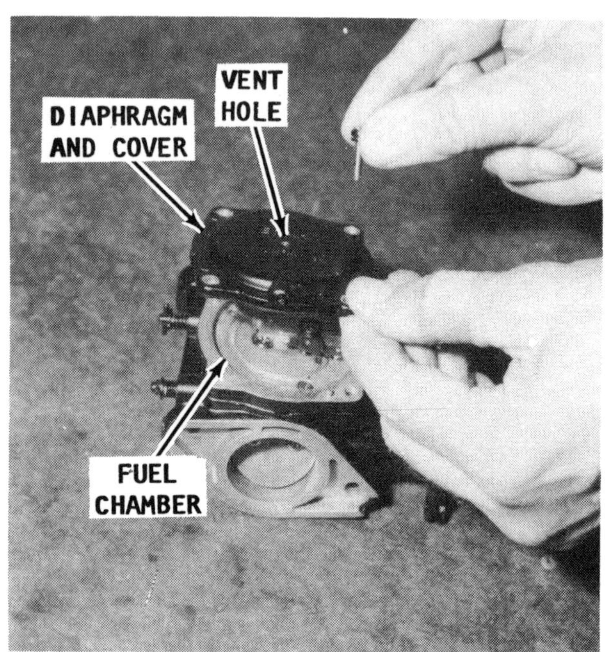

The vent hole for the fuel chamber above the diaphragm, on any diaphragm type carburetor, must be kept clear to the atmosphere at all times, for the diaphragm to function efficiently.

With the two pressure differentials, the diaphragm will flex toward the lowest pressure -- towards the fuel chamber. As the diaphragm flexes, it raises the float arm which, because of a return spring, is in constant contact with the diaphragm. The float arm is hinged. As one side goes up, the other side comes down and allows the inlet valve to open. At the same time, the needle valve falls away from the valve seat. As soon as the inlet valve opens, fuel flows under pressure from the fuel pump into the fuel chamber.

When the piston is on its downward stroke, the crankcase, the carburetor throat, and the fuel chamber are all at a higher pressure than atmosphere. The fuel chamber filled with fuel at a higher pressure now flexes toward the side with the lower pressure -- the side vented to atmosphere.

The return spring pushes the float arm down against the diaphragm. The other side

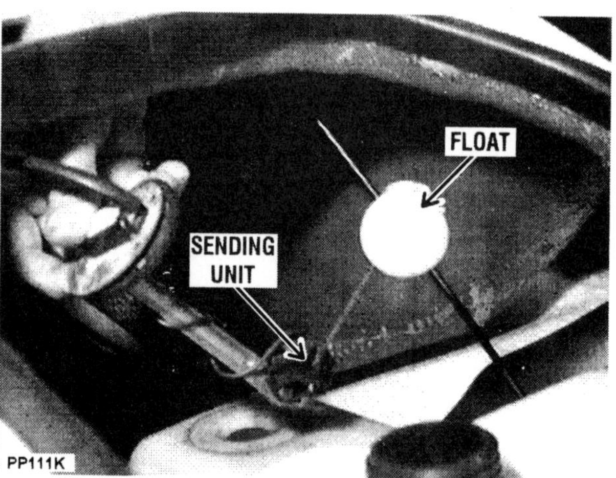

The float sending unit removed from the tank for inspection. The tank can be moved forward without too much trouble to permit withdrawing the sending unit.

of the float arm raises up and pushes the needle valve against its seat closing the inlet valve.

On the next piston downward stroke, the diaphragm flexes back again forcing the fuel in the fuel chamber to be ejected through a jet into the carburetor throat. On the same stroke the fuel chamber is replenished with fresh fuel through the open needle valve. This cycle repeats itself, sending a squirt of fuel into the carburetor throat each downward stroke of the piston.

In order to obtain the proper air/fuel mixture for all engine speeds, some models have high and low speed jets. These jets have adjustable needle valves which are used to compensate for changing atmospheric conditions. In almost all cases, the high-speed circuit has fixed high-speed jets and are not adjustable.

A throttle valve controls the flow of air/fuel mixture drawn into the combustion chambers. A cold engine requires a richer fuel mixture to start and during the brief period it is warming to normal operating temperature. A choke valve is placed ahead of the metering jets and venturi. As this valve begins to close, the volume of air intake is reduced, thus enriching the mixture entering the cylinders.

When this choke valve is fully closed, a very rich fuel mixture is drawn into the cylinders.

6-3 FUEL COMPONENTS AND AVAILABLE GAS

The fuel system includes the fuel tank, fuel pump, fuel filters, carburetor, connecting fuel lines, and the associated parts to connect it all together. Regular maintenance of the fuel system to obtain maximum performance, is limited to changing the fuel filters, cleaning the flame arrestor screens and using fresh fuel.

Even with the high price of fuel, removing gasoline that has been standing unused over a long period of time, is still the easiest and least expensive preventive maintenance possible. In most cases this old gas, even with some oil mixed with it, can be used without harmful affects in an automobile using regular gasoline.

If a sudden increase in gas consumption is noticed, or if the engine does not perform properly, a carburetor overhaul, including boil-out, or replacement of the fuel pump may be required.

Leaded Gasoline

The manufacturer of the units covered in this manual recommend the engines be operated using regular unleaded gasoline with a minimum octane rating of 84 or higher.

In the United States, The Environmental Protection Agency (EPA) phased-out leaded fuel, "Regular" gasoline in 1993. Lead in gasoline boosts the octane rating (energy). Therefore, if the lead is removed, it must be replaced with another agent. Unknown to the general public, many refineries are adding alcohol in an effort to hold the octane rating.

Alcohol in gasoline can have a deteriorating affect on certain fuel system parts. Seals can swell, pump check valves can swell, diaphragms distort, and other rubber or neoprene composition parts in the fuel system can be affected.

Since about 1981, every effort has been made to use materials that will resist the alcohol being added to fuels.

Fuel containing alcohol will slowly absorb moisture from the air. Once the moisture content in the fuel exceeds about 1%, it will separate from the fuel taking the alcohol with it.

A lead substitute additive -- octane booster -- can help prevent detonation when unleaded gasoline is used in a personal watercraft.

This water/alcohol mixture will settle to the bottom of the fuel tank. The engine will fail to operate. Therefore, storage of this type of gasoline is not recommended for more than just a few days.

Air/Fuel Mixture

A suction effect is created in the intake manifold each time the piston moves upward in the cylinder. This suction draws air through the throat of the carburetor. A restriction in the throat, called a venturi, controls air velocity and has the effect reducing air pressure at this point.

The difference in air pressures at the throat and in the fuel chamber, causes the fuel to be pushed out of the metering jet extending down into the fuel chamber. When the fuel leaves the jet, it mixes with the air passing through the venturi. This air/fuel mixture should then be in the proper proportion for burning in the cylinders for maximum engine performance.

Throttle and Choke Valves

A throttle valve controls the flow of air/fuel mixture drawn into the combustion chambers. A cold engine requires a richer fuel mixture to start and during the brief period it is warming to normal operating temperature. A choke valve is placed ahead of the metering jets and venturi. As this valve begins to close, the volume of air drawn in is reduced, thus enriching the mixture entering the cylinders.

When this choke valve is fully closed, a very rich fuel mixture is drawn into the cylinders aiding engine start.

FUEL PUMP

Many times, a defective fuel pump diaphragm is mistakenly diagnosed as a problem in the ignition system. The most common problem is a tiny pin-hole in the diaphragm. Such a small hole will permit gas to enter the crankcase and set foul the spark plugs at idle-speed. During high-speed operation, gas quantity is limited, the plug is not fouled and will therefore fire in a satisfactory manner.

Remote Fuel Pump

At press time, fuel to the carburetors on engines equipped with the Mikuni Super BN, is delivered by a separate fuel pump. Engine models produced from 1992 to 1995 have the pump mounted on the aft end of the air intake/flame arrestor pan. Models produced after 1995, have a remote fuel pump mounted on the starboard outboard bulkhead, just aft of the engine

Troubleshooting, testing and service procedures for the remote fuel pump are presented in Section 6-7, beginning on Page 6-41.

Integral Fuel Pump

As mentioned several times earlier in this chapter, the fuel pump is considered an integral part of the Keihin CDKII carburetor, with one exception. On a triple carburetor installation, as is the case with the engines covered in this manual, the No. 2 -- center -- carburetor does not have the fuel pump. Fuel to this carburetor is delivered as "excess" fuel from the No. 1 -- forward -- and the No. 3 -- aft -- carburetors.

If either fuel pump fails to perform properly, an insufficient fuel supply will be delivered to the carburetors. This lack of fuel will cause the engine to run lean, lose rpm or cause piston scoring.

If the fuel pump diaphragm tears or is punctured, fuel will enter the crankcase through the pulse line causing over-rich mixture. This mixture will foul the plugs and in extreme cases flood the crankcase.

NEVER use liquid Neoprene on fuel line fittings. Always use Permatex when making fuel line connections. Permatex is available at almost all marine and hardware stores.

Detailed illustrated fuel pump service is included in the overhaul procedures in the Disassembling and Assembling portions for the Keihin carburetor in this chapter, beginning on Page 6-27.

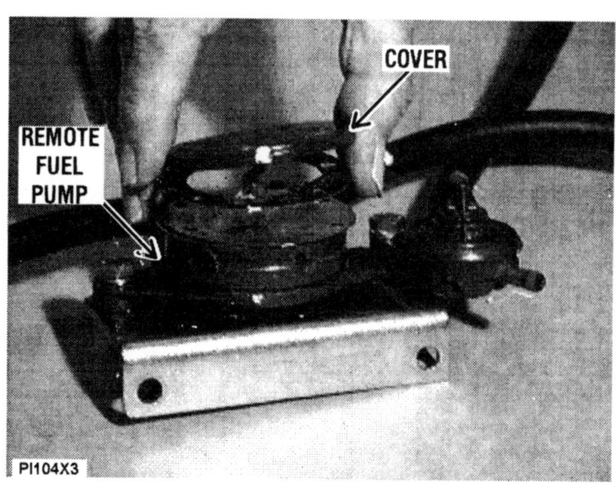

Removing the cover of a remote fuel pump in preparation to servicing the unit. Detailed illustrated procedures to service the remote fuel pump are presented in Section 6-7, beginning on Page 6-41.

Fuel/Oil Mixture

All engines covered in this manual are equipped with an oil injection system, replacing the age-old task of mixing oil with the gas before adding it to the tank.

However, additional oil should be added to the fuel during a "break-in" period following an engine overhaul or following a long storage period.

"Break-in" Lubrication

In order to obtain extra lubrication during the first 10 hours of "break-in", the manufacturer recommends a premix of 25:1 mixture be used in the fuel tank. Therefore, any existing unused fuel in the tank should be removed before adding the premixed solution. This ratio will **ENSURE** adequate lubrication of moving parts which have been drained of oil during the storage period.

Professional mechanics who own oil injected units add an extra ounce of oil per each gallon of fuel. The addition of this extra oil may prevent engine seizure in the event of an oil pump failure or clog in an oil delivery line.

Use only outboard marine oil in the mixture, never automotive oils. Four-stroke automotive engine oil is not formulated to burn completely -- only to lubricate. Therefore, automotive oil, if used in a two-stroke engine will leave an undesirable residue.

Removing Fuel From the System

For many years there has been the widespread belief that simply shutting off the fuel at the tank and then operating the engine until it stops is the proper procedure before storing the engine for any length of time. Right? **WRONG!**

It is **NOT** possible to remove all fuel in the carburetor by operating the engine until it stops. Some fuel is trapped in passages. The **ONLY** guaranteed method of removing **ALL** fuel is to take the time to remove the carburetor, and drain the fuel.

If the engine is operated with the fuel supply shut off until it stops the fuel and oil mixture inside the engine is removed, leaving bearings, pistons, rings, and other parts with little protective lubricant, during long periods of storage.

Proper Procedure

a- Shut off the fuel supply at the tank.

b- Disconnect the fuel line at the tank.

c- Start and operate the engine until it begins to run **ROUGH**.

d- Shut down the engine. Some fuel/oil mixture will remain inside the engine.

e- Remove and drain the carburetor.

By having disconnected the fuel supply, all **SMALL** passages are cleared of fuel even though some fuel is left in the carburetor.

f- Apply a light oil into the combustion chamber through the spark plug opening, as instructed in the Owner's manual.

NOTE

For short periods of storage, simply running the carburetor dry may help prevent severe gum and varnish from forming in the carburetor. This is especially true during hot weather.

6-4 TROUBLESHOOTING

The following paragraphs provide an orderly sequence of tests to pinpoint problems in the fuel systems. It is very rare for the carburetor by itself to cause failure of the engine to start.

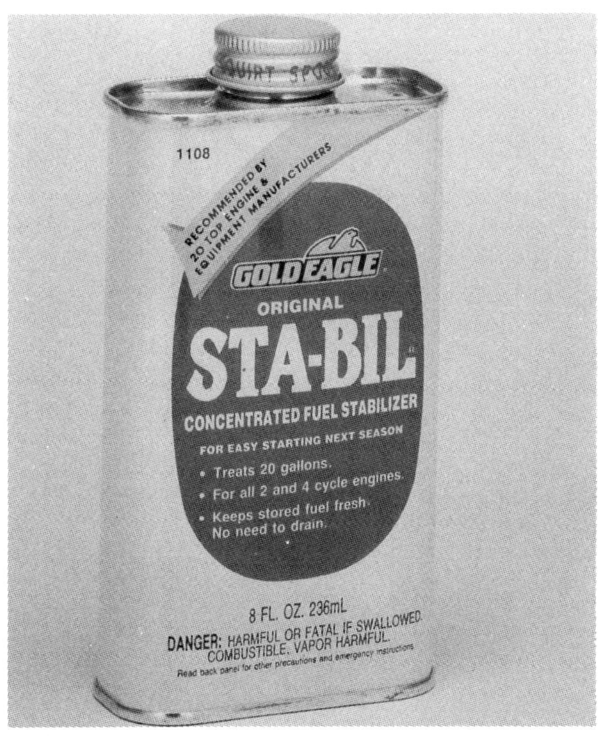

STA-BIL gasoline Stabilizer and Conditioner -- or equal substitute — may be used to prevent fuel from "souring" for up to twelve full months.

Fuel Problems

Many times fuel system troubles are caused by a plugged fuel filter, a defective fuel pump, or by a leak in the line from the fuel tank to the fuel pump. A defective choke may also cause problems. **WOULD YOU BELIEVE**, a majority of starting troubles which are traced to the fuel system are the result of an empty fuel tank, clogged fuel filter, or aged "sour" fuel.

WORDS OF CAUTION

Be sure to loosen the fuel filler cap on the fuel tank **BEFORE** opening any portion of the fuel system, to relieve any pressure built up inside the fuel system. On a hot summer's day, pressure in the fuel system can build up in a short time.

Fuel Filter and Sediment Bowl Service

Have a shop cloth handy to catch any spilled fuel. Un-thread the ring nut and pull down the sediment bowl with the fuel filter inside. A large **O**-ring seals the bowl to the filter cap and ring nut. The function of this **O**-ring is to prevent fuel and air leaks

If the **O**-ring becomes defective, the fuel pump will draw air into the sysetm and the engine will either not start, faulter, or die. Sometimes, this problem is not apparent with a full fuel tank, because the fuel is gravity flow to the carburetor. When the fuel level drops in the tank, the pump takes over, but its effectiveness is impaired by the air leak at the sediment bowl **O**-ring.

Clean out the sediment bowl and rinse the filter screen in solvent every 25 hours of engine operation.

"Sour" Fuel

Under average conditions (temperate climates), fuel will begin to breakdown in about four months. A gummy substance forms in the bottom of the fuel tank and in other areas. The filter screen between the tank and the carburetor and small passages in the carburetor will become clogged. The gasoline will begin to give off an odor similar to rotten eggs. Such a condition can cause the owner much frustration, time in cleaning components, and the expense of replacement or overhaul parts for the carburetor.

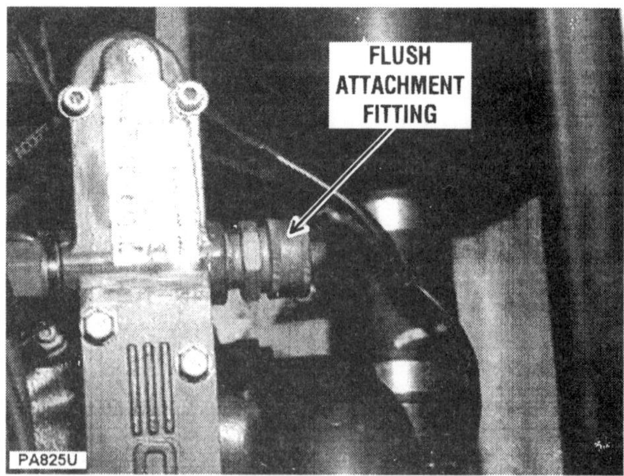

Typical location of the flush attachment fitting on the port side -- aft -- on the water circulating manifold. The location may vary with different models and production years, but the general location is the same.

Even with the high price of fuel, removing gasoline that has been standing unused over a long period of time is still the easiest and least expensive preventative maintenance possible. In most cases, this old gas can be used without harmful effects in an automobile using regular gasoline.

A gasoline preservative additive, such as Stabil Fuel Conditioner and Stabilizer for 2 and 4-Cycle Engines, will keep the fuel "fresh" for up to twelve months. If this particular product is not available in your area, other similar additives are produced under various trade names.

Choke Problems

When the engine is hot, the fuel system can cause starting problems. After a hot engine is shut down, the temperature inside the fuel bowl may rise to 200°F and cause the fuel to actually boil. All carburetors are vented to allow this pressure to escape to the atmosphere. However, some of the fuel may percolate over the high-speed nozzle.

If the choke should stick in the open position, the engine will be hard to start. If the choke should stick in the closed position, the engine will flood making it very difficult to start.

In order for this raw fuel to vaporize enough to burn, considerable air must be added to lean out the mixture. Therefore, the only remedy is to remove the spark plugs; ground the leads; crank the engine through about ten revolutions; clean the plugs; install the plugs again; and start the engine.

Rough Engine Idle

If an engine does not idle smoothly, the most reasonable approach to the problem is to perform a tune-up to eliminate such areas as: fouled or faulty spark plugs.

Other problems that can prevent an engine from running smoothly include: An air leak in the intake manifold; uneven compression between the cylinders; and sticky or broken reeds.

Of course any problem in the carburetor affecting the air/fuel mixture will also prevent the engine from operating smoothly at idle speed. These problems usually include: a weak float arm return spring; leaking needle valve and seat; defective choke; and improper adjustments for idle mixture or idle speed.

Excessive Fuel Consumption

Excessive fuel consumption can be the result of any one of two conditions, or both may apply.

1- Inefficient engine operation.

2- Poor boating habits of the operator.

If the fuel consumption suddenly increases over what could be considered normal, then the cause can probably be attributed to the engine or watercraft and not the operator.

Check the fuel system for possible leaks. A leak between the tank and the pump many times will not appear when the engine is operating, because the suction created by the pump drawing fuel will not allow the fuel to leak. Once the engine is turned off and the suction no longer exists, fuel may begin to leak.

If a minor tune-up has been performed and the condition still exists, then the problem most likely is in the carburetor and an overhaul is in order.

Engine Surge

If the engine operates as if the load on the craft is being constantly increased and decreased, even though an attempt is being made to hold a constant engine speed, the problem can most likely be attributed to the fuel pump, or a restriction in the fuel line between the tank and the carburetor.

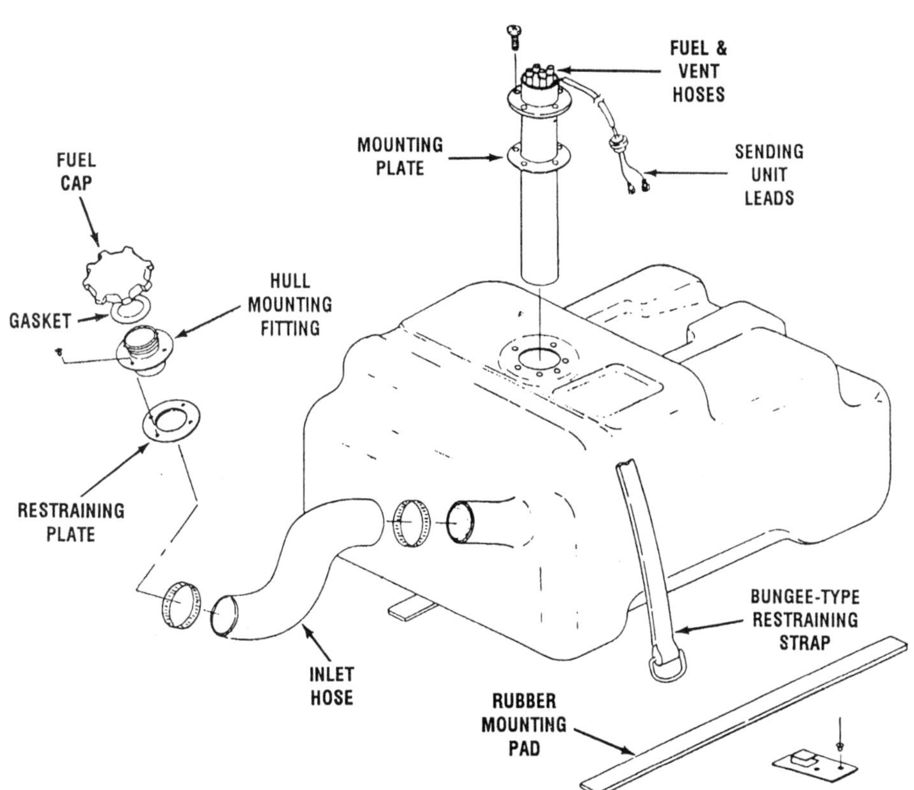

Simple exploded line drawing of a typical fuel tank installation since late 1995. Prior to 1995, the fuel level sending unit was installed in a separate tank opening, as shown on Page 6-4. Actual configuration varies with different models and production years.

6-5 SERVICE MIKUNI SUPER BN CARBURETOR

This section provides complete detailed illustrated procedures for removal, disassembly, cleaning and inspecting, assembling including bench adjustment, installation and operating adjustments for the Mikuni Super BN carburetor.

Good shop practice dictates a carburetor repair kit be purchased and new parts be installed any time the carburetor is disassembled.

Make and earnest attempt to keep the work area organized and to cover parts after they have been cleaned. This practice will prevent foreign matter from entering passageways or adhering to critical parts.

Dual carburetor and triple carburetor installations are removed as a set and may be serviced on the bench **WITHOUT** separating them.

REMOVAL -- 1992-1995

1- Remove the rubber air intake nozzle from the aft end of the flame arrestor. Remove the bolts securing the flame arrestor cover.

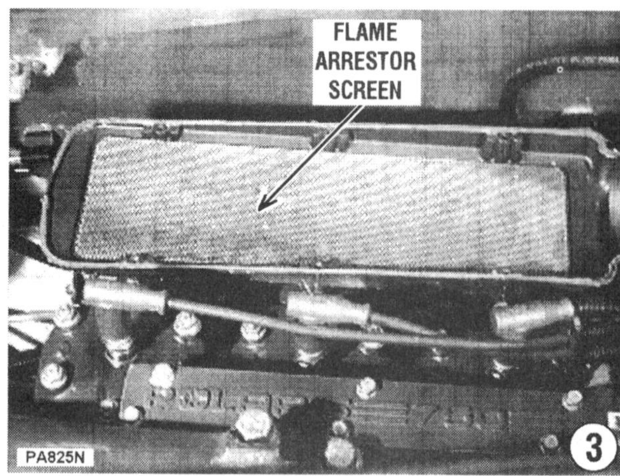

2- Lift off the flame arrestor cover and set it aside.

3- Remove the flame arrestor screen from the pan, and then remove the bolts securing the flame arrestor pan to each carburetor.

4- Disconnect the fuel line:
at the fuel water trap filter,
at the forward end of the fuel manifold, (two hoses on two fittings)
from the return line fitting on the petcock valve.

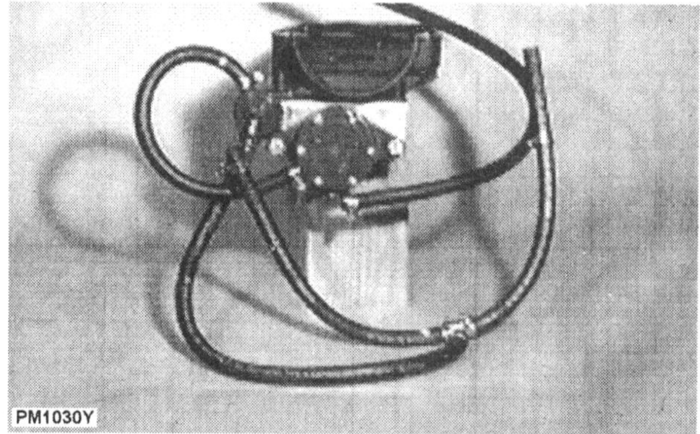

Flame arrestor pan removed with the remote fuel pump still attached and some of the fuel hoses still connected to the pump fittings.

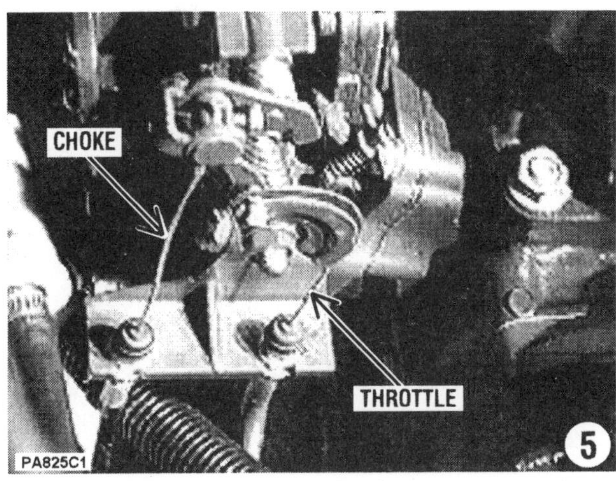

Disconnect the pulse line from the fitting on the back side of the fuel pump.

Lift the flame arrestor pan free of the carburetors with the fuel pump still attached to the aft end of the flame arrestor pan and the remaining hoses still connected to the pump fittings, as shown in the accompanying illustration.

5- Disconnect the throttle and choke cables from the carburetor assembly.

6- Remove the nuts from the underside of the carburetors -- two nuts for each carburetor.

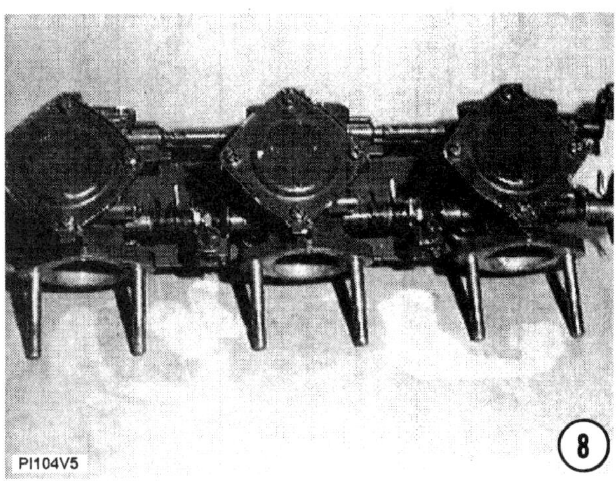

The accompanying illustration identifies the nuts and their location. When the carburetors are installed and the engine is in the craft, the nuts are not visible. The location of these 8mm nuts can only be determined from the photograph and feeling with fingers.

7- CAREFULLY lift the carburetors, as a single unit, straight up and clear of the engine.

8- Move the carburetor assembly to a suitable clean work surface.

REMOVAL -- 1996 AND ON

FIRST, THESE FLAME ARRESTOR WORDS

In 1996, the manufacturer changed the configuration of the air intake/flame arrestor for the different models through the years, equipped with Mikuni carburetors. However, the basic principle and purpose for the unit remained constant. The accompanying line drawings, 1A and 1B, indicate some of these changes, but the removal and installation procedures have remained almost the same. Therefore, read the procedure and check the drawing for the model and production year being serviced.

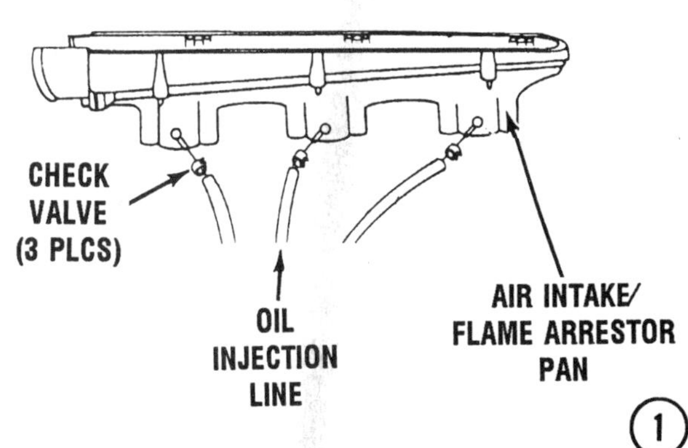

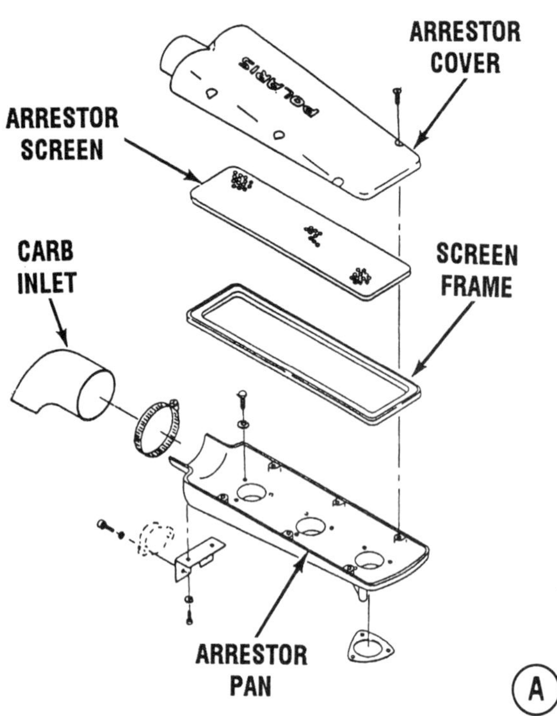

Line drawing to illustrate the air intake/flame arrestor configuration for the Model SLT750 -- 1994 and 1995 installed on the Mikuni carburetors.

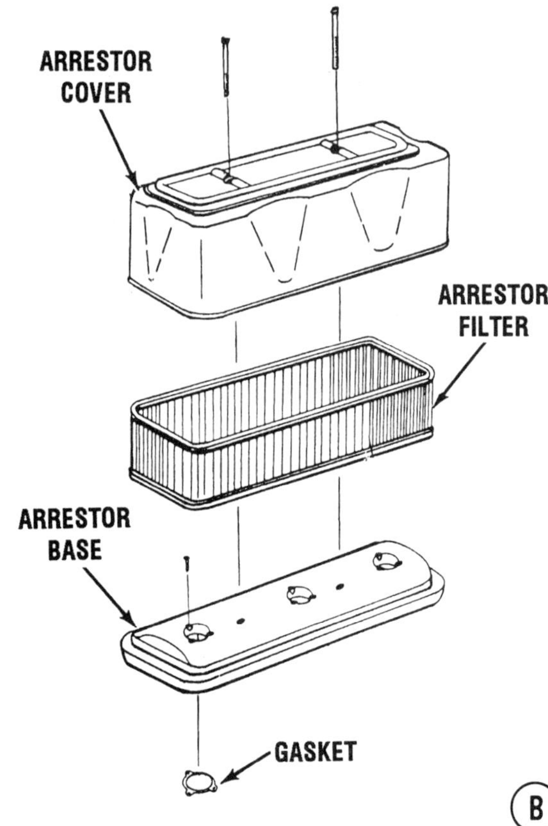

Exploded line drawing of the air intake/flame arrestor used with the Mikuni carburetor on Model SL780 and SLT 780 from 1996 to current.

1- Remove the bolts and associated hardware securing the air intake/flame arrestor cover, and then remove the cover. Lift off the air intake/flame arrestor filter element. If servicing a 1995 model unit, disconnect the oil injection lines from the fittings on the air intake/flame arrestor pan. Remove the air intake/flame arrestor pan.

2- Disconnect the choke cable and the throttle cable, at the carburetor linkage. Disconnect the oil injection pump rod, if the rod is used.

3- Remove the nuts from the underside of the carburetors -- two nuts for each carburetor. The accompanying illustration identifies the nuts and their location. When the carburetors are installed and the engine is in the craft, the nuts are not visible. The location of these 8mm nuts can only be determined from the photograph and feeling with fingers.

4- Lift the carburetor assembly straight up and clear of the engine, and then move the assembly to a suitable clean work surface.

Installation procedures after the carburetor has been assembled, begin on Page 6-26.

DISASSEMBLE MIKUNI SUPER BN CARBURETOR

The following procedures pickup the work after the carburetors have been removed as an assembly; are on a suitable clean work surface, **AND** a carburetor repair has been purchased and is on hand.

The illustrations in this section were developed with a carburetor separated from the rack for photographic clarity.

Remember, the carburetor does have to be removed from the rack in order to perform adequate service work.

As each carburetor is disassembled, keep the parts together. In order to **ENSURE** all parts, except replacement items, are installed back in the original position from which they were removed, it might be best to assemble each carburetor before moving on to the next unit.

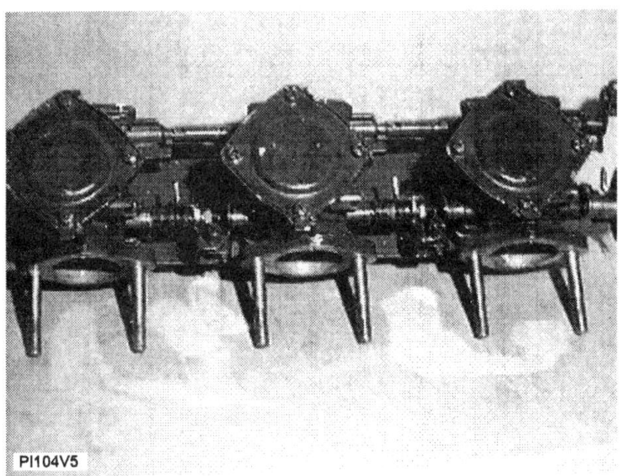

A triple setup of Mikuni Super BN carburetors on the bench ready for disassembly and service. As the text mentions several times, it is not necessary to remove a unit from the "rack" in order to perform a proper overhaul of an individual carburetor. However, if one requires service, it is best to overhaul the others at the same time.

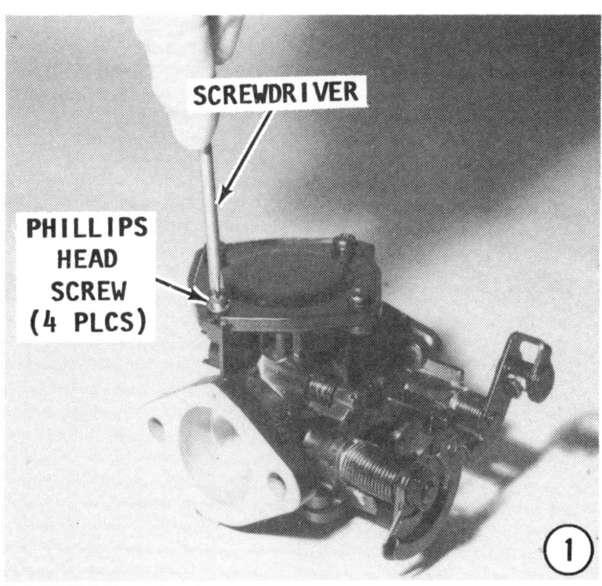

ALWAYS handle internal carburetor parts with care. Keep cleaned, or new parts removed from the package, covered with a clean shop cloth before they are installed to prevent dust and other contaminants from adhering to the cleaned parts or finding it way into internal passages.

Good shop practice dictates a carburetor repair kit be purchased and new parts installed any time the carburetor is disassembled. Such a practice will do much to give "peace of mind" to the owner and contribute to long and efficient service from the carburetor.

1- Remove the four Phillips head screws securing the diaphragm cover to the carburetor.

2- Lift off the diaphragm cover.

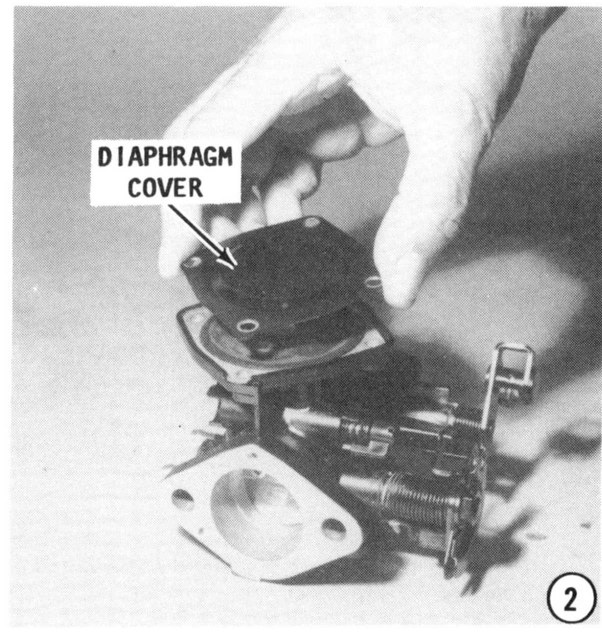

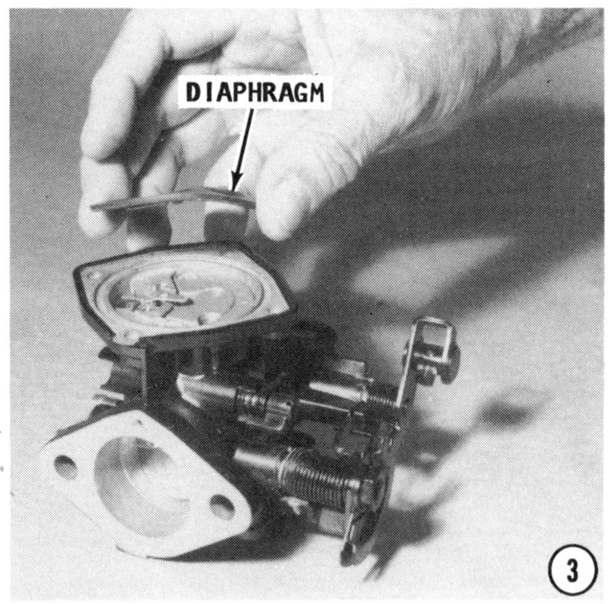

3- Remove the diaphragm from the fuel chamber.

4- Obtain a stubby Phillips head screwdriver and a hammer. Now, tap lightly on the end of the screwdriver while attempting to remove the hinge screw. Do not attempt to remove this screw "cold" -- without tapping the screwdriver -- because the head will almost certainly be stripped!

5- Lift out the float arm and float pin. The needle valve will remain hanging on the

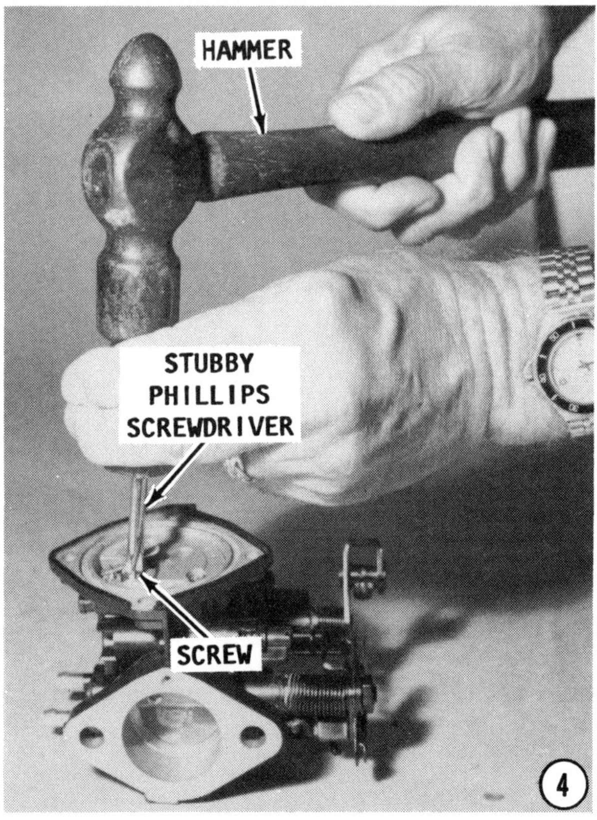

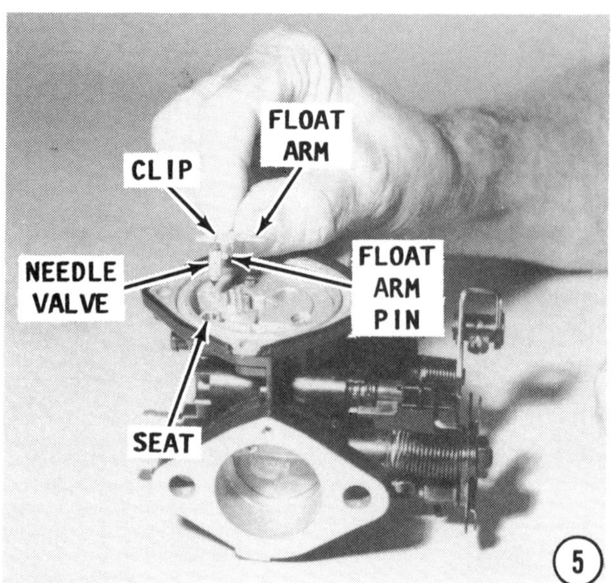

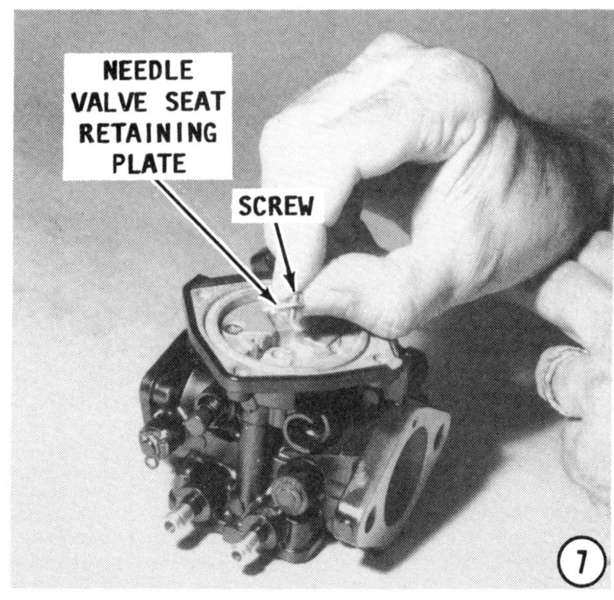

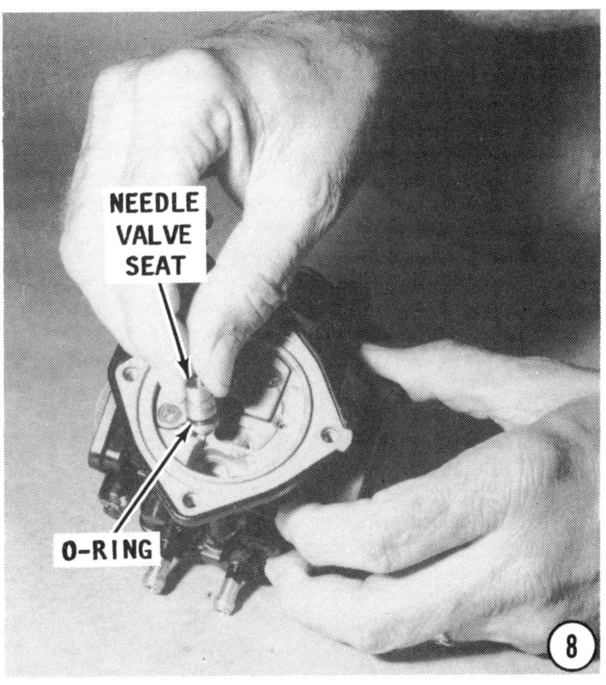

pin by the clip. Lift the valve from the seat.

6- Remove the spring from its recess in the fuel chamber.

7- Use the stubby Phillips head screwdriver and a hammer from Step 15 and lightly tap the end of the screwdriver while attempting to remove the screw securing the retaining plate to the fuel chamber. Remove the screw and plate.

8- Lift out the needle valve seat from the fuel chamber. Remove and discard the O-ring around the seat.

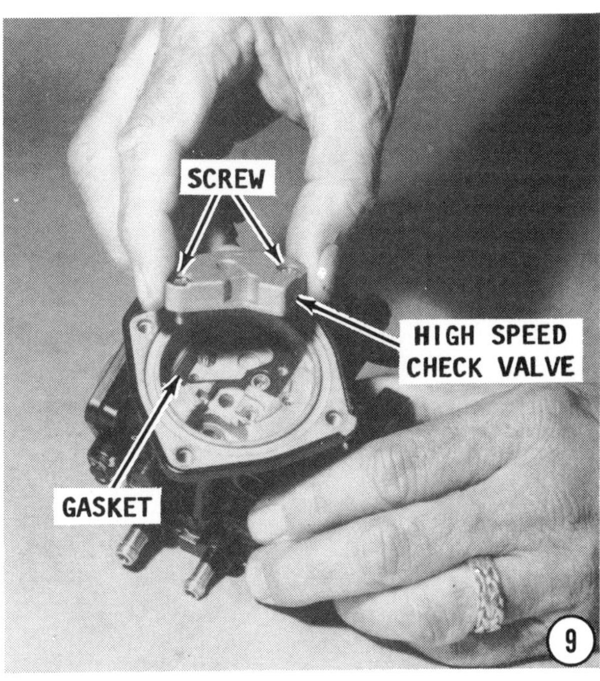

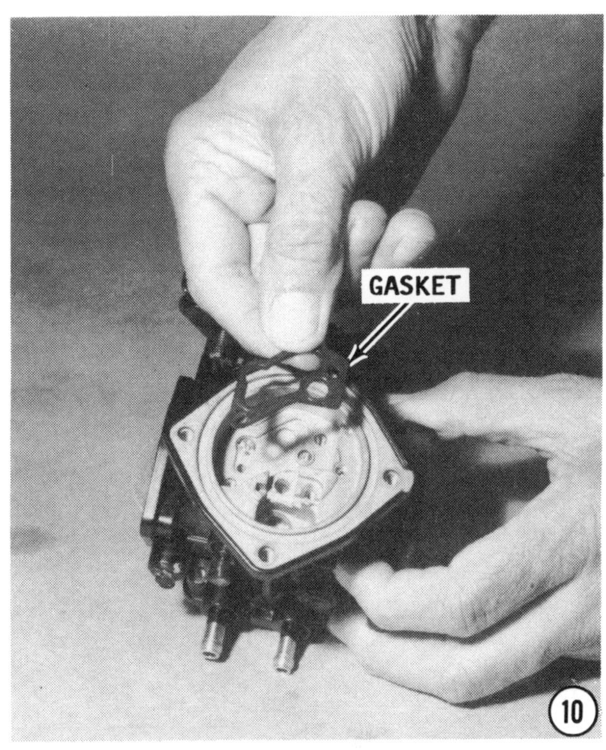

9- If necessary, use the "tapping method" to remove the two small screws securing the high speed check valve housing to the carburetor. Remove the check valve housing.

10- Remove and discard the gasket.

11- Use the correct size screwdriver to remove the main jet (the larger of the two jets), from the fuel chamber.

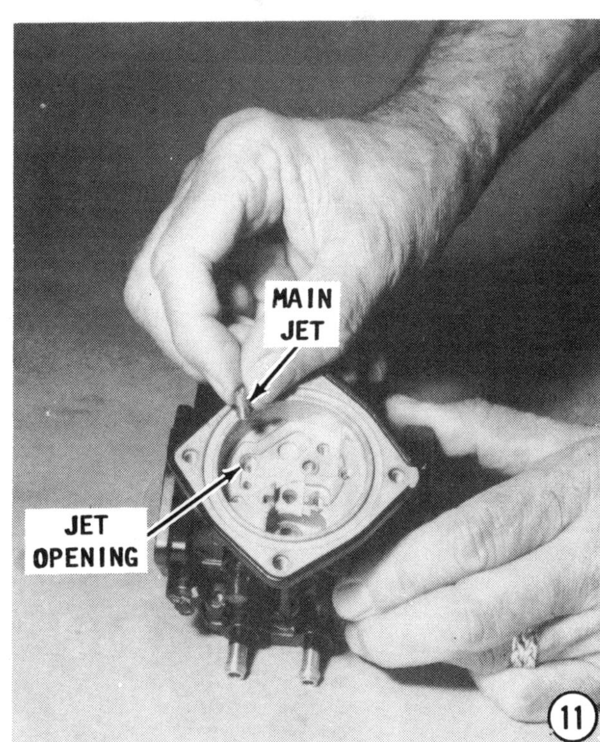

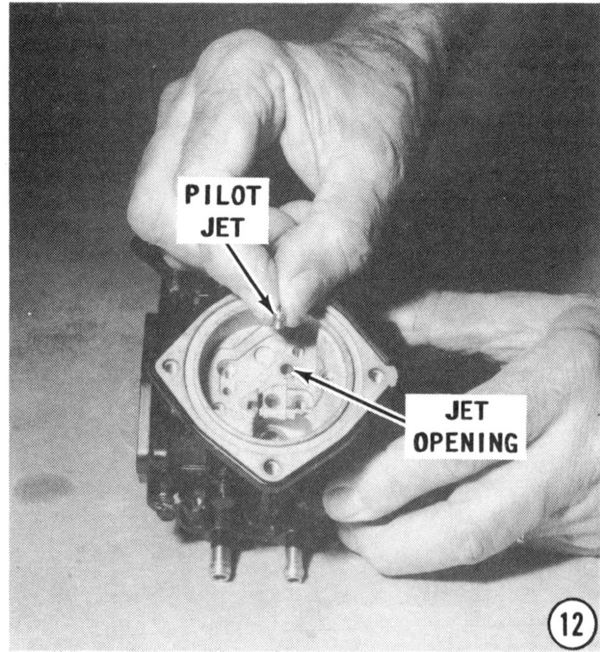

12- Use the correct size screwdriver and remove the pilot jet from the fuel chamber.

13- Remove the small Phillips head screw from the underside of the high speed check valve. Lift off the retainer. Note how a small post on the underside of the retainer indexes into a hole in the housing.

14- CAREFULLY lift off the clear plastic flap. **TAKE CARE** not to lose this item. This flap controls the flow of fuel in one direction and prevents the engine from "faltering" when the throttle is suddenly changed from wide open, to idle, and then back to wide open again.

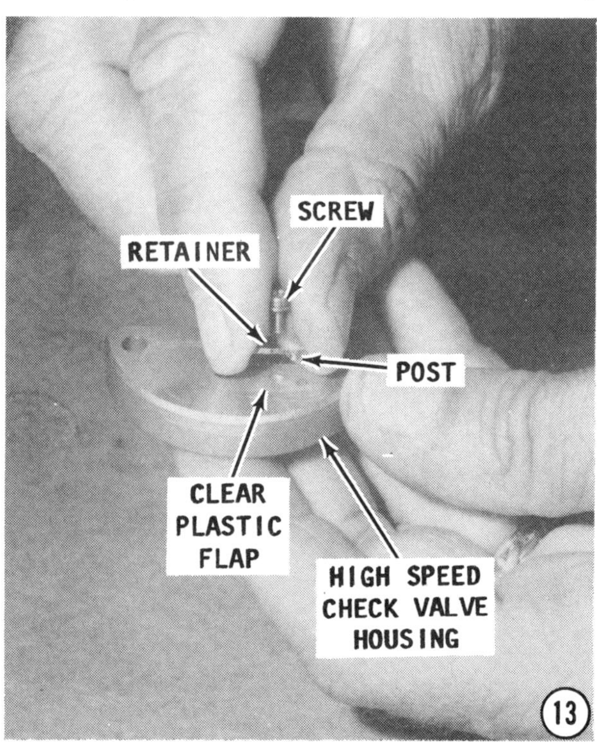

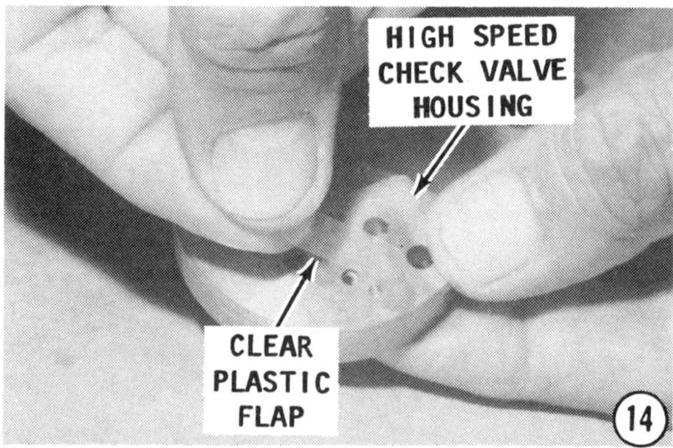

SPECIAL WORDS

Observe the letters **"L"** and **"H"** embossed next to the two mixture adjusting screws. The **"L"** denotes the low speed mixture adjusting screw, and the **"H"** denotes the high speed mixture adjusting screw. Each screw may have a plastic limiter cap installed on the screw head. A tab on the cap indexes into a slot in the collar cast in the carburetor body around the screw. The rotation of the screw is limited by the tab in the slot - hence the name "limiter cap".

15- Pull off the limiter cap on the low speed mixture adjusting screw -- the screw next to the embossed **"L"**.

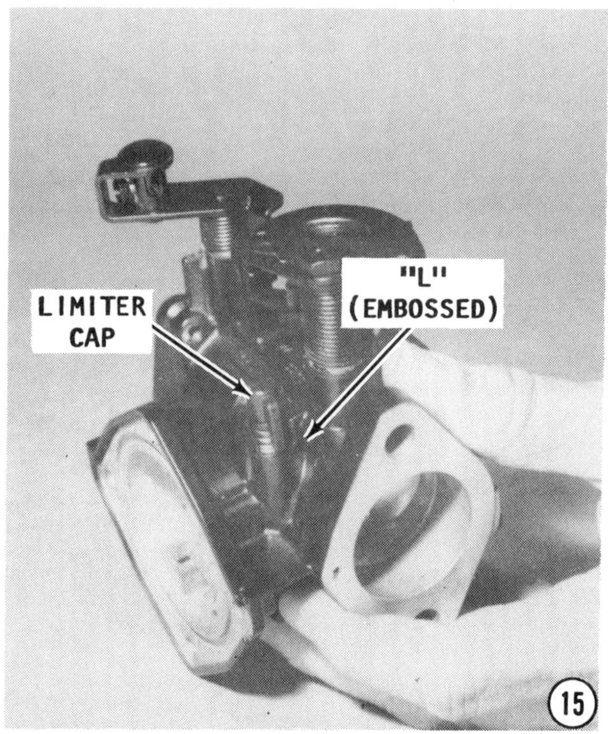

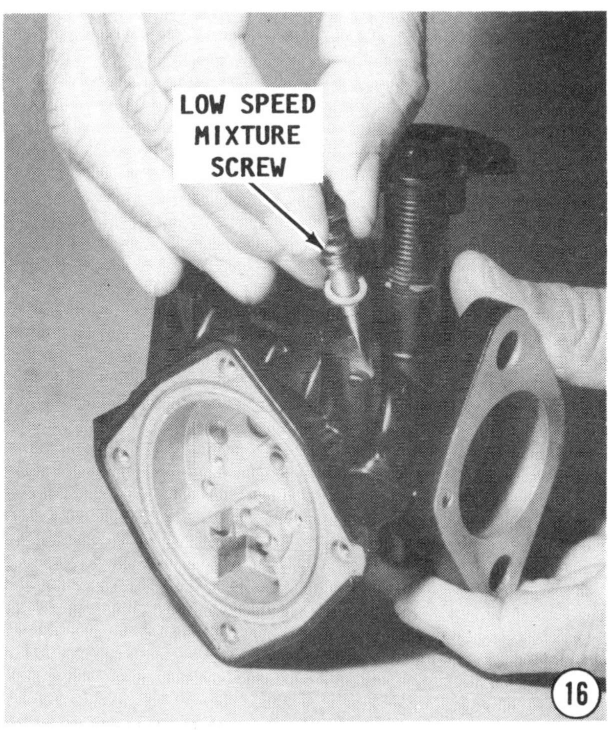

16- Back out the low speed mixture screw.

17- Pull off the limiter cap on the high speed mixture adjusting screw -- the screw next to the embossed **"H"**.

18- Back out the high speed mixture adjusting screw.

CLEANING AND INSPECTING

NEVER dip rubber parts, plastic parts, or diaphragms in carburetor cleaner. These

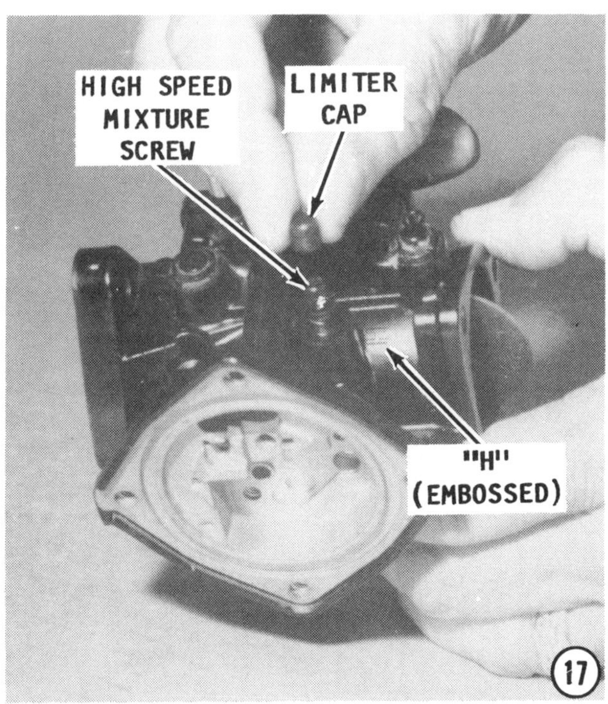

parts should be cleaned **ONLY** in solvent, and then blown dry with compressed air.

Place all metal parts in a screen-type tray and dip them in carburetor cleaner until they appear completely clean, then blow them dry with compressed air.

Blow out all passages in the castings with compressed air. Check all parts and passages to be sure they are not clogged or contain any deposits. **NEVER** use a piece of wire or any type of pointed instrument to clean drilled passages or calibrated holes in a carburetor.

Move the throttle shaft back and forth to check for wear. If the shaft appears to be too loose, replace the complete throttle body because individual replacement parts are **NOT** available.

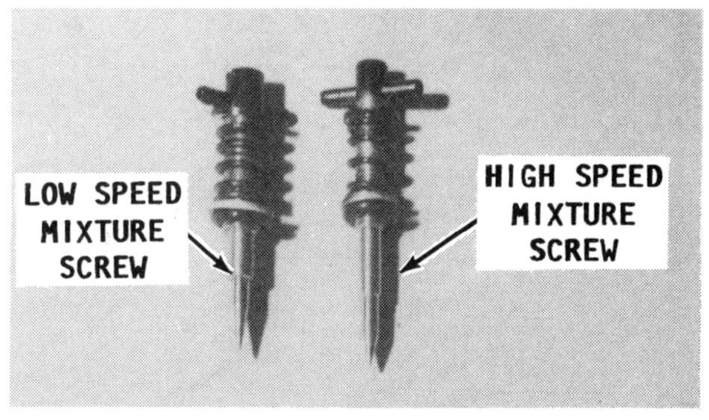

Comparison of the low speed and high speed mixture screws. Notice how the high speed screw is slightly shorter than the low speed screw.

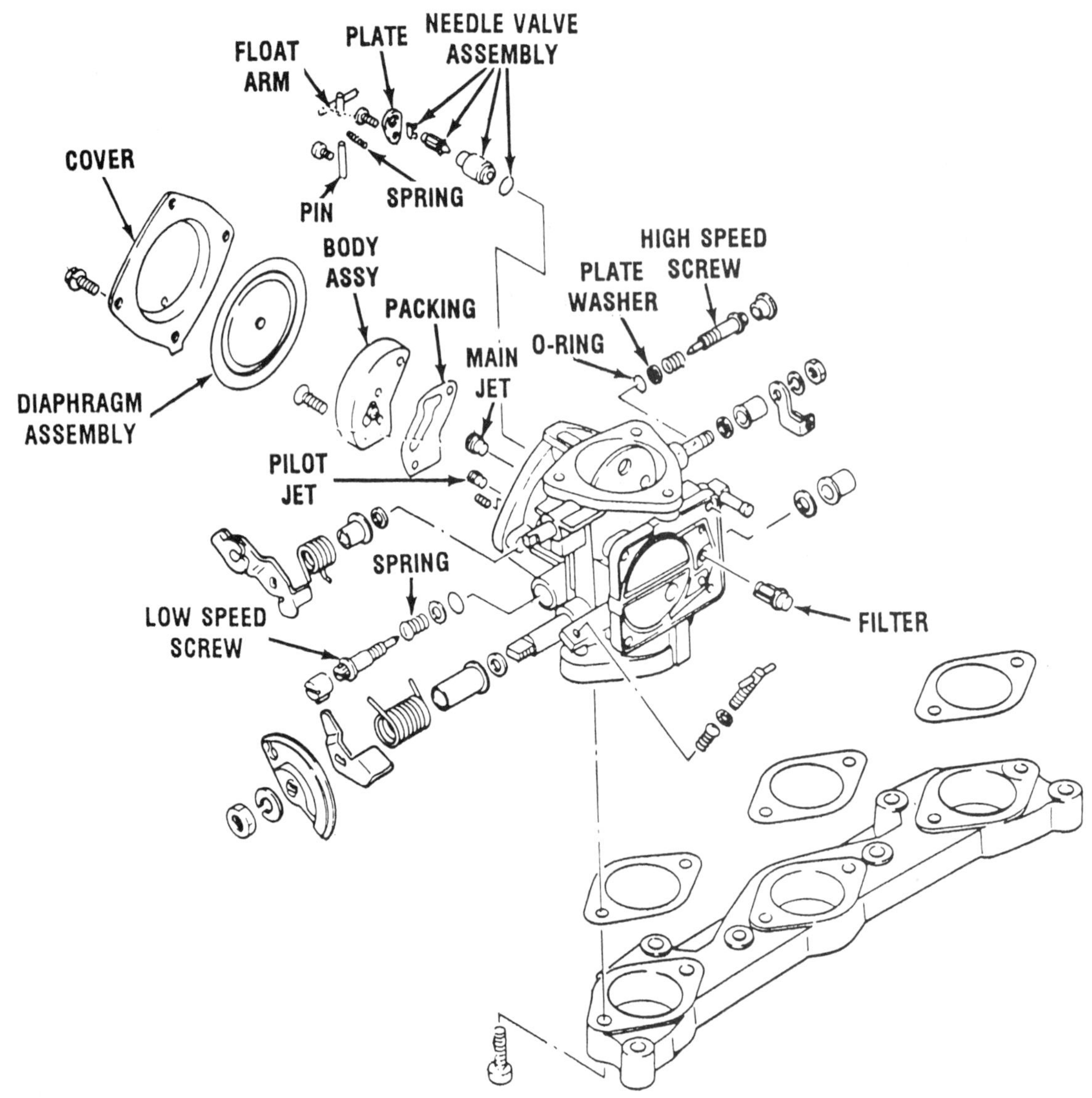

Exploded line drawing of a typical Mikuni super BN carburetor installed on the engines listed in the table on Page 6-2. Major parts have been identified.

CRITICAL AREA

Inspect the condition of both the choke and thottle shafts. Most shaft designs incorporate a cutout portion to enable the butterfly valve to lay flat against the shaft. Two small holes are drilled in this cutout portion to accomodate the small screws securing the valve to the shaft. The weakest point of the shaft is at the screw holes. Small cracks form at the holes due to metal fatigue. These cracks elongate the threaded holes and the small screws shake loose. Once loose, they are sucked into the carburetor, then the intake manifold, and into the crankcase -- eventually finding their way into the combustion chamber. One of these small screws can then cause very expensive damage to pistons, rings and cylinder walls. Therefore, check the cutout in the shaft **CLOSELY** for any evidence of a small crack.

Inspect the main body, airhorn, and venturi cluster gasket surfaces for cracks and burrs which might cause a leak. Check to be sure the float arm return spring has not been stretched. Check the float arm needle contacting surface and replace the diaphragm if this surface has a groove worn in it.

Inspect the tapered section of the low and high speed adjusting needles and replace any that have developed a groove.

As previously mentioned, most of the parts which should be replaced during a carburetor overhaul are included in overhaul kits available from your local marine dealer. One of these kits will contain a matched fuel inlet needle and seat. This combination should be replaced each time the carburetor is disassembled as a precaution against leakage.

Make a thorough inspection of the fuel chamber and fuel pump diaphragms for the tiniest pin hole. If one is discovered, the hole will only get bigger. Therefore, the diaphragms must be replaced in order to obtain full performance from the engine.

Inspect the two vent holes in the first pump body. These holes **MUST** be clean. If these two holes are not perfectly clear, the diaphragm in the fuel pump will not function properly -- full fuel volume will not be pumped.

ASSEMBLING

Special Words on Mixture Screws

Observe the low and high speed mixture screws. They appear to be very similar, but they are **NOT** interchangeable. To identify the correct screw, place them side by side as shown in the lower right illustration on Page 6-17. The low speed screw is longer than the high speed screw. The low speed screw is installed on the linkage side of the carburetor -- next to the embossed "L". The high speed screw is installed on the opposite side -- close to the fuel supply and return fittings -- next to the embossed "H".

Both screws are adjusted later at the completion of the assembling procedures.

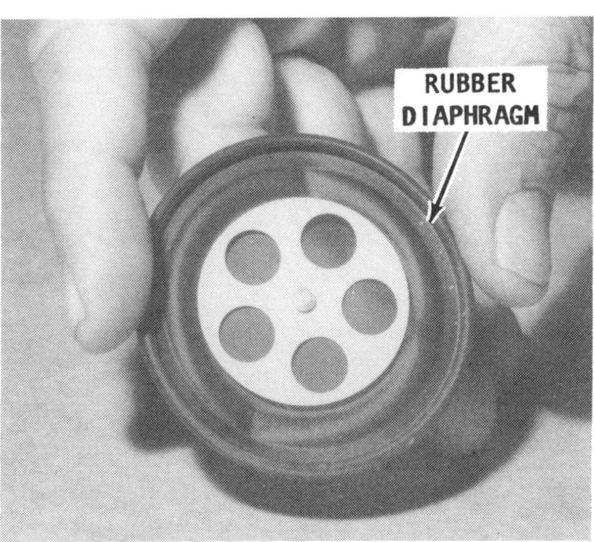

The pump diaphragm must be carefully inspected and replaced if any damage is discovered or even suspected. The tiniest "pin" hole is sufficient cause for replacement.

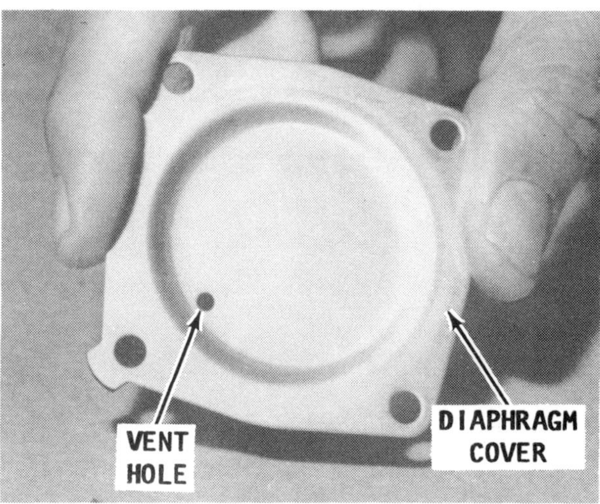

Carefully clean the vent hole in the diaphragm cover. One side of the rubber diaphragm is vented to the atmosphere thru this hole. Diaphragm action is seriously impaired if the hole becomes blocked -- even by the smallest obstruction.

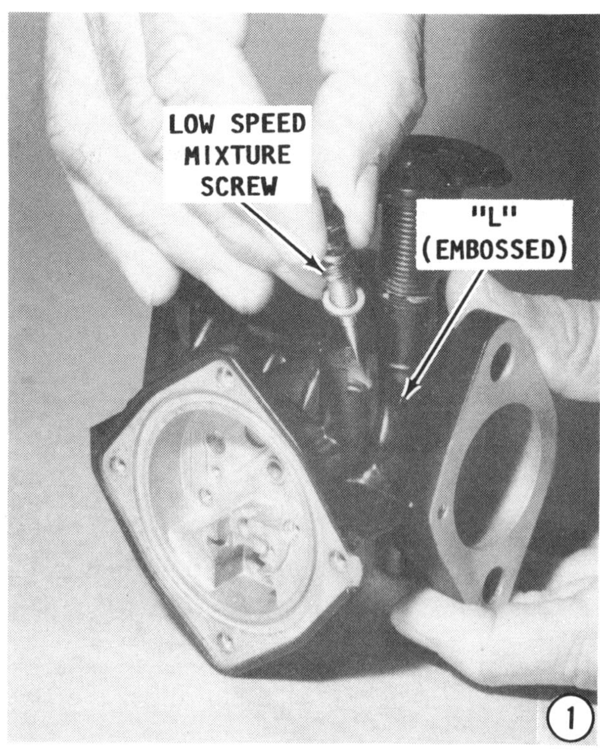

Therefore, thread each screw in just a few turns at this time.

1- Install the low speed mixture screw just a few turns on the linkage side of the carburetor -- next to the embossed "L".

2- Index the tab on the limiter cap into the slot in the collar and snap the cap squarely down over the screw head.

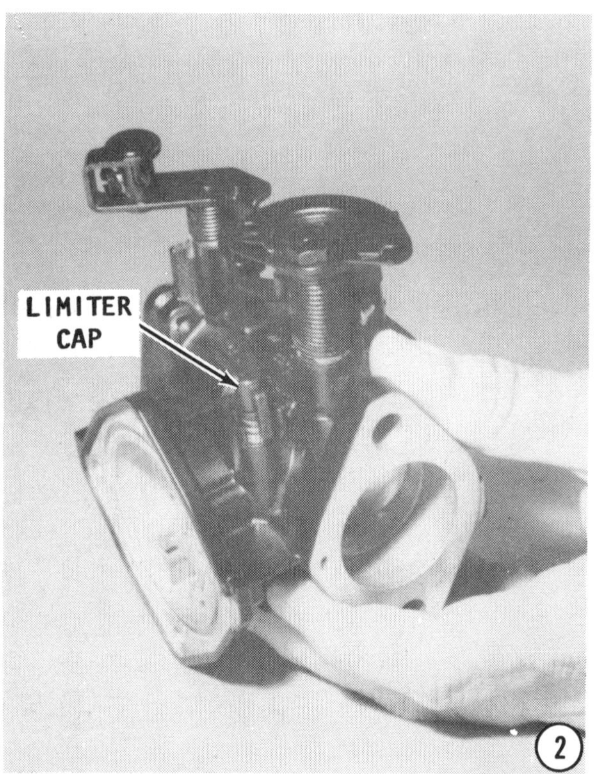

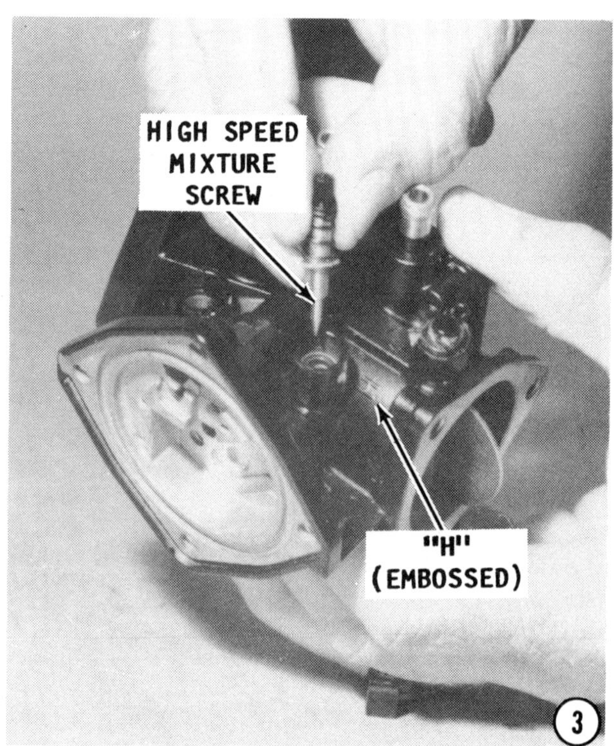

3- Install the high speed mixture screw just a few turns on the same side as the fuel supply and return fittings -- next to the embossed "H".

4- Index the tab on the limiter cap into the slot in the collar and snap the cap squarely down over the screw head.

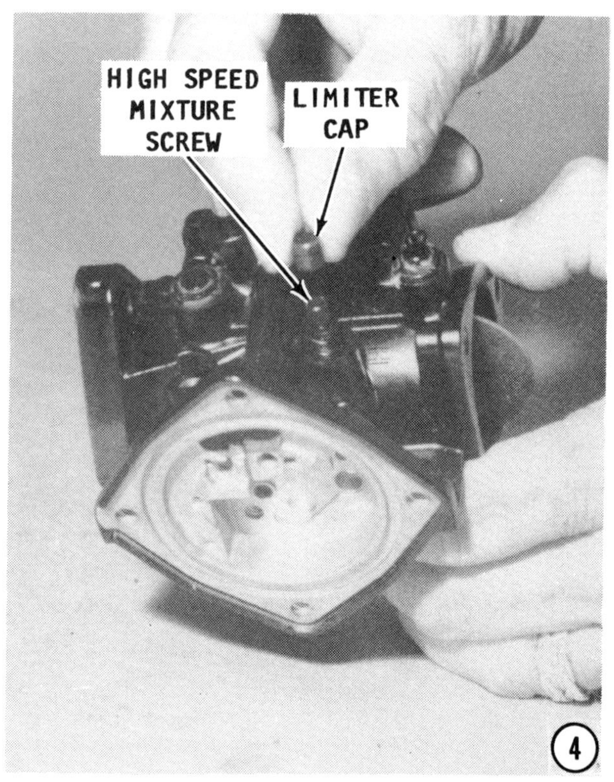

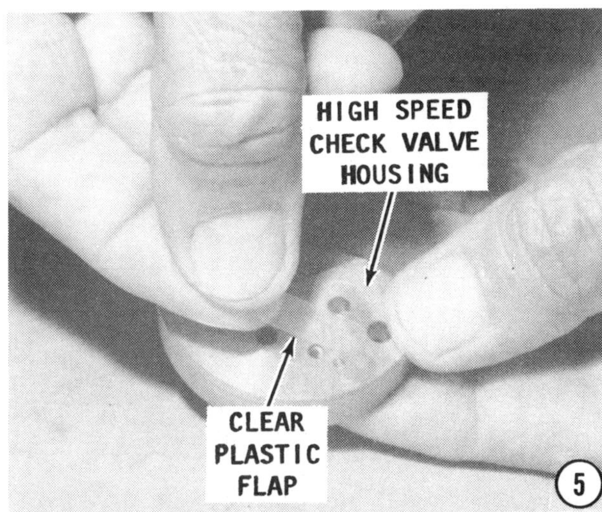

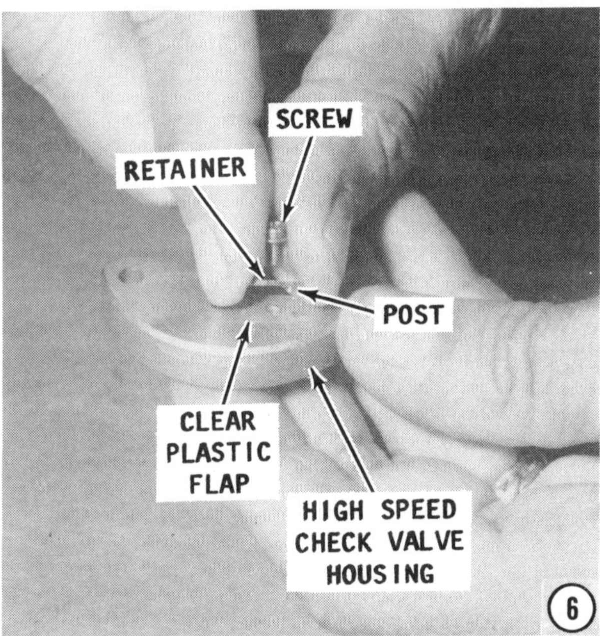

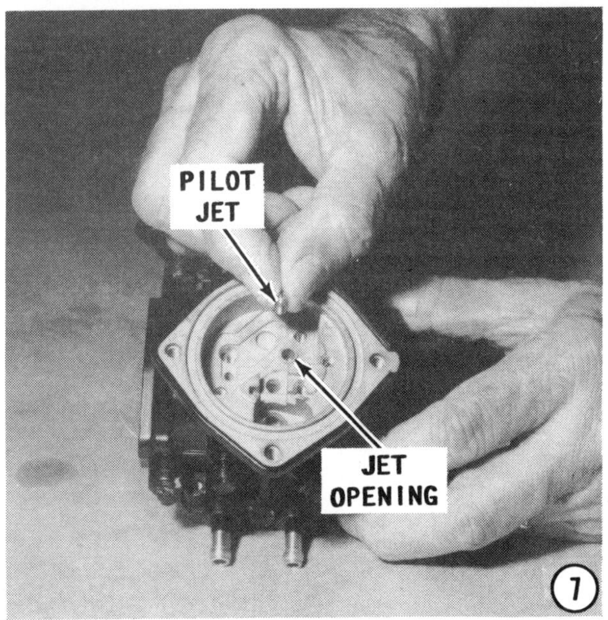

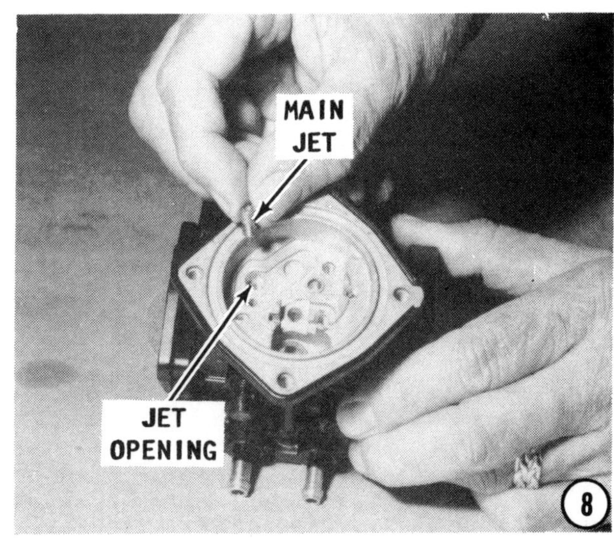

5- Align the two holes in the clear plastic flap with the two holes on the underside of the high speed check valve housing. If the four holes align, the flap is installed properly. If the holes will not align, the flap is upside-down.

6- Position the retainer over the plastic flap, with the post on the underside of the retainer indexed into both the hole in the plastic flap and the hole in the housing. Install and tighten the screw securing both parts to the housing.

7- Install the pilot jet (the smaller of the two), into the jet opening in the fuel chamber. Tighten the jet securely using the proper size screwdriver.

8- Install the main jet into the jet opening in the fuel chamber. Again, using the proper size screwdriver tighten the jet securely.

9- Position the high speed check valve housing gasket into the recess in the fuel chamber.

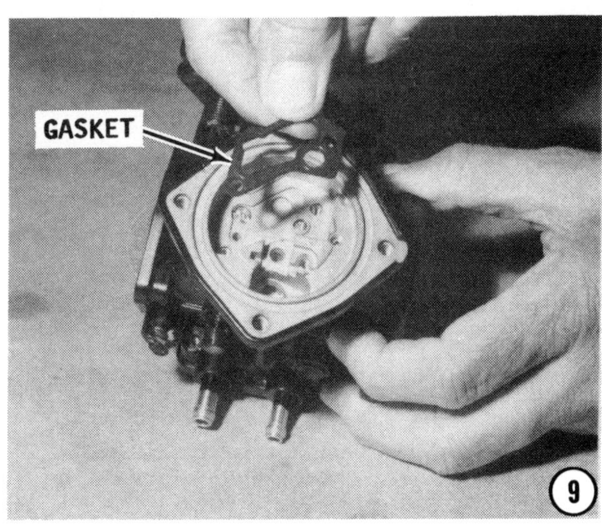

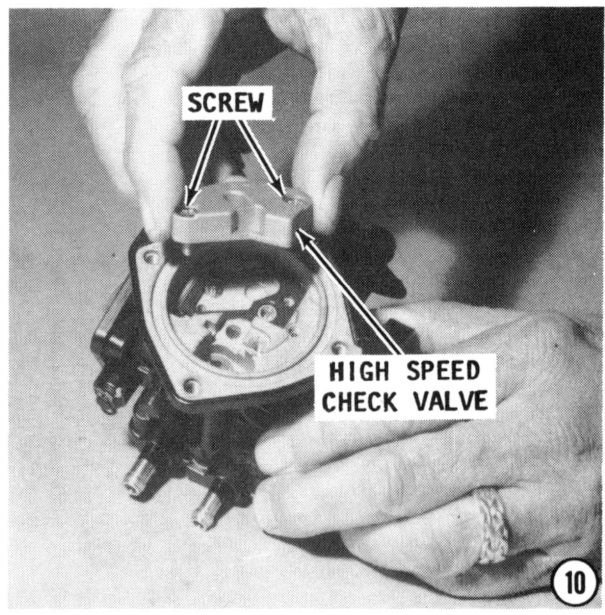

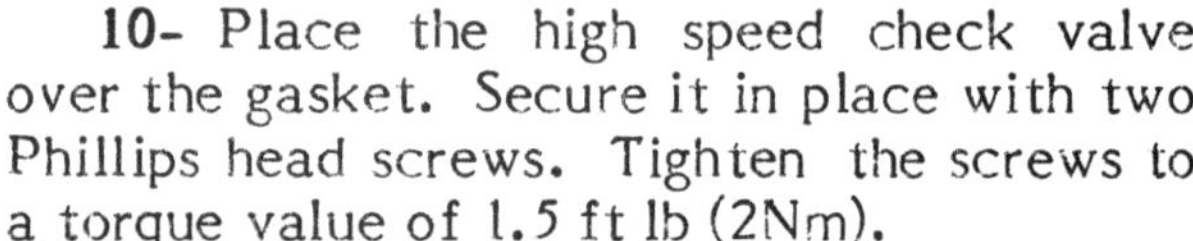

10- Place the high speed check valve over the gasket. Secure it in place with two Phillips head screws. Tighten the screws to a torque value of 1.5 ft lb (2Nm).

11- Install a new **O**-ring around the needle valve seat. Lower the seat into the recess in the fuel chamber.

12- Install the seat retainer over the seat. Secure the seat in place with the Phillips head screw. Tighten the screw to a torque value of 0.7 ft lb (1Nm).

13- Lower the spring into the recess in the fuel chamber.

14- Slide the needle valve clip over the square tab on the float arm. Slide the float arm over the pin, as shown. Lower the assembly down into place and at the same time: guide the needle valve into the needle

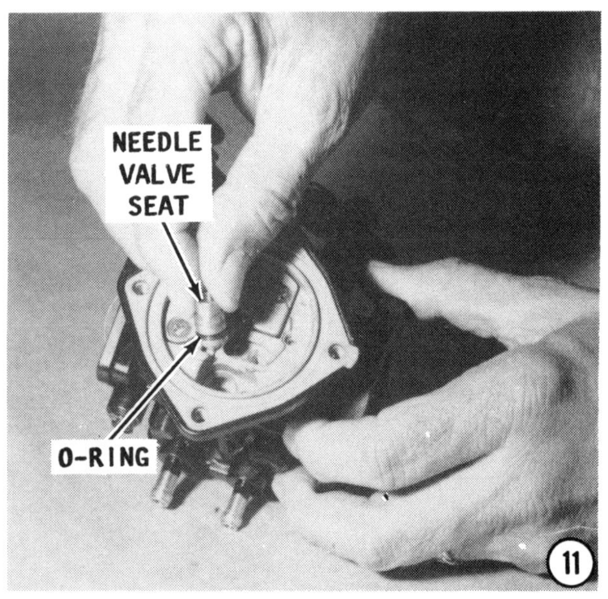

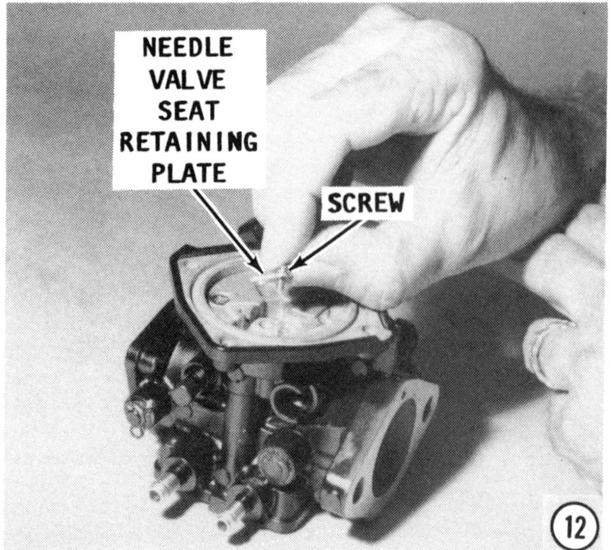

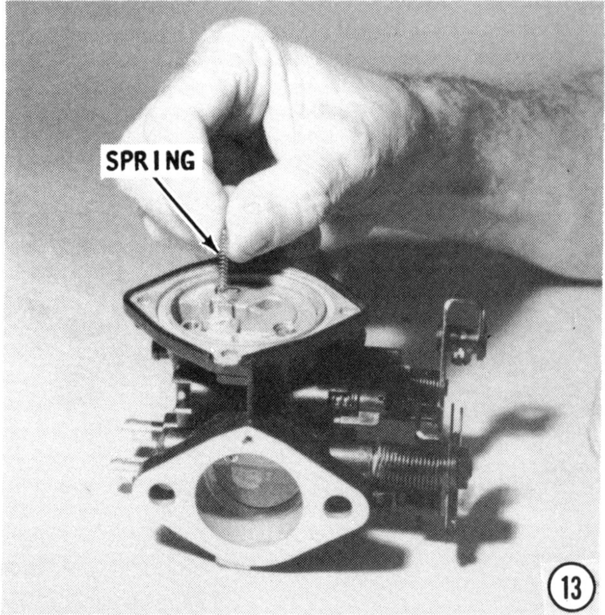

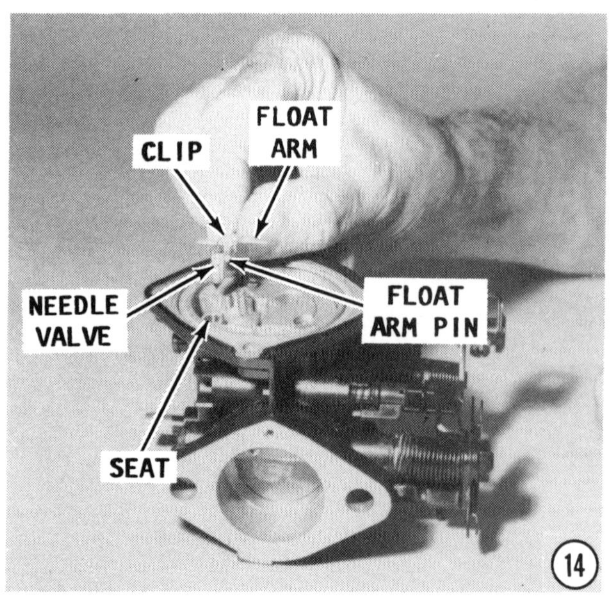

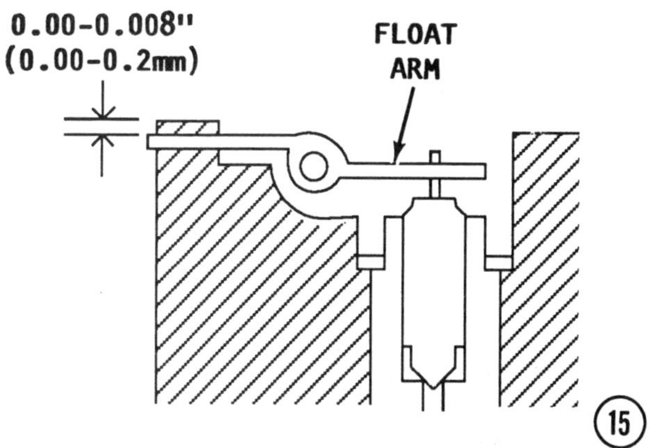

seat; place the float pin into the elongated slot in the fuel chamber; and position the dimple in the float arm over the spring. Secure the pin to the carburetor with the small Phillips head screw. Tighten the screw to a torque value of 0.7 ft lb (1Nm).

Float Arm Adjustment

15- The float arm adjustment is correct when three conditions are satisfied:

a) The top of the float arm is parallel to the fuel chamber surface.

b) The distance between the top surface of the float arm and the diaphragm contact surface of the carburetor is between 0.00-0.008" (0.0-0.2mm).

c) The float arm makes contact with the needle valve, but does not apply any pressure to the valve.

If any of these conditions are not satisfied, then replace the float arm and/or return spring. Do not attempt to bend the

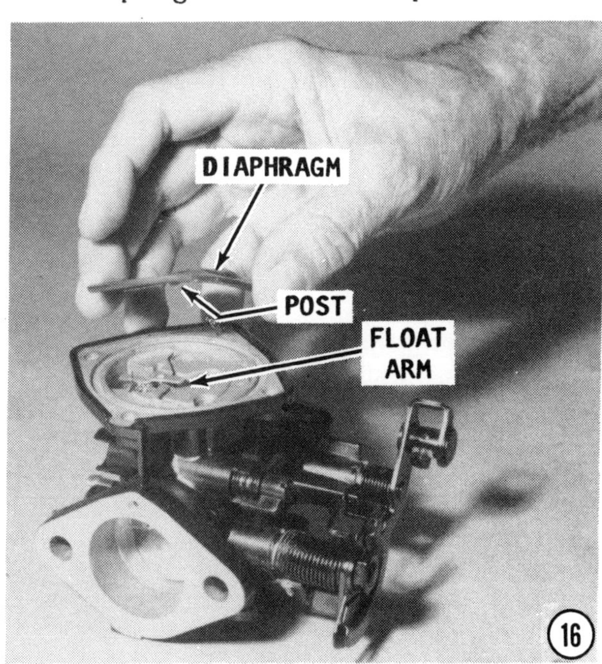

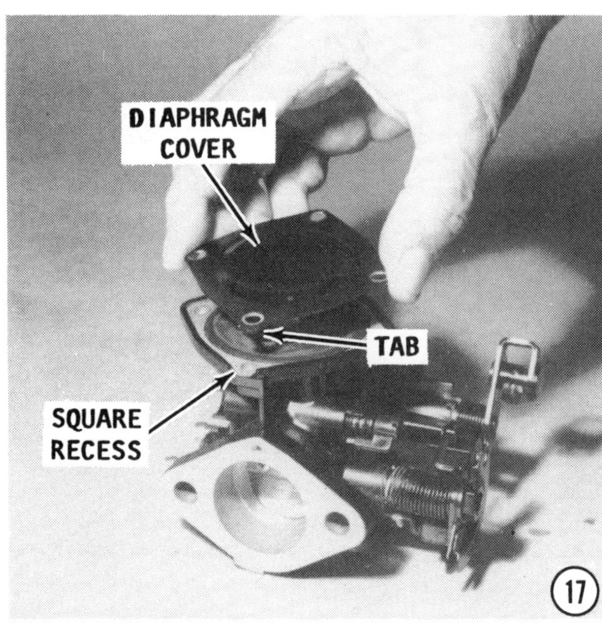

float arm or to change the characteristics of the old return spring.

Depress the rounded end of the float arm. The needle valve should lift smoothly and return inside the seat when the arm is released.

16- Place the diaphragm over the fuel chamber with the float post on the underside facing **DOWN** to rest against the float arm.

17- Position the diaphragm cover over the diaphragm with the tab in the cover corner indexed into the square recess in the carburetor body.

18- Secure the cover to the carburetor with the four Phillips head screws. Tighten

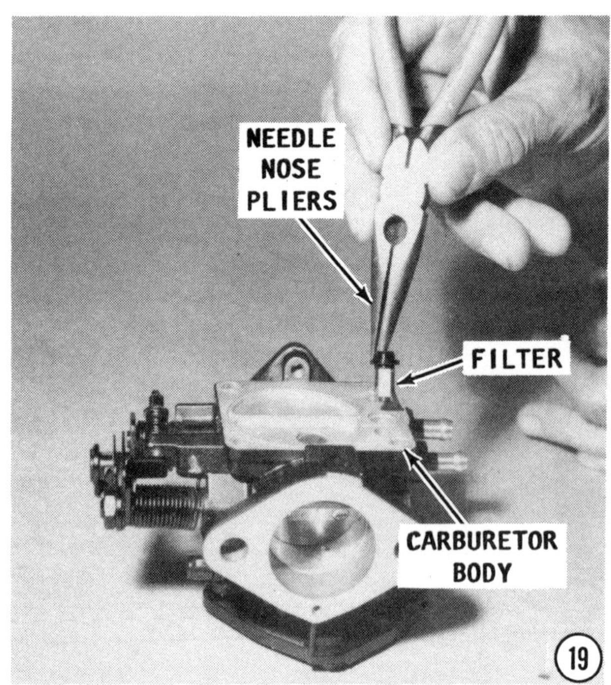

the screws alternately and evenly to a torque value of 3 ft lb (4Nm).

19- Lower the fuel filter into the recess in the carburetor body.

BENCH ADJUSTMENTS

Set the low speed and high speed adjustment screws the correct number of turns "out" from a lightly seated position, as listed in the table in the Appendix.

SPECIAL WORDS

If the following adjustments are attempted while the carburetor/s are installed on the manifold, the flame arrestor cover and the flame arrestor screens must first be removed to enable the choke/throttle (butterfly) valve to be observed.

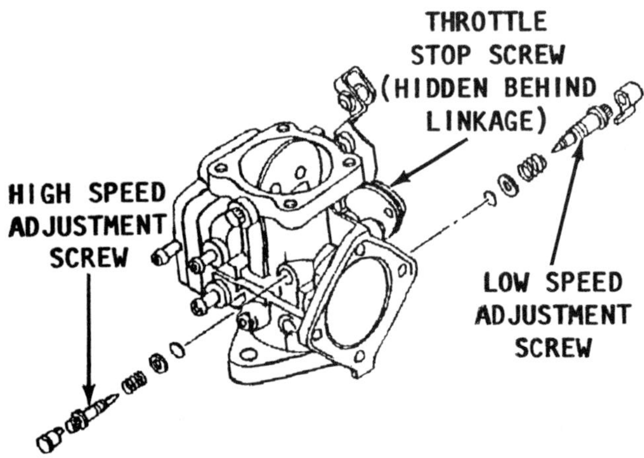

*Exploded line drawing of the Mikuni BN carburetor showing arrangement of parts for the low speed and high speed adjustment screws. Bench adjustment for these two screws is made by seating the screws **LIGHTLY**, and then backing out the number of turns listed in the Appendix for the Model and production year. The forward, center, and aft carburetors all have different settings.*

Low Speed and Idle Mixture Adjustment

Move the craft to a test tank, to a body of water, or connect a flush attachment and hose to the engine. For maximum performance, the idle mixture and the idle rpm should be adjusted under actual operating conditions. Connect a tachometer to the engine.

Lightly seat the low speed adjustment screw. From this position, back the screw out 3/4 to 1 full turn as a preliminary adjustment, at this time.

Start the engine and allow it to warm to operating temperature.

CAUTION

Water must circulate through the jet pump -- to and from the engine, anytime the engine is operating. Circulating water will prevent overheating -- which could cause damage to moving engine parts and possible engine seizure.

NEVER, AGAIN NEVER, operate the engine at high speed with a flush device attached. An engine operating at high speed with such a device attached, might very likely **RUN-AWAY** from lack of a load on the impeller shaft, causing extensive damage.

The throttle stop screw is located on the throttle linkage on the opposite side of the carburetor from the fuel fittings

With the engine operating, slowly rotate the throttle stop screw until the engine idles at 1,200 to 1,300 rpm. Next, rotate the low speed screw inward or outward 1/8th of a turn at a time until the cylinders fire evenly and engine rpm increases. Readjust the throttle stop screw to reduce the rpm to specified idle speed.

Synchronize Throttle Valves Triple Mikuni BN Setup (Illustration "A")

Back off the throttle stop screws on all three carburetors until the end of the screws just barely clear the throttle lever.

Next, back out the throttle stop screws on the No. 1 carburetor until the end of the screw just barely clears the throttle lever. Rotate the throttle stop screw of the No. 1 and No. 2

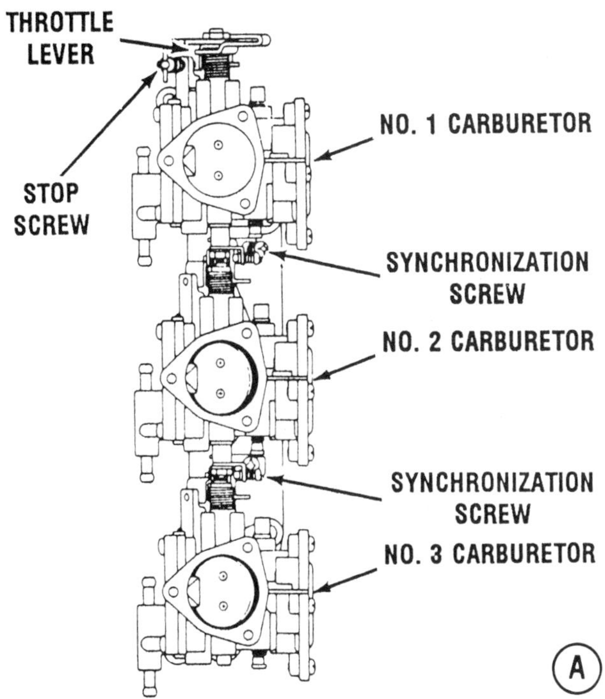

carburetors outward until the ends of both screws just barely clear the throttle lever.

Now, rotate the synchronization screw on the No. 2 carburetor inward until the throttle valves of the No. 2 and No. 3 carburetors are fully closed. **Note:** Turning the synchronization screw inward further than necessary will cause the throttle valve on the No. 3 carburetor to begin to open.

Turn the synchronization screw on the No. 1 carburetor inward until the throttle valve of the No. 1 carburetor is fully closed.

Check to be sure all throttle valves are fully closed. If all valves are not fully closed, repeat this complete adjustment procedure.

Finally, rotate the throttle stop screw to the "set" position noted above.

Synchronize Choke Valves Triple Mikuni BN Setup (Illustration "B")

Begin by backing out the synchronization screw on the No. 1 and No. 2 carburetors until the end of the screw just clears the synchronization lever.

Turn the synchronization screw on the No. 2 carburetor to bring the choke valves of the No. 2 and No. 3 carburetors into a fully closed position when the choke lever is turned to the closed position.

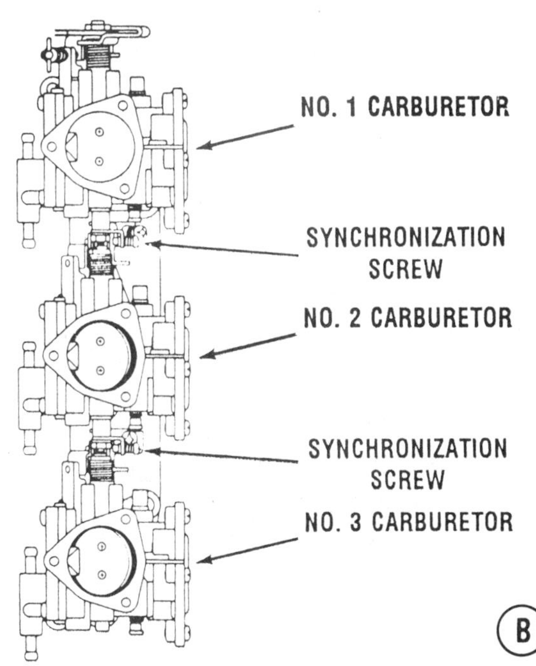

Now, rotate the synchronization screw on the No. 1 carburetor to bring the choke valve on the No. 1 carburetor to the fully closed position when the choke lever is turned to the closed position.

Finally, make a check to ensure all three choke valves are fully closed. If they are not fully closed, repeat the complete procedure.

INSTALLATION -- MIKUNI SUPER BN CARBURETORS TRIPLE SETUP -- 1992-1995

The following procedures pickup the work after the carburetors have been serviced, assembled, and mounted on the "rack".

1- Move the assembled carburetor rack into position on the intake manifold.

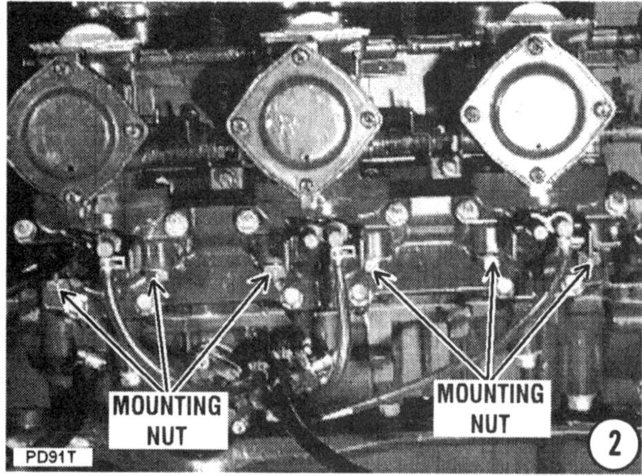

2- Apply just a light coating of Loctite™242, or equivalent, to the threads of the attaching nuts. Start the nuts -- two for each carburetor. Location of the studs should be a little easier than during removal. The studs are not visible and can only be discovered by feeling with a finger and checking the location indicated in the accompanying illustration. Tighten the nuts securely.

3- Install the flame arrestor pan, the filter screen, and then the cover. Secure it all with the attaching hardware tightened securely.

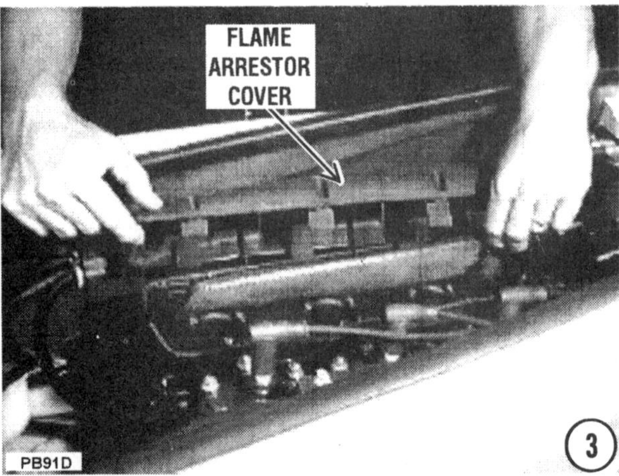

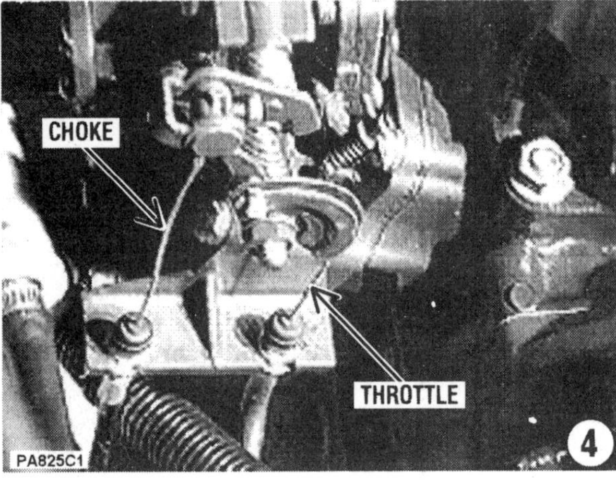

4- Connect the throttle cable, choke cable and the oil pump rod, if used. Connect the fuel hoses to the proper fittings. Connect the oil lines to the carburetor. If servicing a 1995 engine, connect the oil lines to the three fittings on the flame arrestor pan.

INSTALLATION -- MIKUNI SUPER BN CARBURETORS 1996 AND ON

The following procedures pickup the work after the carburetors have been serviced, assembled, and mounted on the "rack".

1- Move the carburetors into position on the intake manifold.

2- Apply just a light coating of Loctite™242, or equivalent, to the threads of the attaching nuts. Start the nuts -- two for each carburetor. Location of the studs should be a little easier than during removal. The studs are not visible and can only be discovered by feeling with a finger and checking the location indicated in the accompanying illustration. Tighten the nuts securely.

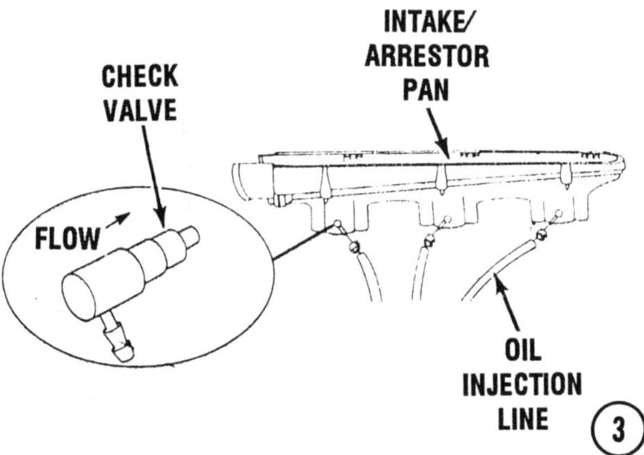

3- Install the flame arrestor pan. If servicing a 1995 model, connect the oil injection lines to the air intake/flame arrestor pan.

4- Connect the choke cable and the throttle cable at the carburetor linkage. Connect the oil pump rod, if the rod is used. The oil pump rod is manufactured as a definite length. No adjustment is necessary or possible. If the rod is bent the least bit, a new rod must be installed. This rod controls a ratio valve inside the pump. This valve allows additional oil to be delivered when the engine is operating under a heavy load at reduced rpm.

5- Position the air intake/flame arrestor filter screen in place, and then the cover. Secure the cover with the attaching hardware.

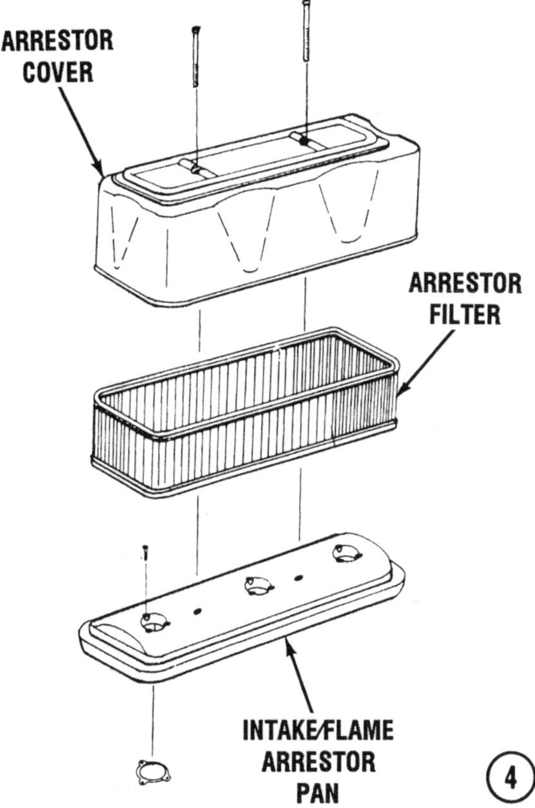

6-6 SERVICE KEIHIN CDK II CARBURETOR WITH INTEGRAL FUEL PUMP

This section provides complete detailed illustrated procedures for removal, disassembly, cleaning and inspecting, assembling including bench adjustment, installation and operating adjustments for the Keihin CDK II carburetor.

Servicing the carburetor may be accomplished while the unit remains mounted to the rack. If a carburetor **IS** removed from the rack, be sure to identify its position to ensure it is installed in the same location on the rack during assembling.

To make identification just a little bit easier -- on a dual setup, only the aft carburetor is equipped with the fuel pump. The forward carburetor is serviced from the aft unit. On a triple carburetor installation -- the center carburetor does not have the fuel pump. The forward and aft units service the center carburetor.

Good shop practice dictates a carburetor repair kit be purchased and new parts be installed any time the carburetor is disassembled.

Make and earnest attempt to keep the work area organized and to cover parts after they have been cleaned. This practice will prevent foreign matter from entering passageways or adhering to critical parts.

Dual carburetor and triple carburetor installations are removed as a set and may be serviced on the bench **WITHOUT** separating them.

The illustrations supporting the removal procedures were made of a dual carburetor installation and the disassembly procedures were made with the carburetor separated from the rack for photographic clarity.

SPECIAL DISCONNECT WORDS

Take time to identify and tag hoses and linkages **BEFORE** making the disconnect.

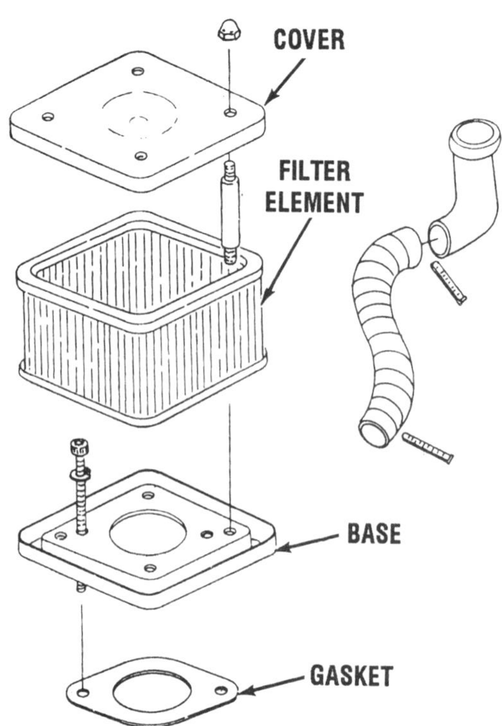

Exploded line drawing of the air intake/flame arrestor arrangement on a dual Keihin carburetor installation on a Polaris Hurricane -- 1996 and on.

Tagging these items now will facilitate speedy and correct connections during installation.

REMOVAL -- KEIHIN DUAL CARBURETOR SETUP

Carburetors from a Hurricane model shown. Differences with other dual Keihin carburetor installation are noted.

1- Disconnect the choke cable from bracket and shaft arm. Disconnect the throttle cable. Disconnect the oil injection pump cable.

2- Remove the attaching bolts securing the air intake/flame arrestor covers. Some models may have a one-piece cover, screen and pan.

Lift off the air intake/flame arrestor covers.

3- Separate the air intake/flame arrestor filters from the pan. Identify and tag, then disconnect the fuel and oil hoses.

4- Obtain the proper size Allen wrench and remove the two screws securing each pan to the carburetor. Actually, these Allen screws could be considered "thru-bolts" because they pass through and secure the carburetor to the intake manifold.

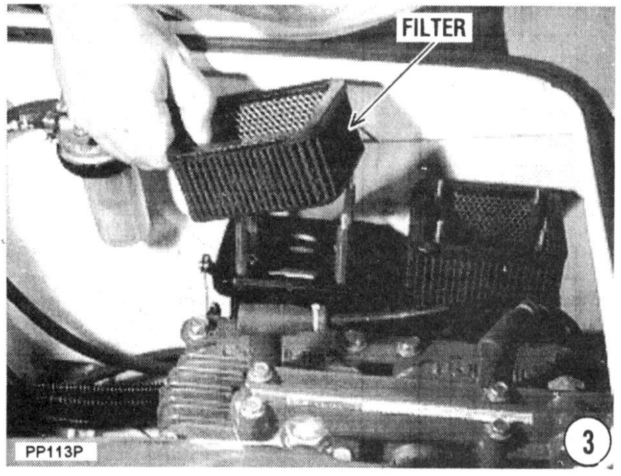

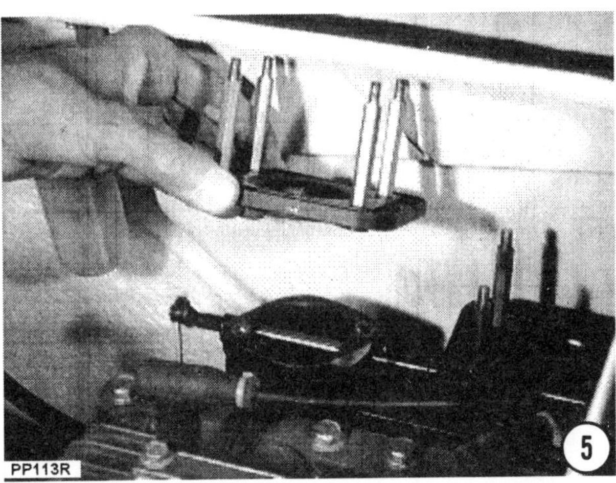

5- Remove each pan from the carburetor. Discard the gasket because a gasket at this location should not be used a second time.

6- Lift the carburetor assembly straight up and free of the intake manifold. Move the carburetor assembly to a suitable clean work surface.

View of the back (outboard) side of the Keihin dual carburetor installation on a twin cylinder engine.

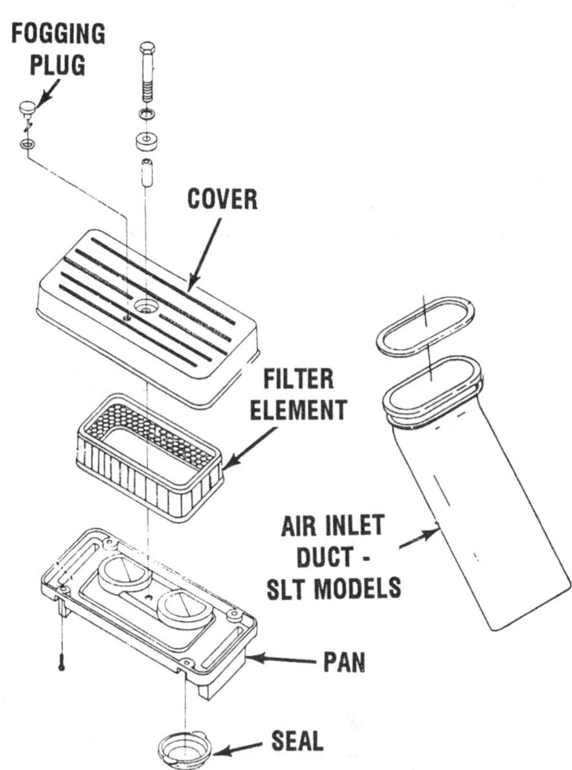

Exploded line drawing of the air intake/flame arrestor arrangement on a twin cylinder SL700 -- 1996; SLT700 -- 1996 and on; and a SL700 Deluxe -- 1997, equipped with a set of twin Keihin CDK II carburetors.

REMOVAL -- KEIHIN II TRIPLE CARBURETOR SETUP

SPECIAL DISCONNECT WORDS

Take time to identify and tag hoses and linkages **BEFORE** making the disconnect. Tagging these items now will facilitate speedy and correct connections during installation.

1- Remove the bolts securing the air intake/flame arrestor cover. **TAKE CARE** not to lose the associated hardware with the bolts, as shown in the accompanying illustration. Remove the cove, and then lift off the air intake/flame arrestor filter element.

2- Disconnect the choke cable, throttle cable, and the oil injection pump cable from the carburetor linkage at the throttle shaft arm. The accompanying illustration shows the linkage for a dual carburetor installation. The linkage is basically the same for a triple carburetor setup.

3- Disconnect the fuel supply line and the fuel return line. Disconnect the oil lines from the carburetors. Loosen the clamp and remove the fuel pump impulse line from the carburetor fuel pump of the forward (No. 1) and aft (No.

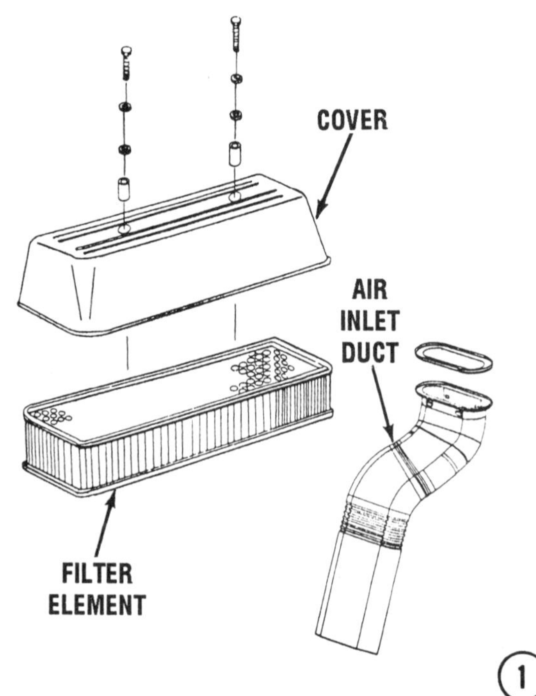

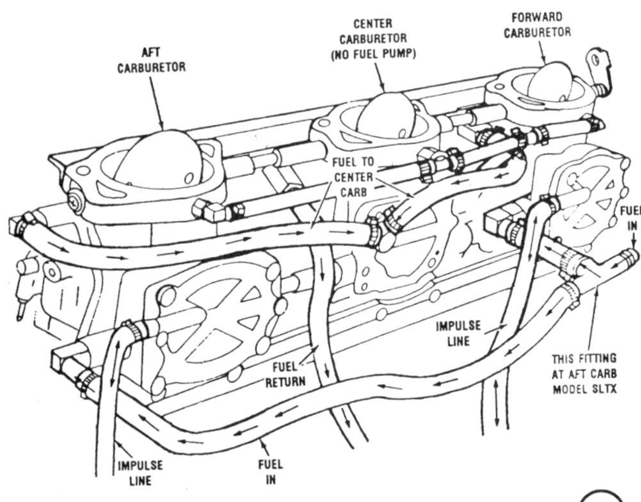

3) carburetor. Obtain proper size Allen wrench and remove mounting bolts at each carburetor securing the pan to the carburetors. Actually, these Allen screws could be considered "thru-bolts" because they pass through and secure the carburetor to the intake manifold.

Lift the carburetor assembly straight up and clear of the engine, and then move it to a suitable clean work surface.

Cover the intake manifold openings with duct tape or other suitable tape material to prevent contamination from entering the crankcase.

DISASSEMBLE KEIHIN CDK II CARBURETOR

The following procedures pickup the work after the carburetors have been removed as an assembly; are on a suitable clean work surface, **AND** a carburetor repair has been purchased and is on hand.

Carburetor "Front" Side

1- Remove the four screws securing the diaphragm cover to the carburetor and lift off

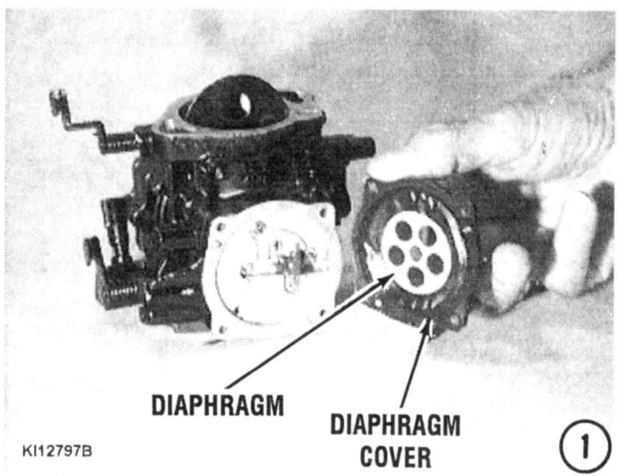

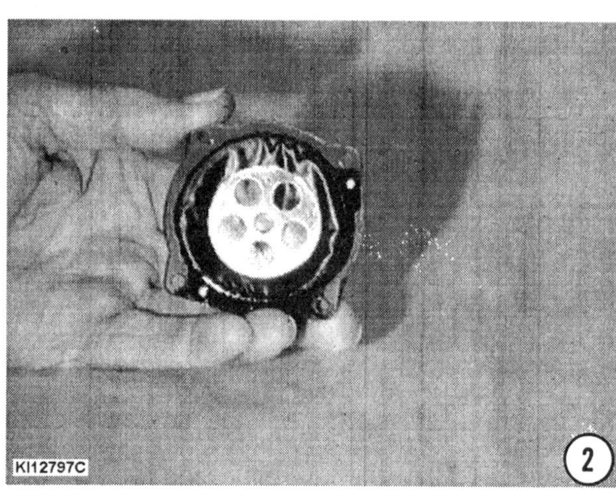

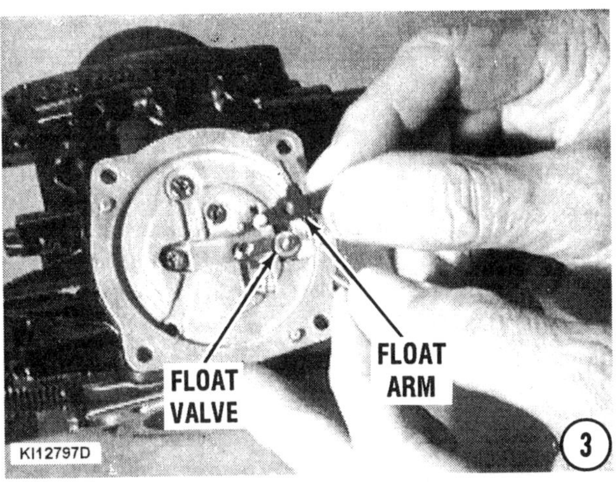

the cover. Remove the screw securing the diaphragm to the carburetor cover.

2- Inspect the diaphragm for any damage, i.e. pinholes or torn surface.

3- Remove the screw securing the float arm into its recess. Now, remove the float arm and float valve.

4- Remove the Phillips head screw securing the jet plate to the carburetor. Remove the jet plate. Inspect the jets to be sure the passageways are clear.

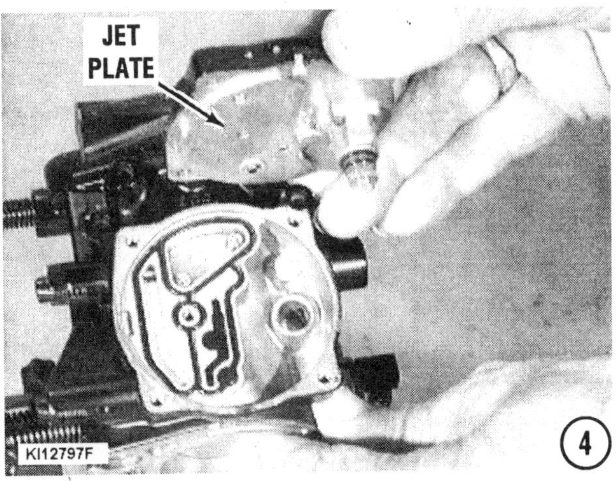

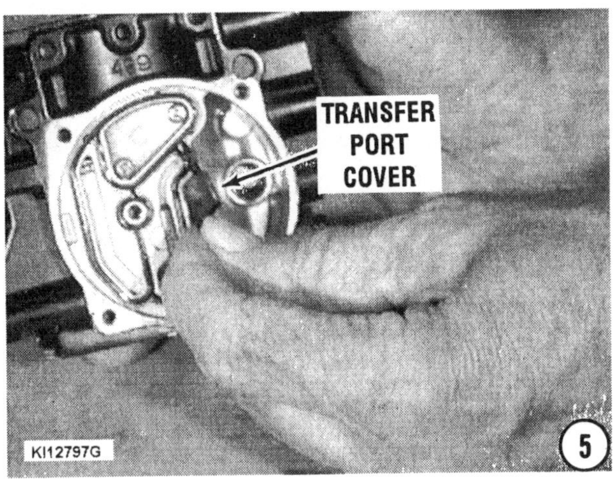

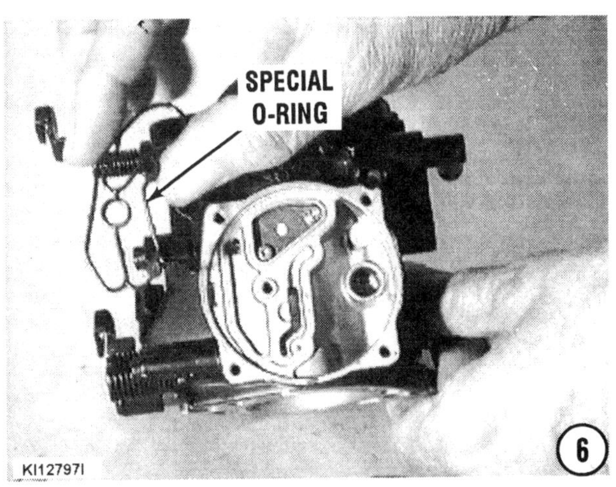

Inspect the fuel screen positioned below the fuel valve. Put the screen up to a light and inspect the screen for any blockage or damage.

5- Remove the transfer port cover. Verify the passageway is clear of debris.

6- Remove and inspect the special "spaghetti-type" O-ring.

7- Remove the two Phillips head screws securing the main chamber check valve assembly to the carburetor. Lift the check valve assembly -- including the metal plate, fiber reed valve, and nylon gasket free of the carburetor. Keep these items in proper order as an aid during assembling.

SPECIAL WORDS
ADJUSTING SCREWS

The low and high speed adjustment mixture screws must **NOT** be removed unless damage to the tapered end is suspected and the screws are to be replaced.

Under normal service, these two screws should remain undisturbed. Each Keihin carburetor is individually adjusted at the factory using a flow meter.

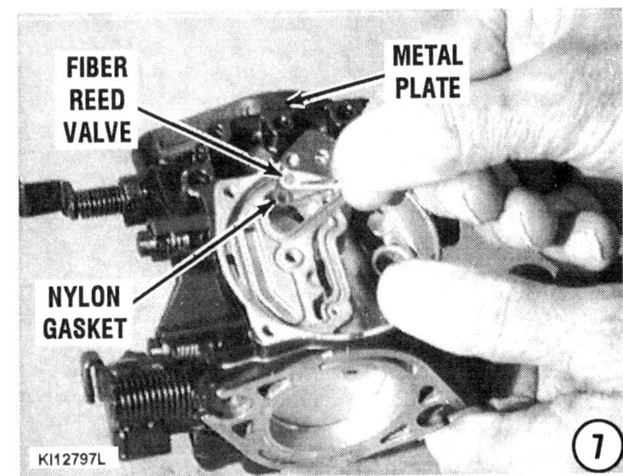

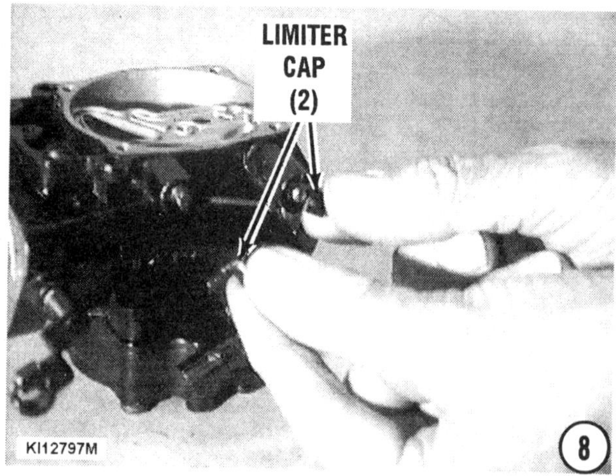

CRITICAL WORDS

Perform Step 8 **ONLY** if the high speed or low speed mixture cannot be adjusted properly. **ALSO** perform the step only if replacement screws have been obtained and are on hand. Failure to adjust the low speed or high speed mixture would suggest the tapered end of one -- but not necessarily both -- of the screws is damaged. In almost all cases, "damage" is caused by the screw being threaded in too tightly and the needle becoming "grooved", because it passed through the orifice. Once the needle is "grooved", proper mixture is not possible.

Each cap has a small flange which indexes into a slot in the carburetor to prevent the adjustment screws from being changed accidentally.

8- Pull off the limiter cap on the low speed mixture adjustment screw -- the screw next to the embossed **"L"**. Now, **CAREFULLY** count the number of turns required to seat the adjustment screw **LIGHTLY** -- in quarter increments, and then back out the low speed adjustment screw. Pull off the limiter cap on the high speed mixture adjustment screw -- the screw next to the embossed **"H"** **CAREFULLY** count the number of turns in quarter turn increments required to seat the screw **LIGHTLY**, and then back out the high speed adjustment screw.

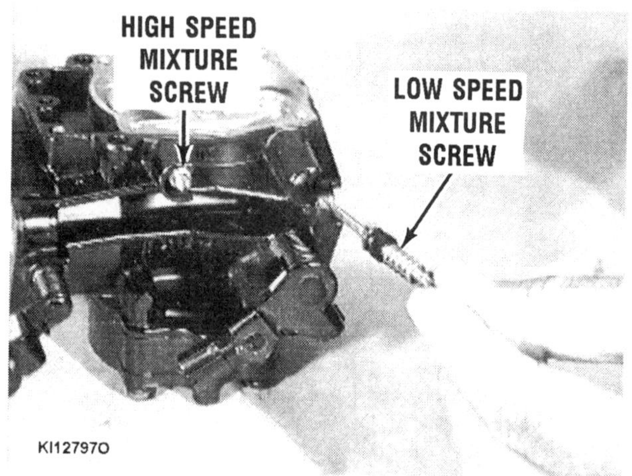

*Before removing either adjustment screw, take time to count the number of turns required to seat the screw **LIGHTLY**, from its present position. Record the number which will be considered a "bench adjustment" during assembling.*

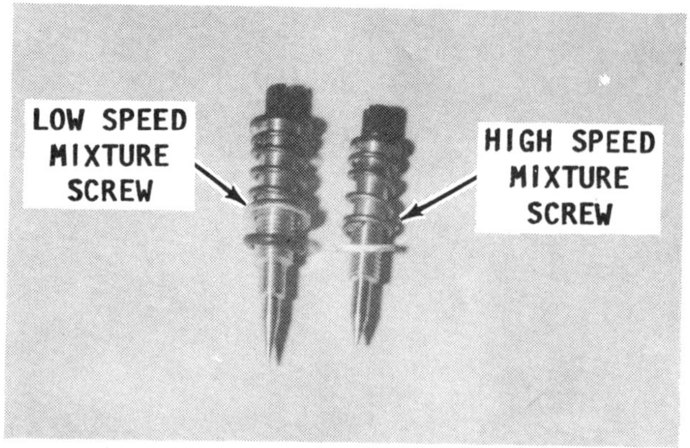

Comparison of the low speed and high speed mixture screws. Notice how the high speed screw is slightly shorter than the low speed screw.

Take time to record the number of turns for each screw. Observe the difference in length and size between the high speed and low speed adjustment screws.

Carburetor "Back" Side (Pump Side)

Rotate the carburetor body around to gain access to the pump side of the carburetor.

9- Remove the screws securing the pump cover to the carburetor body. Lift off the pump cover.

10- Remove the diaphragm and special O-ring. Exercise care when "peeling off" the pump diaphragm to prevent damaging it.

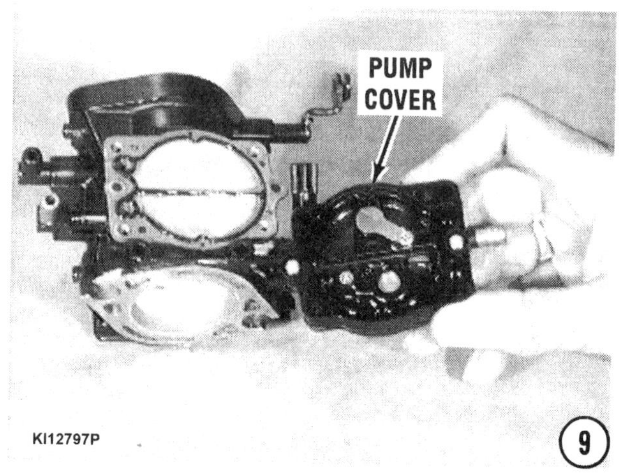

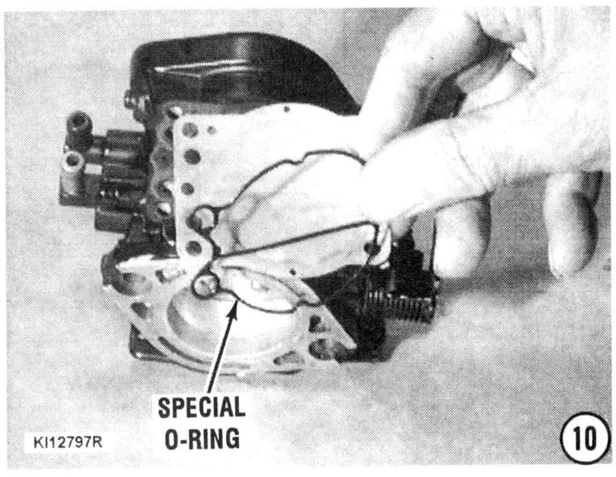

GOOD WORDS

Further disassembling of the fuel pump unit is not necessary, when performing a normal carburetor overhaul. Added to this fact is the requirement for special tools to disassemble the fuel pump.

CLEANING AND INSPECTING CDK-38 AND CDK-40

NEVER dip rubber parts, plastic parts, or diaphragms in carburetor cleaner. These parts should be cleaned **ONLY** in solvent, and then blown dry with compressed air.

Place all metal parts in a screen-type tray and dip them in carburetor cleaner until they appear completely clean, then blow them dry with compressed air. **Do NOT** immerse the carburetor body in a cleaner solution, because such action would surely destroy integral parts of the carburetor assembly.

Instead, use an aerosol can of Choke and Carburetor Cleaner to spray the assembly, as indicated in the accompanying illustration.

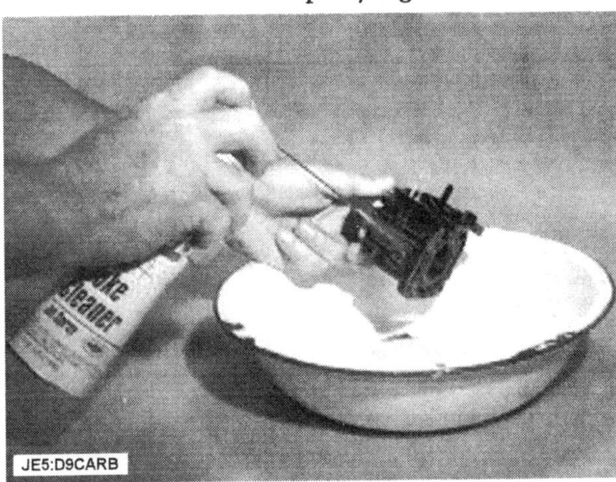

*Using a Choke and Carburetor Cleaner on the carburetor body. **NEVER** use a dip cleaner or hot tank solution, because such action will surely damage the sealing components in the carburetor body.*

Blow out all passages in the castings with compressed air. Check all parts and passages to be sure they are not clogged or contain any deposits. **NEVER** use a piece of wire or any type of pointed instrument to clean drilled passages or calibrated holes in a carburetor.

Move the throttle shaft back-and-forth to check for wear. If the shaft appears to be too loose, replace the complete throttle body, because individual replacement parts are **NOT** available.

CRITICAL AREAS

Inspect the condition of both the choke and throttle shafts. Most shaft designs incorporate a cutout portion to enable the butterfly valve to lay flat against the shaft. Two small holes are drilled in this cutout portion to accommodate the small screws securing the valve to the shaft.

The weakest point of the shaft is at the screw holes. Small cracks form at the holes due to metal fatigue. These cracks elongate the threaded holes and the small screws shake loose. Once loose, they are sucked into the carburetor, then the intake manifold, through the reed valves, and into the crankcase -- eventually finding their way into the combustion chamber. One of these small screws can cause very extensive and expensive damage to reed valves, pistons, rings, and cylinder walls. Therefore, check the cutout in the shaft **CLOSELY** for any evidence of a small crack.

Inspect the main body, airhorn, and venturi cluster gasket surfaces for cracks and burrs which might cause a leak. Check to be sure the float arm return spring has not been stretched. Check the float arm needle contacting surface and replace the diaphragm if this surface has a groove worn in it.

As previously mentioned, most of the parts which should be replaced during a carburetor overhaul are included in an overhaul kit available from the local marine dealer. One of these kits will contain a fuel inlet needle and seat. The needle should be replaced each time the carburetor is disassembled as a precaution against leakage.

Make a thorough inspection of the fuel chamber and fuel diaphragms for the tiniest pin hole. If one is discovered, the hole will only get bigger. Therefore, the diaphragm must be replaced in order to obtain full performance from the engine.

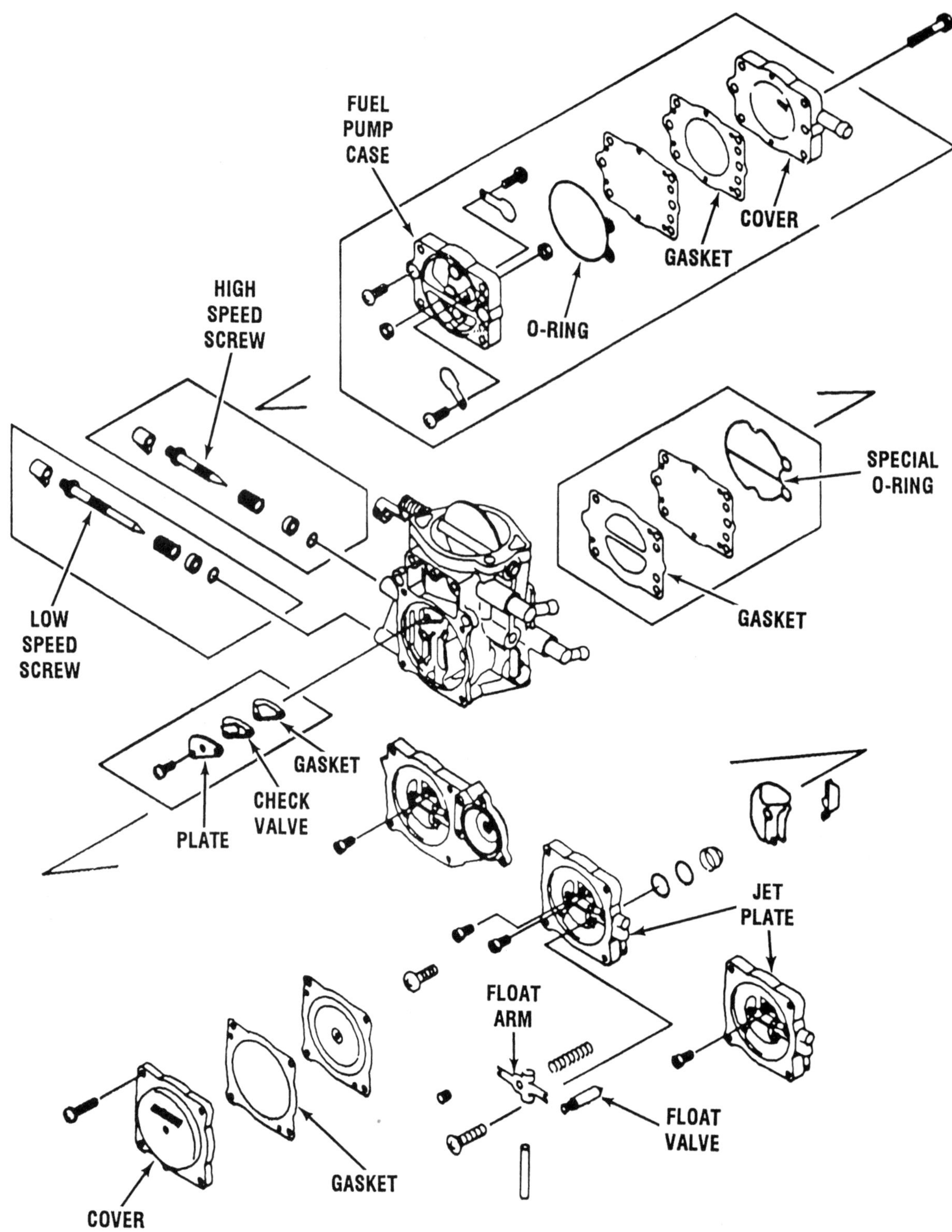

Exploded line drawing of the Keihin CDK II carburetor with integral fuel pump on the "back" side. On a triple carburetor installation the center (No. 2) carburetor does not have the integral fuel pump. On a dual setup, the forward (No. 1) carburetor does not have the fuel pump. Major parts in this illustration have been identified. The manufacturer is constantly making minor changes -- even in the middle of a model year. Therefore, some detailed parts may have changed appearance from the unit being serviced.

ASSEMBLE KEIHIN CDK II WITH INTEGRAL FUEL PUMP

The following procedures pickup the work after the cleaning and inspecting tasks have been completed, as outlined in the previous paragraphs and replacement parts are on hand.

Special Words on Mixture Screws

Observe the low and high speed mixture screws. These screws are **NOT** interchangeable. To identify the correct screw, place them side by side,as shown in the accompanying illustration. The low speed screw is shorter than the high speed screw. Both screws are threaded into the carburetor at the base of the fuel pump. The low speed screw is installed next to the embossed letter **"L"**, and the high speed screw is installed next to the embossed letter **"H"**. Check to make sure both screws are equipped with a washer, a spring, another washer, and an O-ring.

1- Install the low speed screw into the opening embossed with the letter **"L"**. **CAREFULLY** thread the screw into the carburetor body until it is **LIGHTLY** seated. Now, back the screw out the same number of turns recorded during disassembling. If the number of turns out was not recorded, as instructed, back the low speed screw out 7/8 ±1/4 turn as a "bench adjustment" at this time. Install the safety cap over the screw with the flange on the cap indexed into the carburetor slot to prevent the screw from being rotated accidently.

CAREFULLY thread the high speed screw into the opening embossed with the letter **"H"**. until it is **LIGHTLY** seated, and then back it out the same number of turns recorded during disassembling. If the number of turns out was not recorded, as instructed, back the high speed screw out one full turn ±1/4 turn as a "bench adjustment" at this time. Install the safety cap over the adjustment screw, with the flange on the cap indexed into the slot in the carburetor.

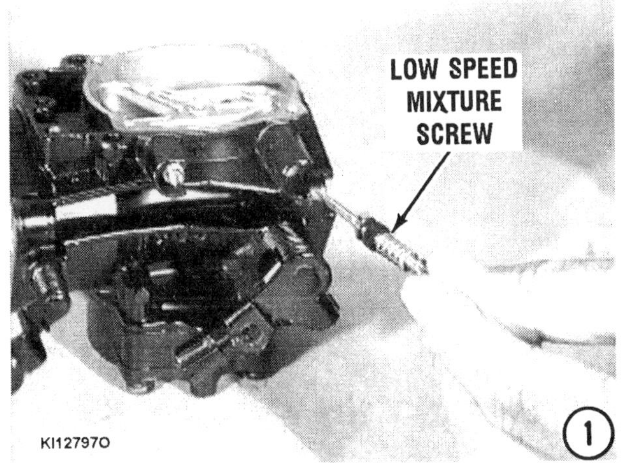

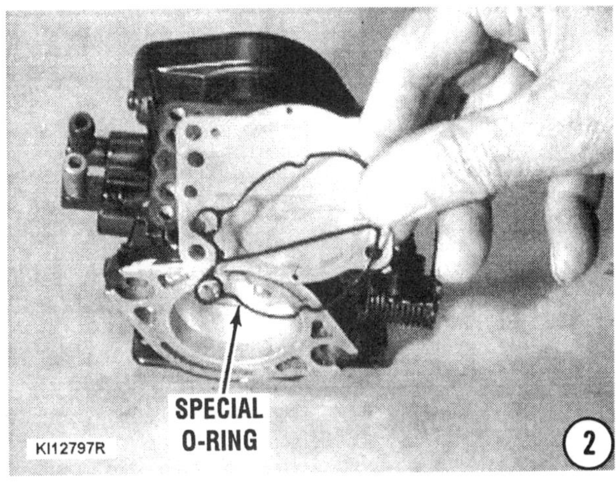

2- Install the pump diaphragm, followed by the special O-ring, into the pump chamber.

3- Position the pump cover over the carburetor body. Install and tighten the screws securely.

Rotate the carburetor body to gain access to the front side -- the cover with the "Keihin" name stamped into it.

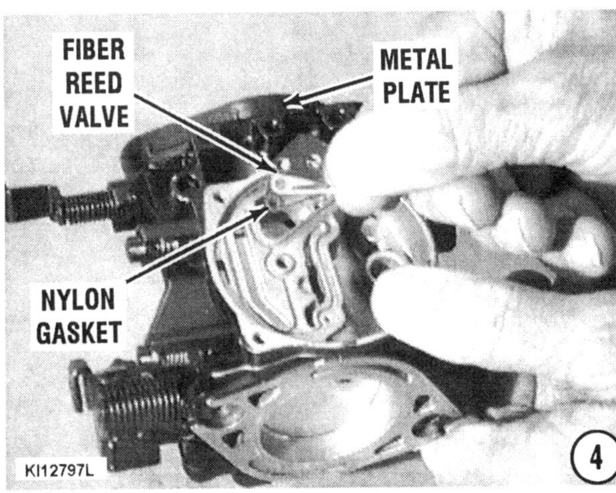

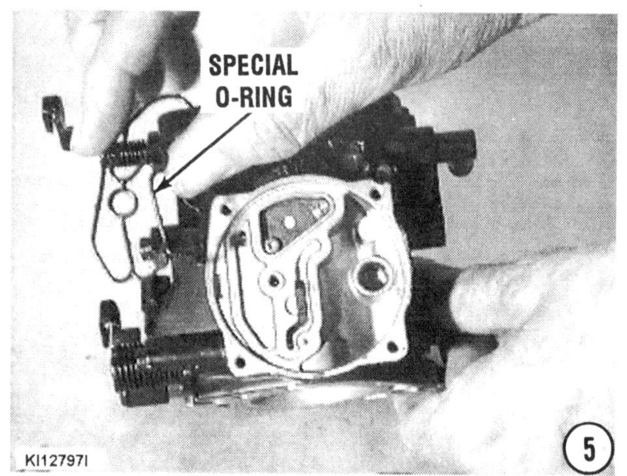

4- Obtain and install **IN SEQUENCE**, the main chamber check valve assembly as follows: First, the **NYLON GASKET** into the gasket recess, next, the **FIBER REED VALVE** into the recess, and finally the metal plate into the recess. Secure the check valve assembly in place with the attaching screws.

5- Insert the special **O**-ring into the groove.

6- Install the transfer port cover.

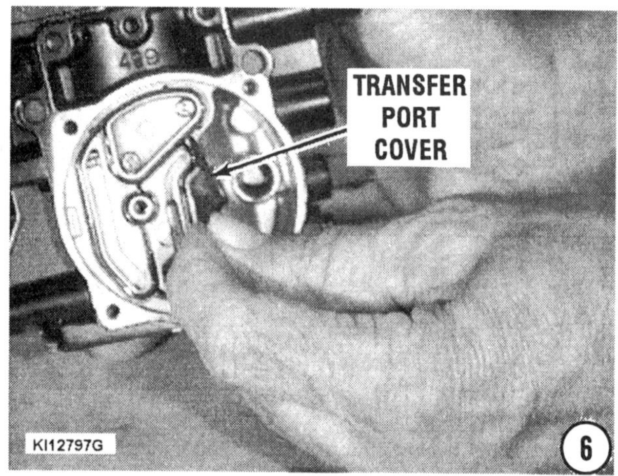

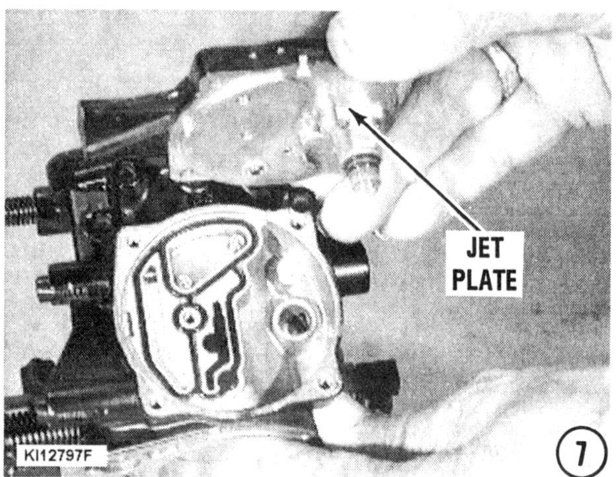

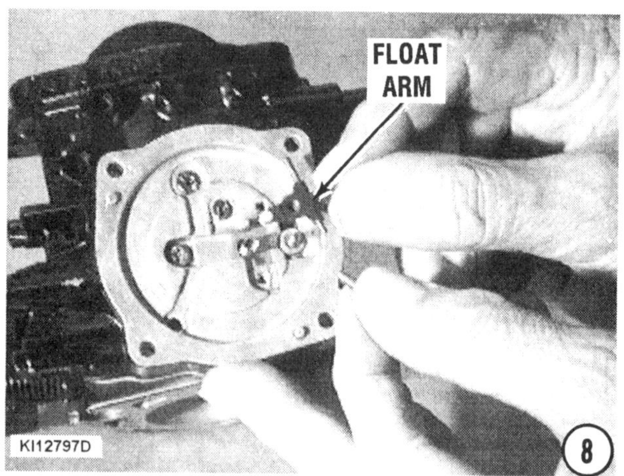

7- Insert the fuel screen into the jet plate below the fuel valve -- if it was removed. Position the jet plate over the carburetor body and secure it in place with the attaching screw.

8- Install the float arm into its recess and secure it with the screw.

9- Position the diaphragm in place on the inside of the diaphragm cover. Tighten the screw just "snug". Install the cover onto the carburetor body. Tighten the screws securely.

10- Place a new carburetor mounting gasket on the intake manifold, with the cutout letters "**U**" and "**P**" facing up -- spelling "UP".

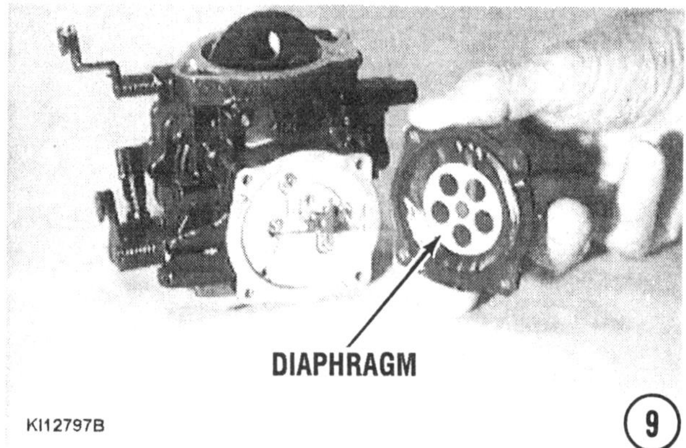

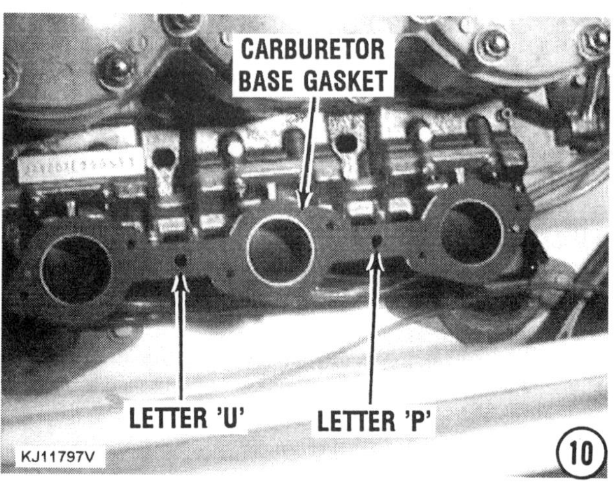

Mixture Screws with Limiter Caps

Since about 1989, the manufacturer has installed small black plastic limiter caps which snap over both the low speed and high speed adjustment screws. The purpose of these caps is to prevent any unauthorized tampering with the factory setting.

Each carburetor is individually adjusted at the factory using a flow meter. After the adjustment is completed, a limiter cap is installed over the head of the screw to prevent accidental tampering.

Each cap has a tab and the tab fits into a slot in the carburetor body at the base of the mixture screw. Therefore, the optimum position of each screw is with the tab down and indexed into the slot.

However, if the mixture screws are removed during service procedures, the low and high speed mixtures must be adjusted and both caps installed over the screws with the tabs on the caps indexed into slots in the carburetor body.

Idle Adjustment Screw

Identify the idle adjustment screw located on the throttle linkage just below the high speed and low speed mixture screws.

Move the craft to a test tank, to a body of water, or connect a flush attachment and hose to the engine. For maximum performance, the idle rpm should be adjusted under actual operating conditions. Connect a tachometer to the engine.

Start the engine and allow it to warm to operating temperature.

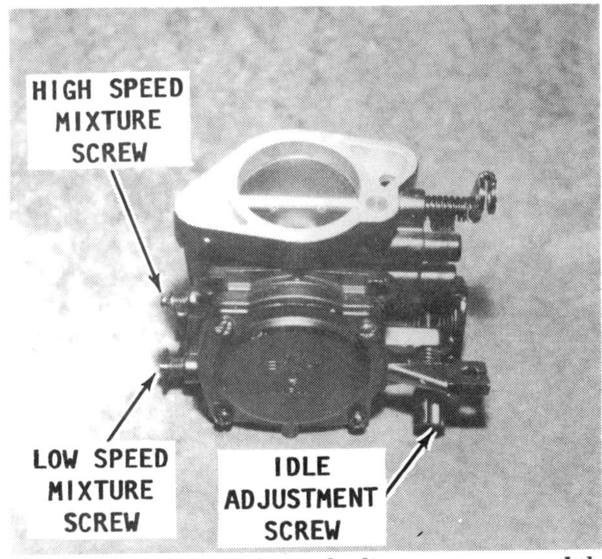

Location of the high speed adjustment screw and the low speed adjustment screw on a Keihin carburetor used on some 650 Series Models.

CAUTION

Water must circulate through the jet pump -- to and from the engine, anytime the engine is operating. Circulating water will prevent overheating -- which could cause damage to moving engine parts and possible engine seizure.

NEVER, AGAIN NEVER, operate the engine at high speed with a flush device attached. An engine operating at high speed with such a device attached, would **RUNAWAY** from lack of a load on the impeller shaft, causing extensive damage.

With the engine running, slowly rotate the idle adjustment screw until the engine idles at 1,150 to 1,350 rpm.

If the engine idle is set too low, the self-circling radius -- described in Chapter 1 -- increases and the craft may lose its self-circling feature.

If the engine idle is set too high, the craft may circle too fast for a displaced rider to climb aboard the craft.

Low and High Speed Mixture Adjustment

Each carburetor is individually adjusted at the factory using a flow meter and should not need any adjustment. Hovever, if replacement screws were installed the manufacturer suggests a starting point from which fine tuning on the water may be necessary.

Lightly seat both mixture screws. From this position back out the low speed screw 1-1/8 turns. Back out the high speed screw 5/8 of a turn.

High Altitude Operation

The idle system should never be changed when operating the craft at higher altitudes -- the low speed systems are not affected enough by changes in altitude to make an adjustment practical.

The idle adjustment just made is the optimum setting for sea level. When operating at higher elevations, the required leaner mixture is obtained by rotating the high speed screw the specified number of turns **CLOCKWISE** according to the following list.

3,300 ft (1,000 m) -- 1/8
6,600 ft (2,000 m) -- 1/4
9,900 ft (3,000 m) -- 3/8

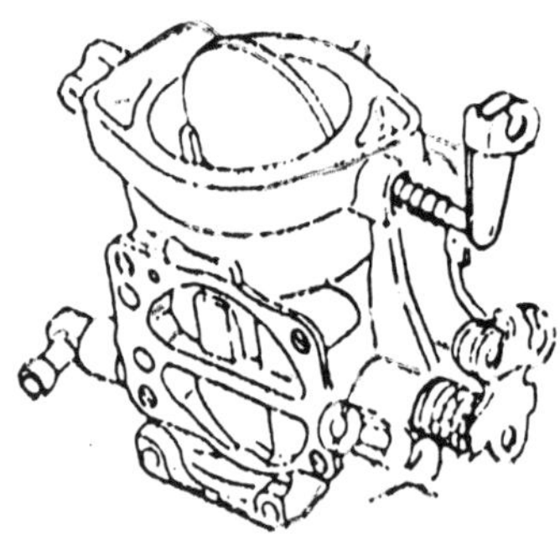

Simple line drawing of the Keihin carburetor to depict the choke lever position moving the choke plate to the fully open position when the choke knob is pushed to the full IN position.

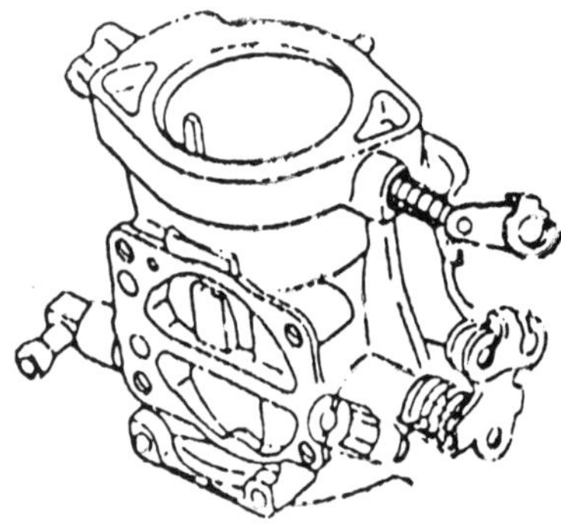

Line drawing of the Keihin carburetor to depict the choke lever position moving the choke plate to the fully closed position when the choke knob is pulled out.

Choke Cable Adjustment

Rotate the choke knob to the full counterclockwise position. Observe the choke pivot arm on the side of the carburetor. The arm should be in the position shown in the accompanying illustration, with a minimum of slack in the choke cable, with the choke butterfly valve fully open.

If the choke needs adjustment: Rotate the choke knob to the full clockwise position. Loosen the two locknuts, one on each side of the cable support bracket. Set the choke pivot to the fully closed position, as shown. Tighten the two locknuts on the support bracket to hold this position, allowing a small amount of slack in the cable.

If the choke cable has stretched, or has become kinked or frayed, or tends to stick, the entire cable must be replaced.

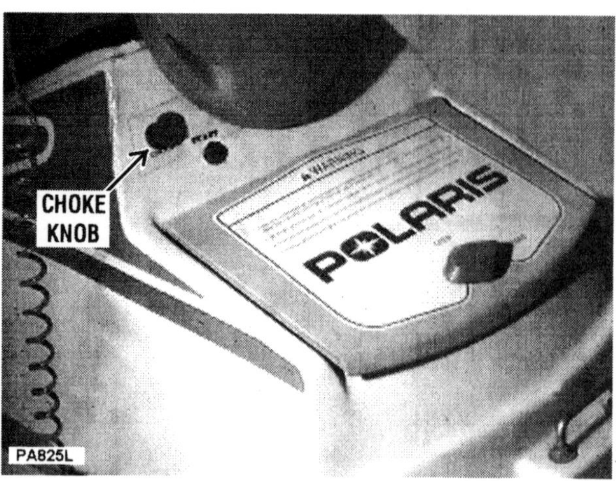

Clear view showing location of the choke knob. Some units may be equipped with a choke lever arrangement, as described in the text. Choke action should be smooth when moving the knob out or in OR when moving the choke lever. Routing of the cable should be checked for kinks, or damage if operation of the knob becomes stiff.

Close view of the throttle and choke cable connected to the throttle shaft arm. The oil cable from the oil injection pump is also visible and identified. Choke and cable adjustments are described in this section's text. Adjustment of the oil injection cable -- when necessary -- is covered in detail in Section 6-8.

Throttle Cable Adjustment

The idle speed must be adjusted **BEFORE** adjusting the throttle lever free "play".

With the throttle lever released and the throttle closed, the idle adjustment screw should rest against the stopper arm of the throttle linkage, with a slight amount of slack in the throttle cable, as shown in the accompanying illustration.

With the throttle lever in the WOT position, the stopper arm of the throttle linkage should rest against the stop on the carburetor.

If the throttle cable needs adjustment: Loosen and rotate the two locknuts around

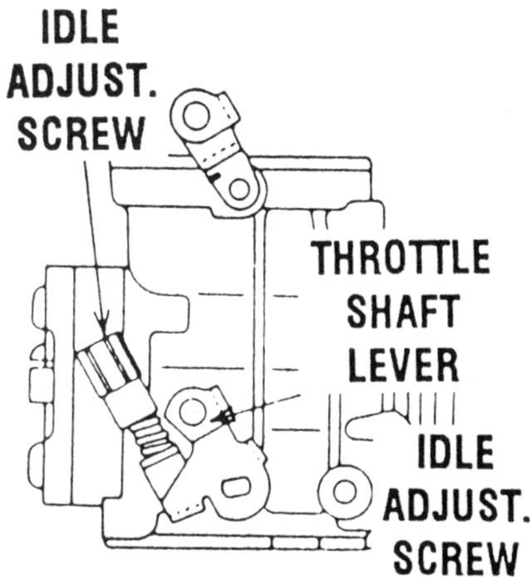

Simple line drawing to depict position of the throttle shaft lever with the throttle released -- at idle speed.

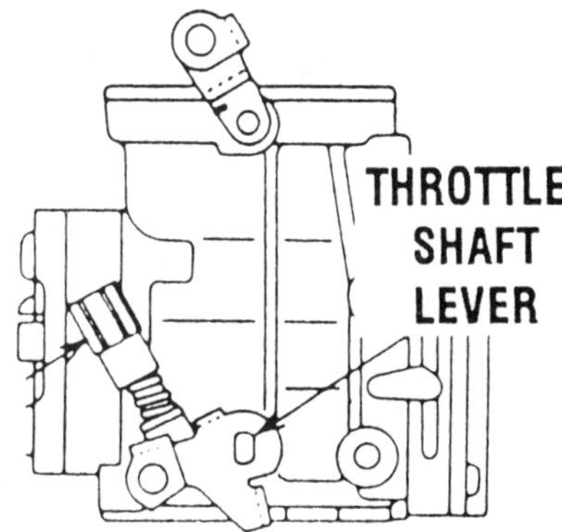

Simple line drawing to depict position of the throttle shaft lever with the throttle at wide open -- "WOT".

the throttle cable at the cable support bracket until, with a minimum amount of slack in the cable, the idle adjustment screw rests against the throttle stopper with the throttle released. Tighten the two locknuts to hold this new adjustment.

If the throttle cable has stretched, or has become kinked or frayed, or tends to stick, the entire cable must be replaced.

Start the engine and check the completed work.

CAUTION

Water must circulate through the jet pump -- to and from the engine, anytime the engine is operating. Circulating water will prevent overheating -- which could cause damage to moving engine parts and possible engine seizure.

NEVER, AGAIN NEVER, operate the engine at high speed with a flush device attached. An engine operating at high speed with such a device attached, would **RUNAWAY** from lack of a load on the impeller shaft, causing extensive damage.

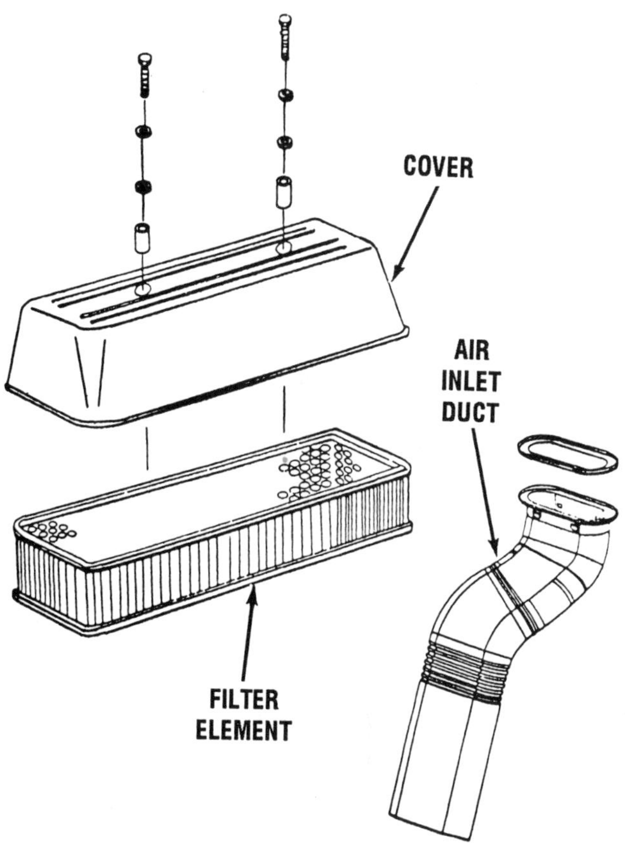

Exploded line drawing of the air intake/flame arrestor filter arrangement on the SLTX - 1996 and on, also the SL900 and SL1050 1997 and on. Check to be sure the associated parts with the bolts are installed in the proper order.

INSTALLATION -- Keihin Carburetor Triple Unit Setup

If the carburetor of a multiple installation was removed from the rack, install the carburetor back onto the rack in the same position from which it was removed. Remember, on a triple installation, the center carburetor does **NOT** have an integral fuel pump.

A- Place a new carburetor mounting gasket on the intake manifold, with the cutout word **"UP"** on the gasket facing up and readable. Some model gaskets may not have the word "UP".

B- Lower the carburetor rack onto the base gasket atop the intake manifold, with the bolt holes aligned.

C- Position the air intake/flame arrestor pan in place on the carburetors. Apply a light coating of Loctite™ 242, or equivalent, to the mounting Allen screws. Secure the carburetors in place with the Allen "thru"-bolts secured to a torque value of 108 in lbs (12.2Nm).

D- Connect the fuel impulse hose to the fitting at the forward (No. 1) and aft (No.3) carburetors. Connect the fuel supply hose to

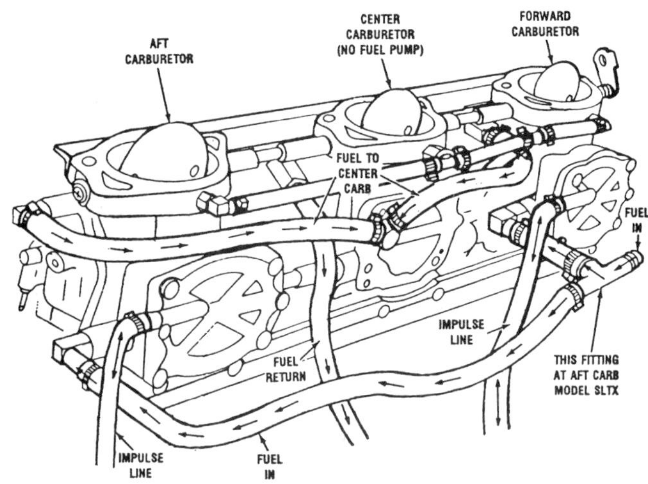

Line drawing to depict the fuel hose identification and routing on an engine with a triple Keihin CDK II carburetor installation. Note the fuel fitting at the forward (No. 1) carburetor in this illustration is at the aft (No. 3) carburetor pump on the Model SLTX - 1997.

the forward and aft carburetors. Connect the fuel line from the forward and aft carburetors to the center carburetor.

E- Connect the throttle cable, the choke cable, and the oil injection cable to the throttle shaft arm. Check Section 6-8 -- Oil Injection -- to ensure the oil injection cable is properly adjusted. The cable was not disturbed while the carburetors were removed, the connection and adjustment should be correct.

F- Place the air intake/flame arrestor filter in place. Slide the associated hardware onto the bolts in the proper order, and then install the cover and tighten the attaching bolts securely.

INSTALLATION -- Keihin Carburetor Dual Unit Setup

1- Arrange two **NEW** carburetor base gaskets on the intake manifold with the bolt

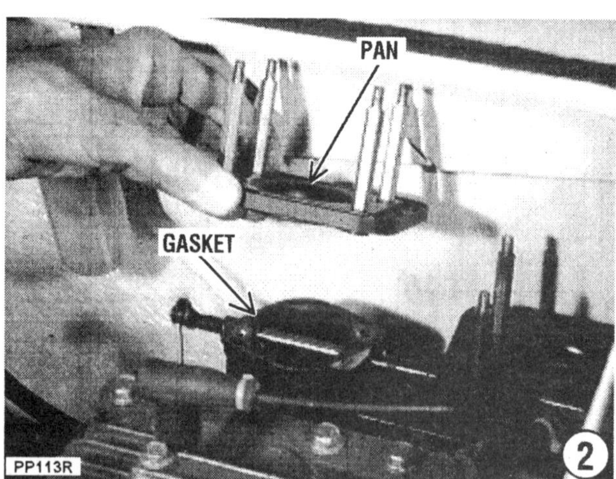

holes aligned. (For photographic clarity the accompanying illustration was taken with the engine out of the craft -- part of the engine disassembling work.) Lower the carburetors attached to the "rack" down into position on the intake manifold.

2- Set two **NEW** air intake/flame arrestor pan gaskets on top of each carburetor Place the air intake/flame arrestor pan in position on each carburetor.

3- Apply a light coating of Loctite™242, or equivalent, to the threads of the attaching Allen

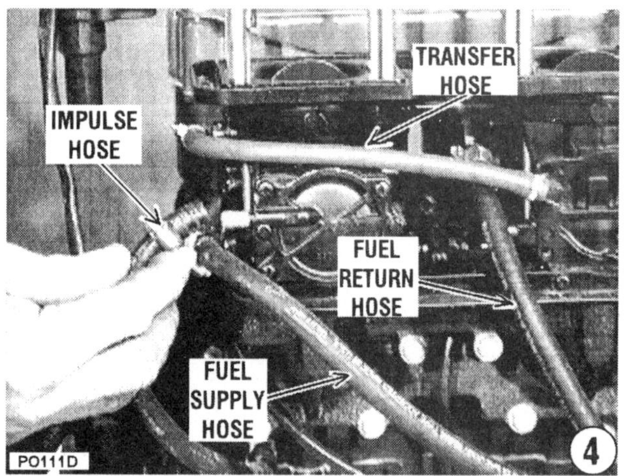

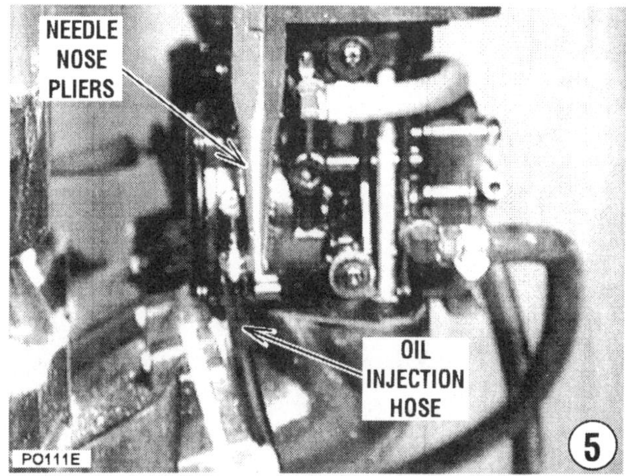

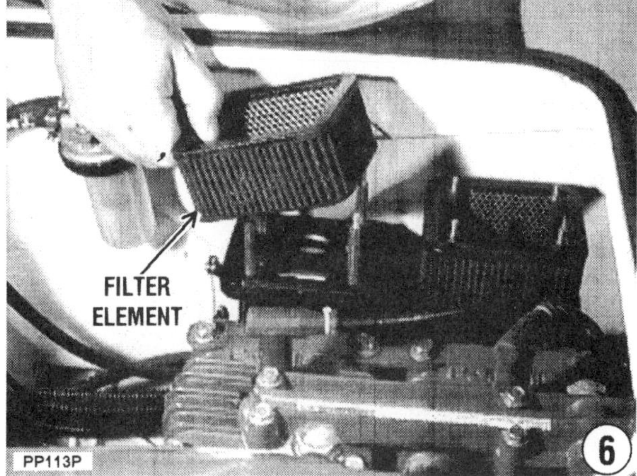

screws. Secure the pan and the carburetor in place with the Allen screws tightened to a torque value of 108 in lbs (12.2Nm). Repeat the procedure for the second carburetor.

4- After the carburetors have been installed, connect the fuel hoses to the proper fittings according to the tags attached during carburetor removal.

5- Connect the oil injection hose to each carburetor. Secure the hose to the fitting using a pair of needle nose pliers and spring clips.

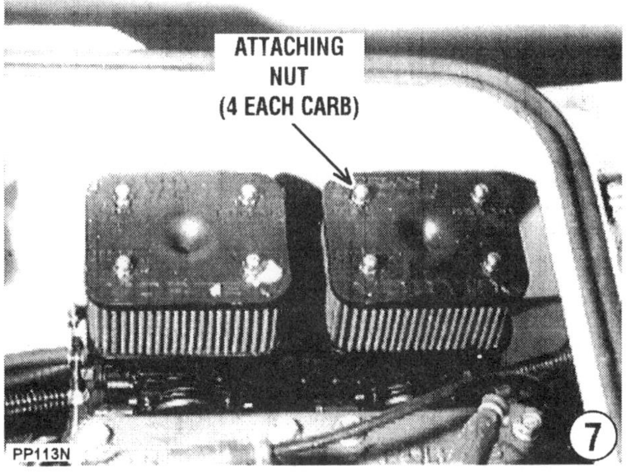

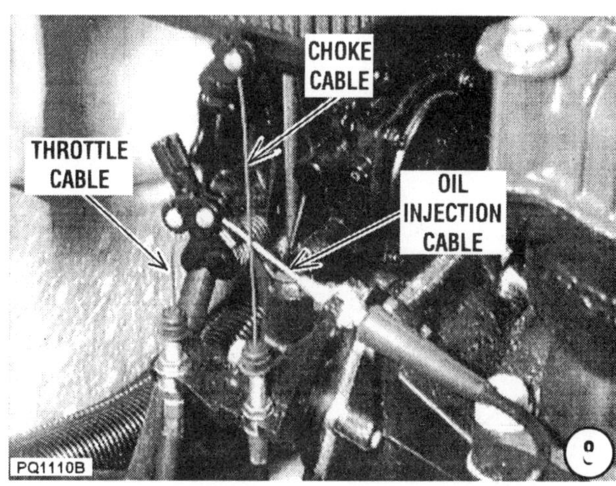

6- Install the two air intake/flame arrestor filters on the pan.

7- Secure the filters in place with the covers and attaching bolts tightened securely.

8- Connect the choke cable, the throttle cable, and the oil injection cable to the throttle shaft arm. Check Section 6-8 -- Oil Injection to ensure the oil cable did not lose its adjustment.

6-7 REMOTE FUEL PUMP

This section presents complete procedures to service the remote fuel pump.

From 1992 through the middle of 1995 the pump was installed on the aft end of the carburetor rack. Since mid 1995 the pump has been mounted on the aft outside bulkhead in the engine compartment.

When the Keihin CDKII carburetor is installed, the remote pump is not used because the fuel pump is considered an integral part of the aft -- dual carburetor set up **OR** aft and forward carburetors -- triple carburetor installation as mentioned earlier.

The remote fuel pump installed on the Fuji and Polaris engines is different in external appearance only. The pump used with the Fuji engine is round whereas the pump used with the Polaris engine is octagon in shape. Operation and internal parts are identical except for the shape, as mentioned.

THEORY OF OPERATION

The next few paragraphs briefly describe operation of the remote fuel pump used on personal watercraft covered in this manual. This description is followed by detailed procedures for testing the pressure, testing volume, removing, servicing, and installing the fuel pump. As mentioned earlier, from 1992 to mid

1995, the fuel pump was mounted on the aft end of the flame arrestor vase plate. Since mid 1995 the remote pump has been mounted in the engine compartment just aft of the engine on the outside bulkhead -- starboard side.

The fuel pump is a diaphragm displacement type. A hose from a crankcase fitting at the aft end of the crankcase -- starboard side -- supplies crankcase impulses from the No. 3 cylinder to the pump. After initial engine cranking to fill the pump and after engine start, the pump supplies adequate fuel to the fuel rail for all three carburetors to meet engine demands all speeds and conditions.

The pump is activated by impulses from the No. 3 cylinder. If this cylinder indicates a wet fouled condition, as evidenced by a set fouled spark plug, be sure to check the fuel pump diaphragms for possible puncture or leakage. Even the tiniest pin hole in a diaphragm will cause loss of engine performance.

If the pump is removed, opened, and disassembled, extreme care **MUST** be exercised when handling the internal components. Installation of the gaskets and diaphragms **MUST** be performed in the proper order or the pump will fail to give maximum performance.

Maintenance

Regular scheduled maintenance of the remote fuel pump is not necessary. However, if the pump is suspected of not functioning properly -- supplying adequate fuel to the carburetors -- a few simple tests and checks may uncover the problem.

1- Operation of the pump may be checked by simply disconnecting the fuel supply line from the one of the carburetors. Place the end of the delivery hose in a suitable container. Start the engine and operate it only at idle speed. Fuel in the manifold and in the carburetors will allow the engine to operate for a short time. While the engine is operating at idle speed a steady flow of fuel should be observed discharged into the container.

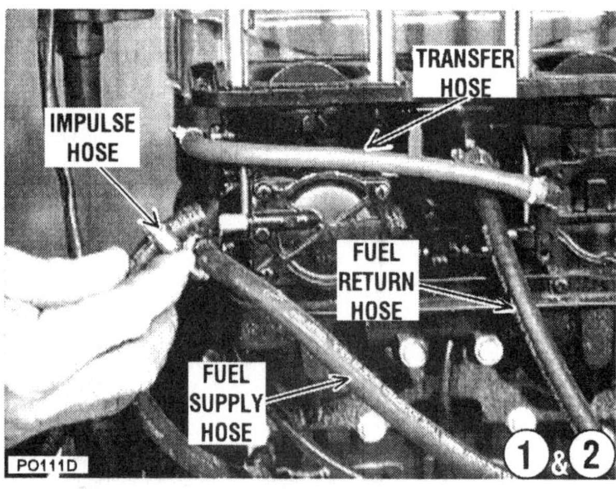

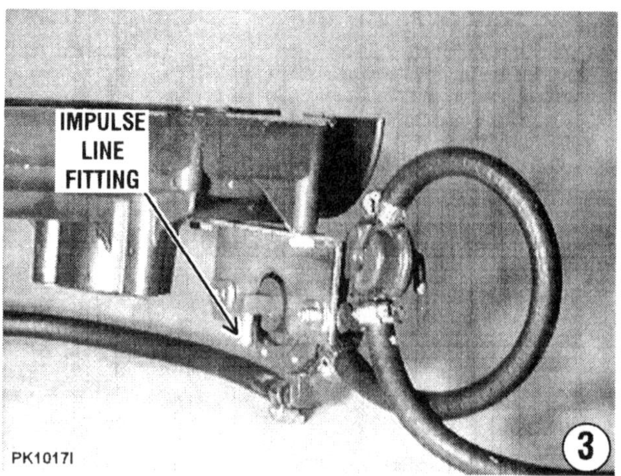

2- If engine performance indicates a fuel delivery problem, or if water is present in the fuel beyond the separator, the pump should be removed, disassembled, cleaned and assembled. The impulse line from the crankcase should also be removed and cleaned to ensure foreign material has not entered the line.

If the pump is disassembled a very close inspection should be made of internal parts to ensure a diaphragm has not sustained a pin hole, crack, or other damage. The check valves should also be checked. If even a slight suspicion exists as to the condition of any component, all diaphragms, check valves and gaskets should be replaced.

3- The pump can be given a pressure test by applying 5 psi to the impulse line fitting. The accompanying illustrations identifies the impulse line fitting on the back side of the pump. The pump diaphragm should hold pressure indefinitely.

REMOTE FUEL PUMP REMOVAL AND DISASSEMBLING

The exploded drawing on the next page will prove most useful to **ENSURE** all parts are installed in the proper sequence.

The pump consists of fragile diaphragms which are easily damaged. The pump should not be disassembled unless replacement diaphragms and gaskets are on hand.

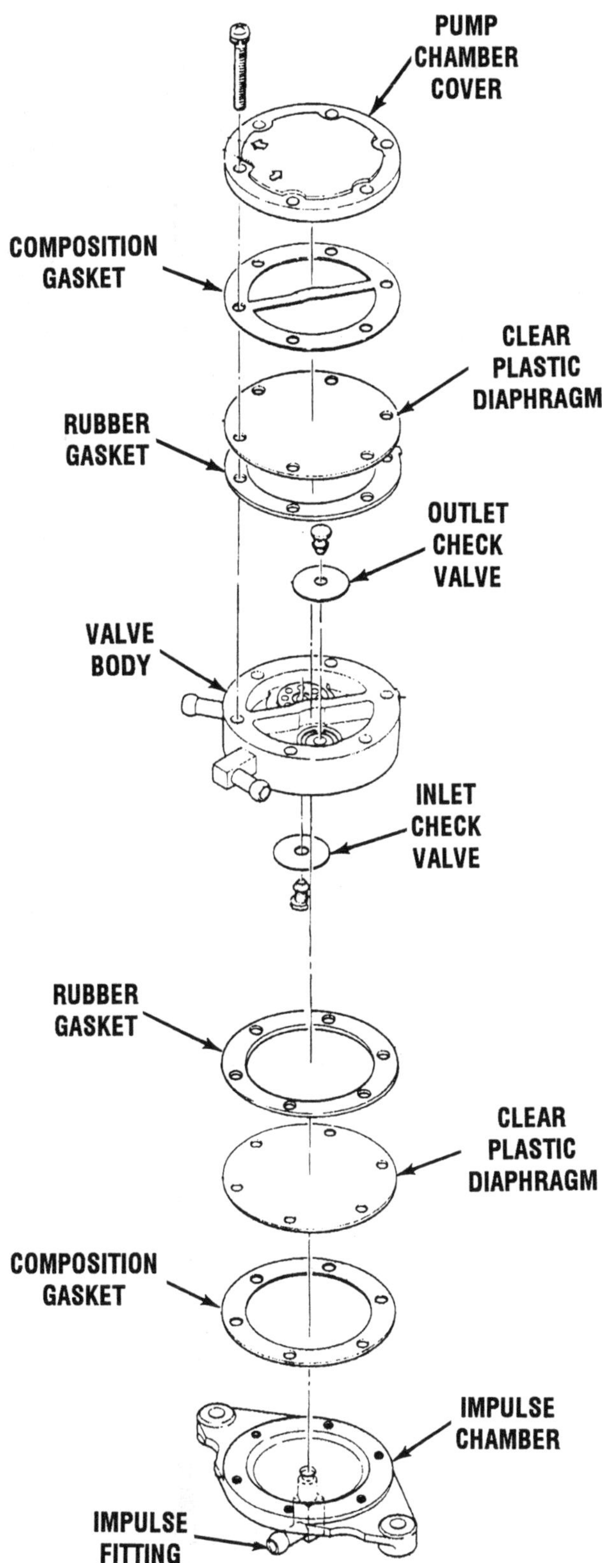

Exploded line drawing of the remote fuel pump. Two configurations have been used on the engines covered in this manual. One pump has an octagon pump body and associated parts, as shown in this illustration. The other pump and associated parts are round. Both pumps operate exactly the same and have the same number of internal parts. The only difference is the shape.

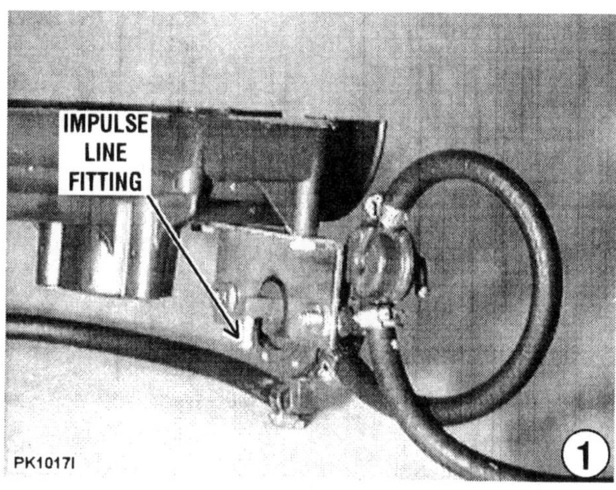

1- Disconnect the impulse hose from the back side of the pump, as shown.

2- Disconnect the inlet hose and the outlet hose from the pump. Take time to plug the disconnected ends of all three hoses to prevent contaminants from entering the crankcase and the fuel system. The plug in the inlet hose will prevent fuel from seeping into the bilge.

3-If the pump is mounted to the aft end of the flame arrestor pan, remove the two mounting bolts securing the pump to the bracket. If the pump is mounted to the bulkhead in the engine compartment, simply remove the two screws. In both cases, remove the pump and move it to a clean work surface with adequate light.

Use "white out" or other similar material and make a mark down one side of the pump as an assist in assembling the internal parts in the same position from which they were removed.

Remove internal parts and keep them in order: the cover, gaskets, diaphragms, valve body and place them in order on the bench. Take note of how the diaphragms are clear plastic and two of the gaskets are rubber and two

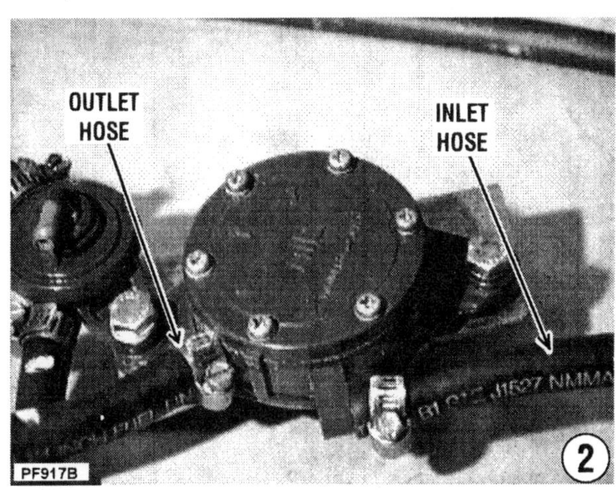

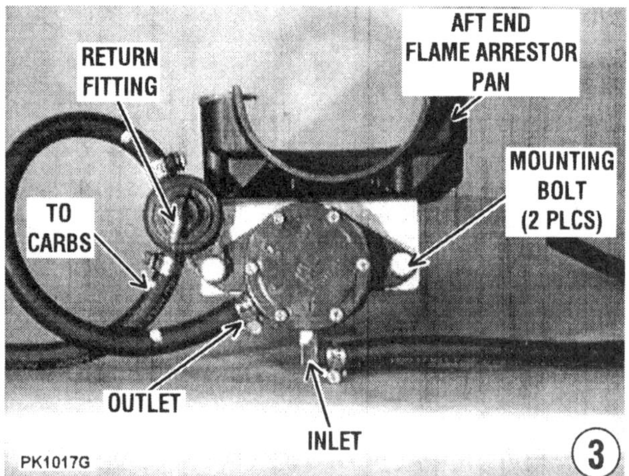

others are composition material. All items **MUST** be assembled in the proper order, as indicated in the accompanying exploded drawing, or the pump will fail to operate properly. Both the round pump and the octagon pump have the same internal parts in the same order. The only difference is their shape.

CLEANING AND INSPECTING

Good shop practice dictates a pump kit be purchased and internal parts replaced any time the pump is disassembled. However, if the parts must be used a second time, Wash them **CAREFULLY** and thoroughly in solvent, and then blow them dry with compressed air from a little distance. **DO NOT** hold the nozzle to close on the diaphragms and check valves, because these items could be damaged by an excessive blast of air.

Inspect each part for wear and damage. The check valves should allow air to pass through in one direction, but not in the opposite direction.

Check the diaphragms for pin holes by holding it up to the light. If pin holes are detected or if the diaphragm is not pliable, it **MUST** be replaced.

ASSEMBLING AND INSTALLATION

Proper operation of the fuel pump is essential for maximum engine performance. Therefore, always use **NEW** gaskets and diaphragms.

NEVER use any type of sealer on fuel pump gaskets.

TAKE CARE not to damage the very fragile flat surface of the check valves. Secure each check valve in place with the screw tightened securely.

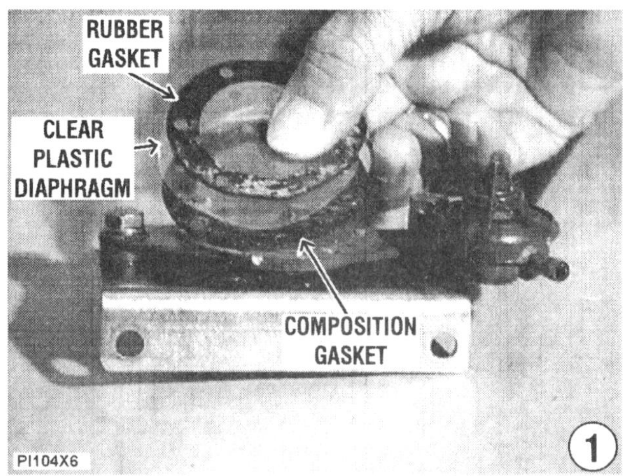

After all parts have been thoroughly cleaned and/or replacement parts have been obtained and are on hand, the pump is ready for assembling.

Use the accompanying drawings to **ENSURE** all parts are installed in the proper sequence. The order for the two fuel pumps mentioned, is the same -- only the shape differs and the arrangement of the fuel fittings.

1- Install the first gasket -- composition gasket, followed by the clear plastic impulse

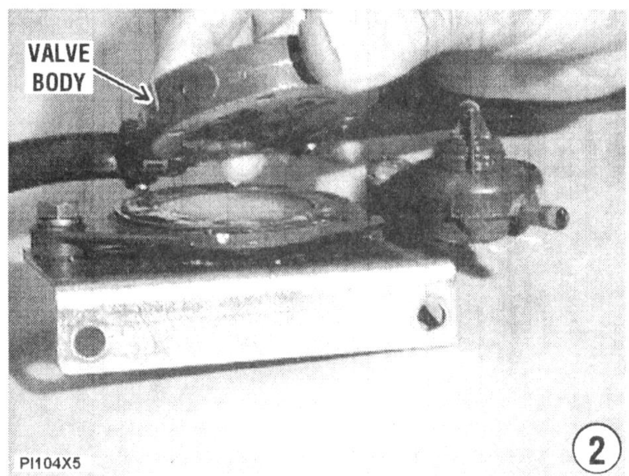

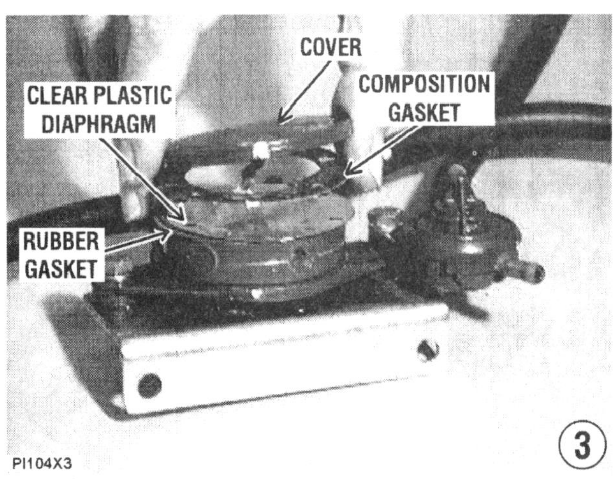

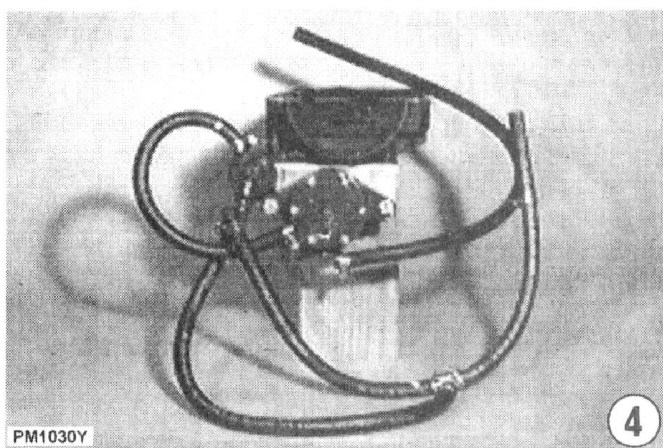

diaphragm, and then the rubber gasket. Use the marks made before disassembling to align all parts in the same position from which they were removed. If new gaskets and diaphragms are being installed, naturally they will not have the marks.

2- If the inlet and outlet check valves were removed, install them into the valve body, as indicated by the exploded drawing. Position the valve body on top of the assembled parts on top of the impulse chamber with the bolt holes aligned with the holes in the chamber.

3- Install the rubber gasket, then the second clear plastic diaphragm, another gasket -- composition, and finally the pump cover. Secure it all in place with the attaching Allen head screws.

4- After the pump is assembled, secure it to the bracket on the aft end of flame arrestor pan or to the bracket on the engine compartment bulkhead. The flame arrestor pan is shown removed from the carburetors for photographic clarity only. Tighten the attaching hardware securely.

Connect the fuel lines, as indicated. The cover has embossed arrows to indicate inlet and outlet. However, there are more arrows than fittings. Therefore, simply use the necessary arrows over the fittings.

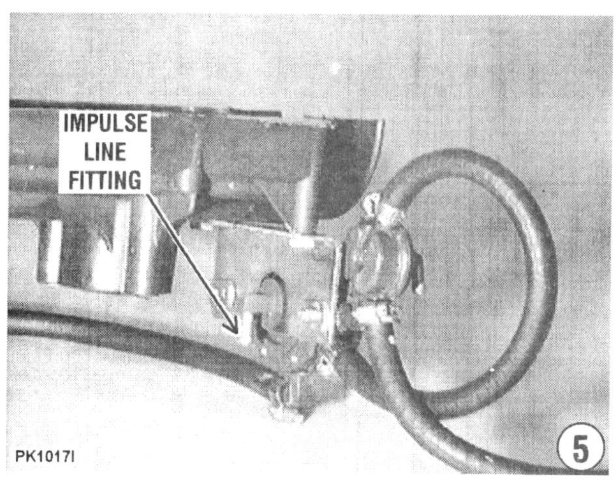

Overall view of the remote fuel pump installed onto the aft end of the intake/flame arrestor pan.

5- The engine pulse line fitting is on the back side of the fuel pump.

CAUTION

Water must circulate through the engine, anytime the engine is operating. Circulating water will prevent overheating -- which could cause damage to moving engine parts and possible engine seizure.

Connect a garden hose to the flush fitting, start the engine and after a few seconds, turn on the water.

Operate the engine and check the completed work. Be sure to inspect all fuel connections for leaks.

NEVER, AGAIN NEVER, operate the engine at high speed with a garden hose connected to the flush fitting. An engine operating at high speed with a hose connected, would possibly **RUNAWAY** from lack of a load on the impeller shaft, causing extensive damage.

6-8 OIL INJECTION

Description

Oil injection is standard equipment on all series engines covered in this manual. Over the years three different model oil pumps have been used. Two are mounted low on the starboard side of the crankcase and are driven by a short shaft and gear indexed with a "worm" type gear on the crankshaft.

One of the two has a linkage rod connecting carburetor synchronizing linkage between the No. 2 (center) and No.3 (aft) carburetors with the pump. This linkage controls an internal

Location of the oil injection pump installed on the lower crankcase -- starboard side on a Fuji engine. Late variable ratio oil pumps are connected to the carburetor throttle shaft arm by a rod which is non-adjustable.

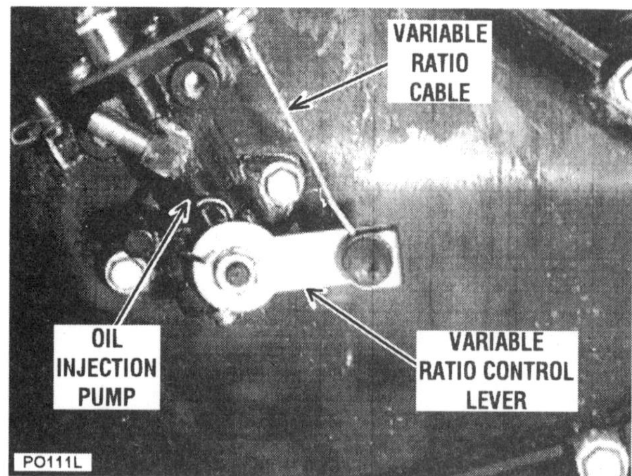

Oil injection pump installed on the flywheel cover on the Polaris engines. The variable ratio cable and lever are clearly visible. The cable must be adjusted correctly for the pump to deliver the proper amount of lubricant.

variable port to ensure the pump will deliver a constant oil/fuel ratio under all engine operating conditions.

The third pump is mounted to the flywheel cover and driven by a short shaft indexed into a slot in the flywheel nut. This pump is also connected to the carburetor synchronizing linkage through a cable arrangement to ensure the pump will deliver a constant oil/fuel ratio under all engine operating conditions.

The oil pump is a positive displacement type unit driven by the short shaft and gear indexed with the "worm" type gear on the crankshaft or by a short shaft indexed into a slot in the crankshaft flywheel nut. With each revolution of the crankshaft, the short shaft moves a plunger cam up and down, pumping oil through hoses to the intake manifold/s, or directly to the carburetor/s, depending of the engine series and model year.

OIL PUMP REMOVAL AND INSTALLATION

REMOVAL -- Pump Mounted Low -- Starboard Side Crankcase

Removal of the oil pump mounted low on the starboard side of the crankcase is not a quick and easy task with the engine in place in the craft. First the flame arrestor/s must be removed, then the carburetors and finally the intake manifold to gain working access to the pump. A mirror with a handle will prove useful to see and remove the mounting bolts. **TAKE CARE** not to lose any of the shim material. The **O**-ring should be replaced. Do not attempt to use an **O**-ring a second time.

Once the pump is free of the crankcase, it can be moved out a bit and the oil hoses disconnected and plugged to prevent contamination from entering.

If the oil pump has a rod connected to the carburetor linkage, disconnect the rod. The rod is a definite length and is not adjustable. If the rod is distorted the least amount, it **MUST** be replaced in order for the pump to function adequately. Remove the oil pump from the craft.

As mentioned earlier, the pump cannot be serviced -- replacement parts are not available. Therefore, if the pump has proven defective, it **MUST** be replaced.

END "PLAY" AND INSTALLATION -- Pump Mounted Low -- Starboard Side Crankcase

If a new oil pump is to be installed, the oil pump bushing end "play" must be determined prior to installing the pump. To determine the proper end "play" proceed as follows:

1- Check to be sure the bushing is fully seated in the block. With a micrometer scale, measure the distance from the surface of the oil pump mounting flange on the crankcase down to the bushing, as indicated in the accompanying illustration. Record this measurement as "A".

2- Next, measure the distance from the oil pump mounting flange up to the oil seal flange, as indicated in the accompanying illustration. Record this measurement as "B".

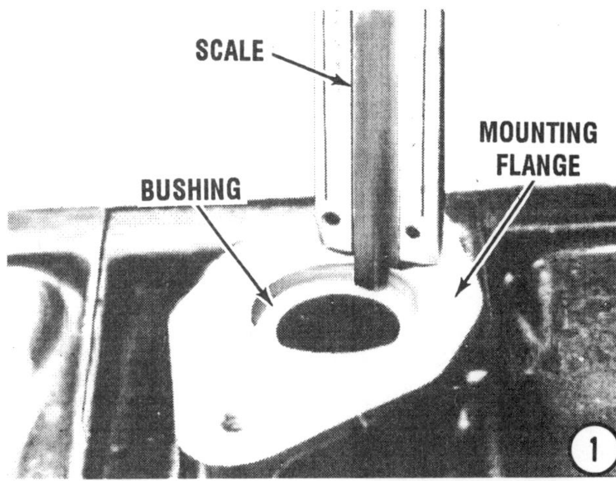

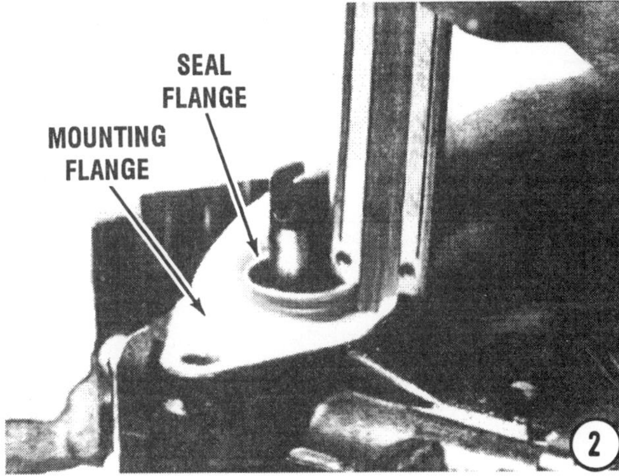

Now, subtract measurement "B" from measurement "A". The answer is the required total bushing end "play".

3- With a dial indicator or other suitable measuring device, measure the thickness of the shim material saved when the old pump was removed. The difference between the total end "play" determined in "B" above and the shim material saved, is the amount to be added or subtracted to provide the specified end "play" of 0.012 - 0.024" (0.3 - 0.6mm).

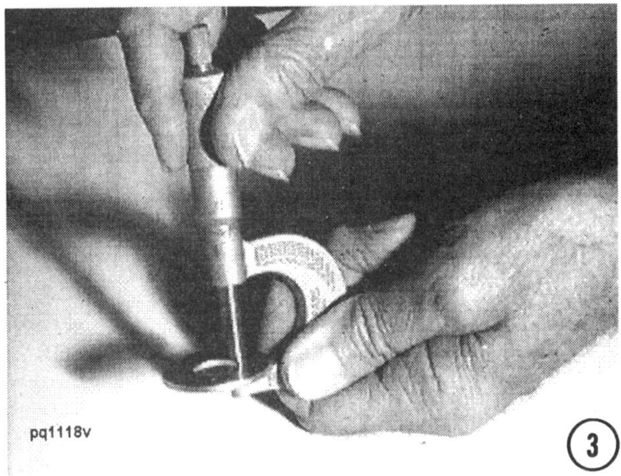

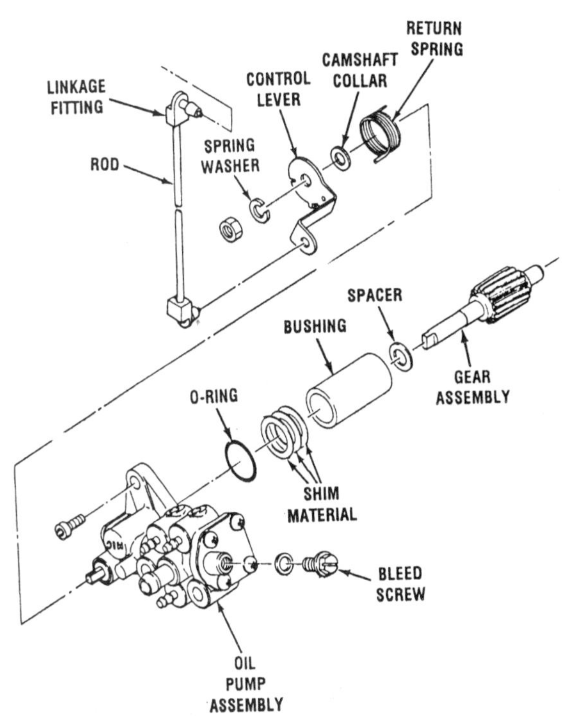

Exploded line drawing of the oil injection pump covered in this short section. Early models do not have the rod connected to the carburetor throttle shaft.

Apply a light coating of grease to a **NEW O-ring** and install it onto the pump. Install the pump using the correct amount of shim material. The slot on the end of the pump shaft must index with the "spade" on the end of the drive gear shaft for the mounting flanges of the pump and the crankcase to make full contact.

Apply Loctite™242 to the attaching bolt threads, and then secure the pump in place with the bolts. Tighten the bolts alternately and evenly to a torque value of 78 in lbs (8.8Nm).

If the pump is equipped with a rod to the carburetor linkage -- connect the lower end of the rod and associated parts to the pump, as indicated in the accompanying exploded line drawing. Connect the upper end to the carburetor linkage. The rod is **NOT** adjustable. The length of the rod will give the required distance between the two joints -- 6.43" ± 0.030" (163.4mm).

REMOVAL -- Pump Mounted On Flywheel Cover

Obtain suitable material to plug the oil lines when they are disconnected. Disconnect the fuel lines and plug them to prevent contamination from entering.

Loosen the cable adjustment nuts slightly -- just enough for the cable to clear the fitting.

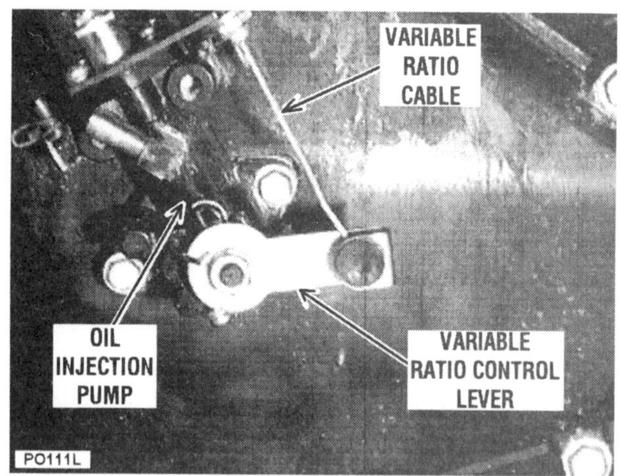

Close view of the oil injection pump mounted to the flywheel cover. Some major parts are identified.

Make an attempt not to lose the adjustment. Disconnect the oil inlet hose and the outlet hoses, and then plug the hoses immediately.

Remove the two attaching bolts securing the pump to the flywheel cover. Lift the pump straight out and clear of the cover. Discard the **O**-ring. Do not attempt to use an **O**-ring a second time.

As mentioned earlier, the pump cannot be serviced -- replacement parts are not available. Therefore, if the pump has proven defective, it must be replaced.

INSTALLATION -- Pump Mounted On Flywheel Cover

Apply a light coating of grease to a **NEW O**-ring, and slide it into place in the oil pump.

Move the pump into place on the flywheel cover with the "spade" on the pump shaft indexed into the slot in the flywheel nut.

Coat the threads of the attaching bolts with Loctite™242. Tighten the bolts alternately and evenly to a torque value of 60 in lbs (6.8Nm).

Slide the oil pump cable into the fitting and then **CAREFULLY** hand tighten the adjustment nuts without affecting the adjustment. Connect the oil lines to the pump and to the carburetors.

Cable Adjustment

Check and adjust the oil cable as follows: The idle speed and throttle lever free "play" **MUST** be properly adjusted before making the oil pump cable adjustment.

Observe the mark on the oil pump lever and the mark on the oil pump body. These two marks should align with each other when the throttle is released and the engine is operating at idle speed. When the cable is adjusted properly, the oil pump lever should move as soon as the throttle is opened -- the slightest amount.

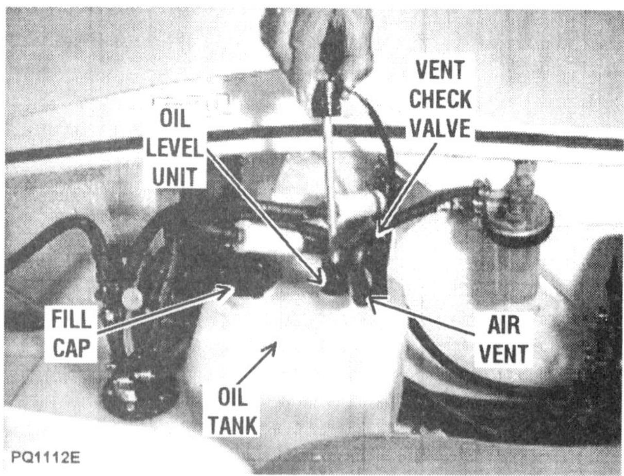

Oil tank installed in a Hurricane model PWC. The shape of the oil tank changes with almost each model due to size and shape of the hull along with size and shape of the fuel tank and other factors. However, they all have basically the same components, cap, oil level sender, vent tube with "roll over" check valve and outlet line to feed the oil pump by gravity.

If the marks are not aligned, simply loosen the two adjustment nuts, move the cable slightly until the marks are aligned, and then tighten the nuts securely to hold the adjustment.

Make a final check to ensure the lever begins to move with the smallest amount of throttle movement.

Oil Pump Output Test

Prepare about a ten minute supply of 40:1 pre-mix using a good grade of two-stroke oil. "Jury-rig" a setup to provide this mixture to the fuel pump during this purging procedure. Check the level of oil in the oil tank and replenish as necessary.

Obtain a transparent measuring cylinder graduated in mL. If such a container is not available, use a measuring cup graduated in ounces. A conversion table on the first page of the Appendix may be used to convert ounces to milli-litres.

Start the engine and allow it to warm to operating temperature.

CAUTION

Water must circulate through the jet pump -- to and from the engine, anytime the engine is operating. Circulating water will prevent overheating -- which could cause damage to moving engine parts and possible engine seizure.

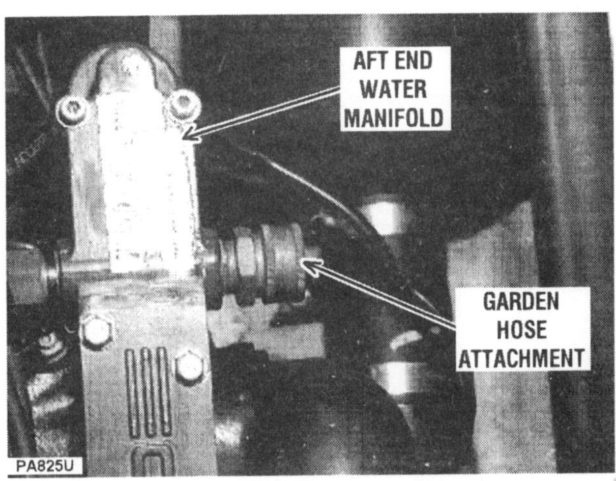

Close view of the garden hose attachment fitting on the aft end of the water exit manifold. Exact location of the fitting may vary, but it is always on the manifold.

Close view of the oil injection pump to be mounted low on the starboard side of the crankcase ready to be installed. The proper amount of shim material has been determined for adequate end "play", as outlined in the text.

NEVER, AGAIN NEVER, operate the engine at high speed with a flush device attached. An engine operating at high speed with such a device attached, would **RUNAWAY** from lack of a load on the impeller shaft, causing extensive damage.

SPECIAL WORDS

During break-in or after long periods of storage, the oil/fuel and oil premix being fed through the carburetor will cause the engine to operate on a heavy mixture of oil.

Using a pair of needlenose pliers, squeeze the clamp on the transparent delivery hose. Push the clamp up the hose. Gently pull the hose from the fitting on the intake manifold and immediately hold it over the container.

Observe the flow of oil for four full minutes to verify the pump is functioning properly.

More Special Words

The oil/fuel mixture prepared earlier will provide adequate lubrication while the system is being tested.

A steady slow pulsing flow with no air bubbles may be expected. Reconnect the line. Shut down the engine and remove the flushing device.

Troubleshooting

Unfortunately, most problems with the oil injection system are only discovered after it is too late. If either of the two hoses becomes clogged, kinked, or restricted, the oil delivery to the cylinders will be reduced.

Insufficient oil delivery to the engine will cause the oil level to drop slower than normal; the engine will overheat due to inadequate lubrication; and moving internal parts will wear more quickly or be severely damaged.

Excessive oil delivery to the cylinders will cause the oil level to drop at a much faster rate than normal. The engine will smoke -- especially at idle speed -- and the spark plugs will become fouled, causing the engine to misfire.

BAD NEWS

No overhaul kits or spare parts are available for the oil pump used on these units. Therefore, a defective oil pump must be replaced.

Systematically check each oil line and connection. A free flow of oil through an unrestricted line is **CRITICAL** for adequate engine lubrication.

CAUTION

Any time the oil tank hose is disconnected, the oil injection pump **MUST** be purged ("bled") of any trapped air. Failure to "bleed" the system could lead to engine seizure due to lack of adequate lubrication.

Instructions for "bleeding" -- purging -- air from the system follows.

Purging Air From Oil Injection System

The following procedures are to be performed any time the oil injection system has

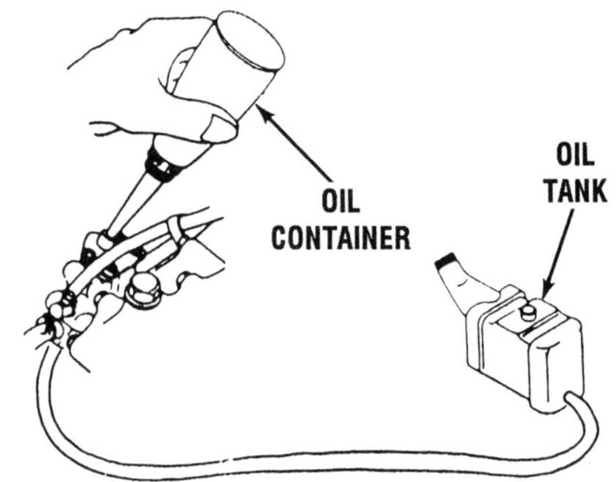

Simple line drawing to depict the ***METHOD*** *only of "bleeding" (purging), air from the oil pump and the line to the oil tank. The container* ***MUST*** *be squeezed in one* ***CONTINUOUS*** *motion. The text explains why.*

been opened (other than to add oil to the tank) and air has entered the system.

Prepare about a ten minute supply of 40:1 premix using two-stroke oil. "Jury-rig" a setup to provide this mixture to the fuel pump during this purging procedure. The oil/fuel mixture prepared will provide adequate lubrication while the system is being purged. Check the level of oil in the oil tank and replenish as necessary.

Start the engine and allow it to warm to operating temperature.

CAUTION

Water must circulate through the jet pump -- to and from the engine, anytime the engine is operating. Circulating water will prevent overheating -- which could cause damage to moving engine parts and possible engine seizure.

NEVER, AGAIN NEVER, operate the engine at high speed with a flush device attached. An engine operating at high speed with such a device attached, would **RUNAWAY** from lack of a load on the impeller shaft, causing extensive damage.

SPECIAL WORDS

Because the fuel and oil premix is being fed through the carburetor, in addition to oil from the oil injection system also being fed directly into the manifold, the engine will be operating on a heavy mixture of oil.

Obtain a small container and have a shop towel handy to catch any oil drips. Loosen the bleed screw on the oil pump a few turns and wait until a slow pulsing flow of oil emerges without any air bubbles. Observe the flow of oil for four full minutes to verify the pump is functioning properly and all air has been expelled from the system. Tighten the screw securely.

Replenish the oil tank with two-stroke oil.

Purging Air From The Oil Pump and Line To The Tank

Obtain a container which is soft enough to permit being squeezed by hand. Obtain some type of nozzle which may be inserted into the oil pump bleed fitting.

Fill the container with roughly 3.5 ounces of recommended two-stroke engine oil.

Remove the air bleeder bolt on the oil pump.

Loosen the oil tank cap slightly to permit air to escape.

Now, inject oil slowly through the oil pump bleed fitting by squeezing the container with one continuous motion. **DO NOT** stop halfway or at any time. This amount of oil is enough to fill the line from the oil pump to the oil tank and means air trapped within the pump and any air in the line will be expelled back into the tank.

CRITICAL WORDS

Squeezing the container intermittently could permit air to enter the oil line and defeat the purpose of attempting to purge air from the pump and the line. Any air left in the pump or the line could cause reduction in oil flow and possible damage to the engine through lack of adequate lubrication.

Install and tighten the air bleeder bolt. Tighten the oil tank filler cap.

7
IGNITION

7-1 INTRODUCTION AND CHAPTER COVERAGE

The less a marine engine is operated, the more care it needs. Allowing a marine engine to remain idle will do more harm than if it is used regularly. To maintain the engine in top shape and always ready for efficient operation at any time, the engine should be operated every 3 to 4 weeks throughout the year.

The carburetion and ignition principles of two-cycle engine operation **MUST** be understood in order to perform a proper tune-up on a marine engine.

The flywheel from an SLT700 Series engine is "pulled" for a closer inspection of the magneto assembly. This service work is difficult to perform if the engine is in the craft. Polaris manufactured engines have an additional magnet holder, between the flywheel and stator plate.

If you have any doubts concerning your understanding of two-cycle engine operation, it would be best to study the operation theory section in the first portion of Chapter 8, before tackling any work on the ignition system.

A Capacitor Discharge Ignition (CDI) is used on all engine models covered in this manual.

7-2 SPARK PLUG EVALUATION

Removal

Remove the spark plug wires by pulling and twisting on only the molded cap. **NEVER** pull on the wire or the connection inside the cap may become separated or the boot damaged. Remove the spark plugs. **TAKE CARE** not to tilt the socket as you remove the plug or the insulator may be cracked.

Examine

Carefully examine all plugs to determine the firing conditions in the cylinder. If the side

The charging ignition coils and charging coils on a Model SLT700 Series are mounted on the stator plate and are well protected behind the flywheel.

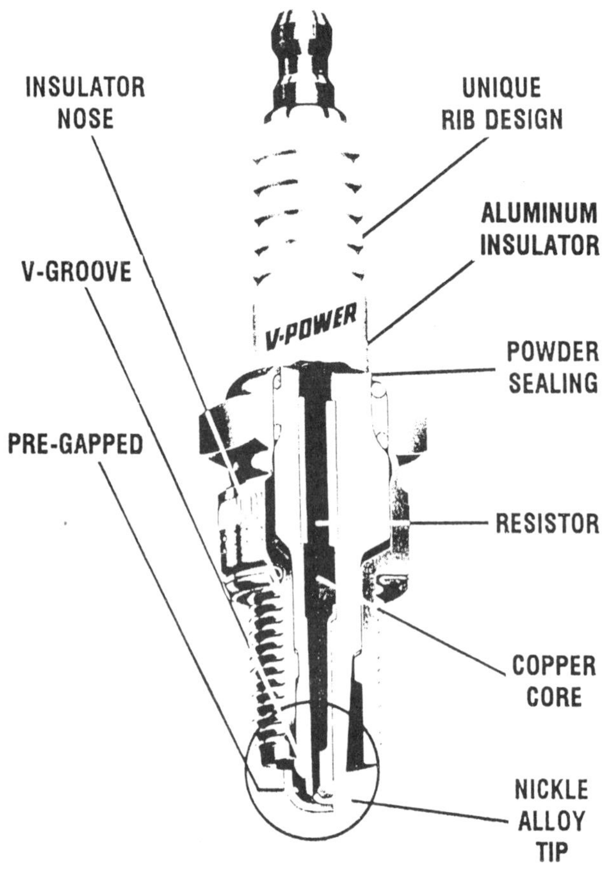

Cross-section drawing of a typical NGK spark plug with some of the features identified.

electrode is bent down onto the center electrode, the piston is traveling too far upward in the cylinder and striking the spark plug. Such

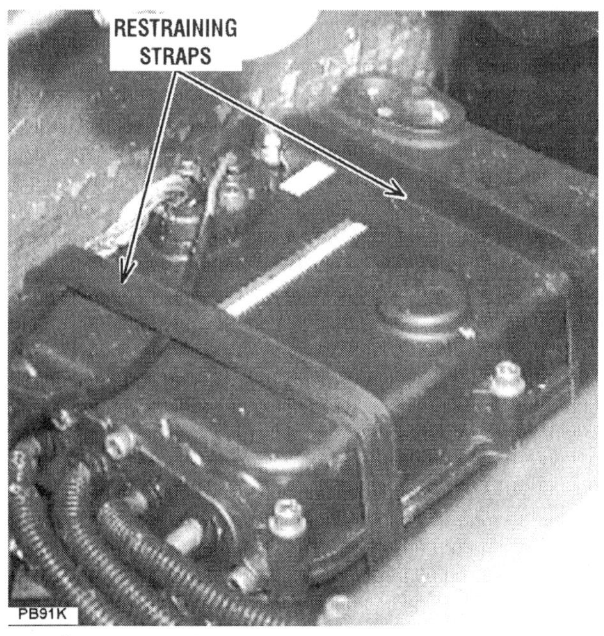

On some Models, the electrical box is secured atop the battery with restraining straps. On other Models, the electrical box is secured to a bulkhead with attaching bolts.

damage indicates the piston pin or the rod bearing is worn excessively.

In most cases, an engine overhaul is required to correct the condition. To verify the cause of the problem, rotate the flywheel by hand. As the piston moves to the full up position, push on the piston crown with a screwdriver inserted through the spark plug hole, and at the same time rock the flywheel back-and-forth. If any play in the piston is detected, the engine must be rebuilt.

Correct Color

A proper firing plug should be dry and powdery. Hard deposits inside the shell indicate too much oil is being mixed with the fuel. The most important evidence is the light gray to tan color of the porcelain, which is an indication this plug has been running at the correct temperature. This means the plug is one with the correct heat range and also that the air/fuel mixture is correct.

Rich Mixture

A black, sooty condition on both spark plug shells and the porcelain is caused by an excessively rich air/fuel mixture, both at low and high speeds. The rich mixture lowers the combustion temperature so the spark plug does not run hot enough to burn off the deposits.

Deposits formed only on the shell is an indication the low-speed air/fuel mixture is too rich. At high speeds with the correct

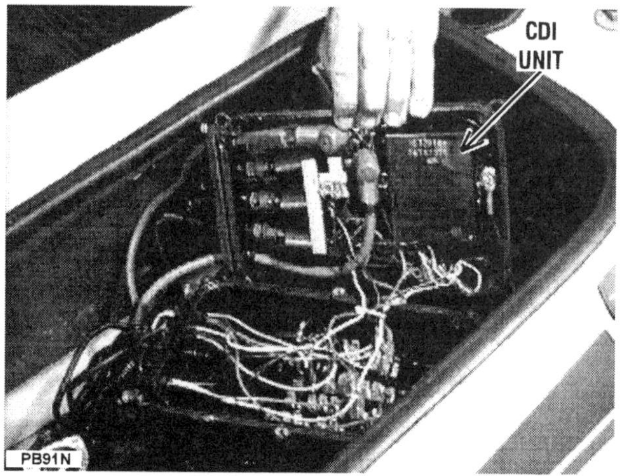

The CDI unit on an SLT700 is mounted inside and well protected by the electrical box. The electrical box also houses the reset button, the cranking motor relay -- to the upper left of the CDI -- the primary charge coils, and the regulator/rectifier.

mixture, the temperature in the combustion chamber is high enough to burn off the deposits on the insulator.

Too Cool

A dark insulator, with very few deposits, indicates the plug is running too cool. This condition can be caused by low compression or by using a spark plug of an incorrect heat range. If this condition shows on only one plug it is most usually caused by low compression in that cylinder. If both plugs have this appearance, then it is probably due to the plugs having a too-low heat range.

Fouled

A fouled spark plug may be caused by the wet oily deposits on the insulator shorting the high-tension current to ground inside the shell. The condition may also be caused by ignition problems which prevent a high-tension pulse from being delivered to the spark plug.

Carbon Deposits

Heavy carbon-like deposits are an indication of excessive oil in the fuel. This condition may be the result of worn piston rings or excessive ring end gap.

Overheating

A dead white or gray insulator, which is generally blistered, is an indication of overheating and pre-ignition. The electrode gap wear rate will be more than normal and in the case of pre-ignition, will actually cause the electrodes to melt. Overheating and pre-ignition are usually caused by overadvanced timing, detonation from using too-

*Damaged spark plugs. Notice the broken electrode on the left plug. The missing part **MUST** be found and removed before returning the engine to service, to prevent serious damage to expensive internal parts.*

This spark plug is foul from operating with an overrich air/fuel mixture, possibly caused by an improper carburetor adjustment.

low an octane rating fuel, an excessively lean air/fuel mixture, or problems in the cooling system.

Electrode Wear

Electrode wear results in a wide gap and if the electrode becomes carbonized it will form a high-resistance path for the spark to jump across. Such a condition will cause the engine to misfire during acceleration. If both plugs are in this condition, it can cause an increase in fuel consumption and very poor performance at high-speed operation. The solution is to replace the spark plugs with a rating in the proper heat range and gapped to specification.

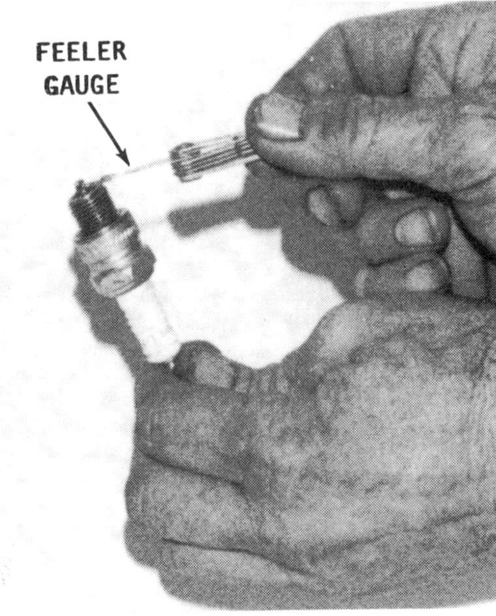

The spark plug gap should always be checked with a wire-type feeler gauge before installing new or used plugs.

Red rust-colored deposits on the entire firing end of a spark plug can be caused by water in the cylinder combustion chamber. This can be the first evidence of water entering the cylinders through the exhaust manifold because of an accumulation of scale. This condition **MUST** be corrected at the first opportunity. Refer to Chapter 8, Engine.

7-3 CDI (CAPACITOR DISCHARGE IGNITION) AND CHARGING SYSTEM

As the stator plate mounted coils rotate, they pass through magnetic fields produced by magnets attached to the flywheel. These coils convert the mechanical energy of motion into electrical energy -- voltage. The charged capacitor discharge system harnesses this electrical energy to provide the necessary high voltage to the spark plugs.

On all models covered, half of the ignition circuit components -- charging, exciter, and pulser coils -- are mounted to the stator plate. The stator is housed and well protected behind the flywheel. The remainder of the ignition components -- ignition coils, CDI unit, and rectifier/regulator -- are located inside an electrical box, within the engine compartment.

Therefore, it is necessary to remove the flywheel to inspect components of the stator plate. If any portion of the stator is determined to be defective, the entire stator assembly must be replaced, because these items are not sold separately.

Elongated slots in the stator plate allow the assembly to be rotated to advance or retard ignition timing. Note the inner magneto cover has been removed in preparation of removing the stator assembly -- Fuji manufactured engines only.

DESCRIPTION AND OPERATION IGNITION CIRCUIT

The following paragraphs describe -- in "layman's" language -- the CDI system and how it functions. Every effort has been made to keep "technical" terms to a minimum.

The CDI system produces high voltage with greater efficiency compared to previous ignition methods. The time necessary to develop the necessary voltage is reduced and timing is very accurate when operating properly. The capacitor discharge ignition system is comprised of two main parts -- the primary circuit and the secondary circuit.

Basically, the primary circuit of the CDI system builds up the voltage in the CDI capacitor. The components of this circuit consist of a series of exciter coils, the trigger coil, and the wiring harness/es from the stator plate to the CDI box and to the ignition coil primary windings.

The secondary circuit is comprised simply of the ignition secondary coil windings, high tension leads and the ground through the spark plug air gap.

The CDI itself, is commonly referred to as the "brain", or "black box" -- it is indeed a sealed black box.

The exciter coil -- ignition generating coil -- and a trigger coil -- pulser coil -- are mounted on the stator plate. The exciter coils provide a source of AC current -- primary ignition voltage. The trigger coil is used to sequentially time the electrical impulses or spark.

The ignition module circuitry converts AC current to DC current. The module stores this

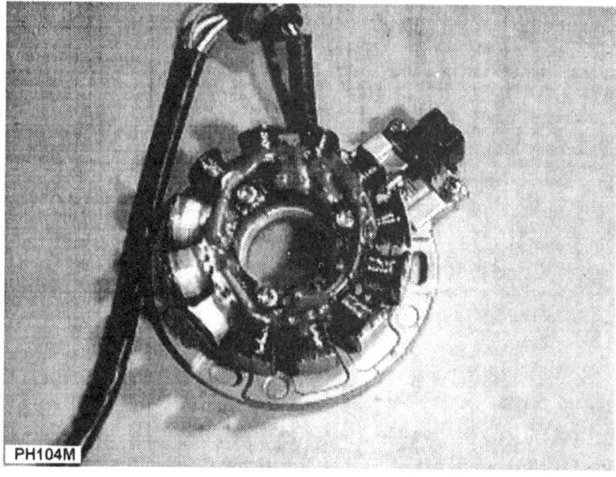

This magneto assembly has been removed from an SLT700 Series engine for further inspection of the stator coils and to check the condition of the stator electrical harness.

DC current and releases it to the ignition coil automatically at the correct time.

A series of diodes, an SCR gate, and a capacitor comprise the basic module circuitry. A diode is a solid state device designed to allow current flow in one direction but prevents flow in the opposite direction. A SCR gate is a solid state electronic switching device which permits voltage to flow only after it has been triggered by another source.

The ignition coil circuitry boosts the DC voltage instantly to approximately 20,000 volts. This high voltage is passed on to the spark plugs through the high tension leads.

Operation

Consider the engine operating and the flywheel rotating. At the point in time when the ignition timing marks align, an alternating coil is induced in the exciter coil. This voltage charges a capacitor in the CDI igniter circuit. As the flywheel continues to rotate, another voltage is induced in the pulser coil.

This voltage flows through a diode and on to the SCR gate. The SCR gate signals the capacitor to discharge the voltage to the primary coils inside the CDI igniter. The voltage in the primary windings of the ignition coil induces a high voltage in the secondary windings and causes a spark to jump to ground across the spark plug electrodes.

In this manner, a spark at the spark plug may be accurately timed by the marks on the flywheel relative to the magnets in the flywheel to provide as many as 100 sparks per second for an engine operating at about 6,000 rpm.

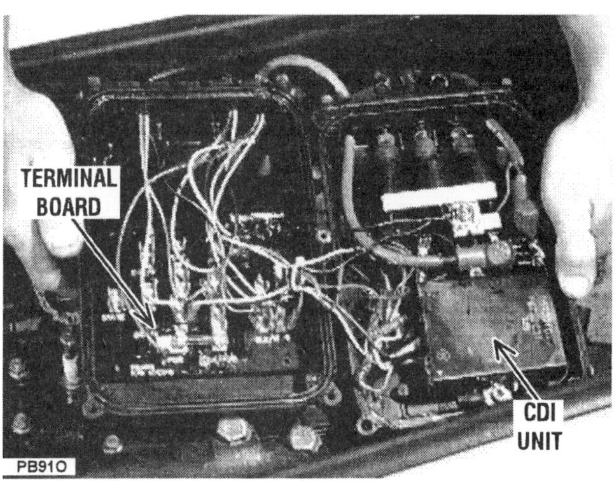

The terminal board shown in the left portion of the electrical box is imprinted with color keys to aid in quick identification of electrical lead connections. The primary charge coils are located "above" the CDI unit.

The stator plate on early Fuji manufactured engines have elongated bolt holes, allowing minor rotation of the plate to facilitate ignition timing.

SPECIAL TIMING WORDS

The stator plate on Fuji manufactured engines has small slots for the mounting bolts. These slots permit the stator to be rotated **SLIGHTLY** in order to obtain the correct degree of timing.

Detailed ignition timing procedures for all models is presented in section 7-5 of this chapter, beginning on page 7-11.

TROUBLESHOOTING CDI IGNITION SYSTEM

Always attempt to proceed with the troubleshooting in an orderly manner. The "shot in the dark" approach will only result in wasted time, incorrect diagnosis, replacement of unnecessary parts, and frustration.

Begin the ignition system troubleshooting with the spark plug and continue through the system until the source of trouble is located.

Spark Plugs

1- Check the plug high tension leads to be sure they are properly connected. Check the entire length of the leads from the plugs to the magneto assembly on the stator plate. If the leads are to be removed from the spark plug, **ALWAYS** use a pulling and twisting motion as a precaution against damaging the connection.

Attempt to remove the spark plug by hand. This is a rough test to determine if the plug is tightened properly. The attempt to loosen the plug by hand should fail. The plug should be tight and require the proper socket size tool. Remove the spark plug and evaluate its condition as described in Section 7-2.

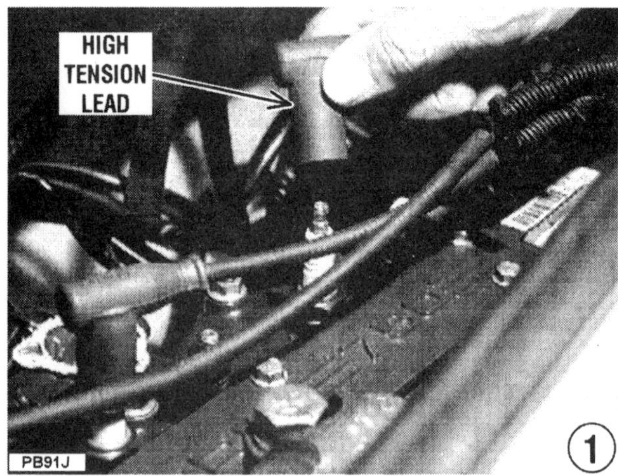

2- Use a spark tester and check for spark. If a spark tester is not available, hold the plug lead about 1/4" (6.4mm) from the engine. Crank the engine through a few revolutions using the cranking motor and check for spark.

A strong spark over a wide gap must be observed when testing in this manner, because under compression a strong spark is necessary in order to ignite the air/fuel mixture in the cylinder. This means it is possible to think a strong spark is present, when in reality the spark will be too weak when the plug is installed. If there is no spark, or if the spark is weak, the trouble is most likely under the flywheel in the stator assembly.

Compression

3- Before spending too much time and money attempting to trace a problem to the ignition system, a compression check of the cylinders should be made. If a cylinder does not have adequate compression, troubleshooting

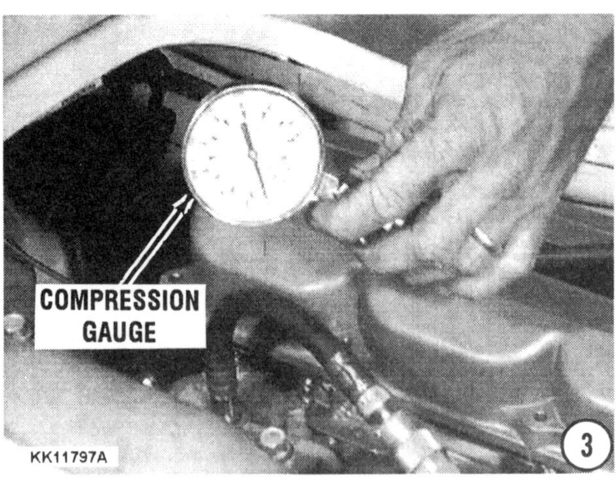

and attempted service of the ignition or fuel system will fail to give the desired results of satisfactory engine performance.

Remove the spark plug lead by pulling and twisting **ONLY** on the molded cap. **NEVER** pull on the wire because the connection inside the cap may be separated or the boot may be damaged. Remove the spark plug. Insert a compression gauge into the cylinder spark plug opening. Ground the spark plug leads to prevent damage to the ignition coil. If a lead is not grounded, the coil will attempt to match the demand created by the spark trying to jump from the electrode to the nearest ground.

Insert a compression gauge into the forward spark plug opening. Crank the engine with the cranking motor, through at least four complete revolutions of the crankshaft, with the throttle at the wide-open position, to obtain the highest possible reading. Repeat the test and record the compression for the other cylinders.

The manufacturer specifies the compression pressure for all engines covered in this manual, should indicate between 100psi - 150psi. (690kPa - 1,034kPa).

GOOD WORDS

On all 1997 1050cc Models, the **CENTER** cylinder will have an approximate 15psi lower reading than the No. 1 (Mag) and No. 3 (PTO) cylinders.

A variation between cylinders is far more important than the actual readings. A variation of more than 5% between the cylinders (service limit) indicates the lower compression cylinder is defective. The problem may be worn, broken, or sticking piston rings, scored pistons or worn cylinders.

7-4 TESTING IGNITION COMPONENTS

General Information

Due to the ever-increasing complex nature of the circuitry and high energy output pulses of micro-second duration, conventional testing devices such as a Volt/Ohm/Ammeter will not measure electrical output with the degree of accuracy required. Very specific testers are required to troubleshoot various subsystems of the CDI ignition system. Therefore, this chapter only includes those tests which may be performed with instruments normally and easily available.

ELECTRICAL SCHEMATICS

When making any tests, refer to the electrical schematic in the Appendix for the engine being serviced. The schematics contain an outline of the top half and the bottom half of the electrical box -- clearly indicating components, wiring, and color code housed inside. All items outside the box are also included.

TESTER WORDS

Resistance tests performed using an analog meter/tester are not as accurate as using a digital multitester. During the following procedures, a Fluke model 73 digital multitester was used to conduct these tests.

Values may vary slightly if using another type of multitester for testing components. Be sure to subtract meter lead resistance, if necessary.

WORDS FROM EXPERIENCE

During the tests, if a reading is slightly different from the specifications, but the engine still operates, then there is no real need to replace the affected component, until it actually fails. Bear in mind, in **MOST** cases electrical components are not returnable, once the item leaves the store. Therefore, make every attempt to avoid the **BUY** and **TRY** method of troubleshooting. Such a practice will lead only to wasted time, unneeded cash outlay, and frustration.

Intermittent or Multiple Problems

Many ignition problems occur only during engine operation -- when the component is subject to vibration and the engine is operating under a load, there is a marked increase in temperature -- from the engine operating.

SPECIAL WORDS ON IGNITION AND CHARGING CIRCUIT TESTING

Many of the components in the ignition and charging circuits have leads with quick-disconnect fittings or harness plugs located inside the electrical box. Therefore, testing begins inside the electrical box.

Some components in the CDI circuitry are housed inside the box -- others are located on the stator plate. If a component inside the electrical box is found to be defective, it may be replaced immediately -- usually through attaching hardware. If a component on the stator -- the excitor coil, charge coil, or trigger coil -- is found to be defective, as mentioned earlier, the flywheel must be "pulled" in order to remove and replace the item. See Chapter 8 Section 8-3, for instructions to pull and install the flywheel on three-cylinder engines. If servicing a two-cylinder engine, refer to Section 8-4.

CRITICAL WORDS

Never attempt to verify the charging circuit by operating the engine with the battery disconnected. Such action would force current -- normally directed to charge the battery -- back through the rectifier and damage the diodes in the rectifier.

__NEVER__ attempt to verify the charging circuit by operating the engine with the battery disconnected. Such action will damage the rectifier/regulator.

Resistance test results are more accurate when performed with a digital ohmmeter as opposed to an analog tester.

Electrical Box Cover Removal

First, these words: Refer to the electrical schematic in the Appendix covering the engine being serviced. An outline indicates components, wiring, and color code housed in the top half and the bottom half of the electrical box.

Due to the number of leads and electrical harnesses entering the electrical box, it is much easier to only remove the box cover and test or remove components. It is not necessary to remove the electrical box for testing purposes. Some leads/harnesses may need to be disconnected to move the box to the top of the engine compartment for easy access.

1- Disconnect the battery cable from the negative terminal of the battery.

2- Remove the electrical box from its secured position. Electrical boxes are mounted either atop the battery, or they may be vertically

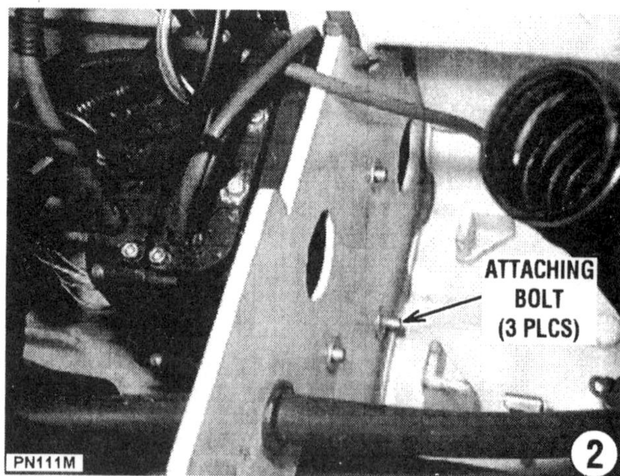

mounted to a bulkhead. If attached to the battery, remove the straps securing the box and battery to the hull. If mounted to a bulkhead, remove the attaching bolts securing the box to the bulkhead.

3- Move the box to a convenient location for testing components -- on top of the engine, for example. Remove the attaching bolts securing the electrical box halves together. Separate the halves carefully, to ensure the seating gasket is not damaged. Certain leads may require disconnecting in order to place the box in a favorable location for servicing.

Stator Assembly Resistance Tests

First, these words: Refer to the electrical schematic in the Appendix covering the engine being serviced. An outline indicates components, wiring, and color code housed in the top half and the bottom half of the electrical box.

It is not necessary to remove the flywheel to test the stator plate components at this time. The stator electrical harness exits the crankcase

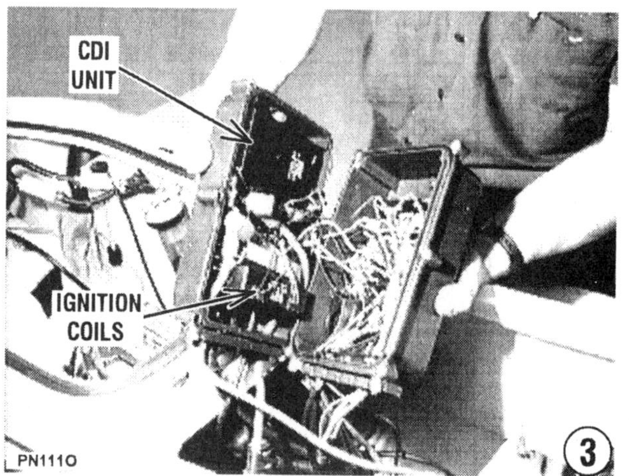

RESISTANCE CHART

FUJI ENGINES		
Part	**Lead Colors**	**Reading**
Alternator Coil	Red/Purple to Yellow	.6 ohms
Exciter Coil	Red/White to Green/Red	490 ohms
Pulser Coil	Blue/Red to Red/White	90 ohms
Trigger Coil	White/Yellow to Black	220 ohms
Ignition Coil	**Connection**	**Reading**
Primary Windings	Black to Black/White	.2-.6 ohms
Secondary Windings	Black to high tension lead (spark plug cap removed)	3.3 K ohms
Spark Plug Cap	Lead inside to the spark plug side	5.0 K ohms

POLARIS ENGINES		
Part	**Lead Colors**	**Reading**
Alternator Coil	Red/Purple to Yellow	.75 ohms (Continuity)
Alternator Coil	Red/Purple or Yellow to Ground	Infinity (No continuity)
Exciter Coil	Black to Ground	0 ohms
Exciter Coil	Black to Grey	80 or 130 ohms; see Electrical Schematic
Exciter Coil	Black to Purple	1200 ohms
Trigger Coil	Refer to test procedure in this Chapter	
Ignition Coil	**Connection**	**Reading**
Primary Windings	Yellow/Brown to Black/White; Black/Green to Black/White; or Black/Blue to Black/White	.35 ohms (Continuity)
Secondary Windings	Black to high tension lead (spark plug cap removed)	1240 or 1550 ohms -- see Coil Description
Spark Plug Cap	Lead inside to spark plug side	5.0 K ohms

near the cranking motor, and then it is routed under the engine and terminates **IN** the electrical box. There are no disconnect fittings between the stator and the harness wire ends in the box.

Therefore, stator coil resistance tests are most easily conducted using the leads in the electrical box. Tested in this manner, the resistance value will be affected by the condition of the conductor -- or lead. Any damaged lead will not accurately indicate **ACTUAL** coil condition. Identify and disconnect the leads from the stator plate to the connections within the electrical box.

Refer to the following table to identify leads and obtain the specific resistance values for those models listed. Use a precision digital multitester to measure the resistance of the stator coils.

Refer to the specific electrical schematic for the model being serviced. If resistance values are not listed on the schematic, use the specifications listed in the following table. Readings may vary ± 10%. Subtract meter lead resistance when measuring small resistance values.

Coil Results

If the meter resistance value is higher or lower than the value given, check the lead for an open circuit or short, respectively.

If the reading is less than expected and the leads are in good condition, most likely a short exists in the coil. No reading (infinity) indicates the coil has an open circuit. Low reading or no reading means the coil is at fault. The flywheel and stator must be removed, and the stator replaced. At press time, these coils were not available on an individual basis -- the entire stator assembly must be replaced.

If the coils are in satisfactory condition, the problem cause is probably a loss of magnetism in the flywheel. The flywheel must be replaced.

For detailed procedures to remove the flywheel and stator plate, see Chapter 8 --Engine.

Hall Effect Sensors
Polaris Engines Only

Stator plate mounted trigger coils on Polaris manufactured engines are referred to as Hall Effect sensors. These sensors become conductive when a magnet in the magnet holder passes by the sensor. This conductivity triggers the release of energy stored in the capacitor, resulting in the ignition spark. The trigger coils --

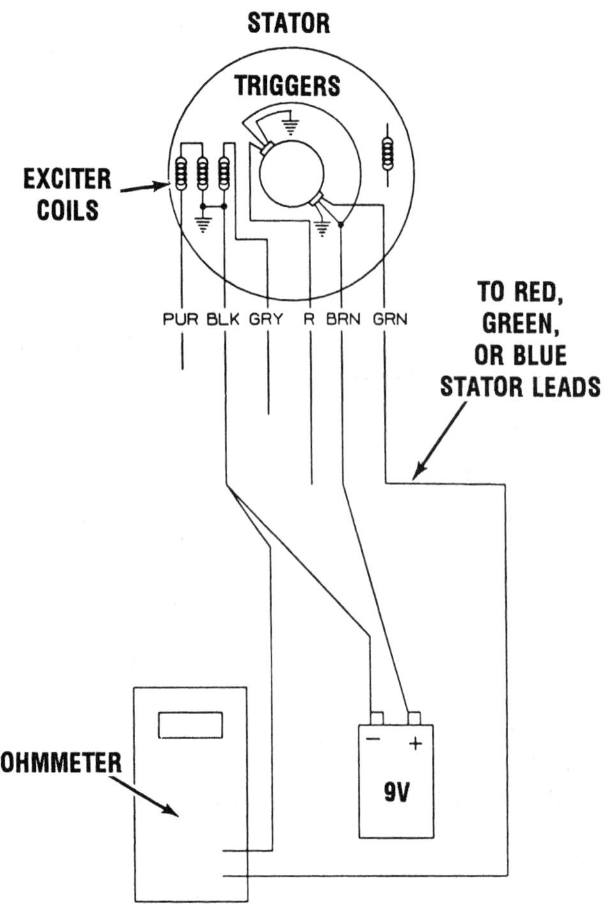

Partial electrical schematic for use with Polaris Static Timing tool P/N 2871745, used in verifying proper operation of Hall Effect sensors -- trigger coils -- Polaris manufactured engines only.

Hall Effect sensors -- receive a constant DC source from the CDI Brown lead. This wire is powered any time the engine is operating.

To test trigger coils, check to be sure the stator harness and all leads are disconnected. If using the Polaris Static Timing Tool -- P/N 2871745 -- the flywheel cover need not be removed. Connect the White tester lead to the stator Red (forward cylinder), Blue (center cylinder), or Green (aft cylinder) stator lead. The tester light should illuminate momentarily if the sensor is working properly.

If the static timing tool is not available, the flywheel cover must be removed to permit rotating the flywheel. Attach a fresh 9V battery to the Black and Brown leads from the stator plate.

Connect a lead from a digital ohmmeter to the Black stator lead. Connect the other tester lead to the Red stator lead.

Slowly rotate the flywheel, observing the ohmmeter reading. The tester should indicate an open circuit -- infinity -- until a magnet in the magnet holder aligns with a Hall Effect sensor -- trigger coil. The tester reading should now indicate low resistance -- under 25 ohms -- until the magnet clears the sensor, and then it should indicate an open circuit as the flywheel continues to rotate.

Repeat this process for all remaining Hall Effect sensors. Again, actual readings may vary depending on the tester. It is important to note that the indication should change from high to low resistance, then back again as the magnet passes a trigger coil.

Ignition Coil Winding Resistance Test

First, these words: Refer to the electrical schematic in the Appendix covering the engine being serviced. An outline indicates components, wiring, and color code housed in the top half and the bottom half of the electrical box.

The ignition coil portion of the CDI igniter can be checked for a broken or badly shorted winding using an accurate ohmmeter.

Primary Winding

Begin by disconnecting the Black/White primary leads from the terminals on the coil. Each coil must be checked individually. Set the meter to the R x 1 ohm scale. Refer to the table for specified values.

Secondary Winding

Begin by rotating the spark plug caps **COUNTERCLOCKWISE** and removing the caps. Next, connect the ohmmeter between the spark plug leads.

Set the meter to the R x 1k scale. Measure the resistance between the high tension cable

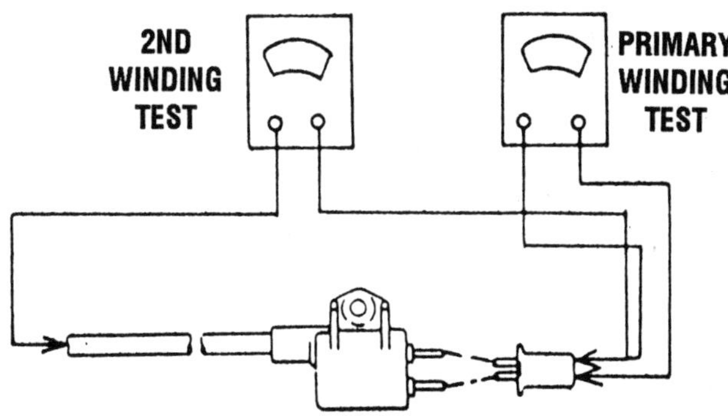

Simple line drawing to depict the primary and secondary winding ignition coil resistance test hookup for all Model engines.

and the coil ground -- small black wire. Measure the resistance at the spark plug cap. Refer to the chart for specified values.

GOOD WORDS

If the ignition coil windings passed the above tests, the coil is probably in good condition. Check the spark plug high-tension leads for visible damage and replace as necessary. **HOWEVER**, if the ignition system still fails to perform properly, and all other components have been checked as satisfactory, replace the coil.

Coil removal and installation procedures are outlined in the following paragraphs.

CDI Igniter Module

The CDI unit is not serviceable. Therefore, if it is found to be defective, the module must be replaced. Remove the two attaching bolts or nuts. **NOTE:** One of these two fasteners may provide a system ground.

7-5 IGNITION TIMING ADJUSTMENTS

All Polaris personal watercraft covered in this manual are equipped with an automatic advance type CDI system. Ignition timing is accomplished electronically and signal output increases proportionately to engine speed increase -- thereby advancing the timing. Such a system has no moving or sliding parts. Therefore, periodic adjustments are not necessary.

The timing is set correctly at the factory and should never need adjustment **UNLESS** the magneto assembly is misaligned or an ignition component -- such as the magneto assembly -- has been replaced.

NOW, THESE CRITICAL WORDS

The two most important items in establishing and maintaining perfect timing on a two-stroke engine are the magneto assembly securing screws and the crankshaft woodruff key.

Magneto Assembly Securing Screws

The magneto assembly is mounted on top of the stator plate and has two elongated slots for the securing screws to pass through and permit rotating the assembly **SLIGHTLY** to achieve proper timing. If the magneto assembly is

Elongated slots on the stator plate allow the magneto assembly to be rotated slightly to correct ignition timing. The attaching screw is removed for photographic clarity. If rotating the plate, the screws should be loosened, but not removed prior to adjusting timing.

allowed to rotate just a small amount, because the securing screws have worked loose, the ignition timing may be off as much as 15°. The symptoms of this misalignment are loss of power and difficult start-up.

Crankshaft Woodruff Key

The same symptoms apply to a sheared Woodruff key allowing the flywheel -- actually the flywheel magnets -- to become out of position on the crankshaft.

During wave jumping or using a wet ramp, when the craft clears the water, the engine rpm increases dramatically -- approaching the limit set by the rpm limiter in an unloaded condition.

Once the craft returns to the water surface, the sudden load on the impeller is transferred along the driveline including the crankshaft.

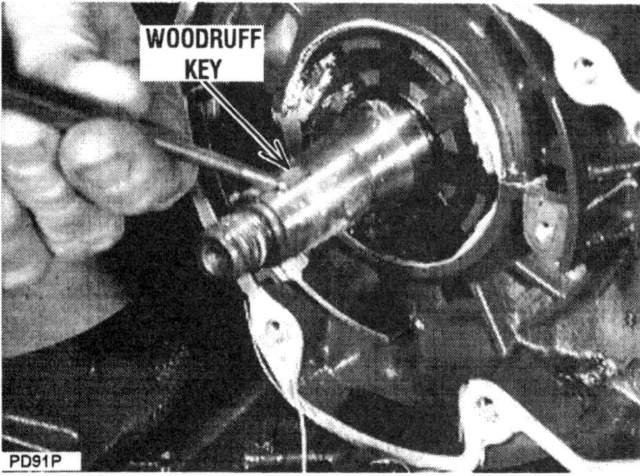

Before removing the Woodruff key, observe its alignment in the crankshaft slot. Carefully remove the Woodruff key and inspect it for signs of wear.

Speed of the rotating pump shaft and the crankshaft is suddenly reduced while the flywheel attempts to continue rotating rapidly due to its mass.

This condition, especially when repeated over a period of time, may shear the Woodruff key and the flywheel becomes repositioned on the crankshaft.

A very small misalignment will result in a change in ignition timing. The craft operator will notice a drastic loss of power. A worse condition would be the flywheel shifting to the degree engine start is not possible.

IGNITION TIMING SPECIAL WORDS

Checking the timing can be an involved process and changing the timing while the engine is **IN** the watercraft may be difficult. This is true of many watercraft manufacturers -- it's the nature of the beast! If a correction must be made to the ignition timing, the flywheel on Fuji engines must be removed. "Pulling" the flywheel without removing the engine from the craft is a difficult task indeed.

All engines are equipped with an ignition timing "viewport" in the stator or magneto cover. Using this viewport for observing timing markings will decrease the number of tasks necessary to check the ignition. The timing of any Polaris engine is recommended as a two-step procedure. The engine should be "statically" checked to verify ignition marks -- out of water -- then, the unit should be dynamically tested -- in the water.

STATIC IGNITION TIMING -- ALL ENGINES

The timing marks should be identified and checked before beginning timing procedures. If using the timing hole located at the top of the flywheel cover, simply remove the plug to observe the timing marks.

Fuji Engines Only

If not using the timing hole, proceed with the following steps.

a- Remove the attaching bolts securing the expansion chamber bracket to the flywheel cover. Pull the bracket away from the cover.

b- Move the fuel tank forward if needed, to allow access for removing the flywheel cover.

The timing viewport is located at the top of the magneto housing. Check to be sure the plug is fastened securely after verifying ignition timing.

Identify and mark any fuel hoses that must be disconnected to move the tank forward.

c- Remove the bolts securing the flywheel cover to the engine crankcase. Some bolts may secure a harness or hose restraining strap to the cover. Mark these bolts and their location for aid in proper assembling.

d- Attach a piece of wire -- a timing marker -- to a forward cylinder base bolt/stud. The wire should be bent **SLIGHTLY** until it points toward the flywheel and must remain secured while the engine is operating.

All Engines

Remove the No. 1 -- forward -- spark plug. Install a dial indicator holder in the spark plug opening. Insert the dial indicator into the holder, but do not lock the dial in place at this time.

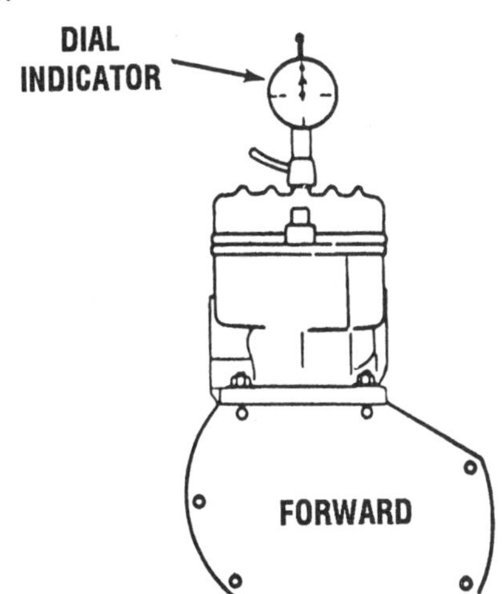

A dial indicator is installed in the spark plug opening as described in the text to determine Top Dead Center (TDC) and Before Top Dead Center (BTDC). These readings are important for accurate timing adjustments.

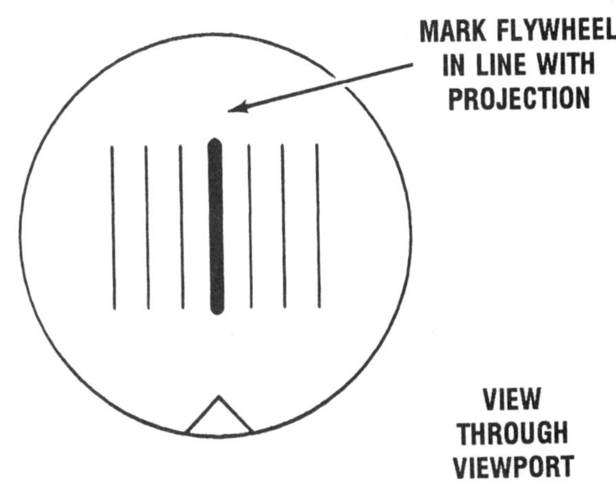

The number of timing procedure steps is greatly reduced if using the timing viewport as described in the text. Use a fast drying white ink to mark the timing line.

Rotate the crankshaft -- using the flywheel holder (Polaris P/N 8700229 and adaptor if servicing a Fuji engine) if available -- until the dial indicator rises up in its holder. The piston is nearing Top Dead Center (TDC). Slide the gauge far enough into the indicator holder to obtain a reading. Tighten the indicator set screw to secure the dial in place in the holder.

Rotate the crankshaft while observing the gauge dial. Find the exact location where the needle stops rotating and begins to reverse direction. This position is Top Dead Center (TDC).

Magneto Cover
Timing Viewport

If using the magneto cover viewing port -- observing the flywheel through the opening should indicate the timing marks -- **CENTERED**. Check to be sure the observation is directly from above the viewport.

If a temporary timing pointer is used, place a temporary TDC mark -- using fast drying white ink or equivalent -- on the flywheel aligned with the timing pointer. The white mark is easier to see than dark ink pens when using a strobe light. Indicate this as being the TDC mark. A second mark shall now be made.

Maintain the position of TDC and then align the "zero" on the dial indicator bezel with the dial pointer. Refer to the Ignition Timing Chart in this section, on Page 7-14. Determine the correct timing specifications for the engine being serviced. Write down the specifications as a handy reference. As an example, a 1996 Hurricane Model has an ignition timing value of 0.1033", (2.55mm) or 20° BTDC.

IGNITION TIMING TABLE

MODEL	Ignition Timing @ 3,000rpm Degrees BTDC	Piston Location BTDC (Ignition Spark)	Piston Location BTDC Static Ref. Location
1992			
SL650 (Early Model)	24° ± 1.5	.139" (3.54mm)	--
SL650 (Late Model)	24° ± 1.5	.139" (3.54mm)	--
1993			
SL650	18° ± 1.5	.079" (2.01mm)	--
SL750	16° ± 1.5	.063" (1.59mm)	--
1994			
SL650	18° ± 2	.079" (2.01mm)	--
SL650/STD	18° ± 2	.079" (2.01mm)	--
SL750	24° ± 2	.139" (3.54mm)	--
SLT750	24° ± 2	.139" (3.54mm)	--
1995			
SL650	18° ± 2	.079" (2.01mm)	--
SL650/STD	18° ± 2	.079" (2.01mm)	--
SL750	24° ± 2	.139" (3.54mm)	--
SLT750	24° ± 2	.139" (3.54mm)	--
SLX780	28° ± 2	.188" (4.78mm)	--
1996			
SLT780 (EC78PWE02 Engine) (EC78PWE03 Engine (EC78PWE05 Engine)	 28° ± 2 26°@1500rpm 26°@1500rpm	 .188" (4.78mm) .139" (3.54mm) .139" (3.54mm)	--
SL780 (EC78PWE02 Engine) (EC78PWE03 Engine)	 28° ± 2 26°@1500rpm	 .188" (4.78mm) .188" (4.78mm)	--
SL700	18° ± 2	.0815" (2.07mm)	.271 - .286" (6.88 - 7.26mm)
SLT700	18° ± 2	.0815" (2.07mm)	.271 - .286" (6.88 - 7.26mm)
SLX780	28° ± 2	.188" (4.78mm)	--
Hurricane	20° ± 2	.1033" (2.55mm)	.271 - .286" (6.88 - 7.26mm)
SL900	18° ± 2	.0815" (2.07mm)	.240 - .260" (6.09 - 6.60mm)
SLTX	18° ± 2	.0815" (2.07mm)	.240 - .260" (6.09 - 6.60mm)
1997			
SL700	18° ± 2	.0815" (2.07mm)	.271 - .286" (6.88 - 7.26mm)
SLT700	18° ± 2	.0815" (2.07mm)	.271 - .286" (6.88 - 7.26mm)
SL780	24° ± 2	.1394" (3.54mm)	--
SLT780	24° ± 2	.1394" (3.54mm)	--
Hurricane	20° ± 2	.1003" (2.55mm)	.271 - .286" (6.88 - 7.26mm)
SL900	18° ± 2	.0815" (2.07mm)	.271 - .286" (6.88 - 7.26mm)
SLTX	18° ± 2	.0815" (2.07mm)	.271 - .286" (6.88 - 7.26mm)
SL1050	18° ± 2	.0815" (2.07mm)	.271 - .286" (6.88 - 7.26mm)

Now, rotate the crankshaft in the **NORMAL** direction until the pointer indicates 0.1033 BTDC -- 1996 Hurricane model. This is the advanced timing mark and should be labeled accordingly, through the timing hole or on the side of the flywheel. The mark **MUST** align with the timing pointer.

Verify the new timing marks by rotating the crankshaft/flywheel. The marks should align properly when the piston is at the correct timing position as shown in the table.

Verify the location of any existing timing marks before making any timing adjustments. If possible, a dynamic timing check should now be performed.

Static Timing
Polaris Engines with
Hall Effect Sensors

First, these words: Refer to the electrical schematic in the Appendix covering the engine being serviced. An outline indicates components, wiring, and color code housed in the top half and the bottom half of the electrical box.

Obtain the Polaris Static Timing Tool -- P/N 2871745. Be sure to use a fresh 9V battery. Open the electrical box and disconnect the stator Brown lead and connect it to the Brown tester lead.

Disconnect the Black stator lead and connect it to the Black tester lead.

Connect the tester White lead to the Red (forward -- No. 1 cylinder) stator lead.

Remove the high tension leads from the spark plugs, and then remove the spark plugs. Install a dial indicator into the No. 1 spark plug opening. Find Top Dead Center (TDC) as previously described in this section.

Slowly rotate the crankshaft counterclockwise -- viewing from the front -- while observing the tester light. Continue slowly rotating the crankshaft as the light illuminates then turns off. Rotate the crankshaft in the **SAME** direction an approximate 1/8 turn, and then **REVERSE** direction -- clockwise -- until the light again illuminates. Stop immediately and take note of the dial indicator reading. Refer to the Ignition Timing Table listed on Page 7-14 for timing specifications.

The static timing is correct if the dial indicator reading is within the tolerances listed for engines equipped with Hall Effect sensors.

If the static timing reading is not within the listed tolerance, the ignition timing must be adjusted. Refer to the following timing adjustment procedures for Polaris manufactured engines.

Dynamic Timing Check

Ignition components are temperature sensitive. Therefore, it is highly recommended the ignition timing should be checked at room temperature -- 68° F (20° C). If not tested at this temperature, the timing may be different than if the timing were set at room temperature.

Remove the dial indicator and install the spark plugs. Tighten the plugs to a torque value of 18 ft. lbs. (24Nm).

Attach a timing strobe light to the lead of the No. 1 -- forward -- spark plug. Connect a tachometer as per the manufacturers instructions.

Connect a garden hose to the engine cooling water supply fitting, or flush fitting connected to the water outlet manifold located atop the cylinder heads.

Start the engine and allow the rpm's to stabilize at idle speed **FOR JUST A FEW SECONDS,** and then turn the water on. Even with the flush attachment supplying water, never operate the engine at high rpm with the craft out of the water.

Increase the engine to 3,000 rpm for all engines -- except 780 Series Fuji 03 and 05 engines. These engines should be checked at 1,500 rpm with a timing specification of 26.5°.

Aim the timing light into the timing viewport or on the temporary pointer. If using the flywheel cover viewport, be sure to look straight down into the hole for the most accurate reading. The pointer should align with the timing mark.

When the engine is to be shut down, turn the water off **FIRST** -- raise the aft portion of the hull -- **WHILE THE ENGINE IS OPERATING AT IDLE** -- "rev" the engine just a **COUPLE** of times to clear water from the exhaust system -- and then shut it down. **NEVER** allow the engine to operate without cooling water for more than 15 seconds.

If the mark aligned with the pointer, disconnect the timing light and tachometer. Remove the hose from the water outlet fitting. Be sure to install the viewport plug, if so equipped.

Timing Adjustment
Fuji Engines

If the timing mark does not align with the pointer, observe the position of the flywheel mark in relation to the pointer. If the mark is counterclockwise from the pointer -- when viewed from the front -- the ignition timing is too advanced. If the mark is clockwise from the pointer, the timing is retarded.

The flywheel must be removed before proceeding. Refer to Chapter 8, for special tools necessary to complete the task and detailed procedures to remove the cover.

Once the flywheel is removed, observe the timing marks on the circumference of the stator plate. If the lower mark is aligned with the crankcase parting line, the ignition timing will be more retarded -- closer to TDC -- than if the upper mark is aligned with the parting line.

LOOSEN but do not remove the stator plate attaching screws. This may take a little effort, if the screws were coated with Loctite during installation. Turn the stator plate counterclockwise to advance timing, turn the stator plate clockwise to retard the ignition timing. Tighten the attaching screws. Do not apply Loctite to the screws at this time. Install the flywheel and tighten the 16mm retaining nut to a torque value of 55 ft. lbs. (75Nm). Tighten the 18mm nut to a torque value of 65 ft. lbs. (88Nm).

Be sure to allow sufficient time for the engine to cool before dynamically checking the timing again.

Start the engine and verify the timing as previously described. If the timing is not correct, repeat the steps to adjust the timing by rotating the stator plate. Now check the ignition timing dynamically -- once the engine has cooled sufficiently.

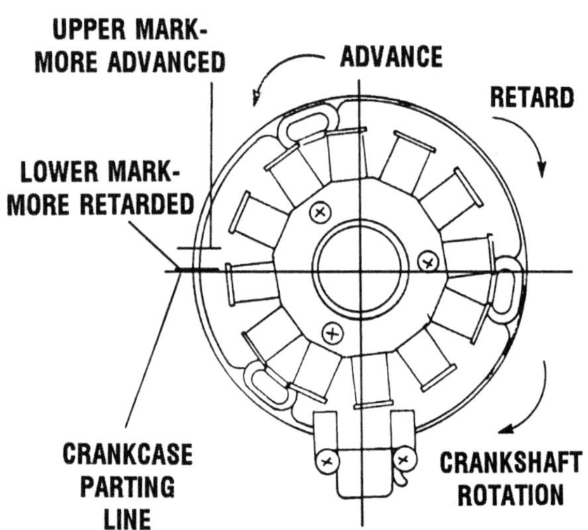

The stator plate is rotated to advance or retard the ignition timing, as shown in this illustration. Be sure to apply Loctite 242 to the attaching screw threads after final adjustment of the plate, to prevent inadvertent rotation due to engine vibration.

Once the timing is correct, the flywheel must be removed. Remove a single stator plate attaching screw and apply a coating of Loctite 242 to the threads. Install the screw and tighten it securely. Repeat the process for each additional screw. In this fashion, the stator plate should **NOT** move, maintaining the proper ignition timing position.

Install the Woodruff key, flywheel, and nut, tightening the retaining nut to the specified torque value.

Remove the timing light and tachometer. Complete all tasks necessary to return the craft to operational readiness.

Timing Adjustment
Polaris

Observe the difference -- distance -- between the timing mark and the pointer.

Remove the oil pump from the flywheel cover. If the oil hoses are not disconnected, the system should not require purging before operating the engine. Remove the attaching bolts securing the flywheel cover to the crankcase. Identify and mark any bolt securing a clamp for electrical harnesses or hoses. Pull the flywheel cover from the crankcase.

Loosen the magnet holder screws. Timing ignition adjustment is accomplished by moving the magnet holder. Rotating the holder **CLOCKWISE** -- when viewed from the front -- will advance the timing and a **COUNTER-CLOCKWISE** movement will retard timing. Rotate the magnet holder in the **OPPOSITE** direction approximately the same amount as the mark must move. Tighten the attaching screws securely.

Perform the dynamic ignition timing check as outlined above. Once the timing is correct, install the flywheel cover and tighten the 6mm attaching bolts to a torque value of 108 in. lbs (12Nm); 8mm bolts to a torque value of 22 ft lbs. (30Nm). Install the oil pump and tighten the attaching bolts to a torque value of 60 in. lbs. (7Nm).

7-6 CHARGING CIRCUIT

The charging coils are installed on the magneto assembly. Therefore, the charging circuit section has been included in this chapter -- Ignition.

Description and Operation

The charging circuit consists of the following items:

Permanent magnets -- attached to the inside perimeter of the flywheel. On some models, these magnets are further protected by a "magnet holder".

Charging coils -- installed on the magneto assembly.

A rectifier/regulator -- located inside the electrical box -- see accompanying illustration.

An external battery

Necessary wiring -- connecting components together.

The negative terminal on the battery must be connected to a good ground on the engine.

As the flywheel magnets rotate, voltage is induced in the charging coil mounted on the magneto assembly under the flywheel. The alternating current passes through a series of diodes inside the rectifier/regulator and is transformed to DC current.

The rectifier/regulator allows a maximum of 14 volts to pass on to the battery. Any excess voltage is passed to an SCR gate and back through the charging coil. In this manner, overcharging the battery is avoided. The life of components in the charging circuit is greatly extended due to the absence of power surges.

The rectifier/regulator is actually a sealed "black box" type solid state unit and cannot be repaired nor can any adjustments be made to the unit. Most Polaris watercraft models use a regulator unit designated as LR-21 or LR-23. If the regulator is found to be defective, the unit must be replaced.

The alternating current generated in the charging coil windings passes to the rectifier/regulator. The rectifier portion changes the alternating current (AC) to direct current (DC) to charge the 12-volt battery. A second function of the rectifier is to smooth the pulses of current to form a steady DC current flow.

The regulator portion of the rectifier/regulator limits the current produced and sends any excess current back through the charging coil windings.

Electronic components in the charging circuit, including the battery, will last much longer if subjected to a low steady voltage. A sudden voltage surge could overheat and damage charging circuit components.

On all models, the magneto assembly is located on the stator plate, behind the flywheel. Therefore, to service charging components, it is necessary to "pull" the flywheel. However, because these components are so well protected behind the flywheel, problems seldom occur in this area of the charging circuit. When problems do occur, they can usually be traced to the rectifier/regulator or to the battery. Again, if either the charging coil/s or the rectifier/regulator fails troubleshooting tests, the defective unit cannot be repaired, it must be replaced.

Charging System Testing (Regulated Voltage)

Obtain a digital multitester and turn the selector to volts DC. Connect the tester leads across the battery terminals.

CAUTION

Do **NOT** operate the engine for more than 15 seconds unless the engine is fitted with a flush fitting and a hose is attached. If the craft is resting in water, it should be secured, in a regular boat "slip". The tester, obviously should be well protected -- grounded -- while performing the check.

Start and operate the engine briefly at 3,000rpm. Observe the measured voltage, which should be approximately 14.5 VDC. A higher reading may indicate a regulator fault or a poor ground at the regulator heat sink.

A lower reading may indicate an excessive system load, an alternator fault, or a regulator problem.

ENGINE TORQUE VALUES

ENGINE FASTENERS	FUJI Engine	POLARIS Engine	FUJI Engine	POLARIS Engine
Air Intake Cover Center Bolt (8mm) Air Intake Cover Screws	REFER TO CHAPTER 8 - ENGINE OR CHAPTER 6 - FUEL			
Crankcase Bolts (8mm)	16 ft lb	22 ft lb	22 Nm	30 Nm
Crankcase Bolts (10mm)	26 ft lb	28 ft lb	35 Nm	38 Nm
Cylinder Head Nuts (8mm)	18 ft lb	-	24 Nm	-
Cylinder Head Bolts (8mm)	-	20 ft lb	-	27 Nm
Cylinder Head Cover Bolts (8mm)	-	22 ft lb	-	30 Nm
Cylinder Base Nuts (10mm)	28 ft lb	40 ft lb	38 NM	54 Nm
Carburetor Mounting Nuts (Mikuni) (8mm)	16 ft lb	-	22 Nm	-
Carburetor Mounting Bolts (Keihin) (6mm)	-	108 in lb	-	12 Nm
Engine Mount Plate (To Engine) Bolts	45 ft lb	50 ft lb	61 Nm	68 Nm
Engine Mount Nuts	45 ft lb	45 ft lb	61 Nm	61 Nm
Engine Mount to Hull (5/16"-18)	14 ft lb	14 ft lb	19 Nm	19 Nm
Exhaust Manifold Bolts (8mm)	16 ft lb	22 ft lb	22 Nm	30 Nm
Exhaust Manifold to Pipe Bolts (12mm)	45ft lb	45 ft lb	61 Nm	61 Nm
Flywheel Nut 14mm	-	90 ft lb	-	122 Nm
Flywheel Nut 16mm 18mm	55 ft lb 65 ft lb	-	75 Nm 88Nm	-
Flywheel Housing Bolt (6mm)	78 in lb	108 in lb	9 Nm	12 Nm
Flywheel Housing Bolt (8mm)	16 ft lb	22 ft lb	22 Nm	30 Nm
Intake Manifold Nuts/Bolts (6mm)	78 in lb	108 in lb	9 Nm	12 Nm
Oil Pump Bolts/Screws (5mm)	48 in lb	60 in lb	5 Nm	7 Nm
Starter Motor Mounting Bolts	108 in lb	108 in lb	12 Nm	12 Nm
Spark Plug	18 ft lb	18 ft lb	24 Nm	24 Nm
Water Outlet Manifold Bolts (6mm)	78 in lb	108 in lb	9 Nm	12 Nm

8
ENGINE

8-1 INTRODUCTION AND CHAPTER ORGANIZATION

The following Polaris Series are covered in this chapter from 1992 through 1997.

SL650 Series	1992 - 1995
SL750 Series	1993 - 1995
SLT750 Series	1994 - 1995
SLX780 Series	1995 Only
SL700 Series	1996 - 1997
SLT 700 Series	1996 - 1997
Hurricane Series	1996 - 1997
SL780 Series	1996 - 1997
SLT780 Series	1996 - 1997
SL900 Series	1996 - 1997
SLTX Series	1996 - 1997
SL700 Deluxe	1997
SL1050 Series	1997

The working sections of this chapter are divided into five main sections as follows:

8-2 Description and Operation -- Two-Cycle Engine
8-3 Complete Service -- Three-Cylinder Engine
8-4 Complete Service -- Two-Cylinder Engine
8-5 Cleaning and Inspecting -- All Engines
8-6 Sealants, Adhesives, Lubricants, and Fuel Stabilizers

Repair Procedures

Service and repair procedures will vary slightly between individual models, but the basic instructions for all 2-cylinder engine are almost identical. The same is true for the 3-cylinder engines. Any variations between models are clearely indicated. Special tools may be called out in certain instances. These tools should be available for purchase from the local dealer. In certain instances, an alternate tool may be obtained or fabricated.

Torque Values

A torque wrench is essential to correctly assemble the engine. **NEVER** attempt to assemble an engine without a torque wrench. All torque values must be met when they are specified. Torque values for various parts of each engine are given in the text.

Attaching bolts or nuts **MUST** be tightened to the required torque value in two progressive sequences, following the specified tightening order. On the first sequence, tighten to 1/2 the torque value. On the second sequence, tighten to the full torque value.

Cleaning

All parts **MUST** be cleaned and thoroughly inspected before they are assembled, installed, or adjusted. Use proper lubricants, or their equivalent, whenever they are recommended.

If a part or assembly is unfit for further service, be sure the proper replacement has been obtained and is on hand, **BEFORE**, the assemblying work begins.

Other Engine Components

Service procedures for the carburetor, fuel pumps, cranking motor, electrical, ignition and other components are given in separate Chapters, as listed in the Table of Contents.

Reed Valve Service

The reeds on all Polaris two-stroke engines are contained in an externally mounted reed block. Therefore, the engine need not be disassembled in order to replace a broken reed.

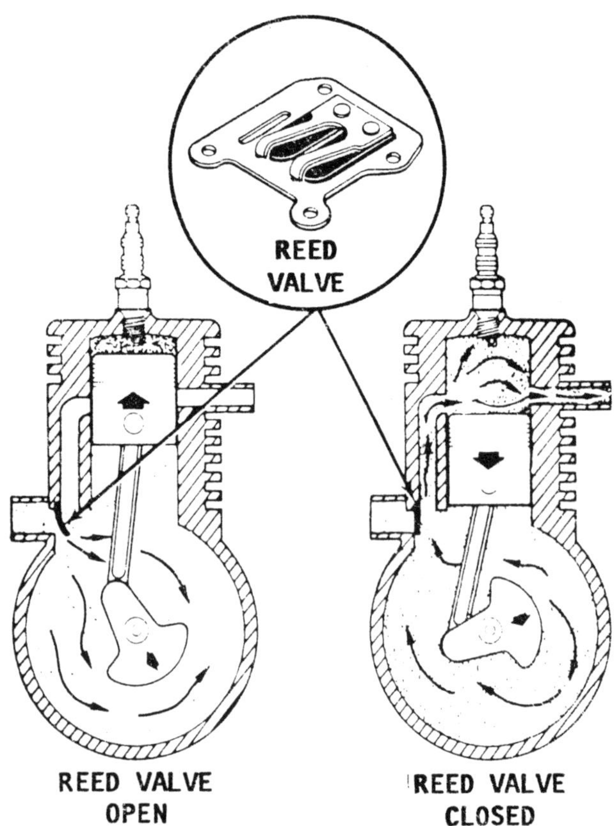

Reed valves are used to control the flow of air/fuel into the crankcase and eventually into the cylinder. As the piston moves upward in the cylinder, the resulting suction in the crankcase overcomes the spring tension of the reed. The reed is pulled free from its seat and the air/fuel mixture is drawn into the crankcase.

8-2 TWO-CYLCE ENGINE DESCRIPTION AND OPERATION

Intake/Exhaust

Two-cycle engines utilize an arrangement of port openings to admit fuel to the combustion chamber and to purge the exhaust gases after burning has been completed. The ports are located in a precise pattern in order for them to be opened and closed at an exact moment by the piston as it moves up and down in the cylinder. The exhaust port is located slightly higher than the fuel intake port. This arrangement opens the exhaust port first as the piston starts downward and therefore, the exhaust phase begins a fraction of a second before the intake phase.

Lubrication

All engines in this manual are equipped with an oil injection system. On the Fuji engines, an oil pump is located low on the starboard side of the block. The oil pump on the Polaris engine

Location of the oil pump on the Fuji engines covered in this manual. This pump is driven by a short shaft with a gear on the end which indexes with a "worm" type gear on the crankshaft. The oil pump on the Polaris engine is mounted on the flywheel cover and is driven by a short shaft indexed into a slot on the flywheel nut.

is mounted on the flywheel cover. See Chapter 6 for detailed oil pump coverage.

Physical Laws

The two-cycle engine is able to function because of two very simple physical laws.

One: Gases will flow from an area of high pressure to an area of lower pressure. A tire blowout is an example of this principle. The high-pressure air escapes rapidly if the tube is punctured.

Two: If a gas is compressed into a smaller area, the pressure increases, and if a gas expands into a larger area, the pressure is decreased.

If these two laws are kept in mind, the operation of the two-cycle engine will be more easily understood.

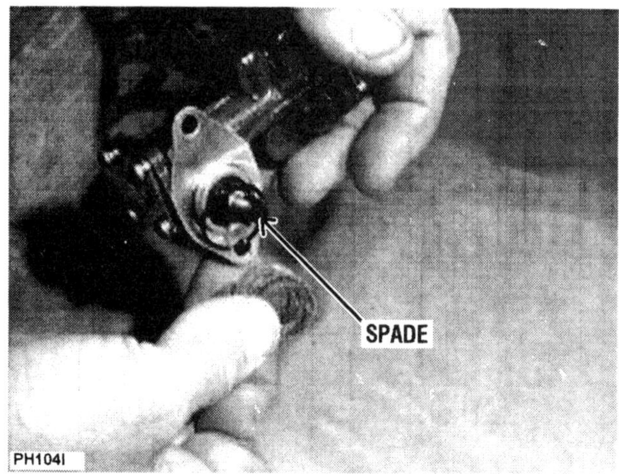

Close view of the oil pump which mounts on the flywheel cover of the Polaris engine and is driven by a short shaft with a "spade" on the end. This "spade" indexes into a slot on the flywheel nut.

ACTUAL OPERATION TWO-CYCLE ENGINE

Beginning with the piston approaching top dead center on the compression stroke: The intake and exhaust ports are closed by the piston; the reed valve is open; the spark plug fires; the compressed air/fuel mixture is ignited; and the power stroke begins. The reed valve was open because as the piston moved upward, the crankcase volume increased, which reduced the crankcase pressure to less than the outside atmosphere.

As the piston moves downward on the power stroke, the combustion chamber is filled with burning gases. As the exhaust port is uncovered, the gases, which are under great pressure, escape rapidly through the exhaust ports. The piston continues its downward movement. Pressure within the crankcase increases, closing the reed valves against their seats. The crankcase then becomes a sealed chamber. The air/fuel mixture is compressed ready for delivery to the combustion chamber. As the piston continues to move downward, the intake port is uncovered. A fresh air/fuel mixture rushes through the intake port into the combustion chamber striking the top of the piston where it is deflected along the cylinder wall. The reed valve remains closed until the piston moves upward again.

When the piston begins to move upward on the compression stroke, the reed valve opens because the crankcase volume has been increased, reducing crankcase pressure to less than the outside atmosphere. The intake and exhaust ports are closed and the fresh fuel charge is compressed inside the combustion chamber.

Pressure in the crankcase decreases as the piston moves upward and a fresh charge of air flows through the carburetor picking up fuel. As the piston approaches top dead center, the spark plug ignites the air/fuel mixture, the power stroke begins and one full cycle has been completed.

TIMING -- TWO-STROKE ENGINES

The exact time of spark plug firing depends on engine speed. At low speed the spark is retarded, fires later than when the piston is at or beyond top dead center. Engine timing is built into the unit at the factory. At high speed, the spark is advanced, fires earlier than when the piston is at top dead center.

Summary

More than one phase of the cycle occurs simultaneously during operation of a two-cycle engine. On the downward stroke, power occurs above the piston while the ports are closed. When the ports open, exhaust begins and intake follows. Below the piston, fresh air/fuel mixture is compressed in the crankcase.

On the upward stroke, exhaust and intake continue as long as the ports are open. Compression begins when the ports are closed and continues until the spark plug ignites the air/fuel mixture. Below the piston, a fresh air/fuel mixture is drawn into the crankcase ready to be compressed during the next cycle.

8-3 COMPLETE SERVICE THREE-CYLINDER ENGINE

ADVICE

Before commencing any work on the engine an understanding of two-cycle engine operation will be most helpful. Therefore, it would be well worth the time to study the principles of two-cycle engines, as outlined briefly in the previous section -- 8-2. A Polaroid, or equivalent instant-type camera is an extremely useful item, providing the means of accurately recording the arrangement of parts and wire connections **BEFORE** the disassembly work begins. Such a record is invaluable during assembling.

Under normal conditions, timing adjustments are not required on the engines covered in this manual. However, if an adjustment is necessary, for any reason, the flywheel must be "pulled" to gain access to the magneto assembly and stator plate, as shown on this Model 750 engine.

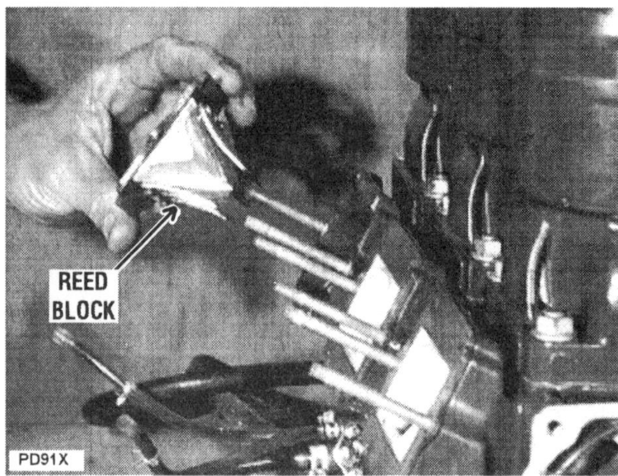

The reed block on most engines can be serviced while the engine remains in place in the engine compartment. The reed blocks must be handled ***CAREFULLY*** *at all times to prevent even the slightest damage. Injury to the block is usually caused while it is out of the engine. Protect it at all times from dust or other contaminants which seem to be attracted to the reeds. The Cleaning and Inspecting section covers service and adjustment of reeds.*

IMPORTANT WORDS

The following series of illustration were developed while servicing a SLT750 Series engine. Other engines differ slightly in appearance, but have the same basic components. Where differences occur, such as bolt tightening patterns or torque values, the differences will be clearly identified.

REED BLOCK WORDS

If the only work to be performed on the engine is servicing the reed block assembly, remove the flame arrestor, carburetors, and perform Disassembling Steps **41 - 45**, beginning on Page 8-12.

Clearance between the reed block and the reed valve is 0.15" (.38mm) maximum.

After the reed block and intake manifold assemblies are installed, install the carburetor and the flame arrestor.

PRELIMINARY TASKS

Engine Overhaul

Move the craft to a suitable work area protected from the elements, if possible. Release the "bungee" type restraining strap securing the electrical box on top of the battery, and move the box aside. Remove the battery cover and disconnect the negative and positive leads. Disconnect the throttle and choke cables from the carburetor. Disconnect the fuel hoses at the fuel tanks.

NOW, THESE WORDS

As mentioned earlier, the following procedures and supporting illustrations were developed while servicing an SLT750 engine. Since some differences may occur between engine models, any differences between other Series engines will be clearly identified.

THREE-CYLINDER ENGINE REMOVAL

Take time to identify and tag the various hoses and electrical leads which are disconnected during the disassembly process. Such a practice will benefit the installation procedures.

1- The engine Serial Number is located atop the water outlet manifold. A warning label may partialy cover the engine ID number. This number **MUST** be available when ordering replacement parts and special tools required for servicing the engine.

2- Remove the bolts securing the flame arrestor cover to the case. Lift the cover and screen clear from the case.

3- Remove the bolts -- three per carburetor -- securing the flame arrestor base plate to the

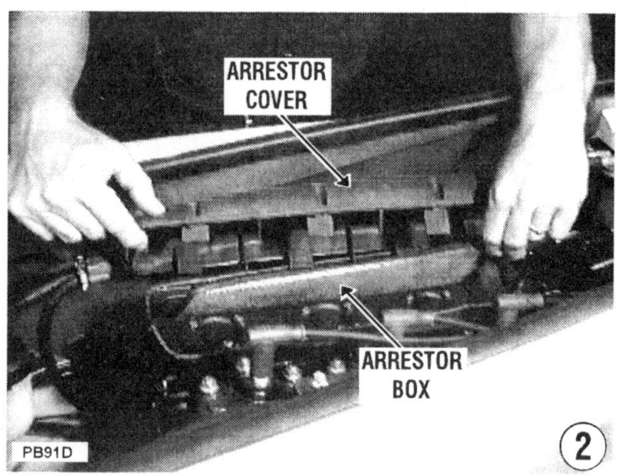

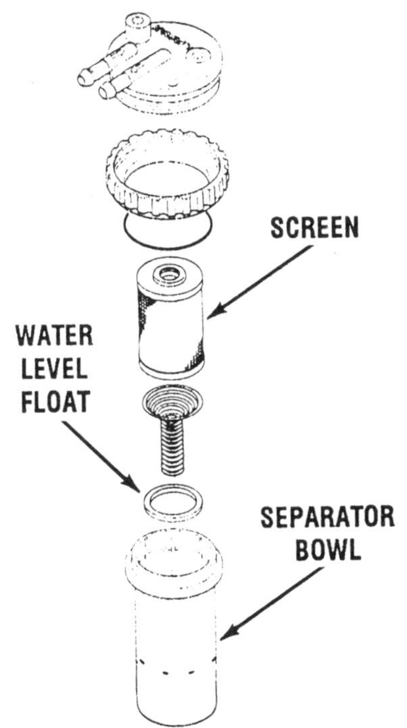

Exploded drawing of the water/fuel separator used since 1995, with internal parts clearly illustrated. The screen/filter assembly improves the separation of water, making visual contamination detection much easier.

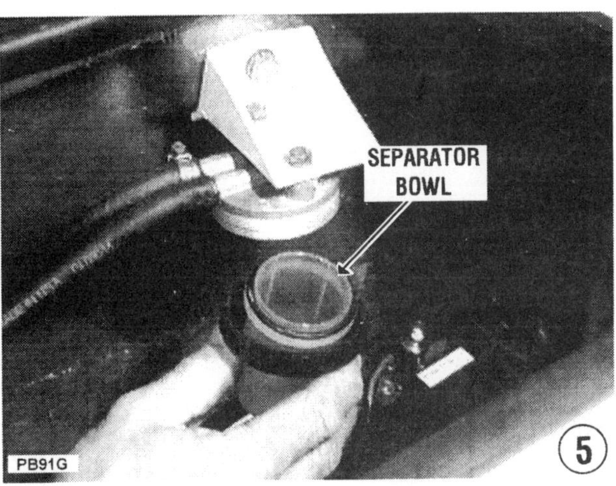

carburetor/s. Lift the base plate carefully. On 1992 to mid 1995 models, the fuel pump is attached to the aft end of this plate.

4- All fuel hoses must be "pinched off", to avoid spilling fuel in the engine compartment. With a set of appropriate clamps, pinch off the fuel filter if it must be replaced. If clamps are not available, use pens or golf tees to plug the ends of hoses.

SPECIAL WORDS

Since 1995, the manufacturer incorporated a screen/filter assembly in the water/fuel separator bowl, as indicated in the accompanying illustration. This screen improves the separation of water, making visual contamination detection easier.

5- Wrap a shop towel around the water/fuel separator cup to prevent spilling any fluids, then loosen the ring nut. Remove the separator bowl from the mounting bracket. In the accompanying illustration, a towel was not used for clarity.

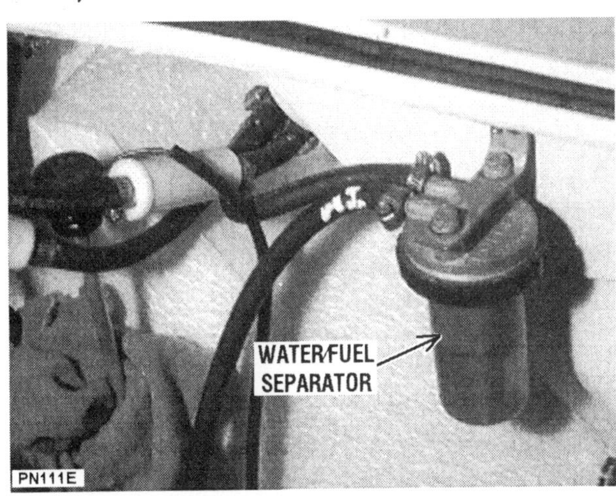

Typical installation of the water/fuel separator illustrated in the adjacent column. The text explains the advantages of the improved separator.

6- On models equipped with the Mikuni Super BN carburetor, a remote fuel pump supplies fuel to each carburetor. Back off the hose clamps on each fuel supply hose, to allow disconnection at the delivery end. Disconnect the hose from the water separator bowl to the fuel pump. If servicing a 1992-mid 1995 unit, lift the flame arrestor base plate, with the fuel pump still attached, up and free of the engine compartment.

7- Disconnect the high tension leads from the spark plugs. **ALWAYS** use a pulling and twisting motion as a precaution against damaging the connection.

HELPFUL WORDS

For simplicity and efficiency of work, the electrical box does not have to be removed to lift the engine out of the watercraft. However, certain electrical leads and harnesses must be disconnected. Therefore, to permit the leads and harnesses to be disconnected, the box must be opened to determine which leads must be disconnected.

8- Remove the restraining straps securing the electrical box and battery to the watercraft.

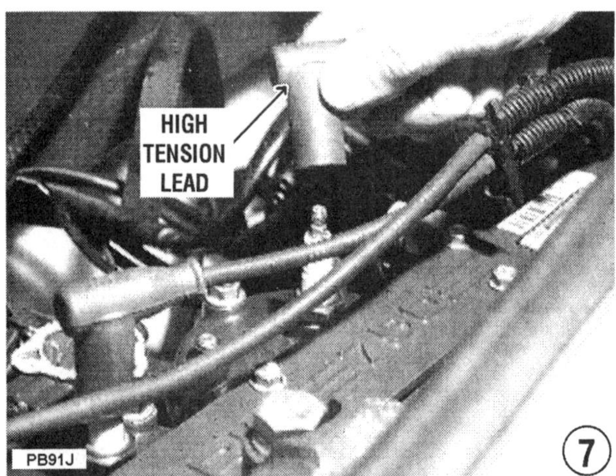

Remove the allen screws securing the electrical box halves together. Separate the box. Take a few moments identifying, tagging and removing the appropriate leads to facilitate moving the box aside. Follow the necessary leads from the engine, to the box. Many of these leads enter the box through a port opening. Trace the leads through the box, and then disconnect them and push the leads through the opening, after removing the grommet.

9- Locate and disconnect the lower exhaust temperature sensor lead in the electrical box.

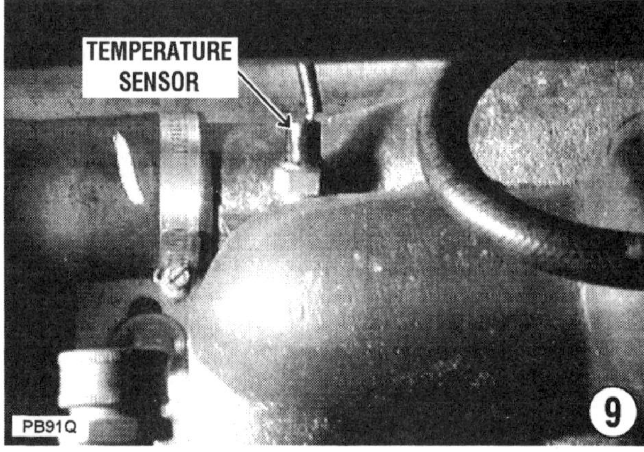

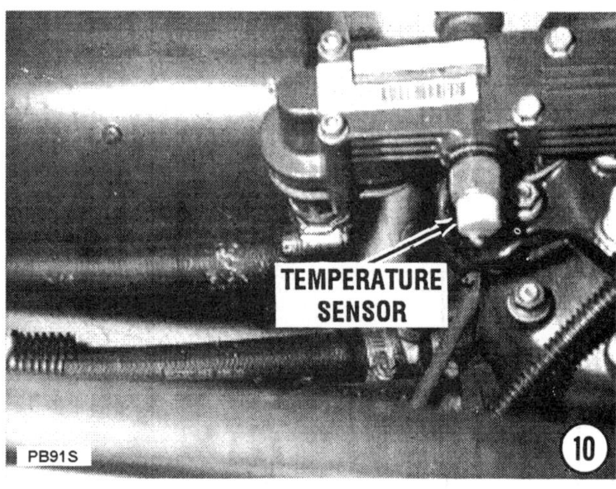

10- Trace the temperature sensor lead from the water outlet manifold, to the electrical box. Disconnect the lead from the electrical box, and then wrap the lead around the engine to keep it out of the way.

Temporarily close the electrical box and secure the two halves together with the attaching screws to keep the box clean and prevent the screws from being misplaced. Set the box aft in a convenient location.

11- Lift the battery cover -- the base for the electrical box -- off the battery. Disconnect the battery leads.

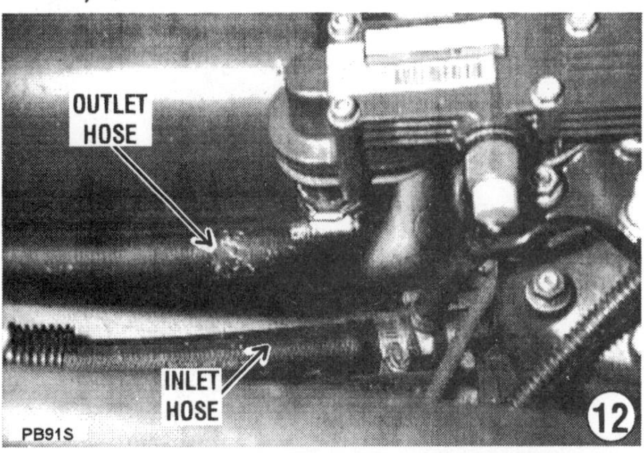

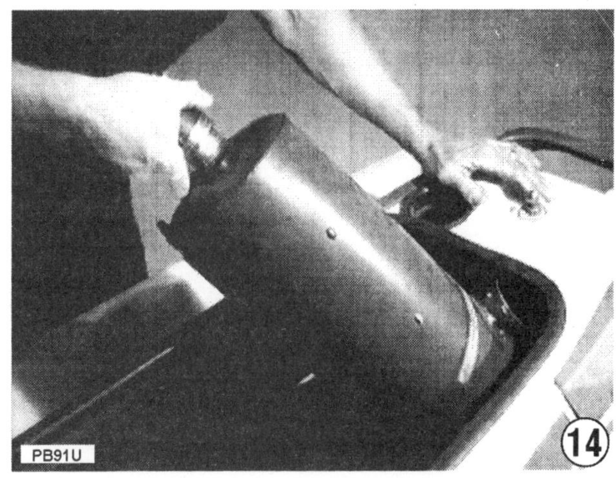

12- Loosen the hose clamps securing the cooling inlet and outlet hoses, and then disconnect the hoses. The lower one is the inlet hose.

13- Loosen the hose clamps securing the exhaust hose to the water box muffler.

14- Work the water box muffler free of the hose fitting and lift it clear of the engine compartment.

15- Remove the two nuts securing the expansion chamber to the hull. Remove the bracket.

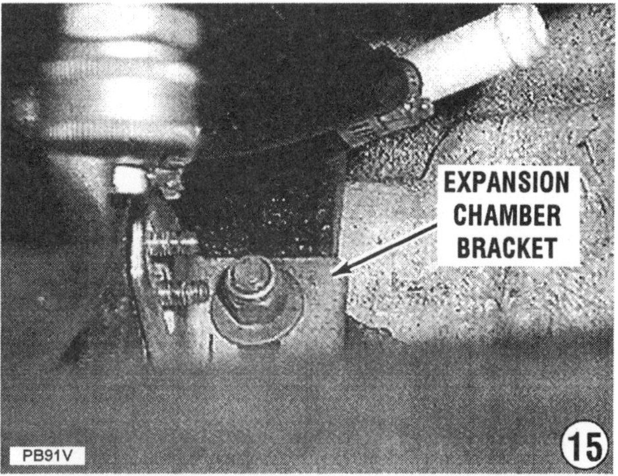

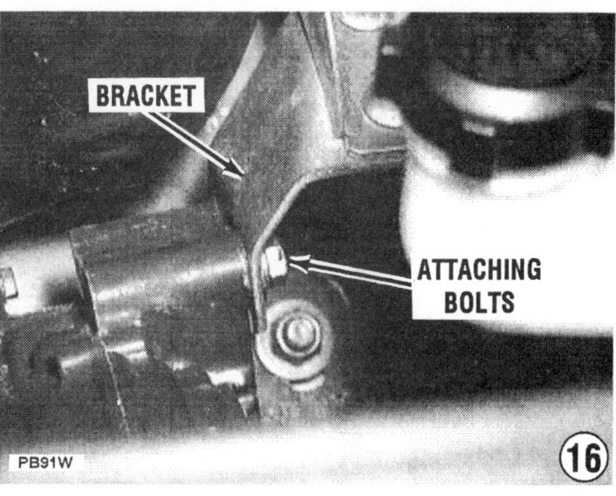

16- Remove the two forward bolts securing the expansion chamber to the flywheel cover, and then remove the brackets.

17- Remove the four attaching bolts securing the expansion chamber to the exhaust manifold.

18- Remove the restraining straps from the fuel and oil tanks. The fuel tank must be moved forward to allow the expansion chamber to be moved forward, rotated and lifted clear of the engine compartment later.

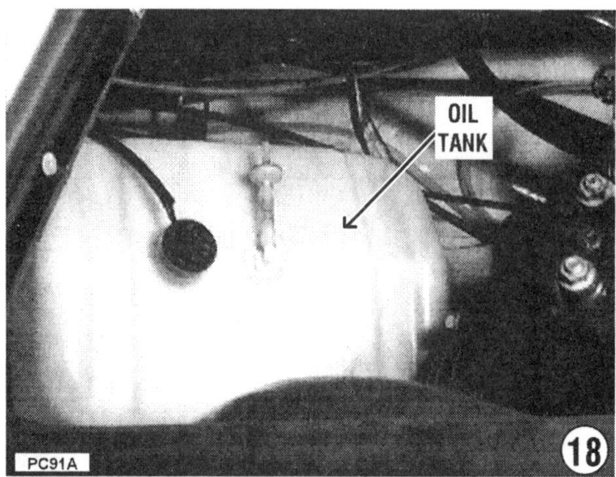

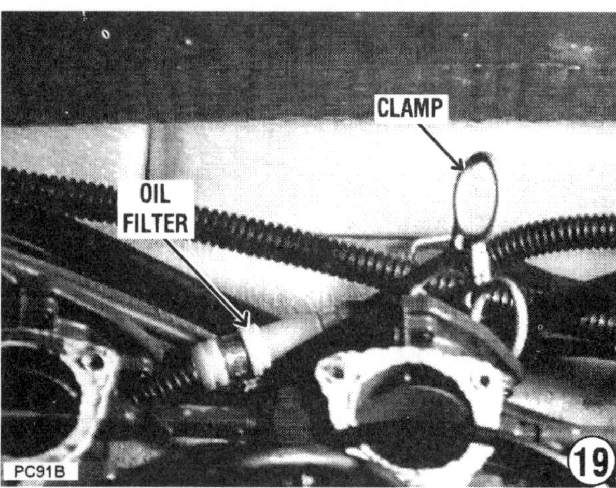

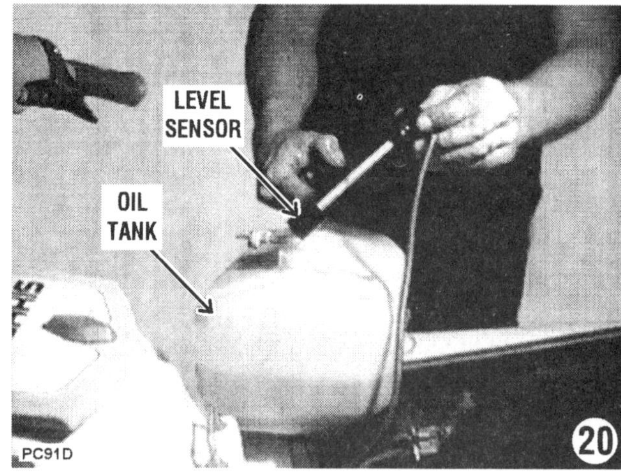

19- "Pinch off" the oil hose at the aft end of the in-line oil filter, and then disconnect the oil filter from the "pinched" hose.

20- Lift the oil tank clear of the engine compartment and set it aside. The oil sensor can be checked at this time.

21- Move the expansion chamber forward, rotate it, and work it up and free of the engine compartment.

22- Loosen the hose clamp securing the coupler cover to the driveshaft coupler, and then remove the cover.

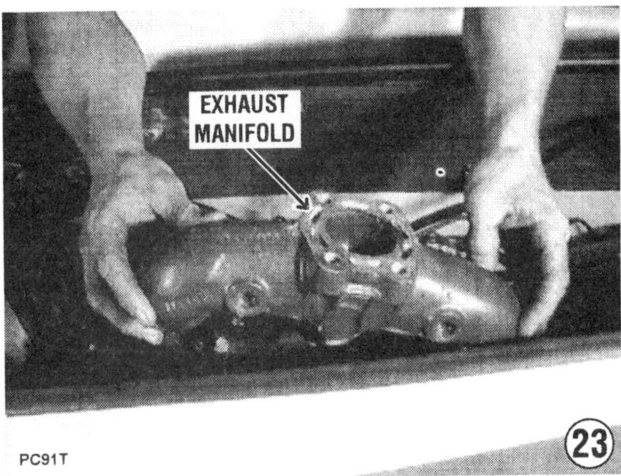

23- Remove the nine bolts securing the exhaust manifold to the engine. Pull the manifold free from the engine, up and out of the engine compartment.

24- Remove the bolt securing the power lead to the cranking motor.

25- Disconnect the throttle and choke cables from the linkage just forward of the No. 1 carburetor.

26- Loosen and remove the four engine mounting plate nuts -- two forward and two aft.

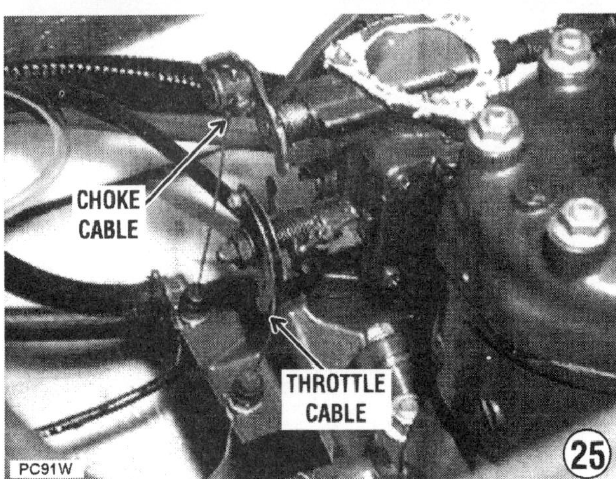

ADVICE

Make one final inspection to ensure all hoses and electrical leads are either out of the way, or wrapped around the engine, in preparation for lifting the engine.

27- Obtain the services of an assistant or a suitable lifting device. If an assistant or lift is not available, the engine may be lifted out by one person, in good shape, if done carefully. Place a board -- just large enough to cover the

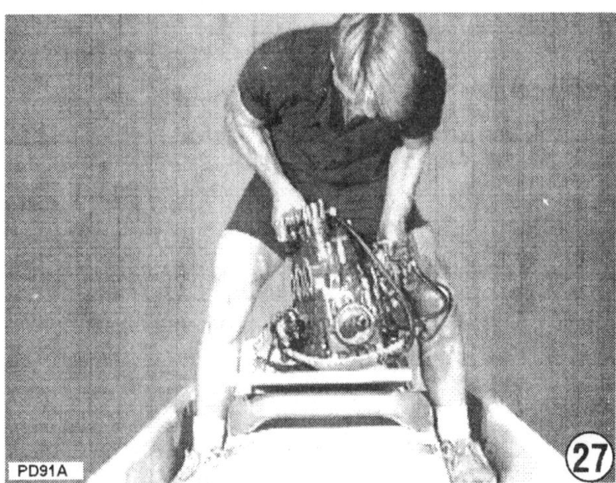

aft portion of the watercraft, as shown. Straddle the engine compartment and move the engine slightly forward, then lift up. Use the legs, not the back! Lower the engine onto the board.

Use care to avoid damaging the check valves on the oil supply lines.

Now, transfer the engine to a suitable work surface for disassembly.

DISASSEMBLING THREE-CYLINDER ENGINE

Flywheel Removal

Take note of any screws or bolts which may have an attached cable or hose clamp. These should be marked with "White-Out", to ensure the proper relocation of such clamps during assembly.

If the engine being serviced has the oil pump secured to the flywheel cover, disconnect and plug the oil lines. Disconnect the oil pump cable, and then remove the oil pump through the two attaching bolts.

28- Remove the bolts securing the flywheel cover to the crankcase. Pull the flywheel cover off the crankcase and magneto cover.

29- Obtain special flywheel holder tool P/N 8700229. If servicing a Fuji engine, an adapter for the special holding tool is also required.

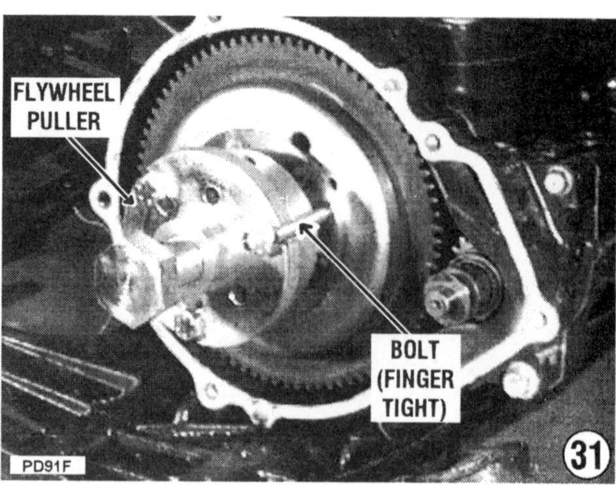

With the pins on the holding tool indexed in the holes in the flywheel, loosen and remove the flywheel nut.

30- Obtain the services of an assistant. Have the assistant use the special tool to prevent the flywheel from rotating and remove the coupler at the aft end of the crankshaft. A crescent type wrench on the coupler flats will do the job nicely.

ALTERNATE METHOD

Remove the No. 3 (aft) spark plug. Estimate the piston position using a screwdriver or similar probe through the spark plug opening. Lower the probe straight down, taking care not to scratch the crown of the piston. Lift the probe in the same fashion.

Rotate the flywheel nut **CLOCKWISE** (when viewed from the front) until the piston is at TDC.

Now, from the TDC position, rotate the flywheel nut approximately 3/4 turn **CLOCKWISE**, which will lower the piston about an inch (25.4 cm) Insert a length of -- say 5/16" rope -- through the spark plug opening to fill most of the space between the piston crown and the head. Once the space is filled with rope, rotate the flywheel nut **COUNTERCLOCKWISE** until the piston crown is "tight" up against the rope. With the crankshaft locked in this position, Loosen and remove the flywheel nut, because the piston against the rope and head will prevent the crankshaft from rotating.

While the rope is still in place on top of the piston to prevent the crankshaft from rotating, remove the coupler at the aft end of the crankshaft. This may be accomplished using a crescent type wrench on the flats provided on the coupler. After the coupler is removed, pull the rope free of the cylinder.

31- Obtain special flywheel puller tool, P/N 8700229. It would be a very difficult, an almost impossible task to remove the flywheel without damaging it if the special tool is not used. The cost of the tool certainly offsets the price of a new flywheel!

CRITICAL WORDS

The bolts supplied with the special tool must be started and tightened **ONLY** fingertight! If over tightened, these bolts will make contact with the stator plate, causing damage to the stator.

32- Obtain two large crescent wrenches -- the first to hold the pulling tool, the other to tighten the center bolt of the tool. Using a ballpeen hammer, lightly tap the end of the center bolt a few times to help jar it loose. Rotate the center bolt **CLOCKWISE** to back the flywheel off the crankshaft.

33- Once the flywheel is loose, pull it free of the crankshaft.

SPECIAL WORDS

The Fuji engines have an inner magneto cover and an outer flywheel cover. These two covers are incorporated into one cover on the Polaris engines. Therefore, if servicing a Polaris engine, disregard the references to an "inner" cover.

34- Remove the seven bolts securing the magneto cover to the crankcase. Six 8mm bolts and one 6mm bolt secure the cover to the engine. An inspection hole is located at the top of the inner cover and is used for ignition timing procedures.

As mentioned in the "Special Words" above, the Fuji engines have an inner magneto cover, attached to the crankcase. This inner cover must be removed if the stator plate is to be removed. During installation, the stator plate is installed first.

Reduction Gear Removal

35- Pull the cranking motor reduction gear free of the crankcase. Inspection and possible service will be outlined later in this section.

Stator Plate Removal

36- Observe the mark embossed on the stator and the line casting on the crankcase edge used for alignment purposes.

37- Loosen and remove the three bolts securing the stator plate to the crankcase.

38- Remove the two screws below the cranking motor reduction shaft securing the electrical harness feed-thru plate to the crankcase.

39- Carefully pull the stator plate free of the crankcase, because the electrical harness is still attached to the stator plate.

Woodruff Key Removal

40- Using a hammer and a rounded-end punch, **LIGHTLY** tap the Woodruff key to jar it loose. Remove the Woodruff key from the

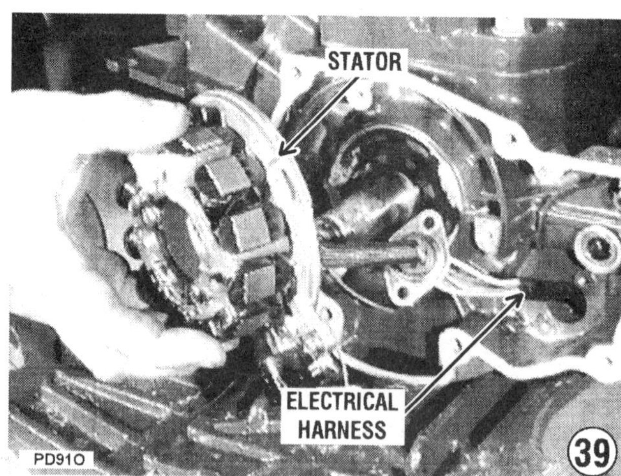

slot in the crankshaft and inspect it for wear. Set the Woodruff key in a safe place or re-install it in the crankshaft after inspection, as a precaution against the key being misplaced.

Carburetor Rack Removal

41- Loosen and remove the bolts securing the carburetor rack to the intake manifold/s -- two bolts per carburetor. Early model engines with Mikuni carburetors may have an individual intake manifold for each cylinder. The carburetor rack assembly may be removed intact. It is not necessary to remove each carburetor at this time.

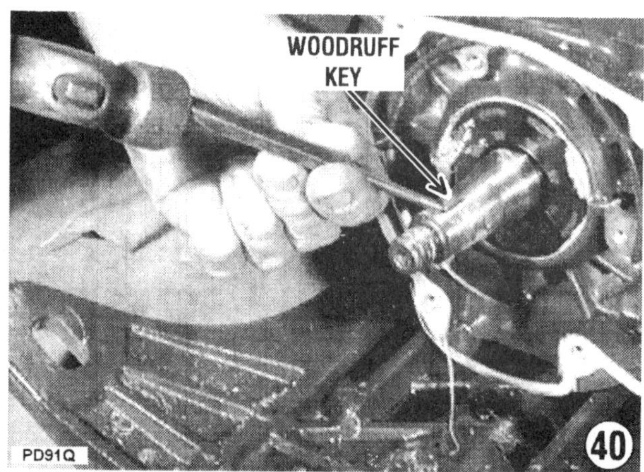

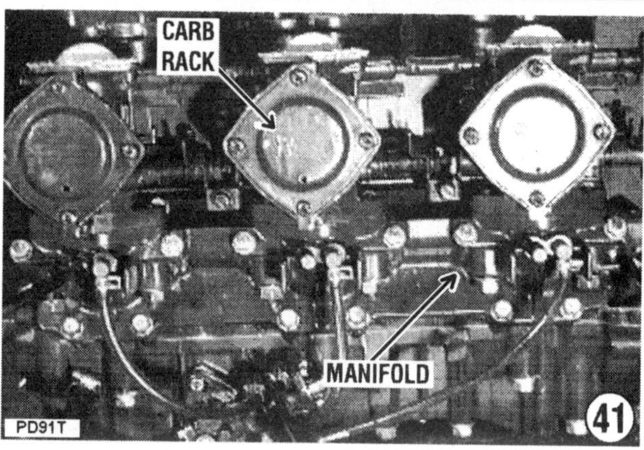

42- Lift the carburetor rack, with the carburetors attached, free of the intake manifold/s. Back off the oil delivery hose clamps from the ends of the hoses. Disconnect these lines. Note the main oil supply line has a larger diameter than the intake delivery hoses.

OIL PUMP WORDS

Three different model oil pumps have been used on the Polaris Watercraft covered in this manual. Two are mounted low on the starboard side of the crankcase and are driven a short shaft and gear indexed with a "worm" type gear on the crankshaft.

One of the two has a linkage rod connecting carburetor synchronizing linkage between the No. 2 (center) and No.3 (aft) carburetors with the pump. This linkage allows the pump to supply added lubrication under heavy load or rpm conditions.

The oil pump on most Fuji engines is located low on the starboard side of the engine. A short pump shaft with a gear on the end, indexes with a "worm" type gear on the crankshaft. An advanced model of the pump also incorporates linkage to the carburetors to enable the pump to deliver additional oil to meet demands of the engine.

The third pump is mounted to the flywheel cover and driven by a short shaft indexed into a slot in the flywheel nut. This pump is also connected to the carburetor synchronizing linkage through a cable arrangement to ensure added oil delivery under adverse engine operating conditions.

OIL PUMP REPLACEMENT

Disassembly and service is not possible on any of the three pumps used. Internal parts are not sold separately. Therefore, if the pump fails to deliver adequate oil, the pump **MUST** be replaced. Chapter 6 -- presents detailed instructions for removal, installation and adjustments.

OIL PUMP REMOVAL

To remove the oil pump, mounted on the engine block, simply disconnect the hoses, plug the lines to prevent contaminants from entering, back out the two attaching bolts and remove the pump.

The oil pump uses an **O**-ring on the pump, pieces of shim material, a bushing, and thrust washer on the pump shaft. The bushing, thrust washer, and short shaft with the gear almost always remain in the block. **TAKE CARE** not to loose or misplace any shim material, because it is critical for bushing end "play" during installation. The **O**-ring will remain in place on the pump.

To remove the oil pump mounted on the flywheel cover, disconnect and plug the oil lines. Disconnect the control cable from the pump. Remove the attaching hardware and the pump.

ATTACHMENT WORDS

Late model series craft may be designed with an oil delivery hose attached -- with a check valve -- to each carburetor or to the air intake. On early Model engines, the hoses attach -- with a check valve -- to the intake manifold/s.

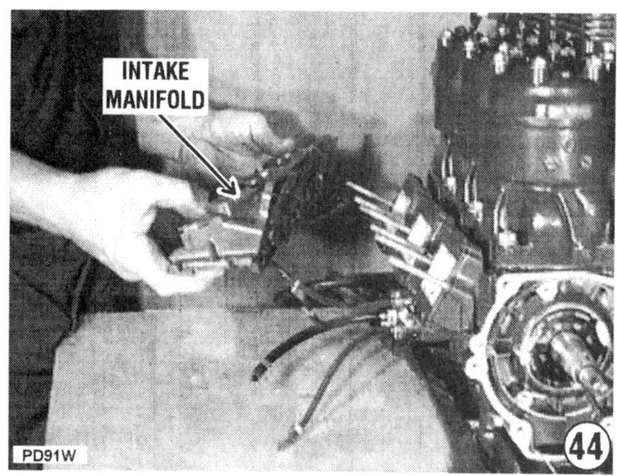

Intake Manifold Removal

43- Remove the 12 mounting bolts and 6 mounting nuts securing the intake manifold to the crankcase.

44- Remove the intake manifold. The reed blocks are now exposed for removal and service.

45- Remove the attaching hardware, and then pull each reed block assembly out of the intake port.

GOOD WORDS ON WATER CIRCULATION

As stated previously, cooling water is introduced into the lower crankcase half and is then routed through the engine to the water outlet manifold, located on top of the cylinders. Circulated water travels through this manifold and is finally channeled out the craft through an exhaust port. In this fashion, water is preheated before it reaches the cylinders. Cold water being introduced directly to the cylinder would certainly result in seizure of the engine.

46- Remove the eight bolts attaching the water exhaust manifold to the cylinders.

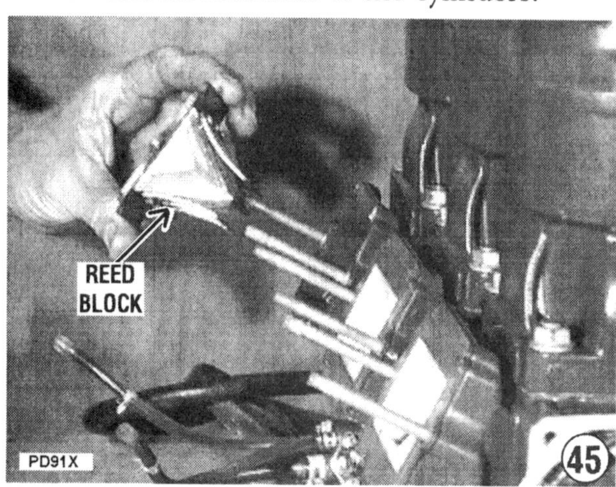

47- Lift the water outlet manifold free of the cylinders. Leave the water temperature sensor in place in the manifold. Use care to protect the sensor lead which was disconnected from the electrical box. This lead does not have a disconnect fitting at the sensor.

SPECIAL CYLINDER AND CYLINDER HEAD WORDS

Polaris has a unique head design in the watercraft industry. Each cylinder is a separate entity and has its own individual head. This design has been used in many reciprocating aircraft engine applications for many years.

The following procedures outline the tasks to remove one head. The other two are removed in the same manner.

It is not necessary to remove a head in order to "pull" a cylinder. The heads should be removed in case of a suspected blown head gasket. In order to lift the crankshaft from the crankcase, all cylinders must be removed.

Cylinder Head Removal

48- Remove the six retaining nuts securing the cylinder head to the cylinder. Some engines

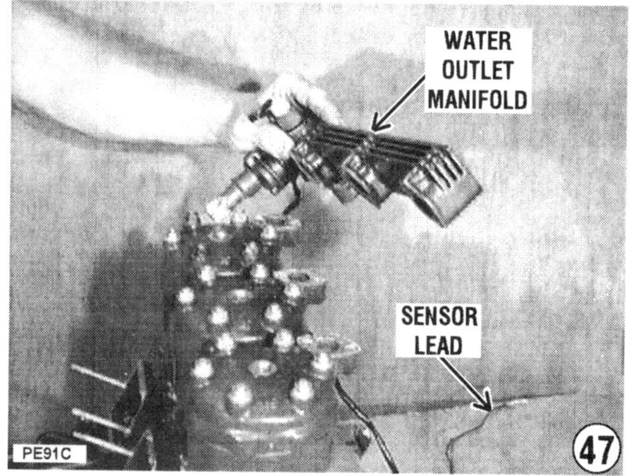

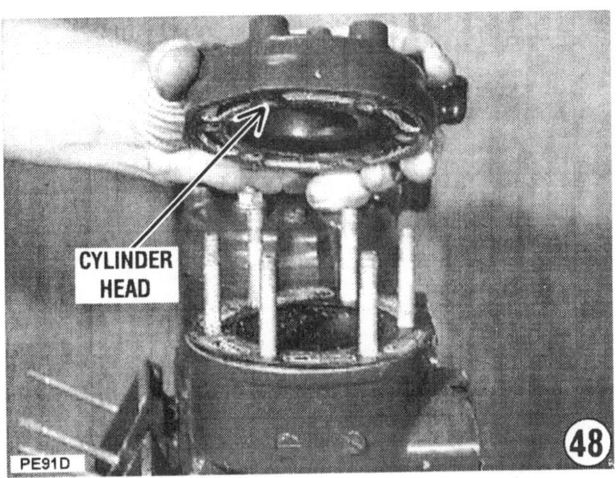

may have a separate head **COVER**, plus the cylinder head. Remove the head from the cylinder.

49- Loosen and remove the four bolts securing the cylinder to the crankcase. **CAREFULLY** lift the cylinder straight up, free of the piston and crankcase. If other cylinders are to be removed, remove cylinder No.1 (forward), and No.3 (aft), in order to gain access to the bolts securing the No.2 (center) cylinder. Perform the same procedure for each cylinder.

SAFETY WORDS

The piston pin C-lockrings are made of spring steel and may pop out of the groove with considerable force. Therefore, warn others in the area and **WEAR** eye protection glasses or a shield while removing the piston pin C-lockrings.

50- Remove the C-lockring from both ends of the piston pin using a small screwdriver inserted into the small groove next to the piston pin opening. Discard the C-lockrings. These rings stretch during removal and must **NEVER** be used a second time.

51-Obtain piston pin remover tool, P/N 2870386. Some "domestic" -- Polaris manufac-

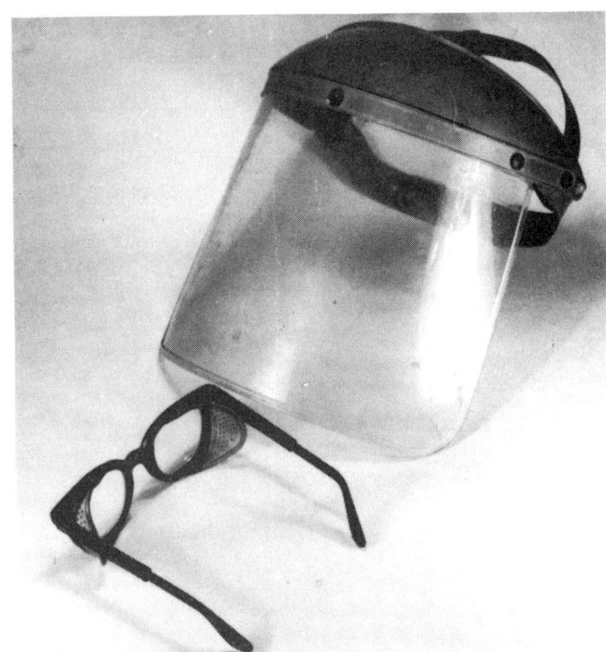

Safety glasses or a safety shield should always be worn when working with C-lockrings. The rings are made of spring steel and may pop out of the groove quite suddenly and with tremendous force.

tured engines -- require adaptor P/N 2871445 for use with the special tool. Push the piston pin out -- free of the piston, or use the special tool. If the special tool is not available and the pin refuses to dislodge, use a short piece of

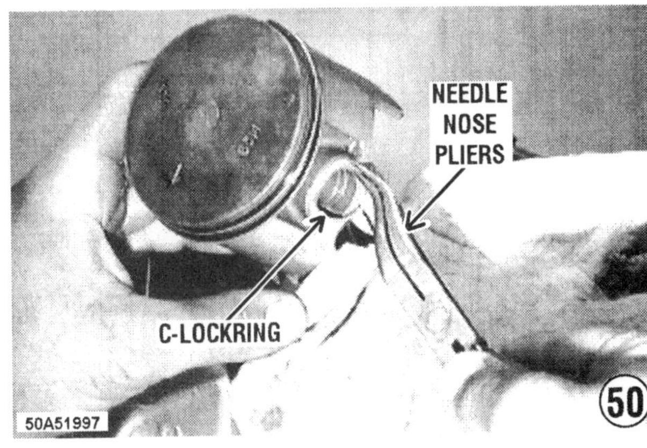

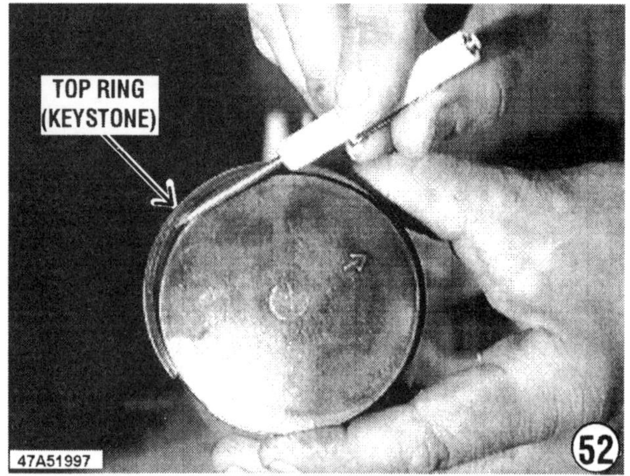

wooden doweling and light taps with a hammer to start it moving. Remove the caged needle bearings from the small end of the connecting rod.

ADVICE

New needle bearings should be installed in the connecting rods, even though they may appear to be in serviceable condition. New bearings will ensure lasting service after the overhaul work is completed. If it is necessary to install the used bearings, keep them separate and identified to **ENSURE** they will be installed onto the same connecting rod from which they were removed.

CRITICAL WORDS

The rod is an integral part of the crankshaft. The two are manufactured together and **CANNOT** be separated. Slowly rotate the bearing races. If rough spots are felt, a new crankshaft must be purchased. The bearings are installed at the factory and are not replaceable.

The engine is almost completely disassembled at this point. The only tasks remaining now are to remove the engine mounting bracket, separate the two crankcase halves, and lift the crankshaft free, as described in the following five steps.

PROFESSIONAL WORDS

Good shop practice dictates to replace the rings during an engine overhaul. However, if the rings are to be used again, expand them **ONLY** enough to clear the piston and the grooves, because used rings are extremely brittle and break very easily.

52- Gently spread the top piston ring enough to pry it out and up over the top of the piston. No special tool is required to remove the piston rings. Remove the lower ring in a similar manner. These rings are **EXTREMELY** brittle and **MUST** be handled with care if they are intended for further service.

NECESSARY PISTON WORDS

Scribe a mark inside the piston skirt, indicating from which cylinder the piston was removed. All associated hardware for each piston **MUST** be kept together. Turn the piston over and carefully place the matching C-lockrings, piston pin, rings and caged needle bearings inside the piston to **ENSURE** proper matching during assembling.

Engine Mounting Bed Bracket Removal

53- Carefully invert the engine block and lower it onto a couple 4" X 4" blocks of wood, as shown, to prevent the connecting rods from taking the weight of the assembly.

54- Remove the four bolts securing the bed bracket to the engine block, and then remove the bracket.

55- Remove the bolts securing the two crankcase halves together. There are up to 21 bolts, three different sizes -- 10mm, 8mm and 6mm. Take note of the locations for each size bolt.

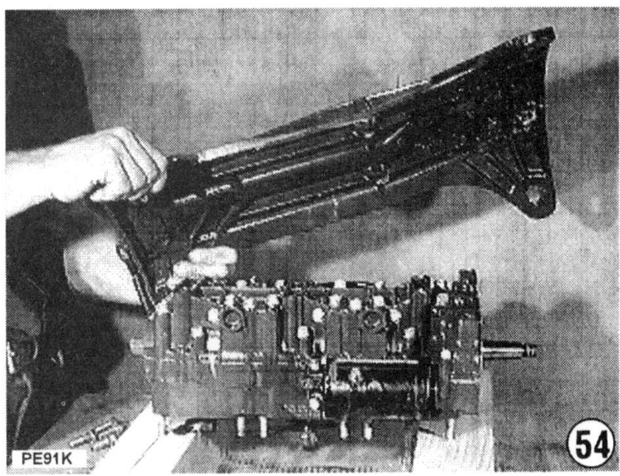

56- Separate the crankcase halves. It may be necessary to jar one or both of the halves with a soft head mallet. **DO NOT** attempt to drive any kind of tool between the halves, because such action would surely damage the mating machined surfaces.

Remove the oil pump gear assembly.

57- Lift the crankshaft clear of the upper crankshaft halve.

The engine can now be considered completely disassembled.

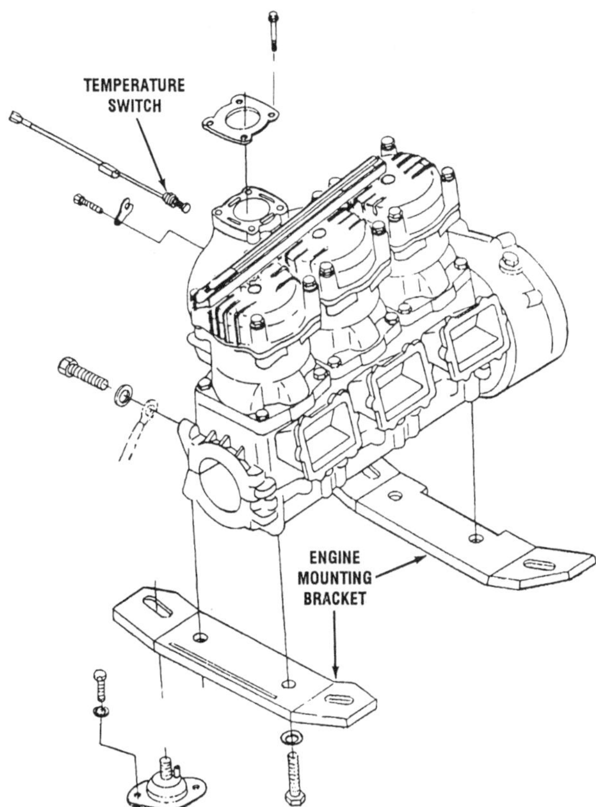

Exploded drawing the engine mount bracket arrangement for the SL900, SL1050 and the SLTX models from 1996 and on.

CLEANING AND INSPECTING

Detailed illustrated procedures to clean and inspect virtually all disassembled parts are presented in a separate "Cleaning and Inspecting" section beginning on Page 8-63.

ASSEMBLING AND INSTALLATION

TEXT CONTINUES ON PAGE 8-22

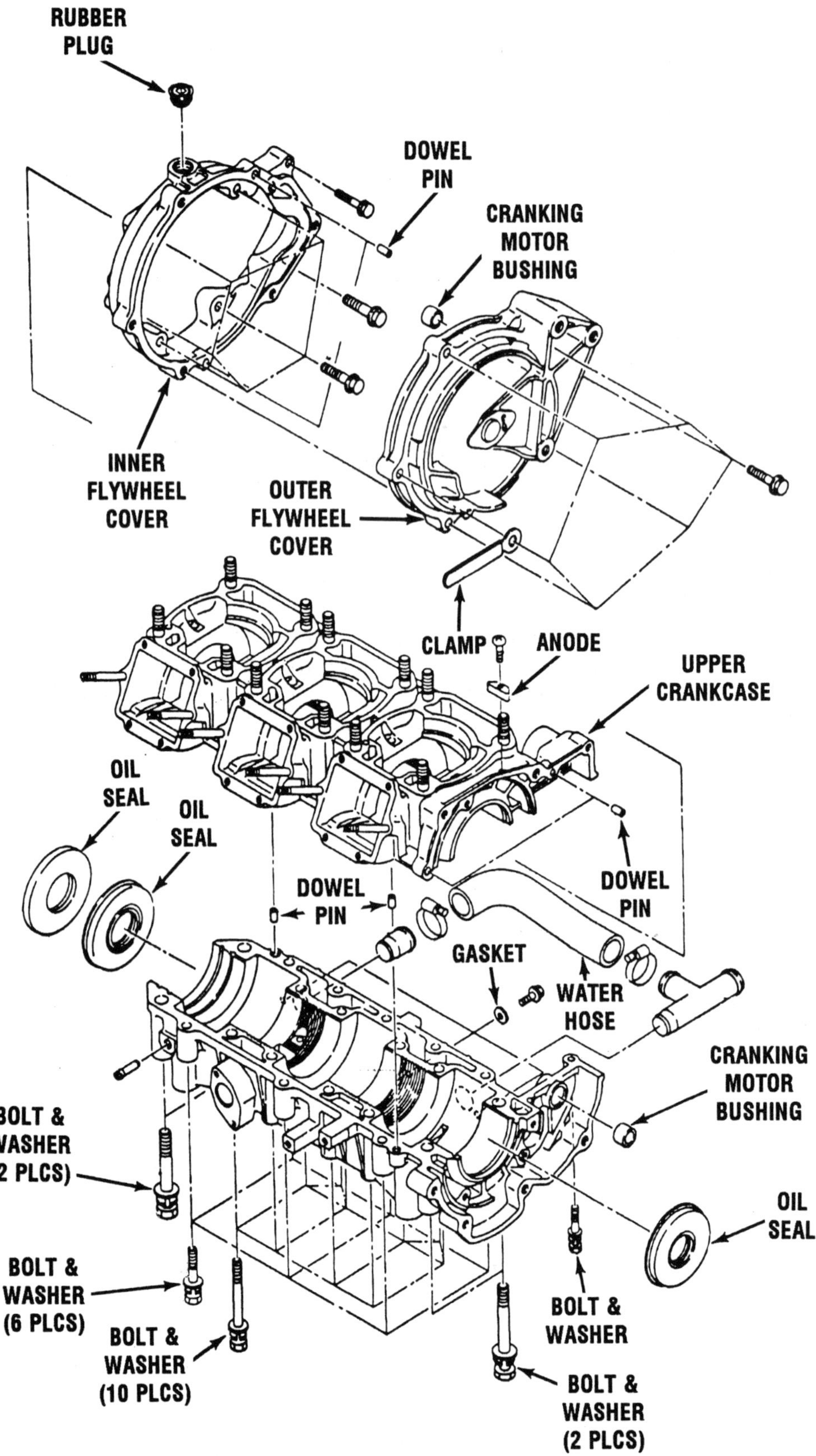

*Exploded drawing of the upper and lower crankcase separated on an SL650 or SL750 for the **1992-1993** model years. Major parts have been identified.*

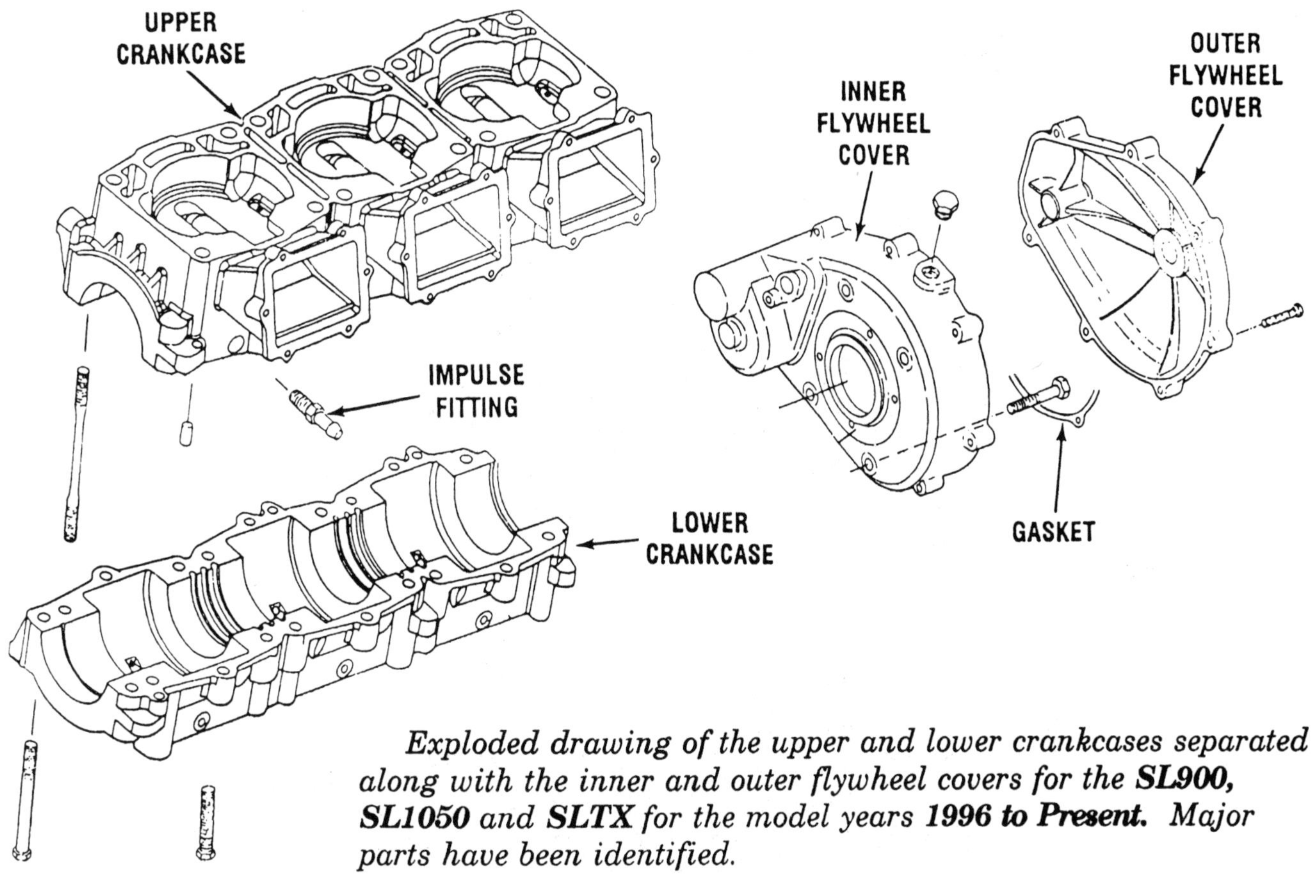

UPPER CRANKCASE

INNER FLYWHEEL COVER

OUTER FLYWHEEL COVER

IMPULSE FITTING

LOWER CRANKCASE

GASKET

Exploded drawing of the upper and lower crankcases separated along with the inner and outer flywheel covers for the ***SL900, SL1050*** *and* ***SLTX*** *for the model years* ***1996 to Present.*** *Major parts have been identified.*

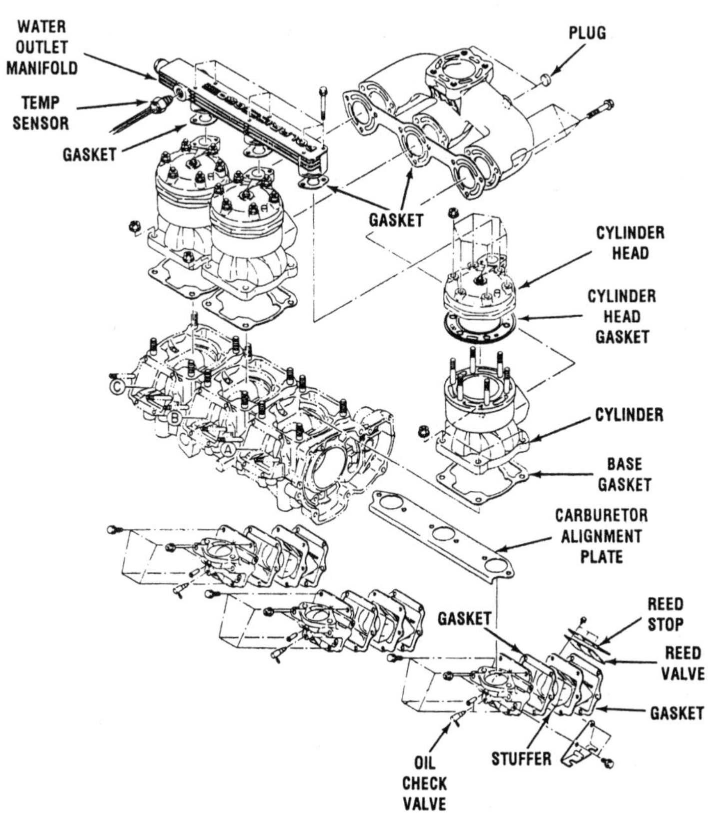

Exploded drawing of the cylinders and associated parts for the ***SL650 and SL750*** *for the model years* ***1992 - 1993.*** *Major parts have been identified.*

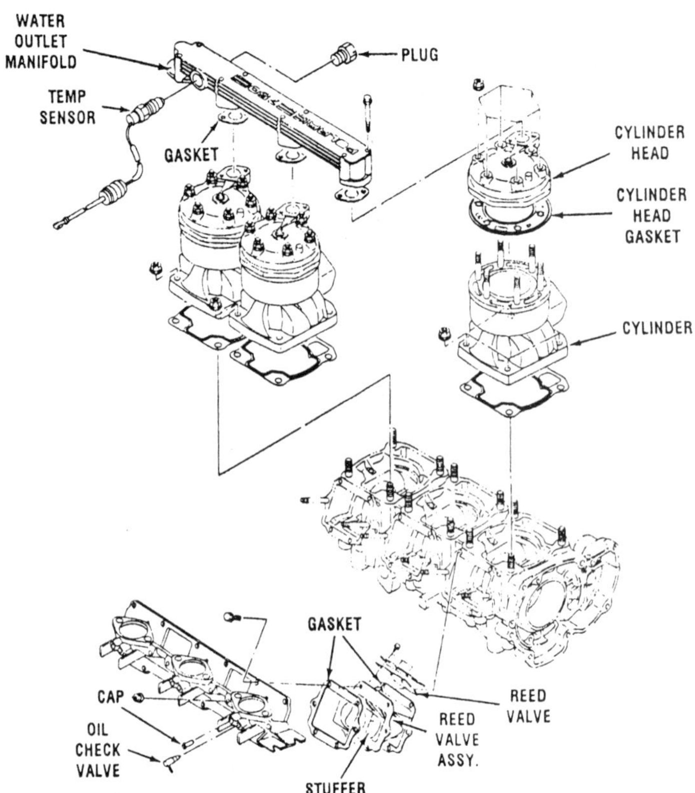

Exploded drawing of the cylinders and associated parts for ***all Models 1994-1995 EXCEPT SLX780.*** *Major parts have been identified.*

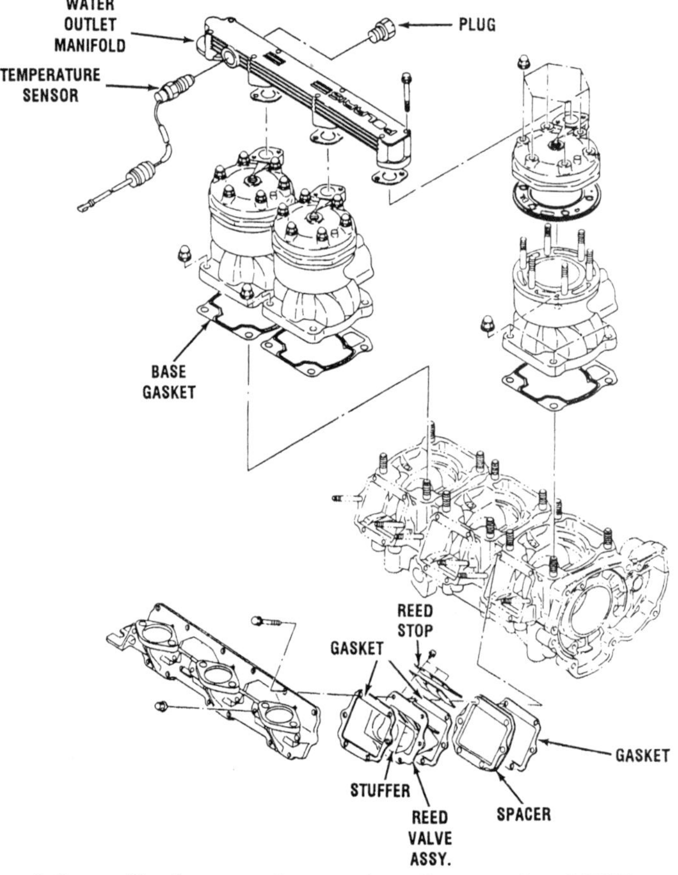

Exploded drawing of the cylinders and associated parts for ***SLX780 -- 1995*** *and* ***SL780*** *and* ***SLX780 -- 1996.*** *Major parts have been identified.*

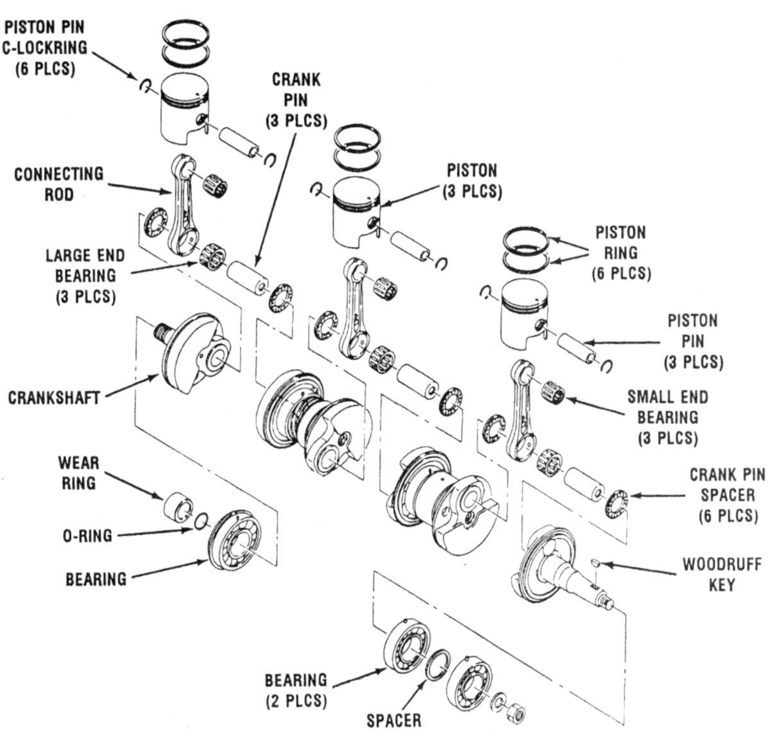

Exploded drawing of the crankshaft and associated parts for the ***SL650 and SL750 -- 1992-1993.*** *Also* ***ALL*** *Models --* ***1992-1995*** *and* ***SL780, SLT780 & SLX780 1996 and ON.*** *Major parts are identified.*

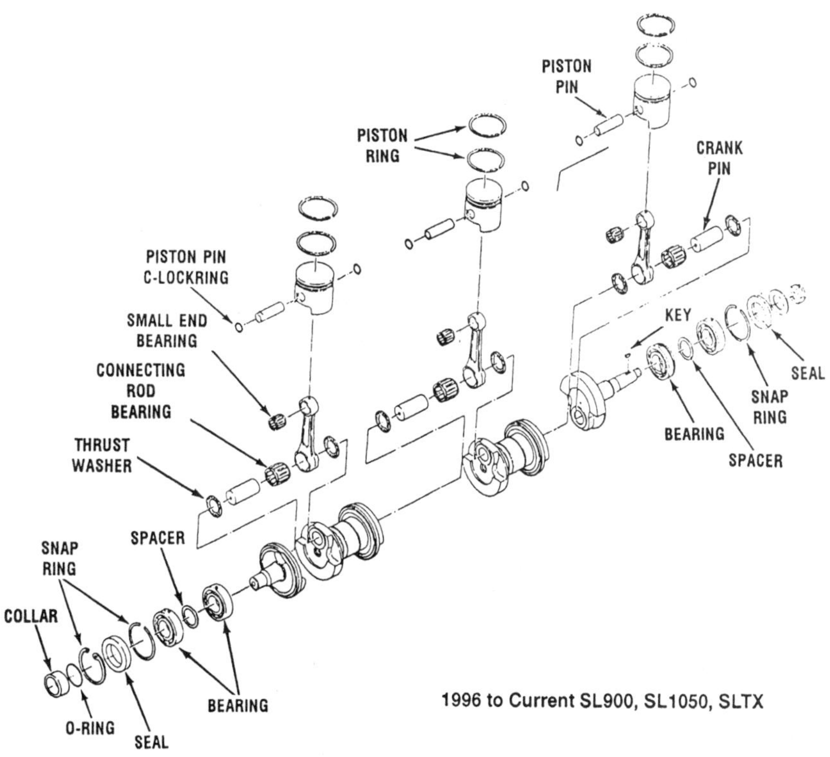

Exploded drawing of the crankshaft and associated parts for Models ***SL900, SL1050,*** *and* ***SLTX -- 1996*** *to* ***Present.***

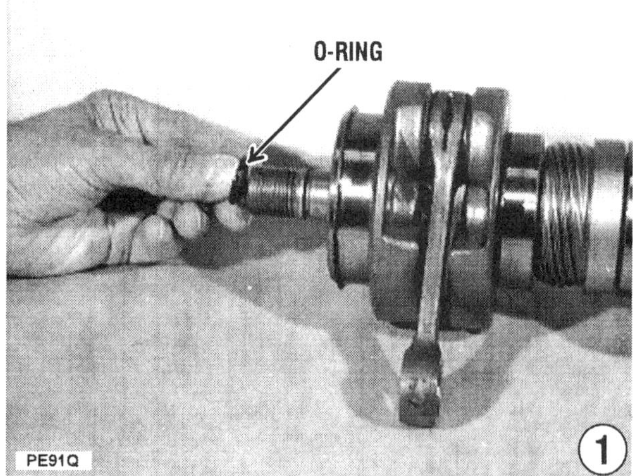

ASSEMBLING AND INSTALLATION -- THREE-CYLINDER ENGINE

The following procedures pickup the work after all disassembled parts have been cleaned and inspected according to the instructions outlined in Section 8-5 **AND** the necessary replacement parts including all gaskets, seals, and **O**-rings have been purchased and are on hand.

Assembling Lower Half Crankshaft Assembly

1- Install a **NEW O**-ring into the groove on the aft end of the crankshaft.

2- Slide the crankshaft collar over the end of the crankshaft and move it forward -- over the previously installed **O**-ring -- until it seats. Note the bearing will ride on this collar.

GOOD OIL SEAL WORDS

Both seals **MUST** be packed liberally with grease before installation.

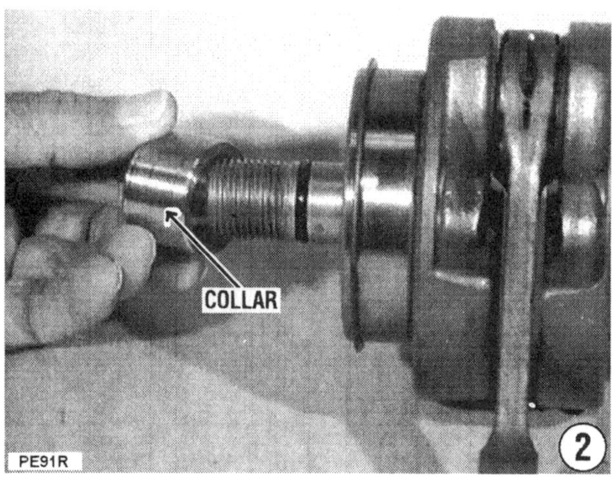

3- Pack a **NEW** oil seal with grease and then push the seal -- over the crankshaft end with the lip going on first. Slide the seal forward until it seats against the collar.

4- Pack a **NEW** second oil seal with grease. For photographic purposes, the accompanying illustration shows the seal installed in Step 2 without grease. Check to be sure the seals face the proper direction.

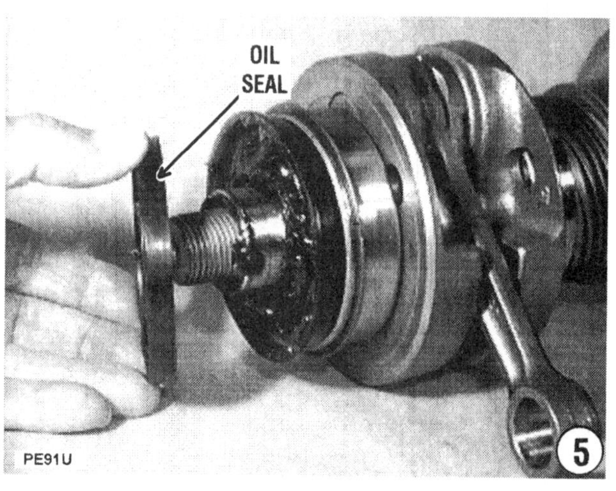

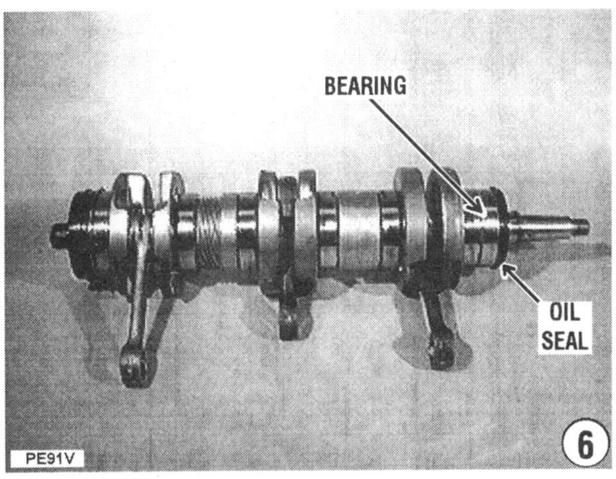

AUTHOR'S WORDS

The seals were now removed, packed with grease, as described in the text, and then permanently installed

5- Install the second seal in place over the crankshaft end and slide it forward until it seats against the first oil seal.

6- Pack a **NEW** seal with grease, and then install it onto the crankshaft, with the lip of the seal going on first. Slide the seal aft on the crankshaft until it makes contact with the crankshaft bearing, as shown.

7- Place the upper crankcase half on two blocks of 4"x4" wood -- to prevent placing a load on the connecting rods. Lower the crankshaft into position in the crankcase.

GOOD CRANKCASE ASSEMBLY WORDS

Before joining the two crankcase halves together, check to be sure **ALL** bearing locating pins are properly positioned in their respective slots. Ensure the crankshaft is set in place properly -- "square" in the upper crankcase half. Due to the worm gear driving the oil pump, once the halves are mated, check to be sure the gear for the oil pump drive aligns with the worm gear on the crank. **BEFORE** bolting the two halves together, rotate the crankshaft to verify the worm gears mesh properly and the oil pump shaft rotates smoothly.

8- Apply Loctite Anaerobic Sealant to the crankcase bolts. This sealant helps prevent moisture from building up on the threads, thus preventing the bolt from "locking up" due to corrosion.

Check to be sure **ALL** bearing locating pins -- forward, center and aft -- are positioned prop-

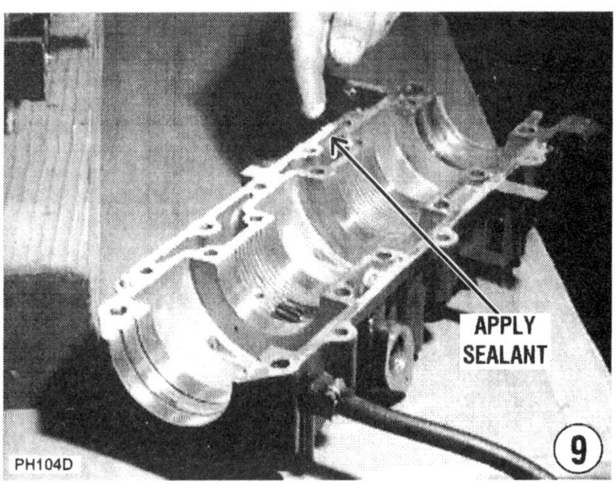

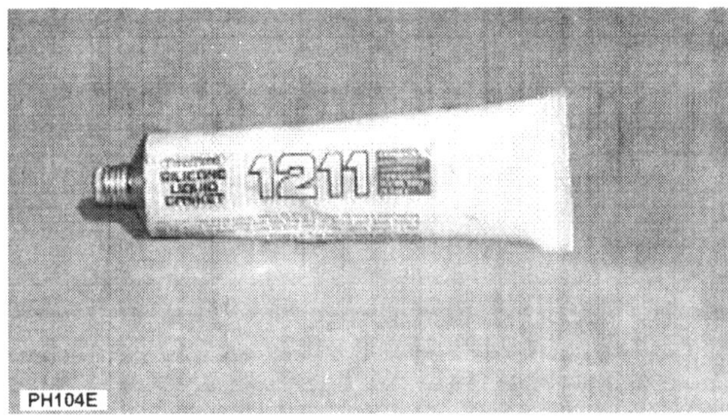

Close look at a tube of "1211" Silicone Liquid Gasket. This material is a heat and chemical resistant substance.

erly to permit each pin to index into the corresponding pin slot. These pins prevent the bearings from rotating and the crankcase halves will not seat properly if any one pin fails to index correctly into the slot. The accompanying illustrations -- No. 8 and 8A -- show the locating pins at the forward end and in the center of the crankshaft properly indexed into the slots.

9- Apply 3-Bond 1211 sealant, or equivalent, to the mating surface of the lower crankcase half. This particular sealant "sets" slowly, allowing time to accomplish other necessary tasks. Therefore, the work does not have to be done at a rapid pace.

10- Check to be sure the mating surfaces of both crankcase halves are free of any old material. Bring the two crankcase halves together. Check to be sure both halves seat together properly.

11- Apply a small amount of Locktite 242, or equivalent to the crankcase bolts. Place the crankcase bolts into the proper position on the lower crankcase. Three sizes are used. The following paragraph describes a method to ensure the bolts are inserted into the proper bolt hole.

Slip each bolt into the estimated opening. Once all bolts are in place, check the exposed

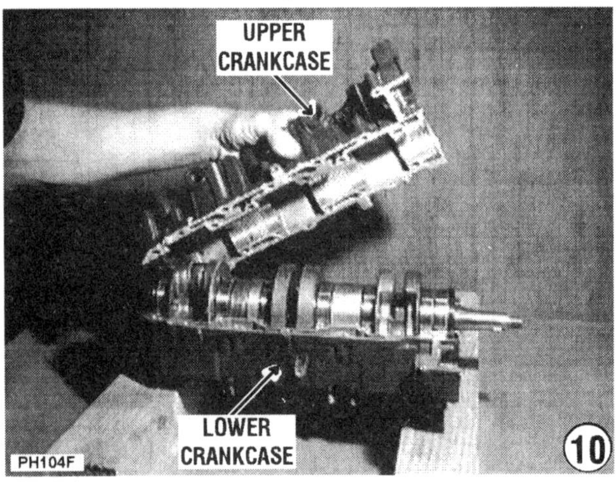

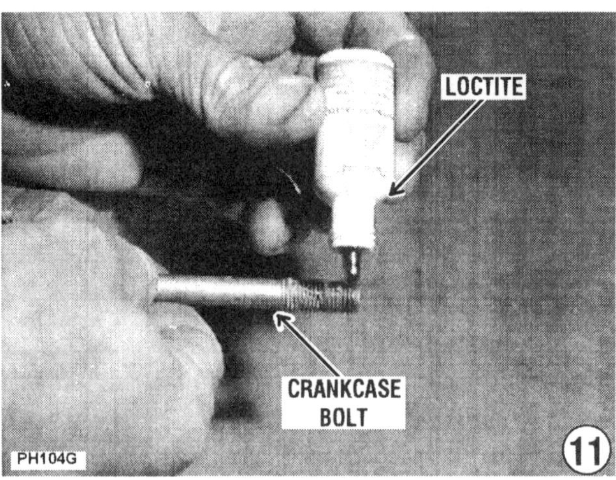

length of each bolt. If they were placed in the correct position, the exposed section of all bolts should be approximately the same height. If a bolt is misplaced, it is generally significantly higher or the head of the bolt may sit flush against the crankcase surface. If a bolt is misplaced, and then tightened to the required torque value, it is very possible such action will crack the crankcase. **BAD NEWS** and an expensive mistake.

12- Tighten the 6mm bolt to a torque value of 8 ft lb; 8mm bolts to a torque value of 16 ft lbs, and 10mm bolts to a torque value of 26 ft lbs. Be sure to begin the torque tightening sequence in the center and work toward both ends alternately and evenly.

OIL PUMP WORDS

As explained in the "removal" section, three different model oil pumps have been used on the Polaris Watercraft covered in this manual. Two are mounted low on the starboard side of the crankcase and are driven a short shaft and gear indexed with a "worm" type gear on the crankshaft.

One of the two has a linkage rod connecting carburetor synchronizing linkage between the

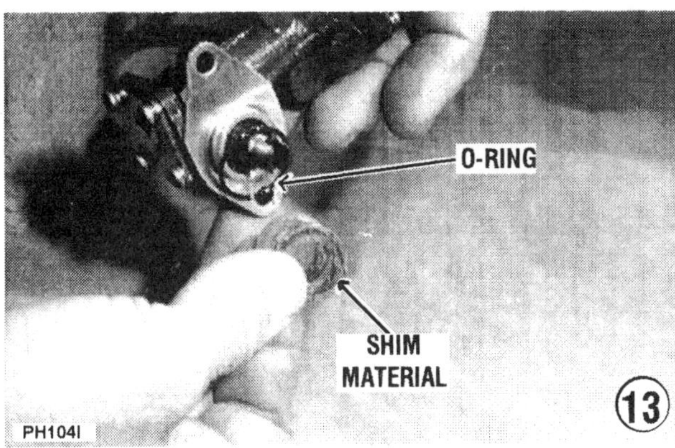

No. 2 (center) and No.3 (aft) carburetors with the pump. This linkage ensures the pump will deliver a constant oil/fuel ratio under all engine conditions.

The third pump is mounted to the flywheel cover and driven by a short shaft indexed into a slot in the flywheel nut. This pump is also connected to the carburetor synchronizing linkage through a cable arrangement to ensure the pump will deliver a constant oil/fuel ratio under all engine operating conditions.

Crankcase Mounted Oil Pump

13- Install the **O**-ring and gasket onto the oil pump shaft. If the pump is equipped with a linkage rod to the carburetor synchronizing linkage, this rod will be attached to the pump just before the engine is lowered into the craft. It is almost impossible to connect this rod after the engine is installed.

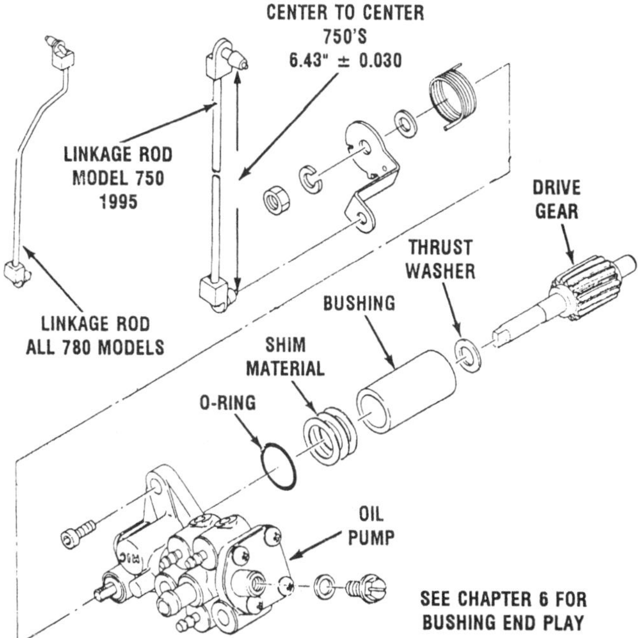

Exploded drawing of the crankcase oil pump with linkage. Major parts are identified.

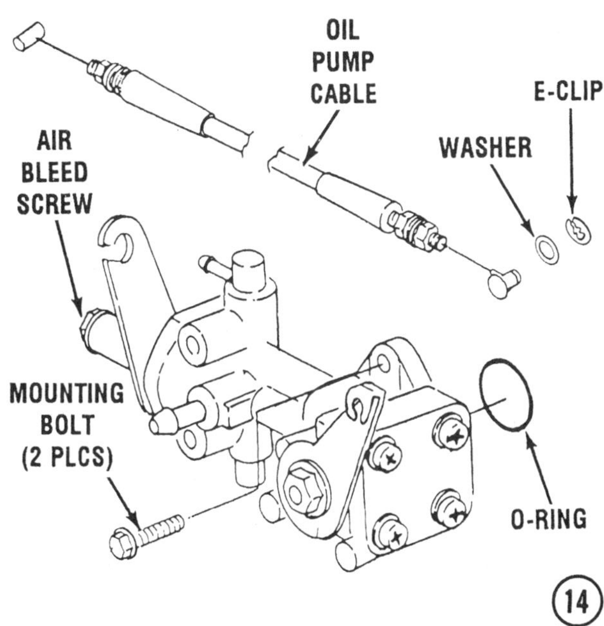

Shaft end "play" must be determined. See Chapter 6, Page 6-46.

Move the oil pump into position with the "spade" on the end of the pump shaft indexing with the internal slot in the pump. It may be necessary to rotate the crankshaft slightly to allow the shaft to index and permit the pump mounting flange to make full contact with the mounting area of the block. Secure the pump in place with the attaching hardware.

Tighten the bolts to a torque value of 48 in bls (5Nm) for a Fuji engine and 60 in lbs (7Nm) for a Polaris

Flywheel Cover Mounted Oil Pump

14- Check to be sure the mating surfaces of the oil pump and flywheel cover are clean. Install a **NEW O**-ring into the oil pump. Move the oil pump into position with the "spade" on

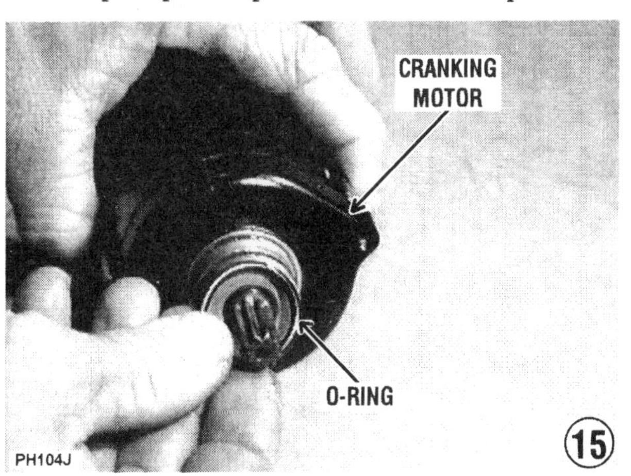

the oil pump shaft indexed into the slot in the flywheel nut. It may be necessary to rotate the pump shaft slightly to permit the "spade" to index fully and permit the two mounting surfaces to make full contact.

Apply Loctite 242 to the threads of the attaching bolts. Secure the pump in place with the bolts tighten to a torque value of 60 in lbs (6.8Nm).

Connect the oil lines to the pump fittings. Slip the oil pump cable into the slot in the bracket, and then connect the cable end to the lever arm with the washer and "E' clip.

The lever should move as soon as the throttle is opened slightly. This cannot be determined until the engine is in the craft and the carburetors are installed. See Oil Pump section in Chapter 6 to make an adjustment.

15- Install the **O**-ring onto the shaft end of the cranking motor. Check to be sure it seats properly.

16- Slide the cranking motor into position in the forward end of the crankcase. Guide the cranking motor shaft through the opening and

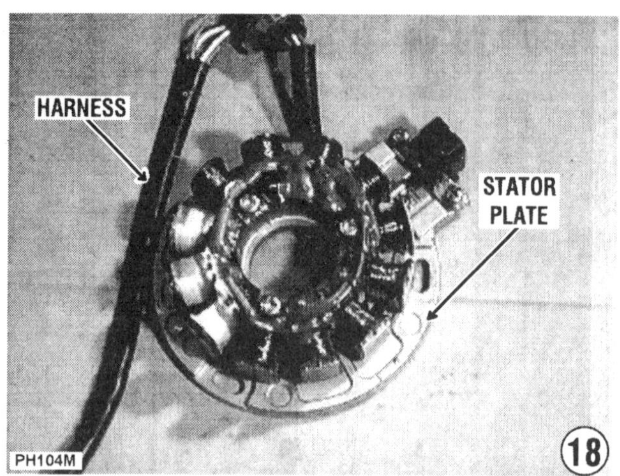

seat the motor in place. The unit should seat against the crankcase firmly.

17- Tighten the cranking motor mounting bolts to a torque value of 108 in lb (12Nm).

18- Make a final inspection of the stator assembly before installation. Verify all components are clean and free of any foreign matter. Check the entire length of the harness one more time to ensure all leads are in good condition without any breaks in the insulation or kinks in the harness.

19- Feed the stator harness through the opening in the crankcase.

20- Continue feeding the harness through the opening and seat the weather proof grommet in place in the opening. Install the attaching screws and tighten them securely.

21- Install the stator assembly over the crankshaft end. Note the stator plate has slots which allow the stator to be rotated **SLIGHTLY** for proper timing alignment. Check to be sure the stamped index mark aligns with the crankcase parting line, as shown.

22- Stators installed on Fuji engines in 1993 only, have two marks -- an upper mark and a lower mark. If installing a new stator on Model SL650 Series, align the upper mark on the stator plate -- more advanced -- with the crankcase parting line, as shown. If servicing a Model SL750 Series, align the lower mark on the stator plate -- more retarded -- with the crankcase parting line.

The mounting openings are actually elongated to permit a very slight amount of timing adjustment.

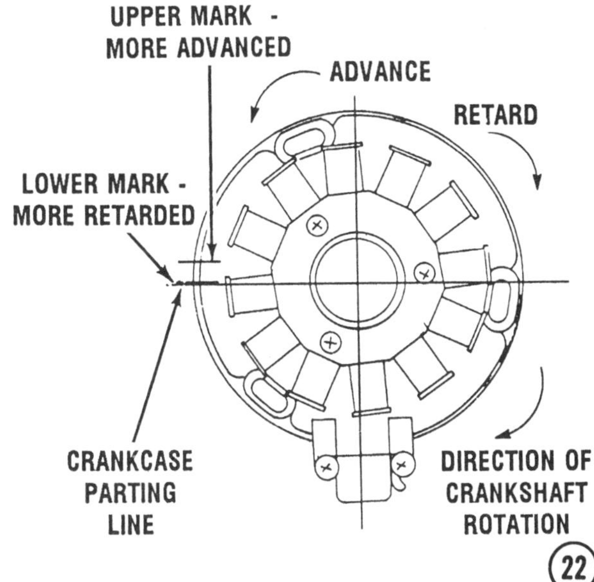

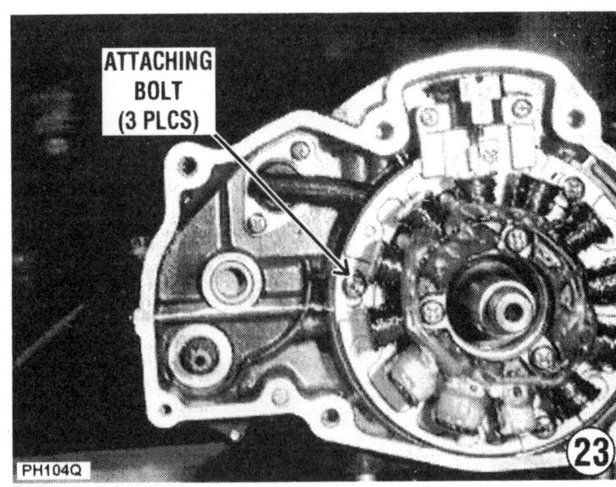

23- Apply Loctite 242, or equivalent, to the stator plate attaching bolts. Install the bolts and tighten them to a torque value of 50 in. lbs (n). Check to be sure the stator electrical harness grommet is secure in the crankcase opening.

One Magneto Cover

24- Apply a non-permanent sealant to the mating surface of the magneto cover. Secure the cover in place with the attaching bolts tightened to a torque value 14 ft lbs (19Nm).

Two Magneto Covers

Some engines have an inner magneto cover and an outer cover. If servicing such an engine, apply a non-permanent sealant to the crankshaft side of the inner cover. The outer cover does not require the sealant. Apply Loctite 242 to the inner magneto cover attaching bolts.

Position the inner cover on the crankcase and secure it in place with the attaching bolts. Tighten the 6mm bolts to a torque value of 78 in lb (9Nm) and the 8mm bolts to a torque value of 14 ft lb (19Nm).

Apply a non-permanent sealant to the mating surface of the outer magneto cover. Install

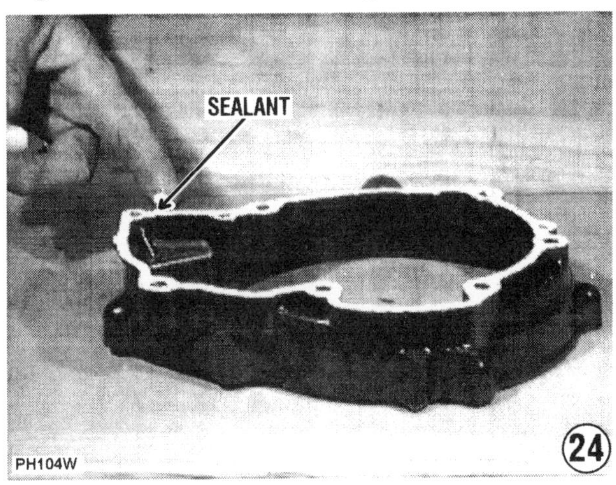

the outer cover on the inner magneto cover. Apply Loctite 242 to the attaching bolts. Install the bolts and tighten them to a torque value of 14 ft lbs (19Nm).

25- Install the drive shaft coupler to the aft end of the crankshaft. Tighten the coupler just hand tight at this time. Be careful when installing the engine to avoid damaging the coupler and crankshaft during lowering and positioning.

Flywheel Installation.

26- Apply a light coating of grease to both ends of the reduction gear. Slide the washer onto the gear shaft. Insert the reduction gear into the recess in the crankcase. Note: The magneto cover/s **MUST** be installed first, otherwise the reduction gear and flywheel cannot be installed.

27- Apply a coating of Loctite 262 to the tapered threads on the forward end of the crankshaft. Verify the Woodruff key is fully seated in the crankshaft slot.

28- Move the flywheel onto the crankshaft, with the keyway in the flywheel hub indexed over the woodruff key.

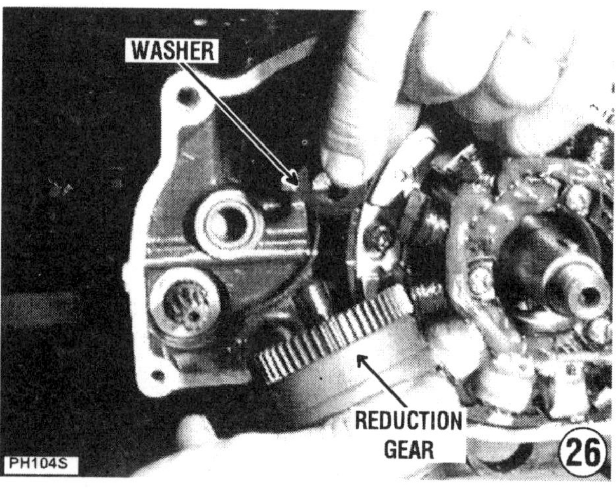

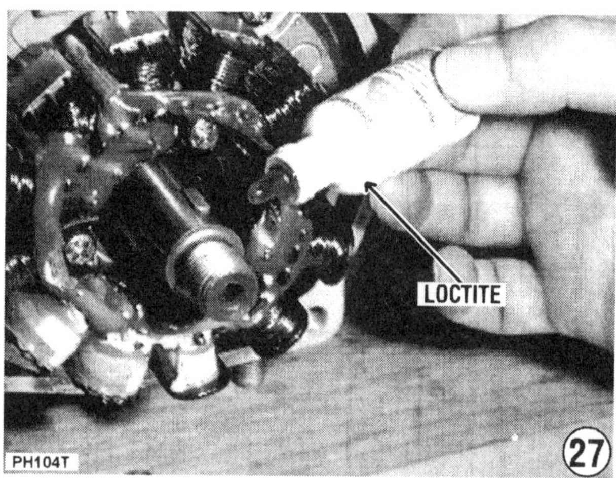

29- Obtain special flywheel holding tool P/N 8700229. If servicing a Fuji engine, an adapter is required -- same part number. Thread the flywheel nut onto the crankshaft, just hand tight. Position the flywheel holding tool over the crankshaft, with the three pins on the tool indexed into the holes in the flywheel. Allow the handle to come down against the workbench surface, as shown.

If servicing a Polaris engine, tighten the flywheel nut to a torque value of 90 ft lb (122Nm). If servicing a Fuji engine, tighten the 16mm

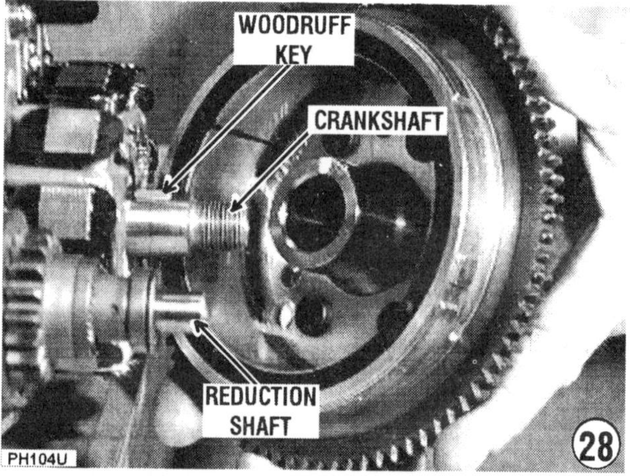

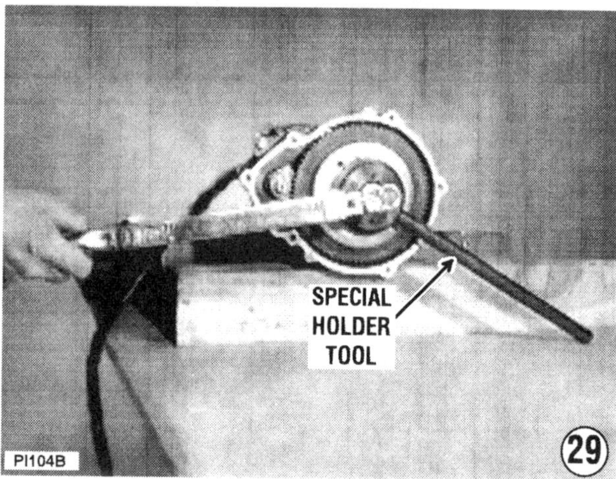

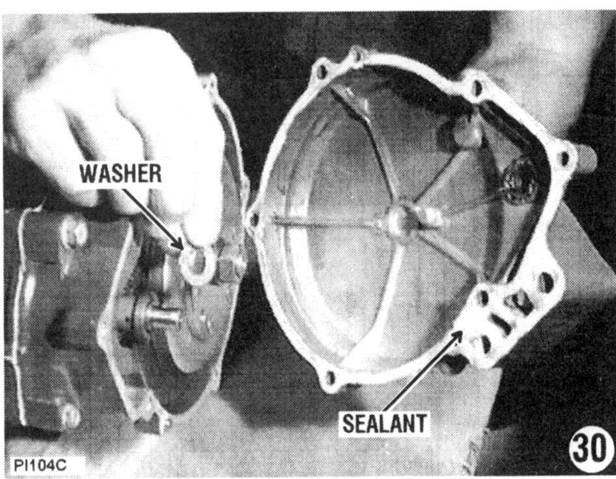

flywheel nut to a torque value of 55 ft lb (75Nm).

30- Apply a coating of sealant to the mating surface of the flywheel cover. Place the washer onto the shaft of the reduction gear. Position the flywheel cover on the crankcase and install the attaching bolts.

If servicing a Fuji engine, tighten the 6mm attaching bolts to a torque value of 78 in lb (9Nm); tighten the 8mm bolts to a torque value of 16 ft lb (22Nm).

If servicing a Polaris engine, tighten the 6mm bolts to a torque value of 108 in lb (12Nm); tighten 8mm bolts to a torque value of 22 ft lb (30Nm).

One of these bolts may secure a clip used for clamping an electrical harness. During disassembly, a mark should have been made to indicate the proper location for this clip. Check to be sure the bolt and clip are installed in the proper location.

31- Rotate the assembled crankcase to position the lower half facing upward. Lower the engine mounting plate onto the crankcase. Check to be sure the plate is positioned properly. The plate can only be installed properly one way -- it is **NOT** reversible.

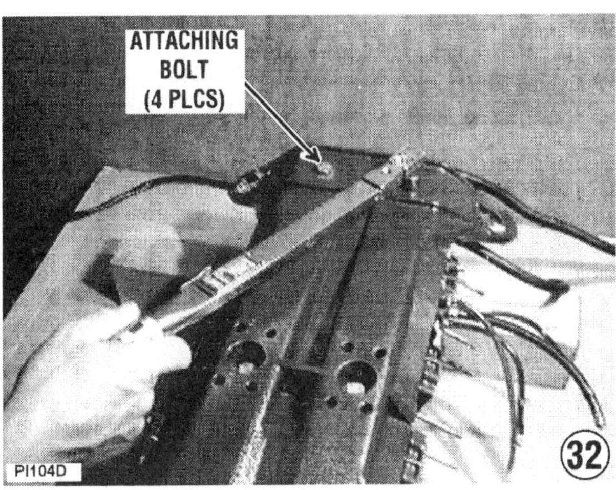

32- Apply a coating of Loctite 242 to the threads of the engine mounting plate attaching bolts. If servicing a Fuji engine, install and tighten the attaching bolts to a torque value of 45 ft lb (61Nm). If servicing a Polaris engine, install and tighten the attaching bolts to a torque value of 50 ft lb (68Nm).

Rotate the engine back to a normal upright position, resting on the engine mounting bracket.

33- Using an oil can, place a "couple squirts" of oil into each oil hole in the crankcase, as shown.

34- Apply a small amount of oil to each lower bearing and to the inside surface of the connecting rod -- small end.

35- Check to be sure the mating surfaces of the cylinders and upper crankcase are clean. Place shop towels or cloths over the crankcase, to prevent contaminants from entering. During piston pin C-lockring installation, a ring may spring loose and could fall into the crank area. **BAD NEWS**! The crankcase would have to disassembled in order to retrieve the lose ring.

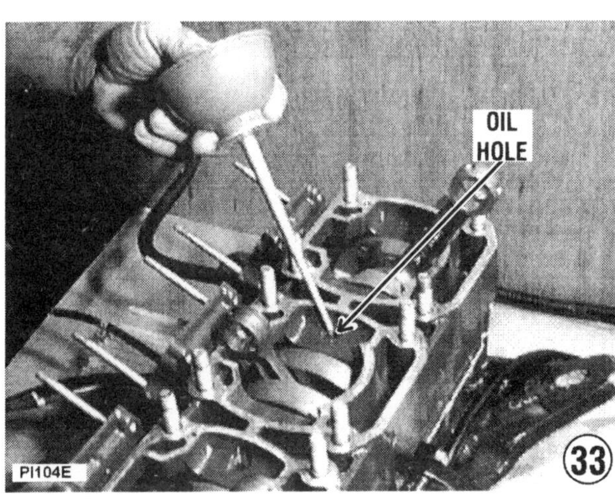

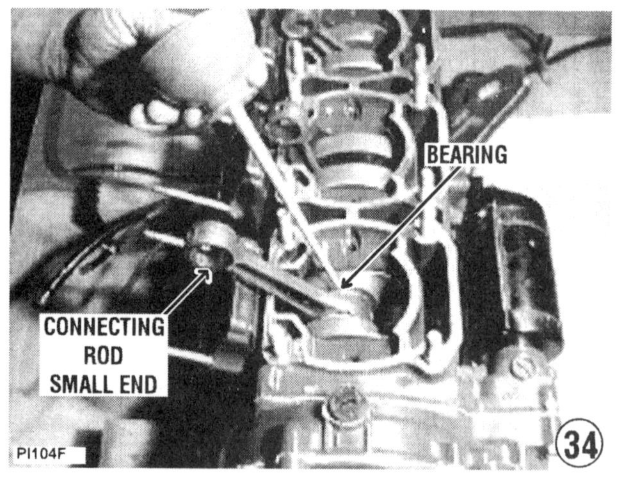

GOOD WORDS ON PISTON ASSEMBLING

During piston assembling, old parts being used a second time **MUST** be matched with the piston from which they were removed. Piston pin C-clips **MUST NEVER** be used a second time. During disassembling, these clips are stretched and lose holding effectiveness. Therefore, they should be discarded. Use only new C-clips.

36- Install a new set of piston rings onto the pistons. No special tool is necessary to install the rings. **HOWEVER**, take care to spread the ring only enough to clear the top of the piston. The rings, especially used rings, are **EXTREMELY** brittle and will snap if spread beyond their limit. Align the ring end gap over the locating pin in each piston groove. The purpose of the pin is to prevent the ring from rotating in the ring groove, during engine operation. If the ring does not straddle the pin in the groove properly, the ring would rotate in the cylinder during engine operation. In a short time an end would catch on a port opening and the ring would break.

37- Coat the caged needle bearing with engine oil and then insert the set into the small end of the same rod from which it was removed. If new bearing sets are to be installed, each new set may be placed in any one of the three rods.

38- Bring the assembled piston down over the small end of the connecting rod. Align the opening through the piston with the opening in the small end of the connecting rod. Insert the piston pin through the piston and needle bearing. The pin is a "push" fit and a special tool is not required.

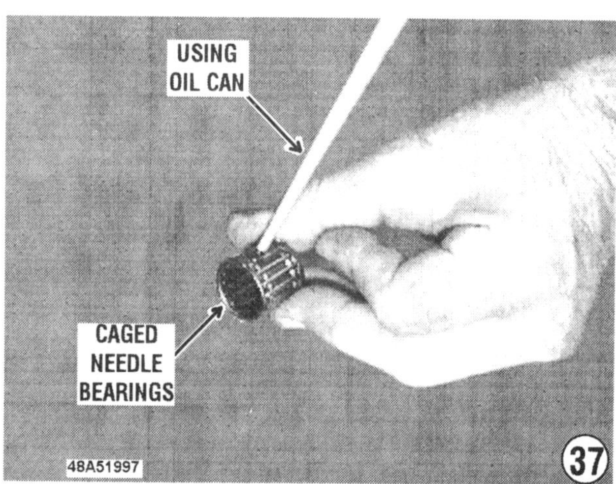

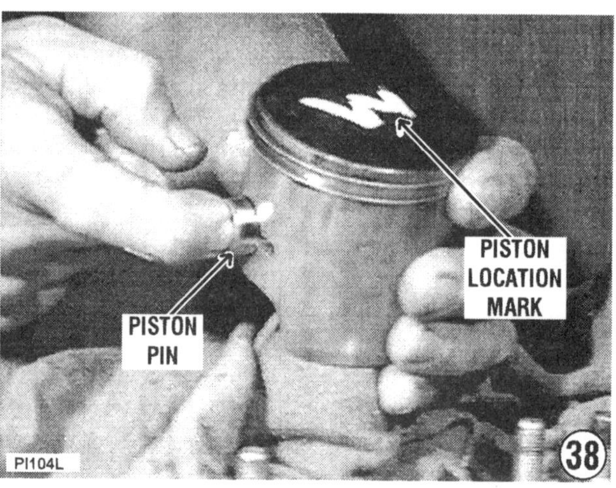

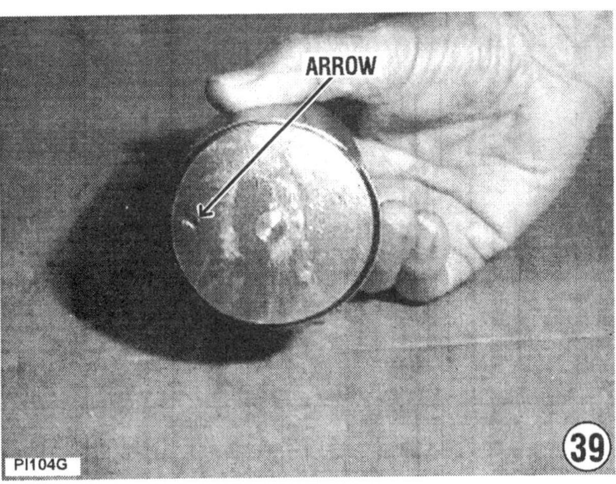

Note: If old parts are being used a second time, the piston must be installed back onto the connecting rod from which it was removed. The visible "M" marking on the piston crown in the accompanying illustration indicates this piston was removed from the middle (No. 2) cylinder and identified during disassembling.

39- Observe the arrow stamped into the piston crown. If servicing a Fuji engine, when installing, the piston onto the connecting rod, the arrow **MUST** point toward the flywheel -- forward end of the engine.

If servicing a Polaris engine, the stamped arrow **MUST** point toward the exhaust side (port side), of the engine.

C-LOCKRING SAFETY WORDS

The piston pin C-lockrings are installed under considerable tension and may snap or pop free. Eye protection should be worn and other personnel in the area warned.

40- The piston pin C-rings **MUST** be installed with the gap facing straight up or straight down -- parallel to the vertical line of

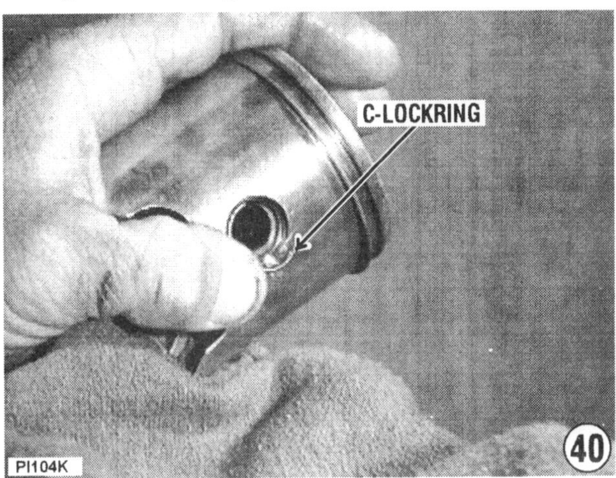

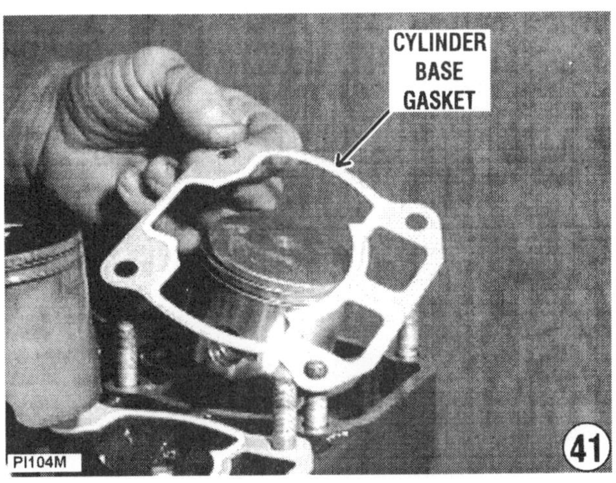

the piston. Install one end of the clip into the recess, then work the clip around, until the other end can be snapped into place.

Assemble the remaining pistons in the same manner. Take time to inspect the piston ring locating pins, to verify each ring gap straddles the pin.

Cylinder Installation

41- The cylinder base gaskets have unique surfaces. One side has a fairly hard silicon bead. The other side is clean -- without any substance. Place a **NEW** gasket over each set of cylinder base studs -- be sure the bead of silicone sealant faces up.

WORDS FROM EXPERIENCE

The best plan to follow when installing the cylinders is to begin with the center cylinder, because with the No. 1 and/or No. 3 cylinders installed, it is extremely difficult to compress the rings. Also, it is much easier to install the center cylinder restraining nuts, without No. 1 and No. 3 cylinders in place.

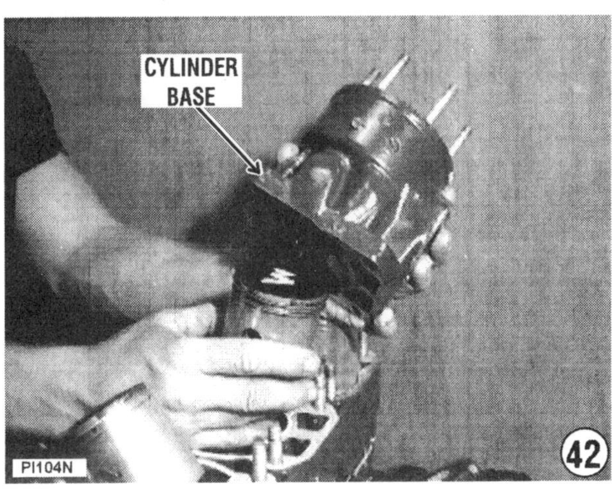

42- Install the center piston first. Apply a small amount of engine oil on the piston sides and rings. Lower the center -- No. 2 cylinder -- down very slowly over the piston and at the same time compress the rings. A ring compressor tool may be used while lowering the cylinder over the piston. If this special tool is not available, the rings may be compressed by hand. Seat the cylinder onto the crankcase.

HELPFUL WORDS

At this time, install the cylinder base nuts and secure them **ONLY** "fingertight". The cylinders must be able to be shifted slightly to allow the exhaust manifold to be properly aligned with the cylinders **BEFORE** installing the engine into the watercraft. After alignment, the cylinder nuts may be tightened to the required torque value. Once the cylinder nuts are tightened, the exhaust manifold is removed and is not installed permanently until after the engine is secured in the craft. It would be difficult indeed, to attempt to align the exhaust manifold with the cylinders once the engine is installed.

The threads in the block for an exhaust manifold bolt were damaged by stripped threads on the bolt. The threads in the block were repaired with the "Tap" tool and the bolt threads were repaired with the "Die" tool from a Tap and Die set.

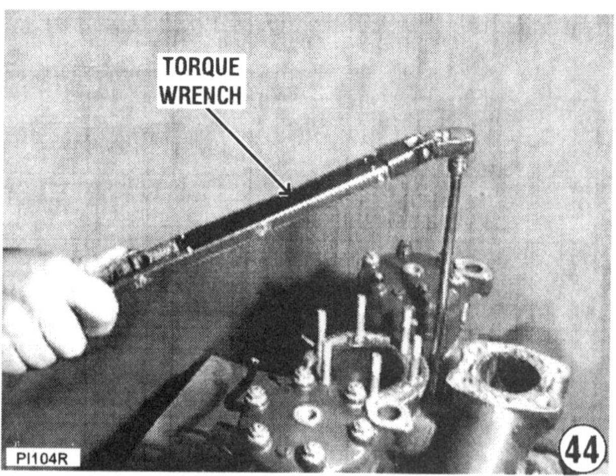

43- Position the exhaust manifold against the cylinders. Install the exhaust manifold bolts and tighten them "just snug" at this time. The manifold will be removed prior to installing the engine in the craft.

44- The manufacturer recommends tightening the cylinder base nuts in a three step sequence, using a criss-cross pattern. If servicing a Fuji engine, tighten the nuts to a final torque value of 28 ft lbs (38Nm). If servicing a Polaris engine, tighten the nuts to a final torque value of 40 ft lbs (54Nm). Once the cylinders are tightened properly, the exhaust manifold may be removed, in preparation for engine installation.

Cylinder Head Installation

45- Lower a **NEW** head gasket onto the cylinder studs. Observe the cylinder head gasket notation. If servicing a Fuji engine, most cylinder head gaskets have a tab, which **MUST**

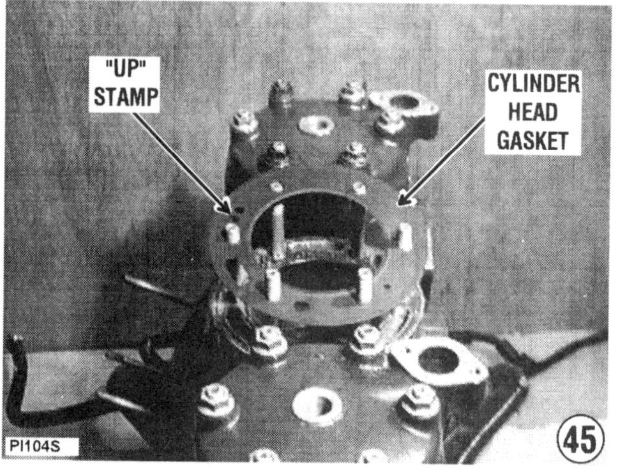

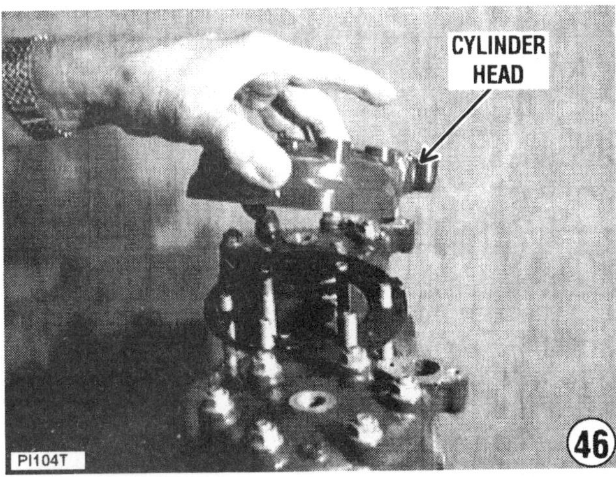

be aligned with the cylinder water outlet, located just below the head mating surface. The gasket **MUST** face with the letters "U" and "P" stamped on the surface. If the gasket is not installed properly, an adequate amount of cooling water will not be delivered through the engine. Furthermore, if the gasket has **"EXUP"** embossed on it, this marking **MUST** be placed on the exhaust side of the cylinder.

GASKET PURCHASE WORDS

According to the manufacturer, different model year engines use different gaskets. When purchasing new gaskets, take an extra minute to verify receipt of the correct gaskets for the engine being serviced. Severe engine damage can occur if the wrong gaskets are mistakenly installed.

46- Lower the cylinder head onto the studs and seat the head against the gasket.

47- Install the cylinder head nuts and tighten them to a value of 18 ft lb (24Nm) -- Fuji engines only. For all Polaris engines, install

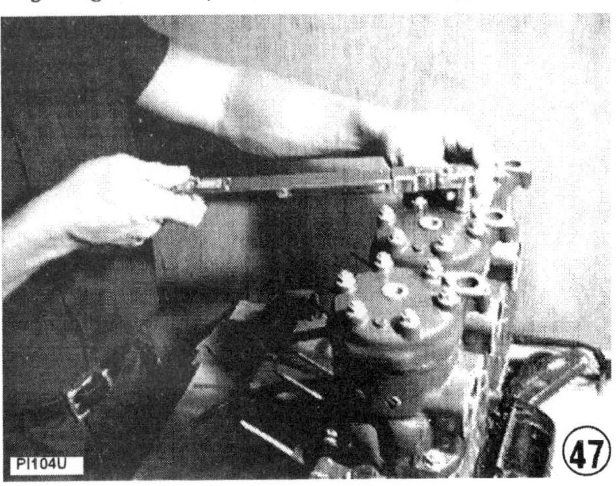

the cylinder head bolts and tighten them to a torque value of 20 ft lb (27Nm) using a three step sequence.

48- Position a **NEW** water outlet manifold gaskets in place on each cylinder head, and then lower the water outlet manifold onto the cylinders. Install the water manifold attaching bolts and tighten them to a torque value of 78 in lbs (9 Nm) -- Fuji engines; 108 in lbs (12Nm) -- Polaris engines.

Reed Block Assembly Installation

FIRST, these special words on Fuji reed block assemblies.

Models manufactured in 1992 and 1993 are equipped with individual carburetor intake manifolds. Since 1994, the Fuji engines have a single manifold for the three cylinders. The old individual manifolds must be aligned properly and installed in the following sequence to prevent air leaks. First, place a **NEW** gasket on the crankcase intake port, followed by the reed block, reed "stuffer", another **NEW** gasket, and then install the intake manifold. Apply Loctite 242 to the four bolts and two nuts. Install the attaching hardware, but **DO NOT** tighten them at this time. Repeat this process for each remaining manifold.

Check to be sure the manifold attaching hardware is loose enough to shift the intake manifolds slightly by hand. Alignment of the intake manifolds is **CRITICAL** to prevent air leaks.

Lower the carburetor mounting plate onto the three manifolds. Install the mounting plate attaching bolts and tighten them securely. The intake manifolds are now aligned.

Tighten the center nuts and bottom bolts of the reed blocks to a torque value of 78 in lb

(9Nm). Once all reed blocks are secured, remove the carburetor mounting plate.

Tighten the top bolts of the reed blocks to a torque value of 78 in lb (9Nm).

1994 to Present
Single Intake Manifold
Fuji and Polaris

49- Place a new gasket on the crankcase intake port, followed by the reed block assembly, then another gasket. Repeat this process for each cylinder.

50- Position the intake manifold in place on the intake port studs. Some engines may use bolts to secure the intake manifold. Seat the manifold against the reed blocks. Apply Loctite to the mounting studs or attaching bolts. Install the attaching nuts/bolts and tighten them, to a torque value of 78 in lb (9Nm), starting in the center and working alternately and evenly toward both ends.

51- Insert an oil delivery hose into each of the oil pickup/check valve fittings on the intake manifold, if applicable. Beginning in 1995, some models inject the oil directly into the carburetor. Secure each hose to the fitting with the clamp.

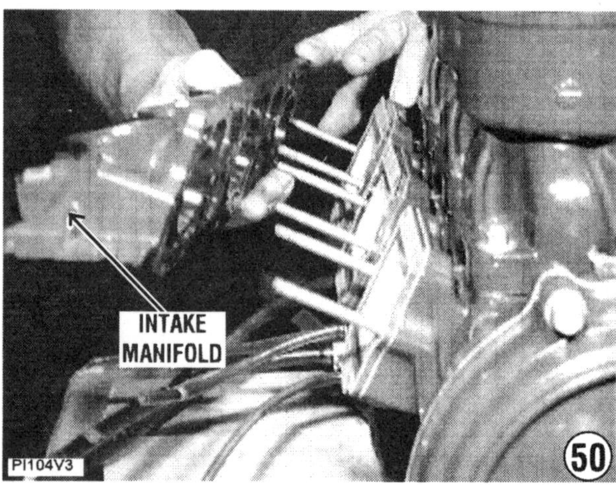

52- Place a **NEW** gasket on each intake manifold port. Lower the carburetor rack assembly onto the intake manifold.

53- Apply Loctite to the studs or attaching bolts. Install the carburetor mounting washers and nuts -- Mikuni carburetors -- and tighten them to a torque value of 16 ft lb (22Nm). Install the carburetor mounting bolts -- Keihin carburetors -- and tighten them to a torque value of 108 in lb (12Nm). Use the standard

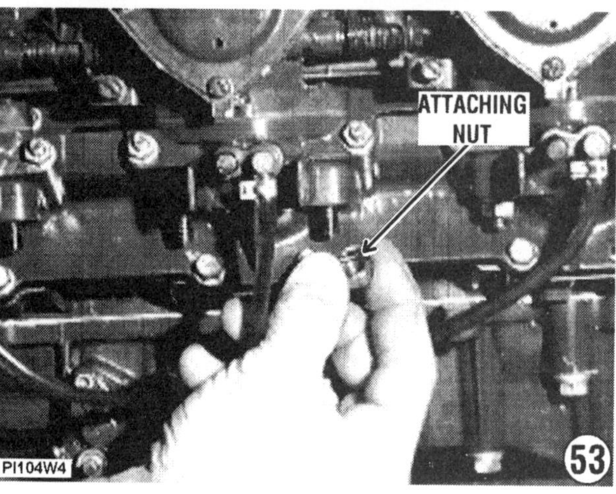

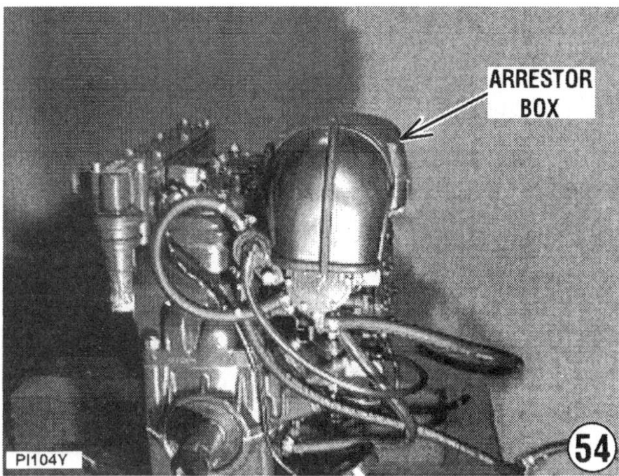

torque sequence of staring in the center and work outward toward both ends, alternately evenly.

54- Position the flame arrestor box in place on top of the carburetor rack. Secure the box with the attaching hardware.

Engine Installation

The engine is basically ready for installation in the watercraft. In final preparation, remove the battery, and slide the electrical box aft in the engine compartment. Gather up the electrical leads and fuel hoses and tuck them against the outboard bulkhead. If an engine lift is not available, the engine may be lowered into the watercraft by one person.

55- If any shim material was used on the mounting studs, they should have remained in place when the engine was removed. Set a small sheet of plywood on the aft end of the watercraft, as shown. Move the engine onto the piece of wood. Straddle the compartment, facing aft. Grasp the engine, lift it -- using your legs, not the back -- and move it "forward", then lower it through the engine compartment

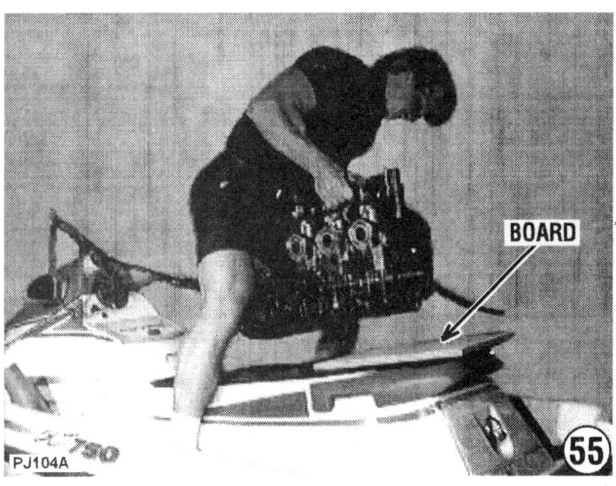

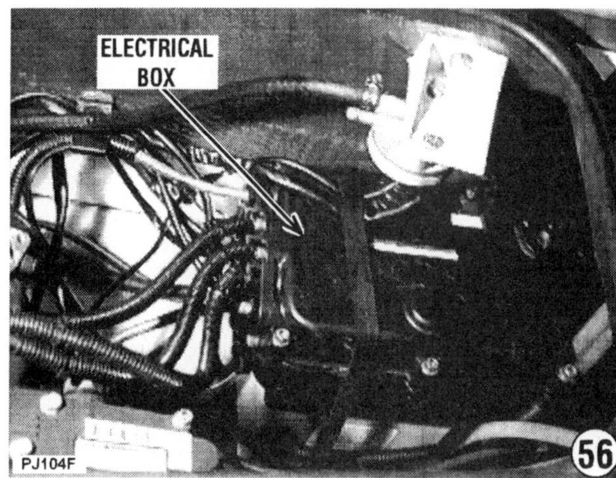

opening and onto the engine mounting bracket studs. Check to be sure the engine has not crimped any hoses or wires

Tighten three of the engine mounting bracket nuts to a torque value of 45 ft lbs (61Nm). Tighten the aft port side nut just "snug", because the nut will be removed later when the aft expansion bracket is installed on the stud.

56- Install the battery and connect the positive and negative cables. Position the battery cover on top of the battery. Connect the stator harness leads to the electrical box. These leads feed through the box, secured by a watertight grommet. It may require referencing the electrical schematic to identify particular lead connections. Once the connections are complete, close the electrical box, checking to be sure no wires are caught between the two halves and tighten the attaching hardware securely. Place the electrical box on the covered battery and secure the "stack" with the two restraining straps.

ENGINE SUPPORT EQUIPMENT INSTALLATION

Exhaust Manifold Installation

57- Insert the attaching bolts into the exhaust manifold. Place a **NEW** gasket over the bolts and seat it against the manifold. Position the manifold over the exhaust ports. Thread the attaching bolts "finger tight", to permit the manifold to be shifted slightly, if necessary. After all bolts are "finger tight", secure the bolts to a torque value of 16 ft lb (22nm) -- Fuji engine. If servicing a Polaris engine, tighten the manifold bolts to a torque value of 22 ft lb (30Nm).

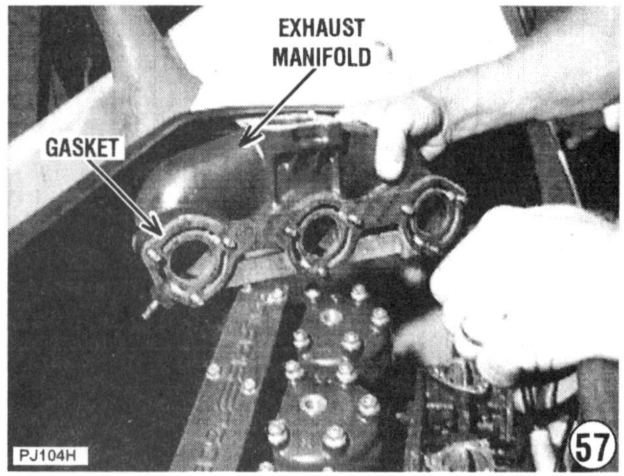

58- Install a **NEW** gasket atop the exhaust manifold. Note any markings indicating the proper positioning of the gasket -- such as "**UP**". The fuel and oil tanks must be moved forward to allow room to lower the expansion chamber into the engine compartment. On some models, it may be necessary to remove the oil tank. With the tank moved forward, lower and rotate the expansion chamber into the compartment, in front of the engine and work it around to position on the port side.

59- Position the expansion chamber on top of the exhaust manifold. Install the attaching bolts, taking note of different lengths. With the bolts properly positioned, tighten them to a torque value of 40 ft lb (54Nm).

60- Secure the stay bracket to the expansion chamber, if it was removed. Install the bolts securing the expansion chamber stay bracket to the flywheel cover. Tighten the bolts securely.

61- Install the aft bracket for the expansion chamber. One of the engine support bracket studs is used to secure the lower part of the

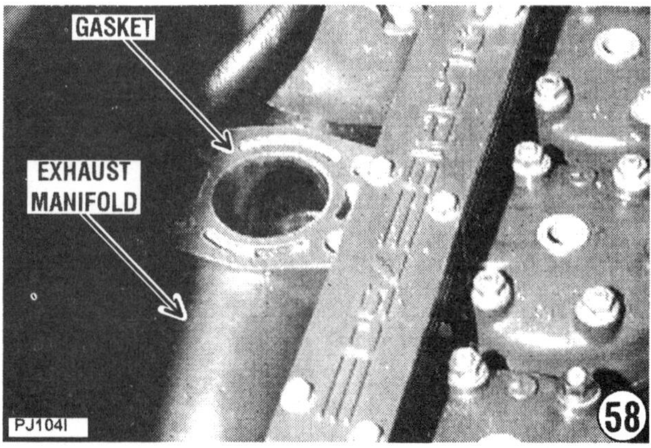

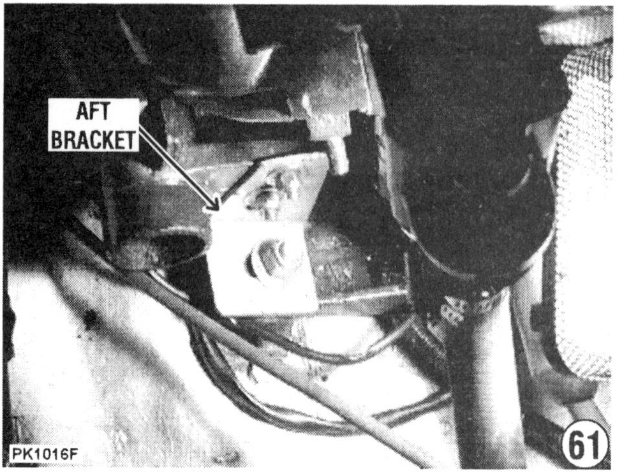

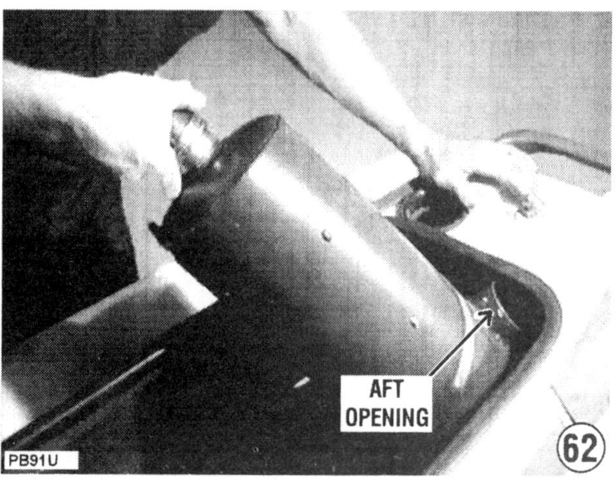

bracket. Tighten the engine mounting bracket nut to a torque value of 45 ft lbs (61Nm).

62- Work the exhaust muffler into the engine compartment and aft into position with the large opening at the aft end facing up and the metal strip on the bottom of the muffler positioned in the groove in the bottom of the hull.

63- Check to be sure a couple of hose

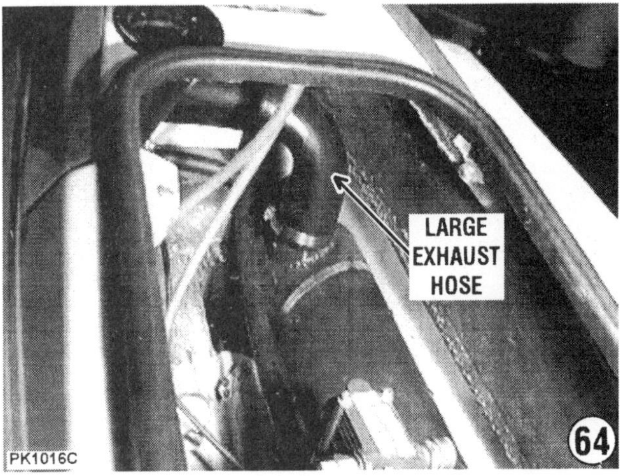

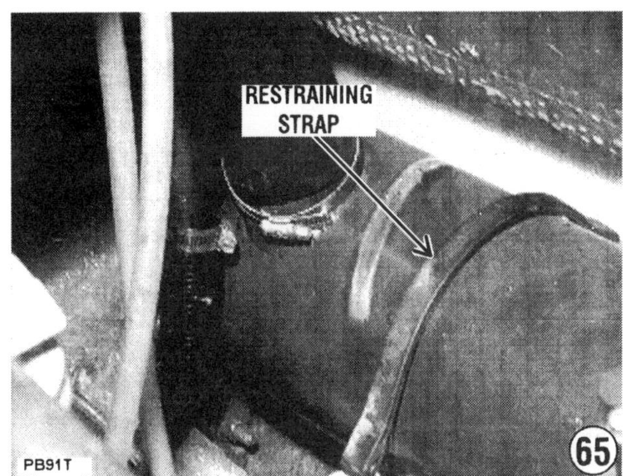

clamps are on the forward muffler hose. Shift the muffler forward and at the same time slide the forward hose over the expansion chamber exhaust fitting. Secure both ends of the hose with the hose clamps.

64- Verify a hose clamp is on the exhaust hose. Slide the large exhaust hose down over the muffler fitting and secure it in place with the hose clamp.

65- Hook one end of the muffler restraining strap onto the fitting outboard down low in the hull. Secure the muffler in place by stretching the "bungee" type restraining strap over the muffler and hooking it in place in the fitting secured in the hull.

CONNECTION WORDS

Connect hoses and linkage items per the tags attached during the disassembling. If the tags were not attached as instructed, in most cases, the shape and length of the hoses and linkage will indicate where the connection is to be made.

8-4 COMPLETE SERVICE TWO-CYLINDER ENGINE

ADVICE

Before commencing any work on the engine, an understanding of two-cycle engine operation will be most helpful. Therefore, it would be well worth the time to study the principles of two-cycle engines, as outlined briefly in Section 8-2. A polaroid, or equivalent instant-type camera is an extremely useful tool, providing the means of accurately recording the arrangement of parts, hoses and wire connections **BEFORE** the disassembly work begins. Such a record will prove most valuable during assembling.

IMPORTANT WORDS

The following series of illustrations were developed while servicing a Hurricane Series engine. Other 2-cylinder engines differ slightly in appearance, but have the same basic components. Where differences occur, such as bolt tightening patterns or torque values, the differences will be clearly identified.

REED BLOCK WORDS

If the only work to be performed on the engine is servicing the reed block assembly, remove the flame arrestor and carburetors by performing Disassembly Steps 22 through 29 on Page 8-42, then go to Page 8-63 in the Cleaning and Inspecting section.

Clearance between the reed block and the reed valve is 0.15" (0.38mm) maximum.

After the reed block and intake manifold assemblies are installed, install the carburetor and the flame arrestor.

Engine Overhaul

Move the craft to a suitable work area protected from the elements, if possible.

NOW, THESE WORDS

As mentioned earlier, the following procedures and supporting illustrations were developed while servicing a Hurricane engine. Since some differences may occur between engine models, any differences between other 2-cylinder Series engines will be clearly identified.

TWO-CYLINDER ENGINE REMOVAL

Take time to identify and tag the various hoses and electrical leads which are disconnected during the disassembly process. Such a practice will benefit the installation procedures.

1- The engine Serial Number is located atop the flywheel cover. This number **MUST** be available when ordering replacement parts and special tools required for servicing the engine.

2- Disconnect the positive and negative leads at the battery. Release the "bungee"-type

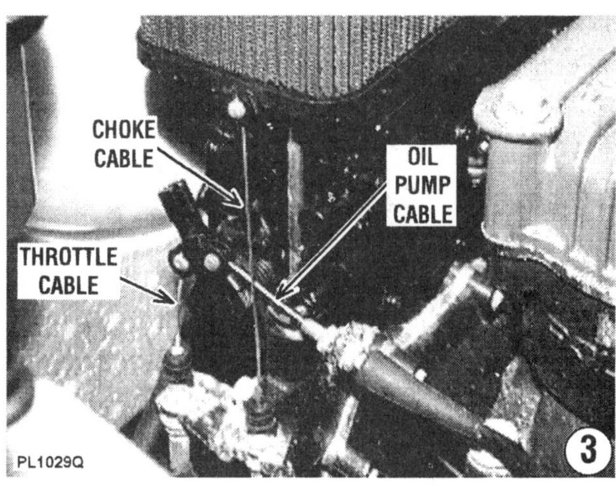

restraining strap and move the battery aft out of the way.

3- Disconnect the choke cable, throttle cable, and oil injection cable from the carburetor linkage arm.

4- "Pinch" the oil supply hose to the pump and then disconnect the hose and the return hose from the oil tank. Back out the screw securing the small bracket to the forward wall of the oil tank. Once the bracket is free, release the tank from the bracket, and then lift the oil tank up and out of the engine compartment.

5- "Pinch off" the inlet hose to the water separator filter. Use some "white-out" or tag to identify the supply hose from the water separator filter, to the fuel pump incorporated in the aft carburetor, and then disconnect the hose from the filter.

6- Identify the supply and return hoses from the fuel pump on the aft carburetor, and mark the return hose at the carburetor. After it is marked, disconnect the supply hose from the fuel/water separator. Disconnect the return hose. Note: Hose at aft carburetor is a supply hose and the hose between the carburetors is the return hose.

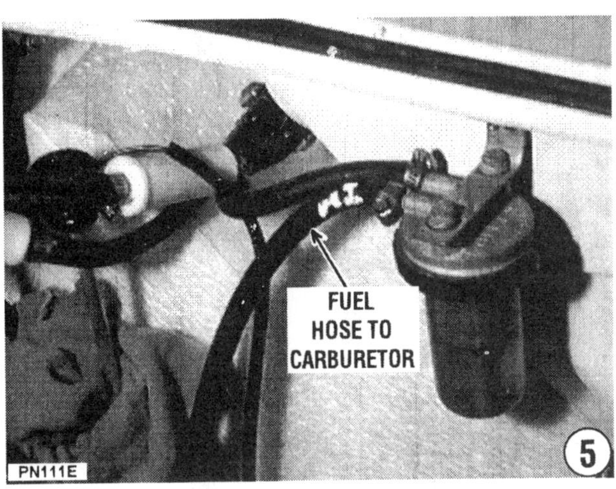

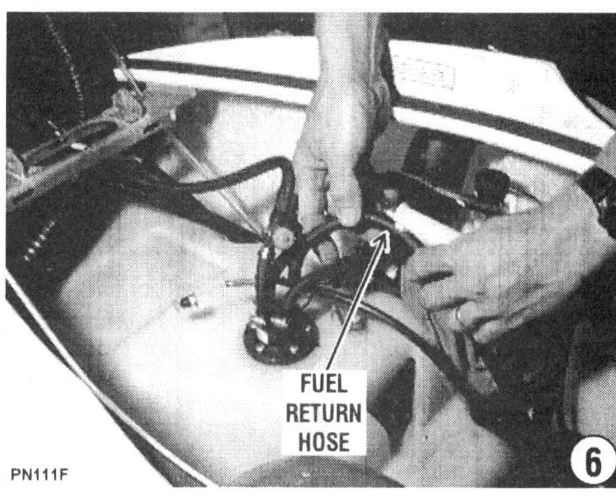

7- Remove the four bolts on top of the expansion chamber. Two of these bolts may be difficult to remove because the expansion chamber is so close to the curved portion of the hull.

8- Remove the forward inboard bolt securing the expansion chamber to the exhaust manifold.

9- Disconnect -- by loosening the hose clamps -- the large cooling hose to the expansion chamber and the smaller hose to the exhaust manifold.

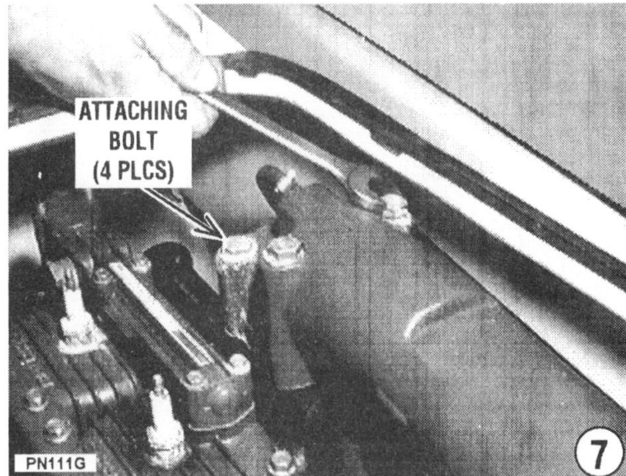

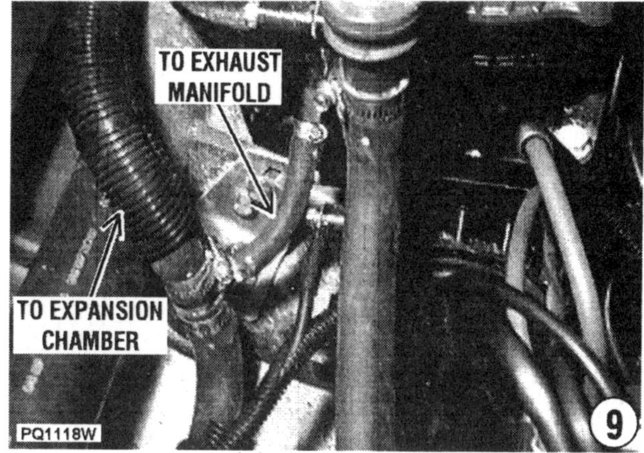

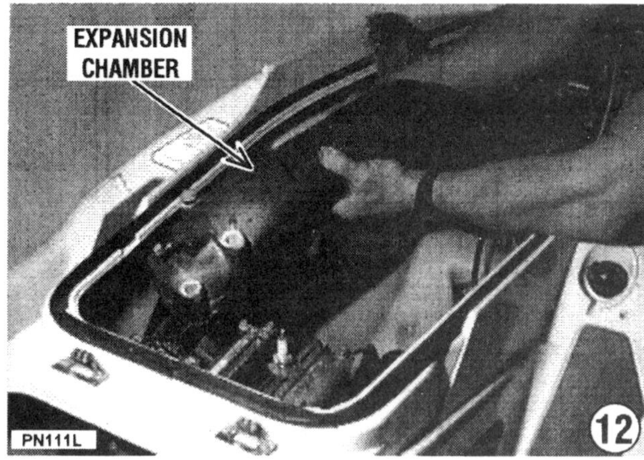

10- Remove the two bolts securing the expansion chamber to the hull through a bracket attached by the aft port side engine mounting bolt.

11- Loosen the hose clamp and disconnect the large exhaust hose from the expansion chamber. This hose does not come off easily.

12- Lift the expansion chamber up, free of the exhaust manifold, and then rotate it slightly, and work it out of the engine compartment.

13- Remove the three Allen head screws securing the electrical box to the small bulkhead.

14- Lift the electrical box up, but not clear of the engine compartment. Now, work the water proof grommet for the temperature sensor lead out of the electrical box opening. Work the water proof grommet out of the opening for the stator harness leads.

15- Remove the bolts securing the two halves of the electrical box together, and then open the box. Disconnect the stator harness leads and the temperature sensor lead from the quick disconnect terminals in the box. Note the leads and the terminals are color identified.

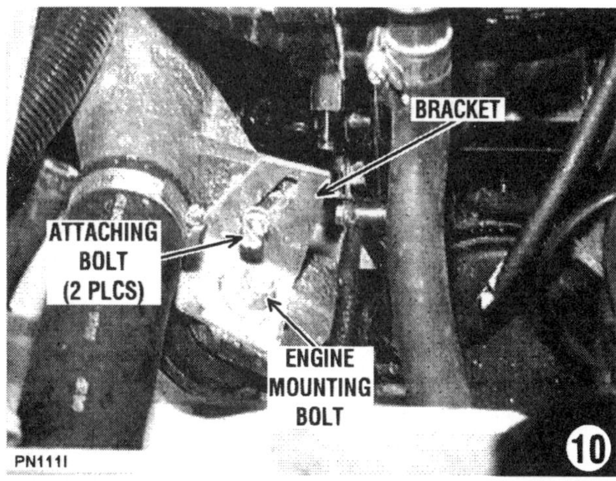

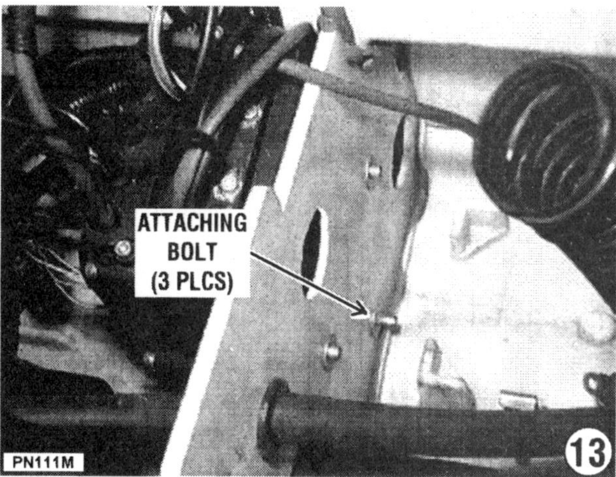

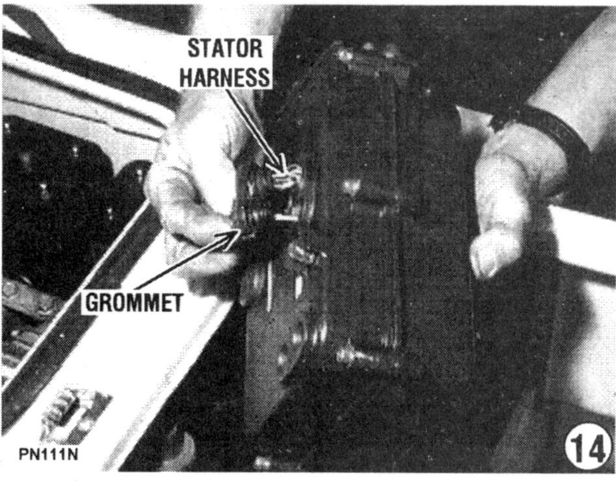

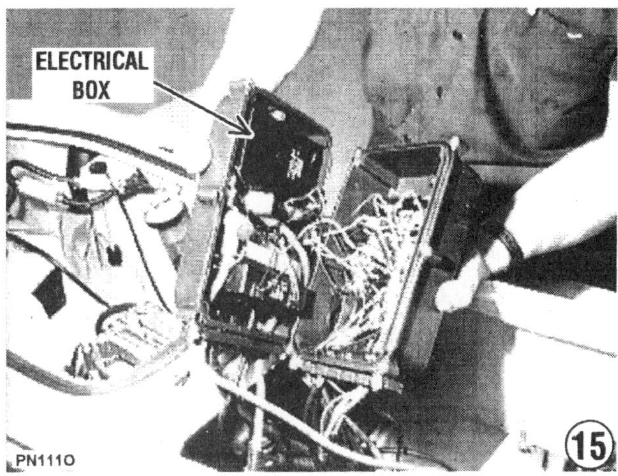

Pull these leads back out through the openings and free of the box.

16- Remove the six attaching bolts, and then remove the exhaust manifold.

17- Disconnect the high tension lead from the cranking motor terminal.

18- Loosen the hose clamps securing the cooling water outlet hose to the water cooling manifold. Remove the bolt securing the cranking motor ground wire to the engine block. Slide the ground wire free.

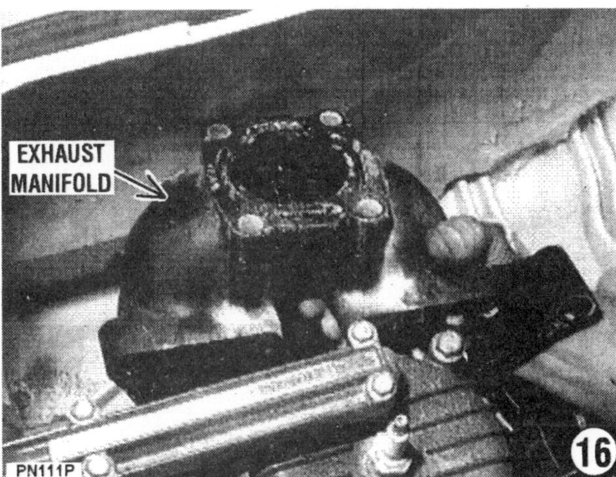

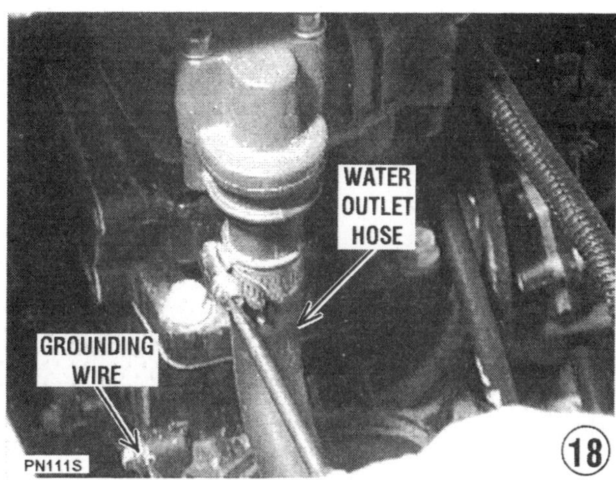

19- Pry the driveshaft coupler cover free. No bolts, screws, or clamps are used. Loosen the hose clamp securing the driveshaft coupler to the driveshaft. Once the clamps are loose, work the coupler aft and free of the driveshaft.

20- Take time to wrap all loose electrical leads and hoses around the engine to ensure they do not hang up when the engine is lifted from the craft. The manufacturer suggests an eyelet be welded to two bolts with the same threads and the proper size to fit into the spark

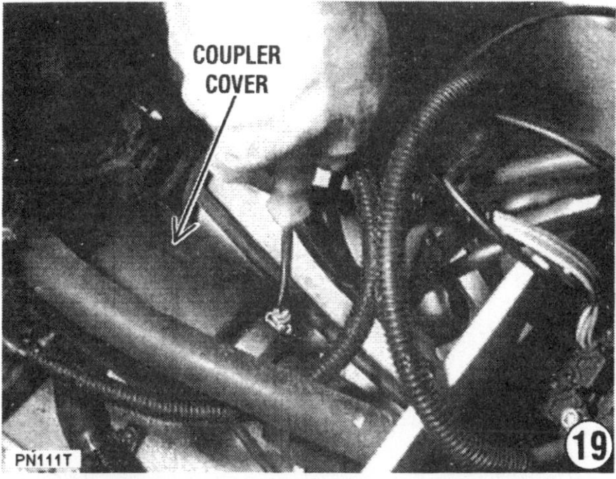

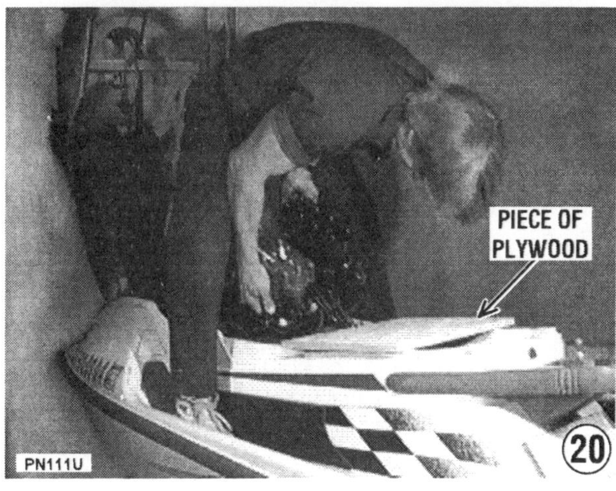

plug openings. Obtain a suitable lifting device. Pass a chain through the two eyelets threaded into the spark plug openings and up to the lifting device. Guide and lift the engine clear of the engine compartment.

Easy Alternate Method

Place an adequate size piece of plywood on the aft portion of the craft, as shown. One man in reasonable good physical shape can straddle the engine compartment, grasp the engine on the port and starboard side, and then using his legs, not his back, lift the engine up and clear of the compartment -- then set in on the piece of plywood.

21- Move the engine to a suitable work surface of convenient height.

HELPFUL WORDS

During the disassembly work, if a component or assembly is determined to be in satisfactory condition for further service, "let a sleeping dog lie", do not disturb the component, but continue with only the tasks required to return

the engine to satisfactory operating performance.

DISASSEMBLE -- TWO CYLINDER ENGINE

22- Remove the nuts securing the air intake/flame arrestor cover to the studs in the carburetors. Lift the cover and individual screen-type filters off the air intake/flame arrestor base plate.

23- Disconnect the pulse hose from the aft carburetor. The forward carburetor does not have a fuel pump -- therefore no pulse hose. The forward carburetor is serviced with fuel from the aft carburetor.

24- Using a pair of needle nose pliers, release the spring clamp on the oil supply hose to the carburetor. Work the oil hose free of the fitting. Repeat the procedure for the other carburetor.

25- Remove the carburetor mounting bolts, two per carburetor.

26- Lift the carburetors up -- as a set -- and free of the intake manifold. Move them to a

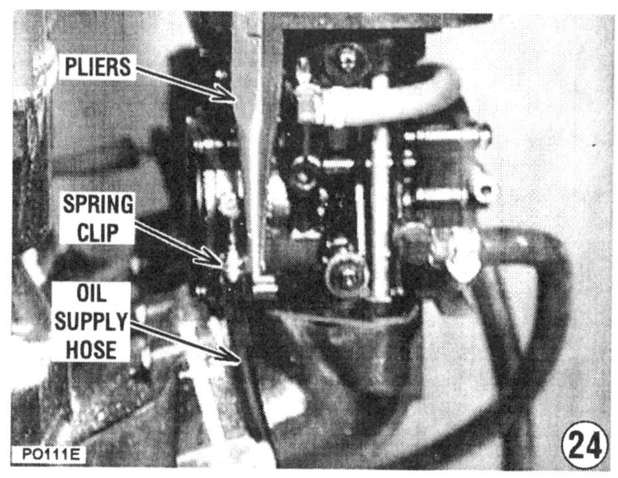

clean surface and cover them with a shop cloth to prevent dust, dirt, or other contaminants from finding their way into the carburetors. See Chapter 6 for detailed illustrated procedures to disassemble the carburetors, including the integral fuel pump.

27- Remove the bolts securing the intake manifold to the upper crankcase.

28- Carefully lift the manifold off the block.

29- With care, remove the reed blocks. These blocks are held in place by the intake manifold. Therefore, no bolts or screws are involved.

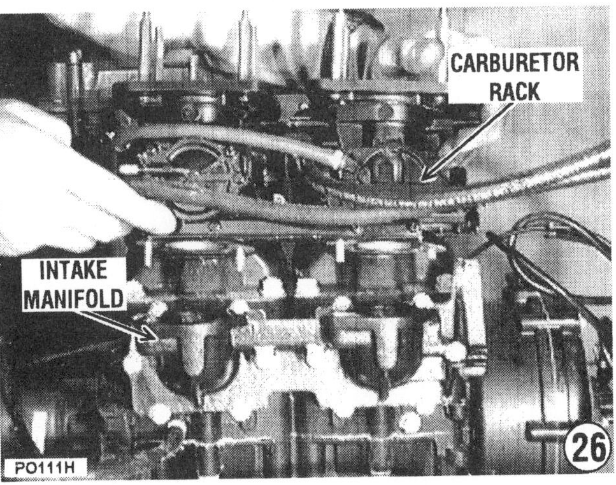

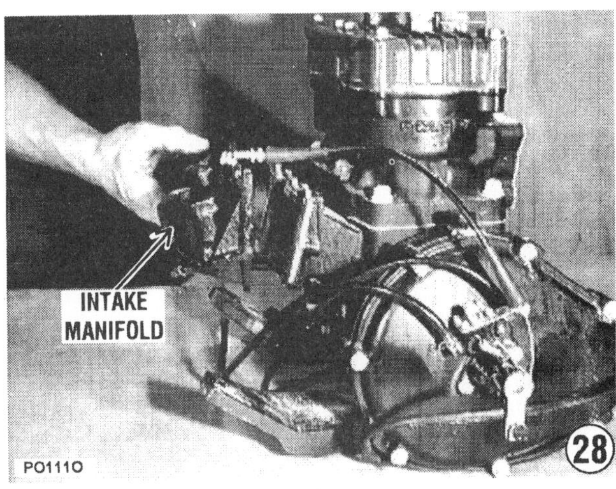

ILLUSTRATION WORDS

The illustrations for the next four steps -- Steps 30 thru 33 inclusive -- were taken while the engine remained in the craft.

Cylinder Head Removal

30- Remove the bolts securing the cylinder head cover bracket, and then remove the bracket.

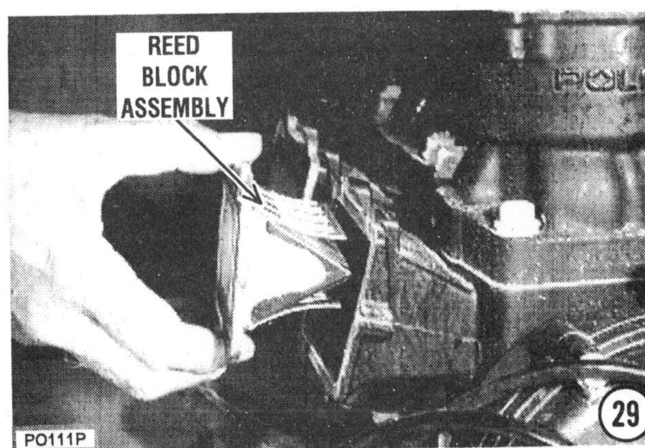

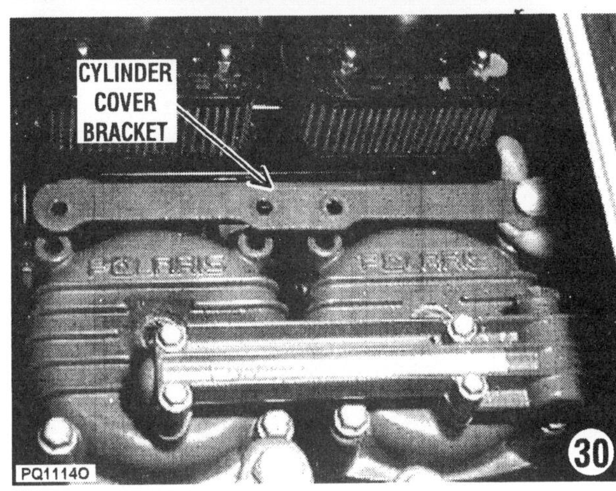

31- Remove the water manifold mounting bolts, and then remove the manifold.

32- Back out the cylinder head cover mounting bolts, and then lift the cover free.

33- Remove the cylinder head bolts, and then the cylinder head.

Repeat the previous four steps for the other cylinder.

ILLUSTRATION WORDS

The illustrations for the next 8 steps were developed while disassembling a Polaris 3-cylinder engine. However, the procedures and techniques involved to dismantle a 2-cylinder engine are basically the same. Therefore, the illustrations are valid for this section covering a two cylinder engine.

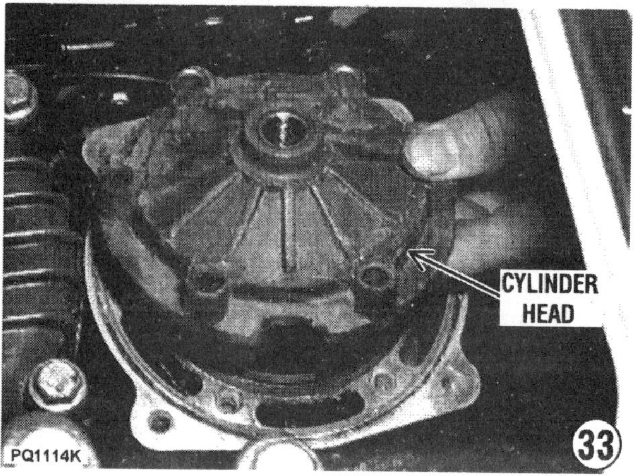

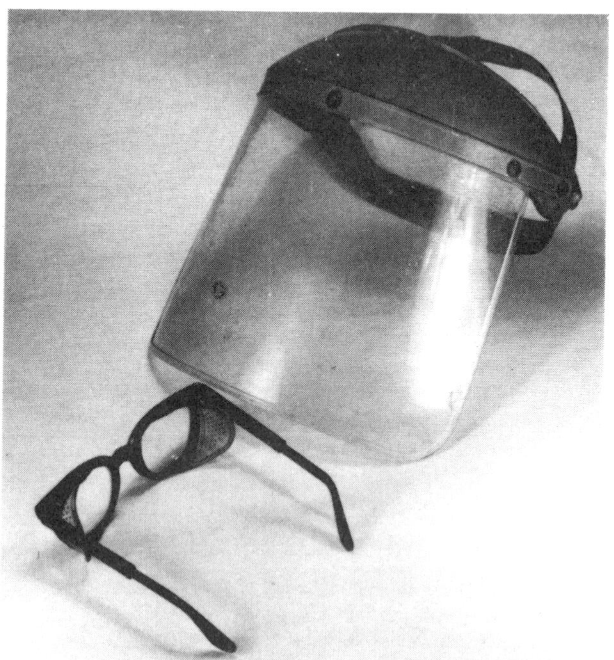

Snap rings are under tremendous tension during removal and installation, presenting a potential hazard. As an eye protection measure, safety glasses or a shield should ALWAYS be worn during work with such snap rings. Warn others in the area such work is in progress.

34- Loosen and remove the four bolts securing the cylinder to the crankcase. **CAREFULLY** lift the cylinder straight up, free of the piston and crankcase. If the other cylinder is to be removed, perform the same procedure.

SAFETY WORDS

The piston pin C-lockrings are made of spring steel and may pop out of the groove with considerable force. Therefore, warn others in

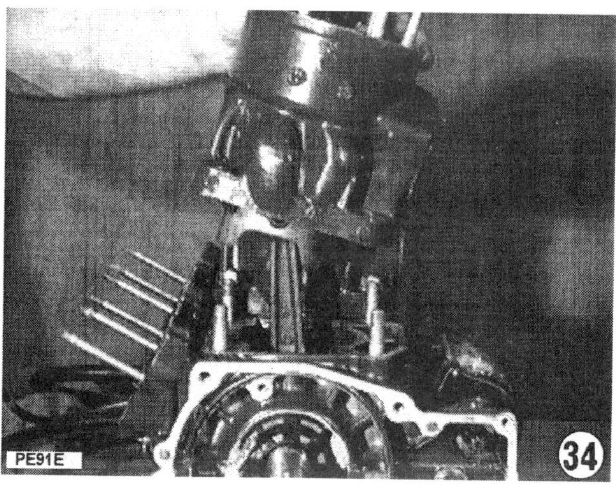

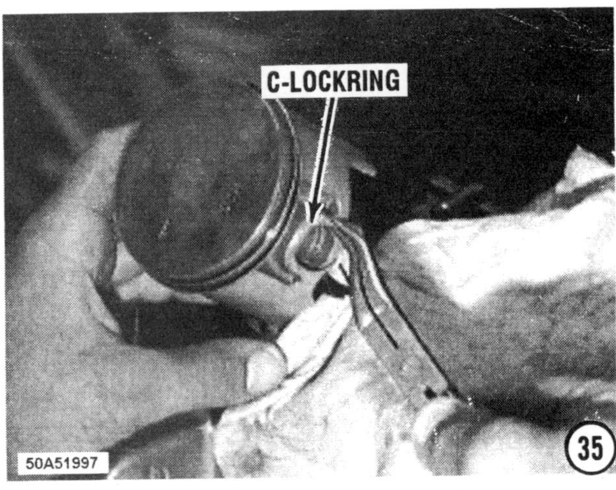

the area and **WEAR** eye protection glasses or a shield while removing the piston pin C-lockrings.

35-- Remove the C-lockring from both ends of the piston pin using a small s;pari of needle nose pliers in the small groove next to the piston pin opening. Discard the C-lockrings. These rings stretch during removal and must **NEVER** be used a second time.

36- Obtain piston pin remover tool, P/N 2870386. Polaris manufactured engines -- require adaptor P/N 2871445 for use with the special tool. Push the piston pin out -- free of the piston, or use the special tool. If the special tool is not available and the pin refuses to dislodge, use a short piece of wooden doweling and light taps with a hammer to start it moving.

Remove the caged needle bearings from the small end of the connecting rod.

ADVICE ON NEEDLE BEARINGS

New needle bearings should be installed in the connecting rods, even though they may appear to be in serviceable condition. New bearings will ensure lasting service after the

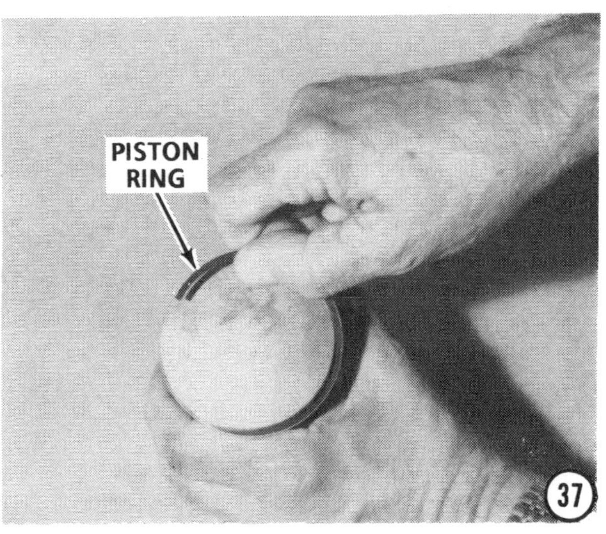

overhaul work is completed. If it is necessary to install the used bearings, keep them separate and identified to **ENSURE** they will be installed onto the same connecting rod from which they were removed.

CRITICAL WORDS

The rod is an integral part of the crankshaft. The two are manufactured together and **CANNOT** be separated. Slowly rotate the bearing races. If rough spots are felt, a new crankshaft must be purchased. The bearings are installed at the factory and are not replaceable.

PROFESSIONAL WORDS

Good shop practice dictates to replace the rings during an engine overhaul. However, if the rings are to be used again, expand them **ONLY** enough to clear the piston and the grooves, because used rings are extremely brittle and break very easily.

37- Gently spread the top piston ring enough to pry it out and up over the top of the piston. No special tool is required to remove the piston rings. Remove the lower ring in a similar manner. These rings are **EXTREMELY** brittle and **MUST** be handled with care if they are intended for further service.

NECESSARY PISTON WORDS

Scribe a mark inside the piston skirt, indicating from which cylinder the piston was removed. All associated hardware for each piston **MUST** be kept together. Turn the piston over and carefully place the matching C-lockrings, piston pin, rings and caged needle bearings

inside the piston to **ENSURE** proper matching during assembling.

Engine Mounting
Bed Bracket Removal

38- Carefully invert the engine block and lower it onto a couple 4" X 4" blocks of wood, as shown, to prevent the connecting rods from taking the weight of the assembly.

Remove the two bolts securing each of the two bed brackets to the engine block, and then remove the brackets. Note the forward bracket is longer than the aft bracket.

39- Remove the bolts securing the two crankcase halves together. There are 6 bolts, two different sizes -- 10mm and 8mm. Take note of the locations for each size bolt.

Separate the crankcase halves. It may be necessary to jar one or both of the halves with a soft head mallet. **DO NOT** attempt to drive any kind of tool between the halves, because such action would surely damage the mating machined surfaces.

Lift the crankshaft clear of the upper crankshaft half.

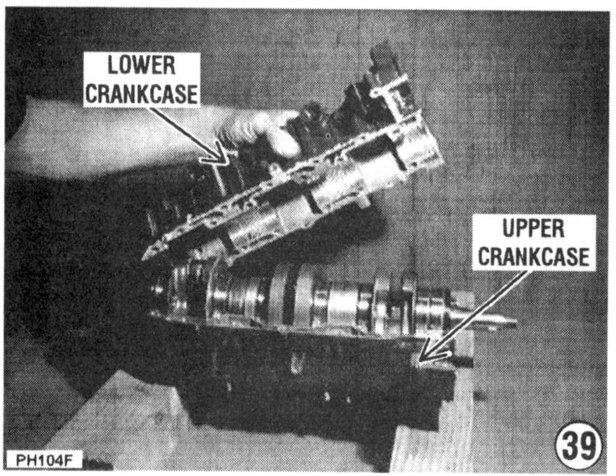

The engine can now be considered completely disassembled.

HOWEVER, the seals at both ends of the crankshaft may be removed if the determination is made they are no longer fit for service. See exploded drawing on Page 8-48. The bearings at both ends are a "press fit". Therefore, special tools will be required to remove and install them.

Exploded drawings of the engine follow on the next two pages.

CLEANING AND INSPECTING

Detailed illustrated procedures to clean and inspect virtually all disassembled parts are presented in a separate "Cleaning and Inspecting" section beginning on Page 8-63.

ASSEMBLE -- TWO-CYLINDER ENGINE

The following instructions cover assembling of virtually all parts of a 2-cylinder Polaris engine **AFTER** all used parts have been inspected and cleaned and replaceable parts have been obtained and are on hand.

Good shop practice dictates **NEW** gaskets, seals and **O**-rings be installed anytime the engine is overhauled.

A complete package of all **O**-rings, seals, gaskets, etc. reqiured for a 2-cylinder overhaul is available at one price from from the local Polaris dealer or directly from the factory.

If certain parts were not removed or disassembled, simply skip the steps involved in those components and continue with the necessary tasks to assemble the engine to efficient operation.

A suitable torque wrench is absolutely necessary duringt installationof parts and assemgblies to ensure the engine will perform at top efficiency, when the work is completed.

ILLUSTRATION WORDS

A few illustrations in this section were actually taken during a 3-cylinder overhaul. However, the procedures are valid and any changes from a 3-cylinder unit to the 2-cylinder engine will be clearly called to the reader's attention.

TEXT CONTINUES ON PAGE 8-49

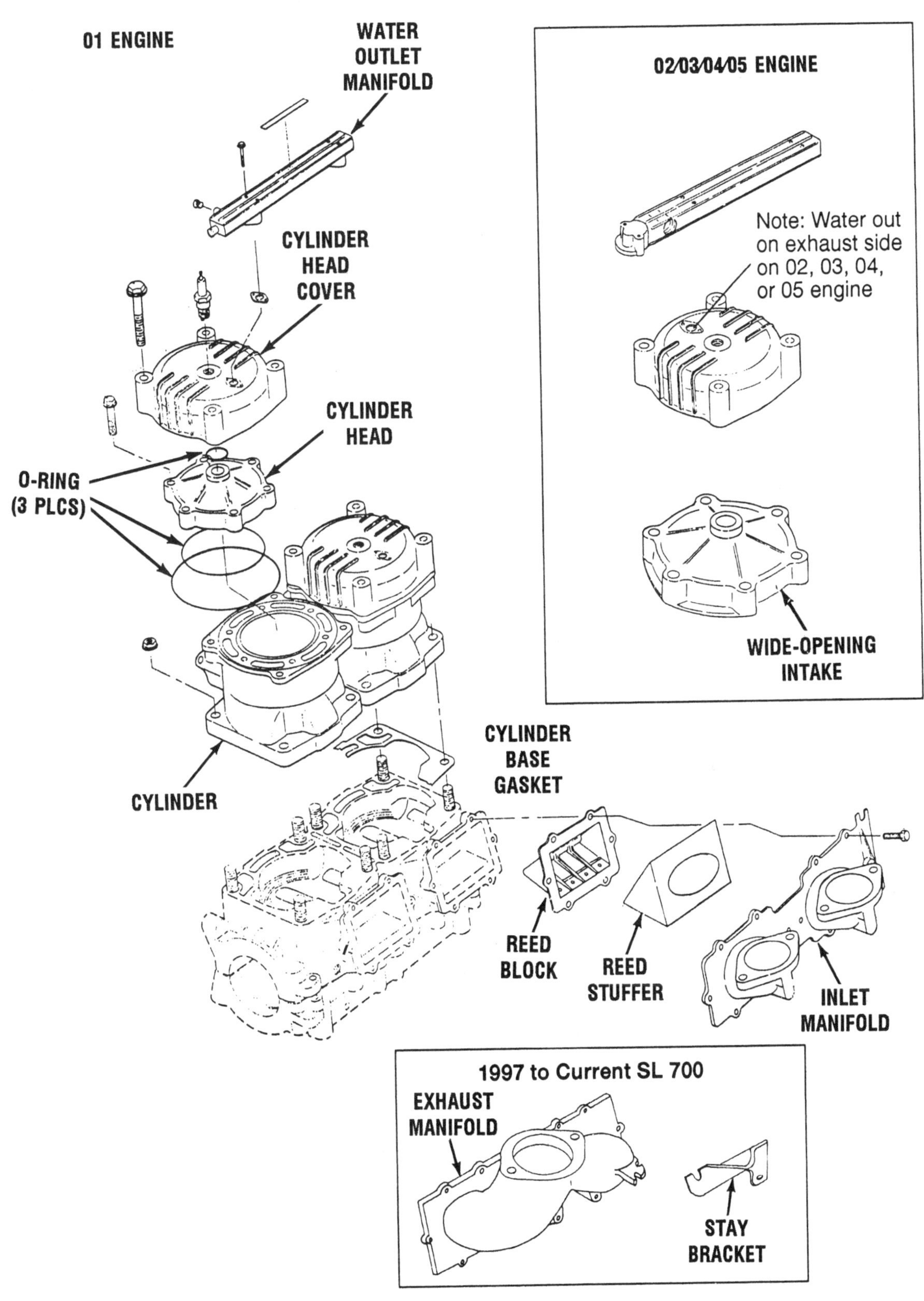

Exploded drawing of 1996 to Current Model SL700, SL700 Deluxe, SLT700 and Hurricane Series engines. Major parts are identified.

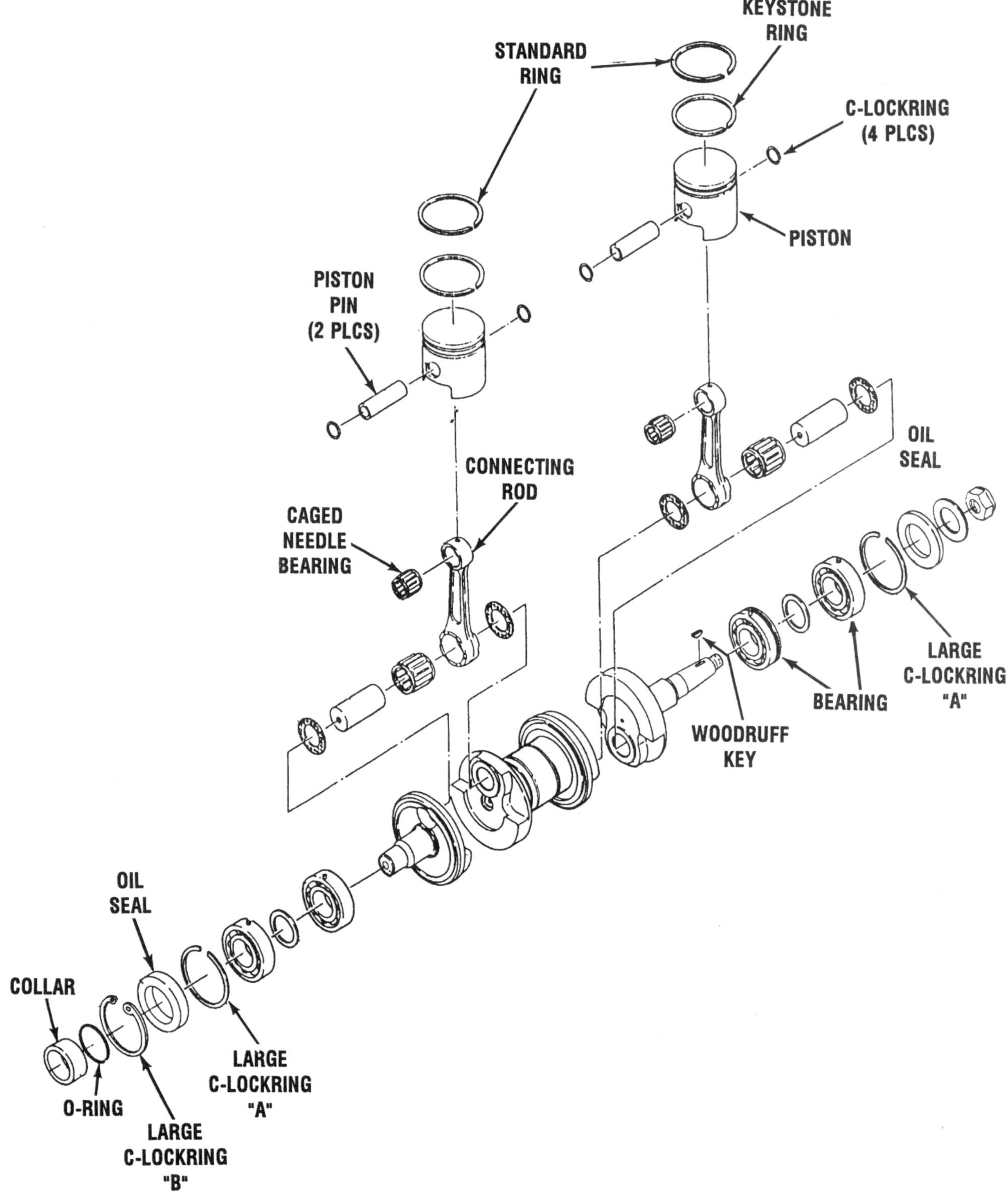

Exploded drawing of the crankshaft assembly on 1996 to Current Model SL700, SL700 Deluxe, SLT700 and Hurricane Series engines. Major parts are identified.

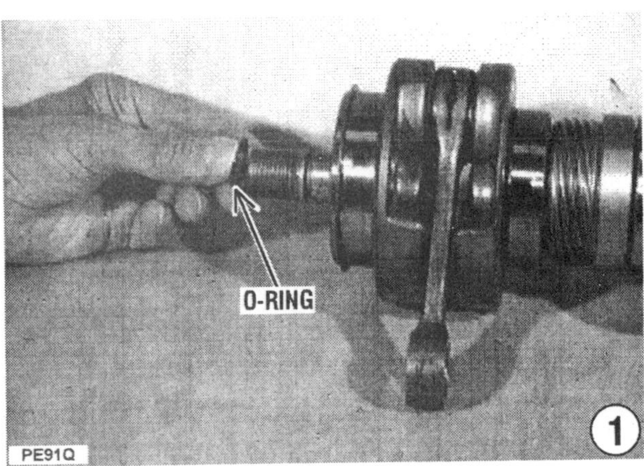

Assemble Lower Half Crankshaft Assembly

1- Install a **NEW O**-ring into the groove on the aft end of the crankshaft.

2- Slide the crankshaft collar over the end of the crankshaft and move it forward -- over the previously installed **O**-ring -- until it seats. Note the bearing will ride on this collar.

GOOD OIL SEAL WORDS

Both seals **MUST** be packed liberally with grease before installation.

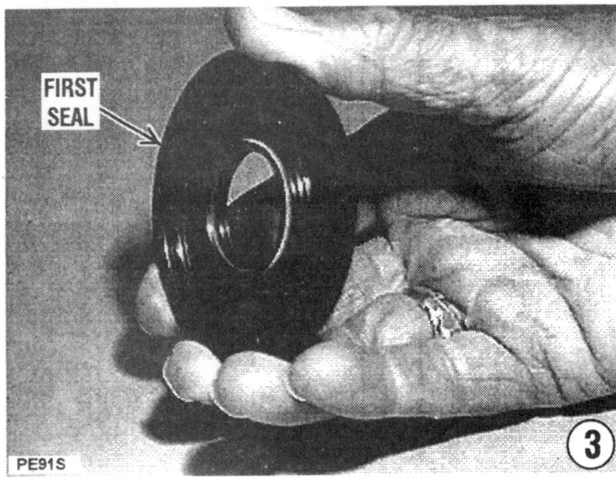

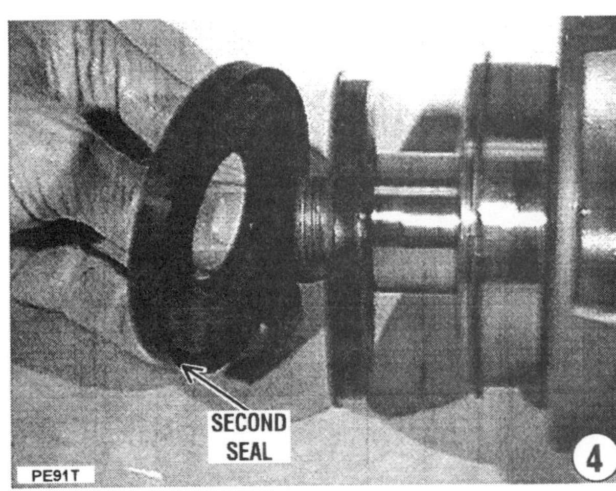

3- Pack a **NEW** oil seal with grease and then push the seal -- over the crankshaft end with the lip going on first. Slide the seal forward until it seats against the collar.

4- Pack a **NEW** second oil seal with grease. For photographic purposes, the accompanying illustration shows the seal installed in Step 2 without grease. Check to be sure the seals face the proper direction.

AUTHOR'S WORDS

The seals are now removed, packed with grease, as described in the text, and then permanently installed

5- Install the second seal in place over the crankshaft end and slide it forward until it seats against the first oil seal.

6- Pack a **NEW** seal with grease, and then install it onto the crankshaft, with the lip of the seal going on first. Slide the seal aft on the crankshaft until it makes contact with the crankshaft bearing, as shown.

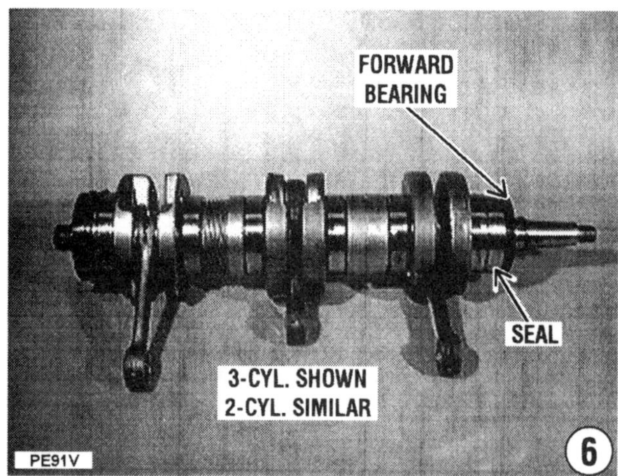

7- Place the upper crankcase half on two blocks of 4"x4" wood -- to prevent placing a load on the connecting rods. Lower the crankshaft into position in the crankcase.

GOOD CRANKCASE ASSEMBLY WORDS

Before joining the two crankcase halves together, check to be sure **ALL** bearing locating pins are properly positioned in their respective slots. Ensure the crankshaft is set in place properly -- "square" in the upper crankcase half. Due to the worm gear driving the oil pump, once the halves are mated, check to be sure the gear for the oil pump drive aligns with the worm gear on the crank. **BEFORE** bolting the two halves together, rotate the crankshaft to verify the worm gears mesh properly and the oil pump shaft rotates smoothly.

8- Apply Loctite Anaerobic Sealant to the crankcase bolts. This sealant helps prevent moisture from building up on the threads, thus preventing the bolt from "locking up" due to corrosion.

Check to be sure **BOTH** bearing locating pins are positioned properly to permit each pin to index into the corresponding pin slot. These pins prevent the bearings from rotating and the crankcase halves will not seat properly if any one pin fails to index correctly into the slot. The accompanying illustrations -- No. 8 and 8A -- show the locating pins at the forward end and in the center of the crankshaft properly indexed into the slots.

9- Apply 3-Bond 1211 sealant, or equivalent, to the mating surface of the lower crankcase half. This particular sealant "sets" slowly, allowing time to accomplish other necessary

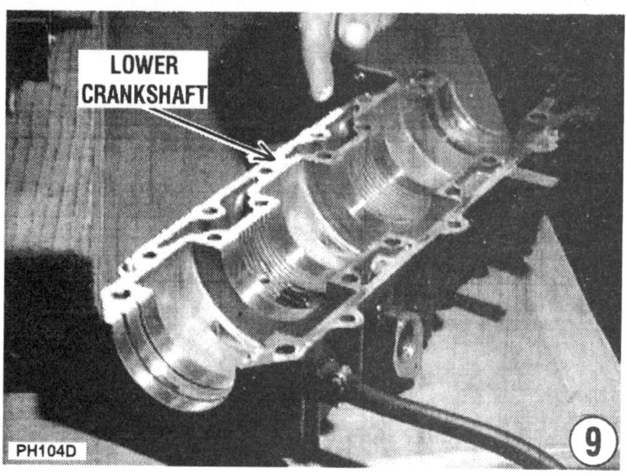

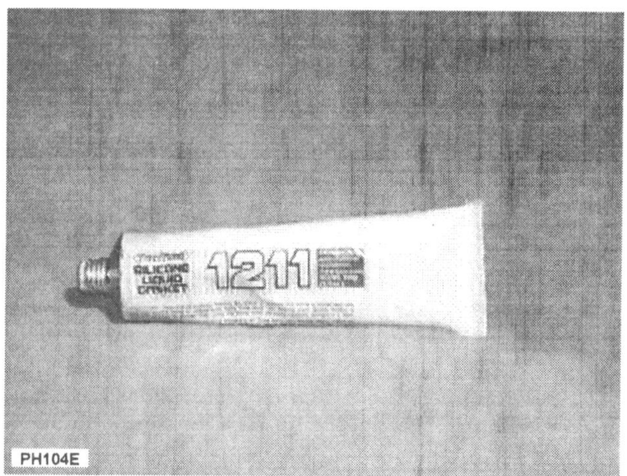

The "3-Bond 1211" sealant mentioned in the text, is a silicone-like substance. This sealant sets much slower than many other sealants, allowing time for other necessary tasks to be accomplished before it sets.

tasks. Therefore, the work does not have to be done at a rapid pace.

10- Check to be sure the mating surfaces of both crankcase halves are free of any old material. Bring the two crankcase halves together. Check to be sure both halves seat together properly.

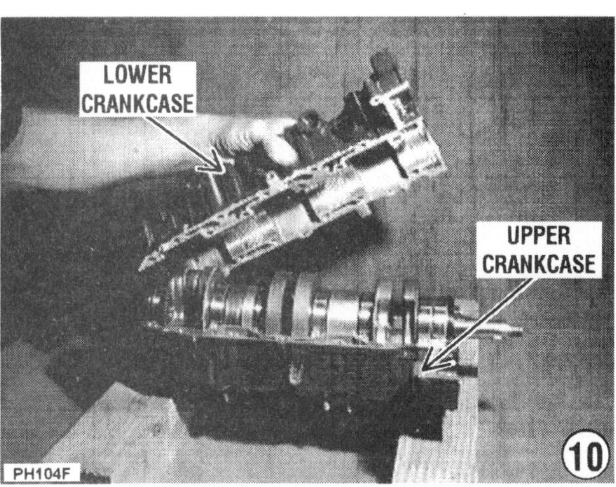

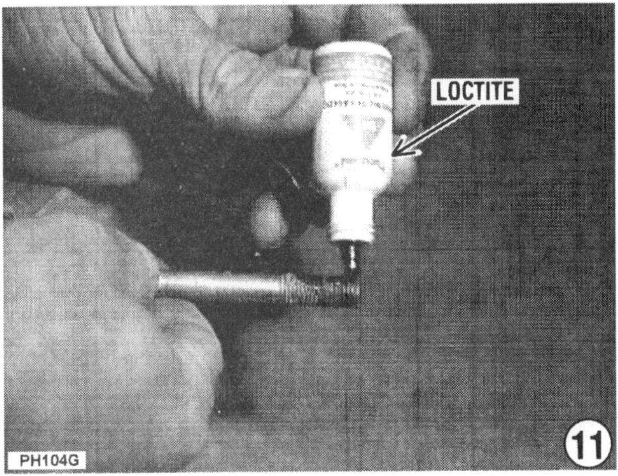

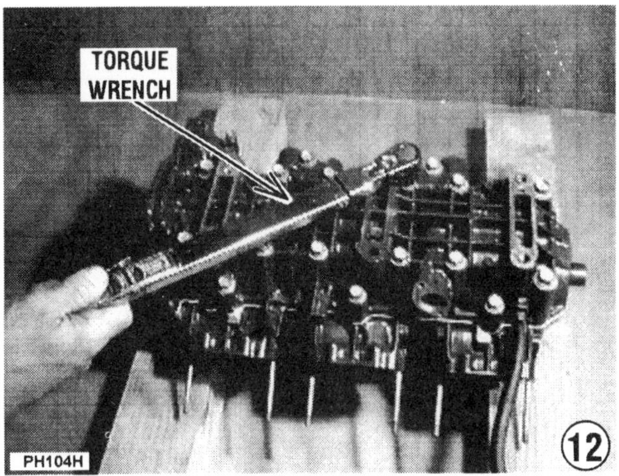

11- Apply a small amount of Locktite® 242, or equivalent to the crankcase bolts. Place the crankcase bolts into the proper position on the lower crankcase. Two sizes are used. The following paragraph describes a method to ensure the bolts are inserted into the proper bolt hole.

Slip each bolt into the estimated opening. Once all bolts are in place, check the exposed length of each bolt. If they were placed in the correct position, the exposed section of all bolts should be approximately the same height. If a bolt is misplaced, it is generally significantly higher or the head of the bolt may sit flush against the crankcase surface. If a bolt is misplaced, and then tightened to the required torque value, it is very possible such action will crack the crankcase. **BAD NEWS** and an expensive mistake.

12- Tighten the 8mm bolts to a torque value of 22 ft lbs, and 10mm bolts to a torque value of 28 ft lbs. Be sure to begin the torque tightening sequence in the center and work toward both ends alternately and evenly.

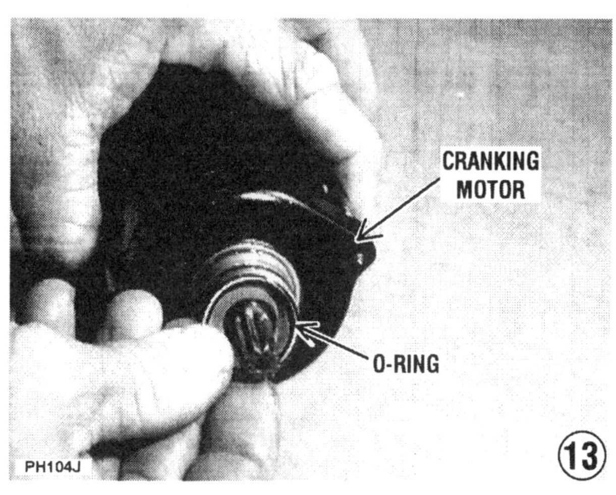

Cranking Motor
Installation

The following three steps are performed with the engine still inverted on the blocks of wood.

13- Install the **O**-ring onto the shaft end of the cranking motor. Check to be sure it seats properly.

14- Slide the cranking motor into position in the forward end of the crankcase. Guide the cranking motor shaft through the opening and seat the motor in place. The unit should seat against the crankcase firmly.

15- Tighten the cranking motor mounting bolts to a torque value of 108 in lb (12Nm).

Stator Installation

16- Make a final inspection of the stator assembly to ensure the harness leads are not crimped, the grommets are intact, the coils are in good condition, and any loose foreign material has been blown clear with compressed air.

17- Press the waterproof grommet into the opening and secure it with attaching hardware. Secure the stator with the bolts tighten to a torque value of 50 in lbs (6Nm).

18- Apply just a dab of grease to the Woodruff key, and then work it into the crankshaft keyway. Install the stator onto the crankshaft by first feeding the electrical harness through the crankcase opening.

19- Install and tighten the drive shaft coupler to the aft end of the crankshaft, just **HAND** tight at this time.

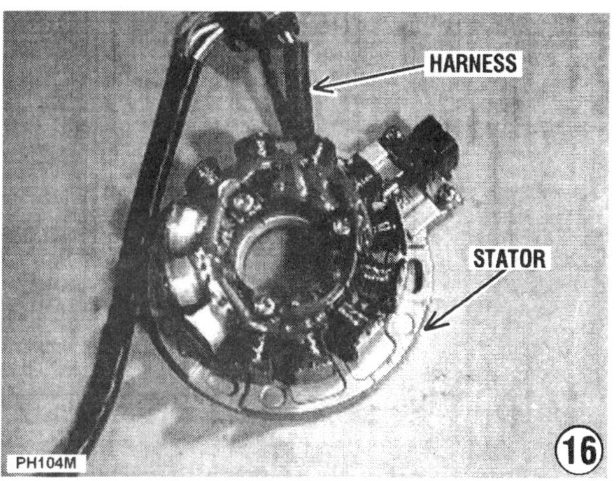

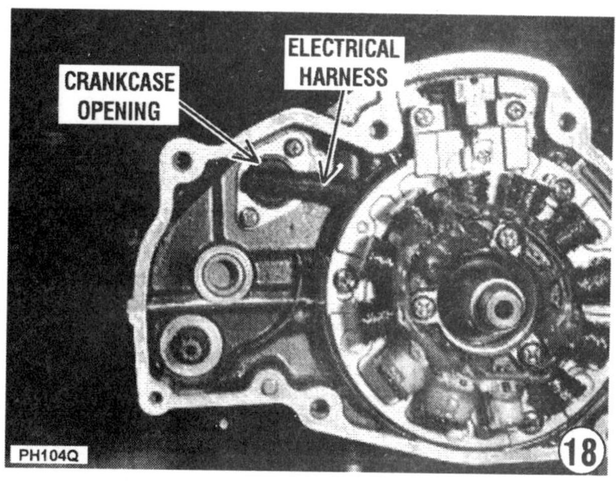

Flywheel Installation

20- Apply a light coating of grease to both ends of the reduction gear. Slide the washer onto the gear shaft. Insert the reduction gear into the recess in the crankcase.

21- Apply a coating of Loctite 262 to the tapered threads on the forward end of the crankshaft. Verify the Woodruff key is fully seated in the crankshaft slot.

22 Move the flywheel onto the crankshaft, with the keyway in the flywheel hub indexed over the woodruff key.

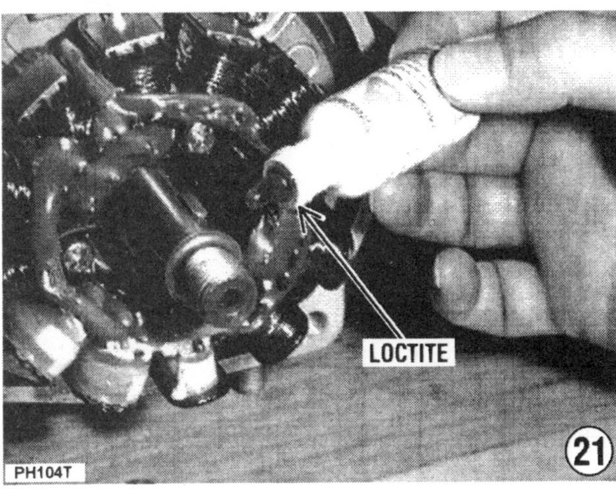

23- Obtain special flywheel holding tool P/N 8700229. Thread the flywheel nut onto the crankshaft, just hand tight. Position the flywheel holding tool over the crankshaft, with the three pins on the tool indexed into the holes in the flywheel. Allow the handle to come down against the workbench surface, as shown.

Tighten the flywheel nut to a torque value of 90 ft lb (122Nm).

24- Apply a coating of sealant to the mating surface of the flywheel cover. Place the washer onto the shaft of the reduction gear. Position

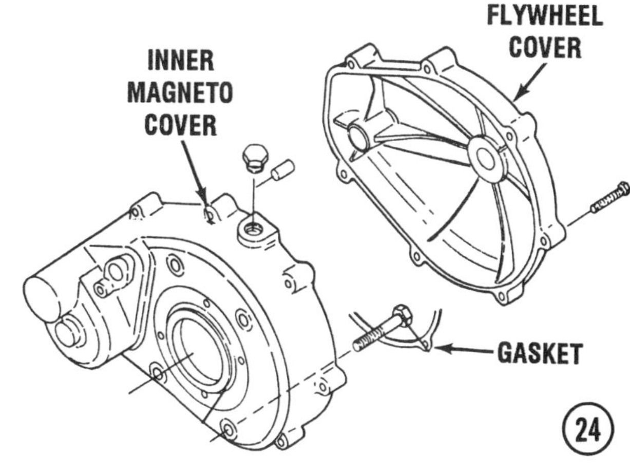

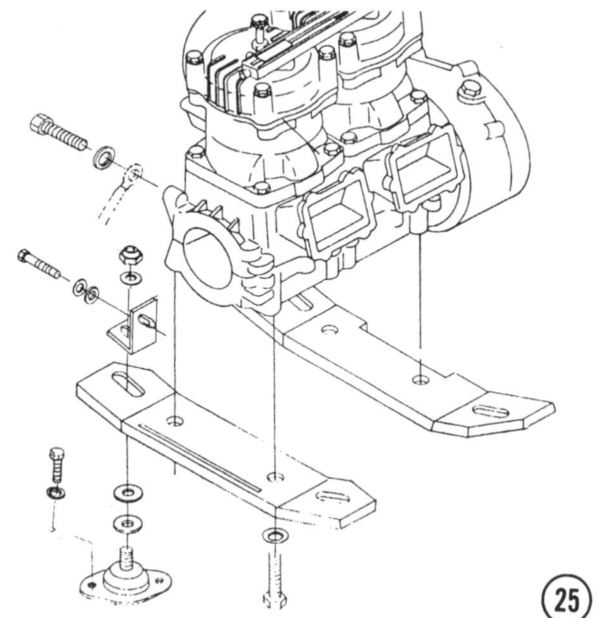

the flywheel cover on the crankcase and install the attaching bolts.

Tighten the 6mm bolts to a torque value of 108 in lb (12Nm); tighten 8mm bolts to a torque value of 22 ft lb (30Nm).

One of these bolts may secure a clip used for clamping an electrical harness. During disassembly, a mark should have been made to indicate the proper location for this clip. Check to be sure the bolt and clip are installed in the proper location.

25- Rotate the assembled crankcase to position the lower half facing upward. Lower the engine mounting brackets onto the crankcase. Check to be sure the brackets are positioned properly. The brackets can only be installed properly one way -- they are **NOT** reversible.

Apply a coating of Loctite® 242 to the threads of the engine mounting plate attaching bolts. Tighten the attaching bolts to a torque value of 50 ft lb (68Nm).

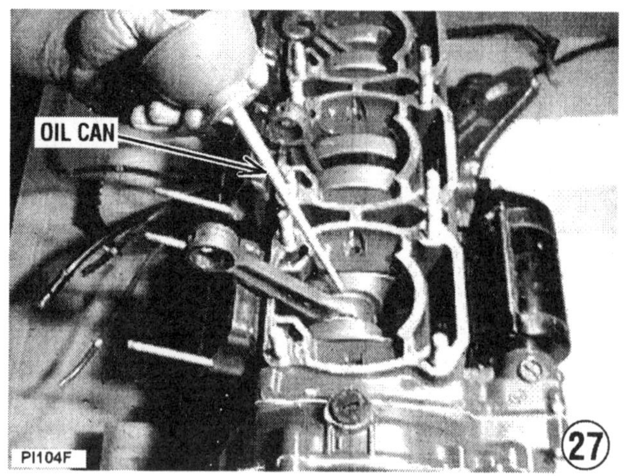

Rotate the engine back to a normal upright position, resting on the engine mounting brackets.

26- Using an oil can, place a "couple squirts" of oil into each oil hole in the crankcase, as shown.

27- Apply a small amount of oil to each lower bearing and to the inside surface of the connecting rod -- small end.

28- Check to be sure the mating surfaces of the cylinders and upper crankcase are clean. Place shop towels or cloths over the crankcase, to prevent contaminants from entering. During piston pin C-lockring installation, a ring may spring loose and could fall into the crank area. **BAD NEWS!** The crankcase would have to disassembled in order to retrieve the lose ring.

GOOD WORDS ON PISTON ASSEMBLING

During piston assembling, old parts being used a second time **MUST** be matched with the piston from which they were removed. Piston pin C-clips **MUST NEVER** be used a second time. During disassembling, these clips are

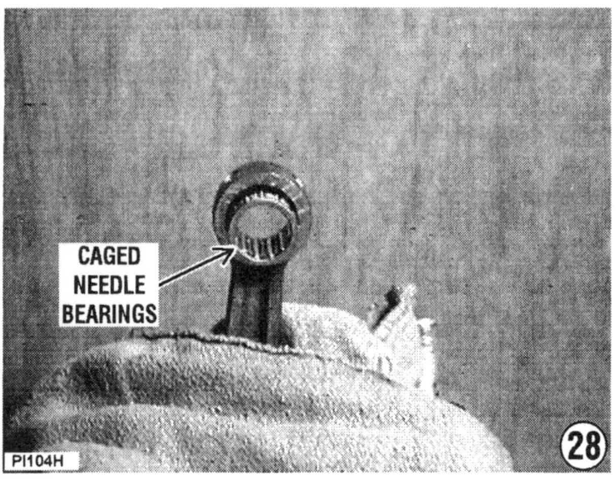

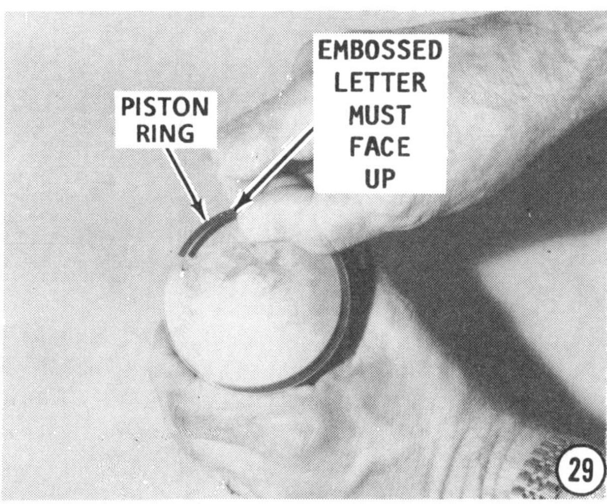

stretched and lose holding effectiveness. Therefore, they should be discarded. Use only new C-clips.

29- Install a new set of piston rings onto the pistons. No special tool is necessary to install the rings. **HOWEVER**, take care to spread the ring only enough to clear the top of the piston. The rings, especially used rings, are **EXTREMELY** brittle and will snap if spread beyond their limit. Align the ring end gap over the locating pin in each piston groove. The purpose of the pin is to prevent the ring from rotating in the ring groove, during engine operation. If the ring does not straddle the pin in the groove properly, the ring would rotate in the cylinder during engine operation. In a short time an end would catch on a port opening and the ring would break.

30- Coat the caged needle bearing with engine oil and then insert the set into the small end of the same rod from which it was removed. If new bearing sets are to be installed, each new set may be placed in either rod.

31- Bring the assembled piston down over the small end of the connecting rod. Align the

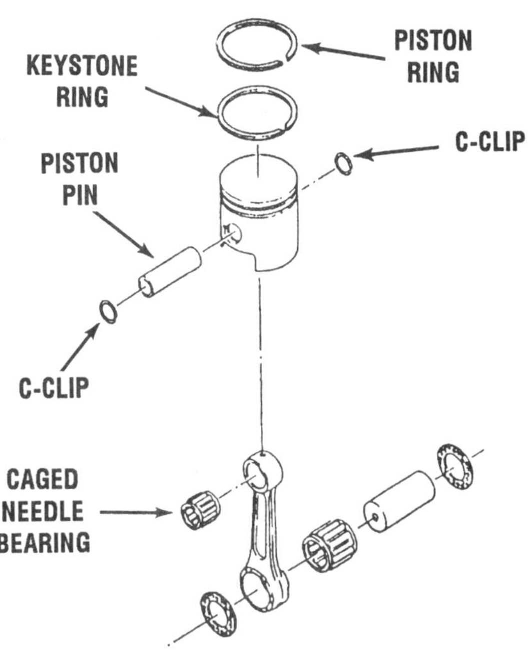

Simple line drawing of the piston installation on a two-cylinder Polaris engine. Because the piston rod is an integral part of the crankshaft and cannot be removed, the lower caged needle bearing cannot be removed.

opening through the piston with the opening in the small end of the connecting rod. Insert the piston pin through the piston and needle bearing. The pin is a "push" fit and a special tool is not required.

Note: If old parts are being used a second time, the piston must be installed back onto the connecting rod from which it was removed. The visible "M" marking on the piston crown in the accompanying illustration indicates this piston was removed from the Magneto (No. 1) cylinder and identified during disassembling.

32- Observe the arrow stamped into the piston crown.

The stamped arrow **MUST** point toward the exhaust side (port side), of the engine.

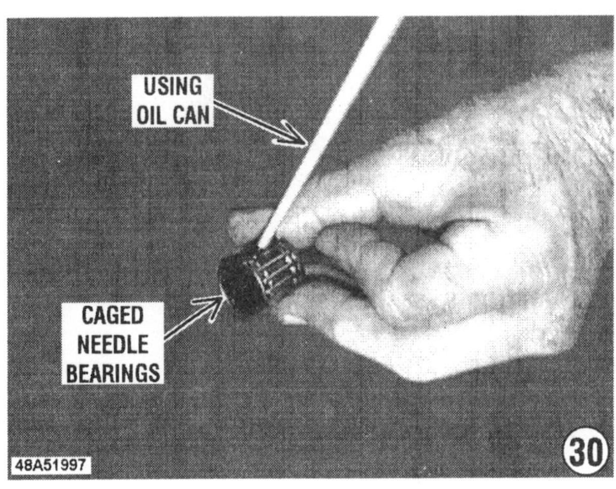

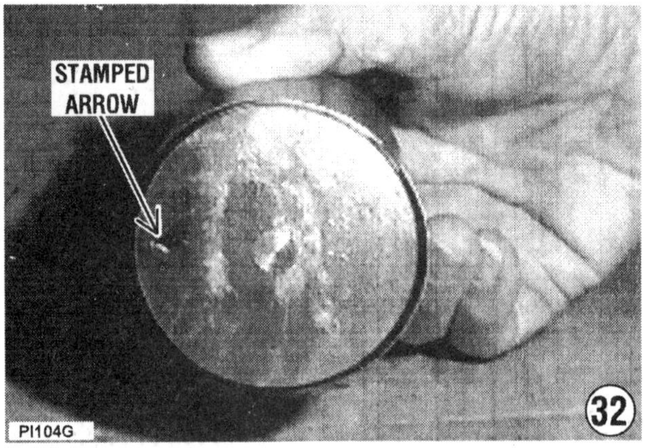

C-LOCKRING SAFETY WORDS

The piston pin C-lockrings are installed under considerable tension and may snap or pop free. Eye protection should be worn and other personnel in the area warned.

33- The piston pin C-rings **MUST** be installed with the gap facing straight up or straight down -- parallel to the vertical line of the piston. Install one end of the clip into the recess, then work the clip around, until the other end can be snapped into place.

34- Assemble the other piston in the same manner. Take time to inspect the piston ring locating pins, to verify each ring gap straddles the pin. The pin prevents the ring from rotating during engine opertion. If the ring was free to rotate, an end would eventually catch on a port opening and break.

Cylinder Installation

35- The cylinder base gaskets have unique surfaces. One side has a fairly hard silicon bead. The other side is clean -- without any substance. Place a **NEW** gasket over each set of cylinder base studs -- be sure the bead of silicone sealant faces up.

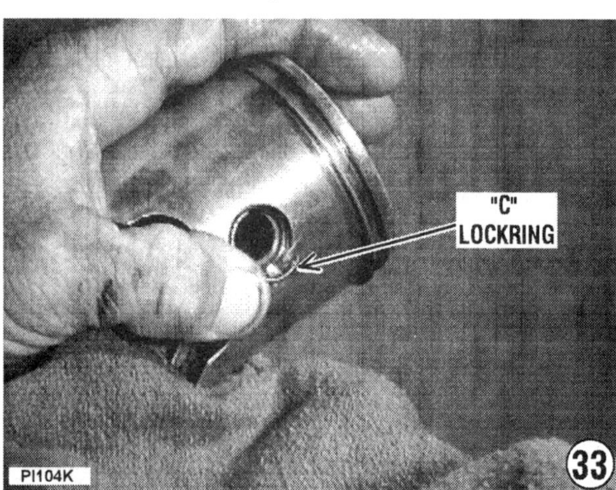

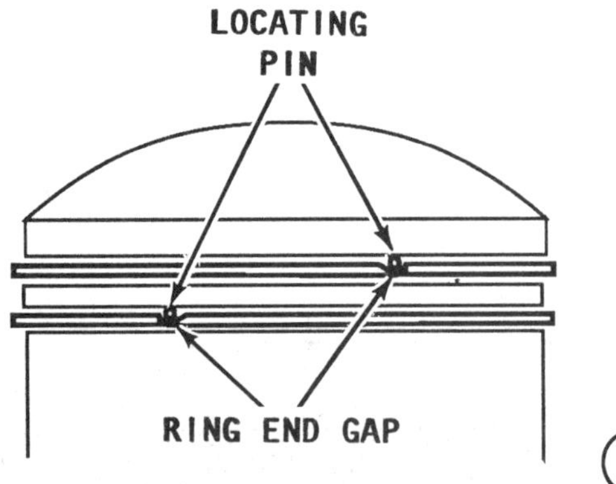

36- Install either piston first. Apply a small amount of engine oil on the piston sides and rings. Lower the cylinder -- down very slowly over the piston and at the same time compress the rings. A ring compressor tool may be used while lowering the cylinder over the piston. If this special tool is not available, the rings may be compressed by hand. Seat the cylinder onto the crankcase.

HELPFUL WORDS

At this time, install the cylinder base nuts and secure them **ONLY** "fingertight". The

cylinders must be able to be shifted slightly to allow the exhaust manifold to be properly aligned with the cylinders **BEFORE** installing the engine into the watercraft. After alignment, the cylinder nuts may be tightened to the required torque value. Once the cylinder nuts are tightened, the exhaust manifold is removed and is not installed permanently until after the engine is secured in the craft. It would be difficult indeed, to attempt to align the exhaust manifold with the cylinders once the engine is installed.

GOOD EXHAUST MANIFOLD INSTALLATION WORDS

The accompanying illustration shows the exhaust manifold being installed in the craft, but actually it is done now **TEMPORARILY** to ensure the cylinders are aligned to permit installation of the exhaust manifold permanently after the engine is in the boat. The task of attempting to shift the cylinders to align with the exhaust manifold after the engine is installed is almost an impossible task.

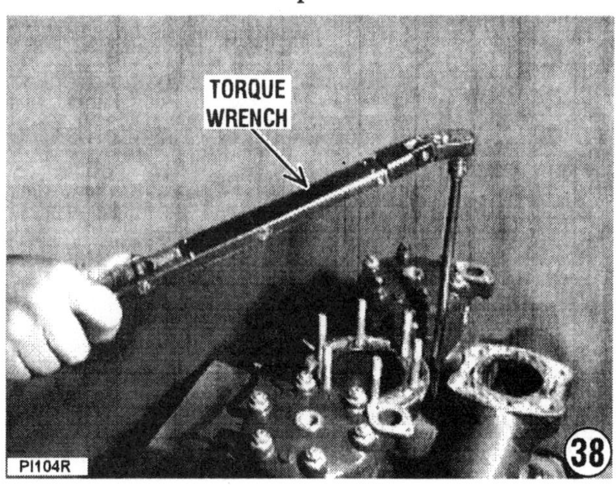

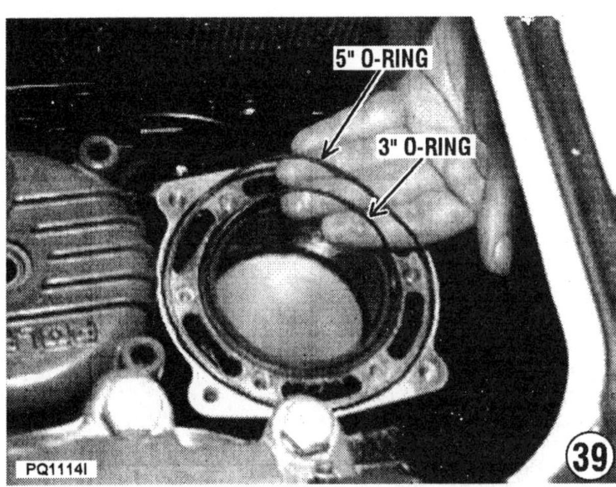

37- Position the exhaust manifold against the cylinders. Install the exhaust manifold bolts and tighten them "just snug" at this time. The cylinders are now aligned to receive the exhaust manifold in the craft. The manifold will be removed prior to installing the engine in the craft, because there is not sufficient room to install the engine with the exhaust manifold secured in place.

38- The manufacturer recommends tightening the cylinder base nuts in a three step sequence, using a criss-cross pattern. Tighten the nuts to a final torque value of 40 ft lb (54Nm). Once the cylinders are tightened properly, the exhaust manifold may be removed, in preparation for engine installation.

Cylinder Head Installation

39- Apply a light coating of grease onto the two new **O**-rings. Fit the **NEW** large -- approx. 5" -- **O**-ring into the outboard groove in the top end of each cylinder. Work the apporx. 3" **O**-ring into the smaller inboard groove, as shown.

40- Lower the cylinder head into place with the small cutout with an embossed **"E"** next to it facing the exhaust manifold. The large cutout

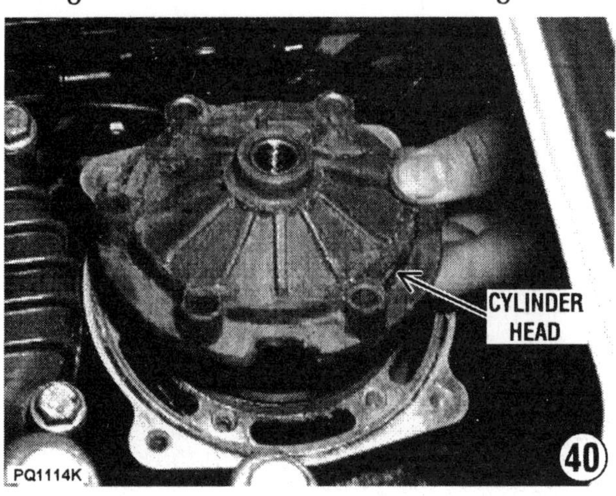

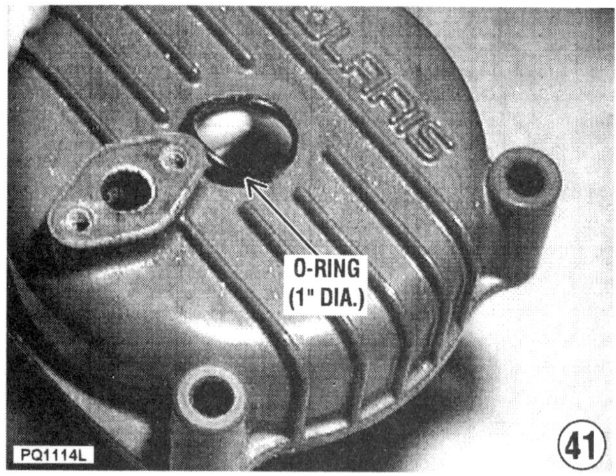

will then be on the intake manifold side. Secure the cylinder head with the head bolts tightened in two sequences to a torque value of 20 ft lbs (27Nm). Tighten the bolts in a cross pattern starting in the center and working alternately towards both sides.

41- Coat a **NEW** -- approx. 1"-- **O**-ring with a light coating of grease, and then force it into the inside groove in the head cover.

42- Position the head cover in place over the cylinder head. Install and tighten the restraining bolts just fingertight, at this time.

Repeat the previous four steps (39 thru 41), for the other cylinder.

43- Position a **NEW** water outlet manifold gasket in place on each cylinder head, and then lower the water outlet manifold onto the cylinder covers. **Temporarily**, install the water manifold attaching bolts, just fingertight, at this time.

44- Align the cylinder cover bracket holes with the bolt holes in the cylinder covers. Install the restraining bolts and tighten them just fingertight, at this time.

NOW, remove the water outlet manifold, and tighten all cylinder head cover bolts on

both sides, alternately and evenly in two sequences to a torque value of 22 ft lbs (30Nm).

Install the water outlet manifold again and this time, tighten the bolts alternately and evenly in two sequences to a torque value of 108 in lbs (12Nm). The water outlet hose will be connected to the manifold after the engine is secured in the craft.

Reed Block Assembly Installation

45- Insert the reed blocks into the crankcase intake openings. Slide the reed stuffers into the

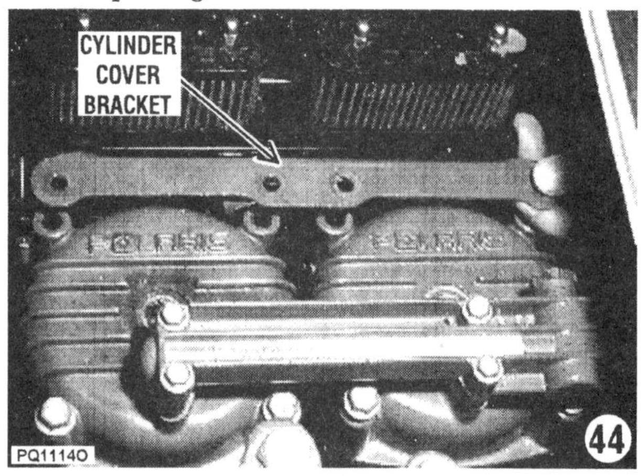

reed blocks, as indicted in the accompanying line drawing.

46- Position the intake manifold in place on the crankcase, with the bolt holes aligned. Seat the manifold against the reed blocks. Apply Loctite® to the attaching bolts, and then install and tighten them, to a torque value 108 78 in lb (12Nm), starting in the center and working alternately and evenly toward both ends.

47- Position **NEW** carburetor mounting gaskets on the intake manifold. Move the assembled carburetors mounted on the "rack"

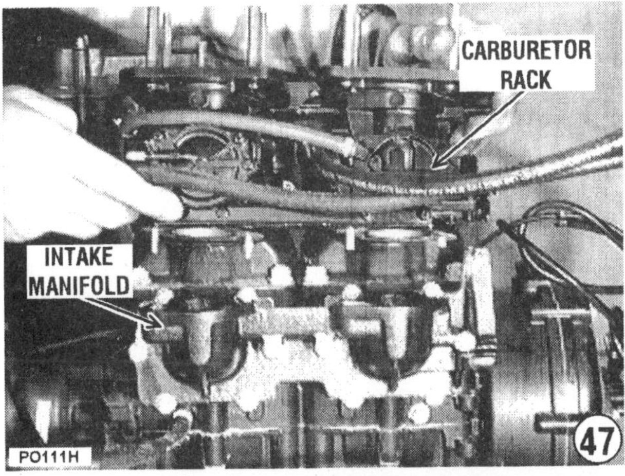

onto the intake manifold. Place the air intake/flame arrestor base pan onto the carburetors. Slide the long Allen "thru" bolts down through the pan and the carburetor, and then thread them into the intake manifold. Tighten the bolts to a torque value of 108 in lbs (12Nm).

48- Set the filter elements on the pan, and then secure them in place with the covers and attaching nuts. (Illustration shows the engine in the craft when actually it was still on the work bench.) Tighten the nuts securely.

Oil Injection Pump

Installation

49- Lubricate the oil pump **O**-ring with light grease, and slide it onto the oil pump sleeve. Bring the pump up to the flywheel cover. Note the position of the slot in the flywheel nut. Now, the "spade" on the end of the oil pump shaft indexes into the slot in the nut. Therefore, rotate the pump shaft slightly to align with the slot in the nut and then move

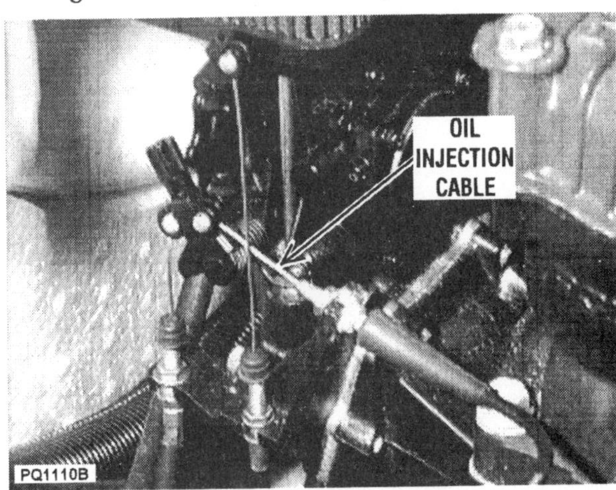

Close look at the oil injection cable hookup at the throttle shaft arm at the carburetors. Proper adjustment is essential to ***ENSURE*** *adequate oil delivery to the engine under all operating conditions, especially with a heavy load at low rpm.*

the pump up against the flywheel cover. Apply Loctite® 242, or equivalent to the threads of the attaching bolts.. Secure the pump with the two bolts tightened to a torque value of 60 in lbs (7Nm). Connect the oil pump cable to the carburetor throttle shaft arm.

Insert an oil delivery hose into each of the oil pickup/check valve fittings on the carburetors. Secure each hose to the fitting with a clamp.

Adjustment
Oil Pump Cable

Check and adjust the oil cable as follows:

The idle speed and throttle lever free "play" **MUST** be properly adjusted before making the oil pump cable adjustment.

Observe the mark on the oil pump lever and the mark on the oil pump body. These two marks should align with each other when the throttle is released and the engine is operating at idle speed. When the cable is adjusted properly, the oil pump lever should move as soon as the throttle is opened -- the slightest amount.

If the marks are not aligned, simply loosen the two adjustment nuts, move the cable slightly until the marks are aligned, and then tighten the nuts securely to hold the adjustment.

Make a final check to ensure the lever begins to move with the smallest amount of throttle movement.

Engine Installation

The engine is basically ready for installation in the watercraft. Gather up the electrical leads and fuel hoses left in the craft and tuck them against the outboard bulkhead. If an engine lift is not available, the engine may be lowered into the watercraft by one person.

50- If any shim material was used on the mounting studs, they should have remained in place when the engine was removed. As mentioned in the removal procedures, the manufacturer suggests an eyelet be welded to two bolts with the same threads and the proper size to fit into the spark plug openings. Thread the eyelet bolts into the spark plug openings several turns, to ensure they do not pull loose. Obtain a suitable lifting device. Pass a chain through the two eyelets threaded into the spark plug openings and up to the lifting device. Lift and guide the engine into the engine compartment.

Easy Alternate Method

Place an adequate size piece of plywood on the aft portion of the craft, as shown. Move the engine onto the piece of plywood. One man in reasonable good physical shape can straddle the engine compartment, grasp the engine on the port and starboard side, and then using his legs, not his back, lift the engine off the plywood, forward and down into the engine compartment. As the engine is lowered, check to be sure the engine mount bolt holes index over the studs in the hull.

51- Tighten three of the engine mounting bracket nuts -- the two forward nuts and the aft starboard nut -- to a torque value of 45 ft lbs (61Nm). Tighten the aft port side nut just "snug", because the nut will be removed later when the aft expansion bracket is installed on the stud.

52- Install the battery and connect the positive and negative cables. Secure the battery in place with the "bungee" type restraining strap/s. Open the electrical box. Connect the stator harness leads inside the electrical box -- color to color. These leads feed through the box, secured by a water-tight grommet.

If the color code cannot be identified, it may require referencing the electrical schematic to

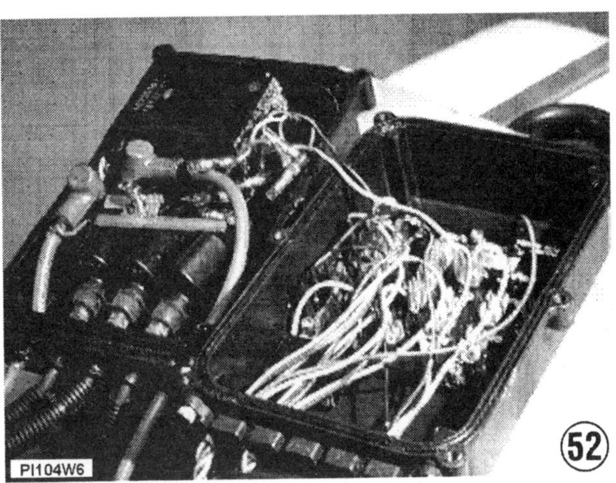

identify particular lead connections. Once the connections are complete, close the electrical box, checking to be sure no wires are caught between the two halves and tighten the attaching hardware securely. Secure the box to the small bulkhead with the three attaching bolts.

ENGINE SUPPORT EQUIPMENT INSTALLATION

Exhaust Manifold Installation

53- Insert the attaching bolts into the exhaust manifold. Place a **NEW** gasket over the bolts and seat it against the manifold. Position the manifold over the exhaust ports. Thread the attaching bolts "fingertight", to permit the manifold to be shifted slightly, if necessary. After all bolts are "fingertight", secure the bolts to a torque value of 22 ft lb (30Nm).

54- Install a **NEW** gasket atop the exhaust manifold. Note any markings indicating the proper positioning of the gasket -- such as "**UP**". It will necessary to move the oil tank -- without disconnecting oil lines -- to allow room to lower

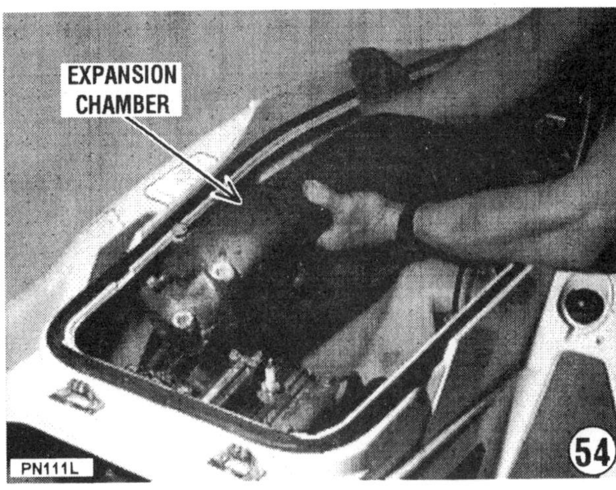

the expansion chamber into the engine compartment. On some models, it may be necessary to remove the oil tank. Lower and rotate the expansion chamber into the compartment, in front of the engine and work it around into position on the port side.

55- Position the expansion chamber on top of the exhaust manifold. Install the attaching bolts, taking note of different lengths. The two outboard bolts will be difficult to tighten because of the chamber being so close to the curved

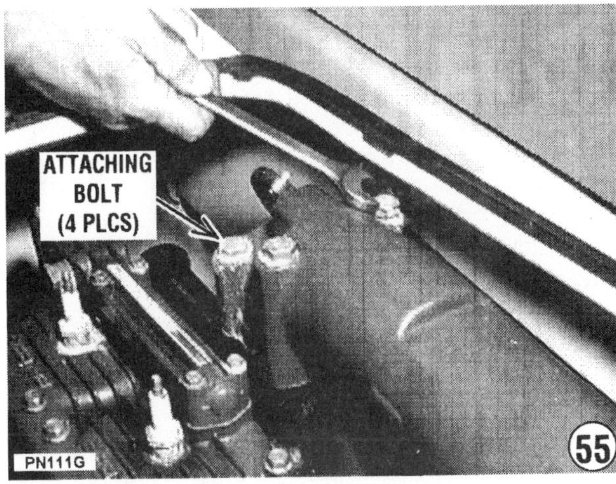

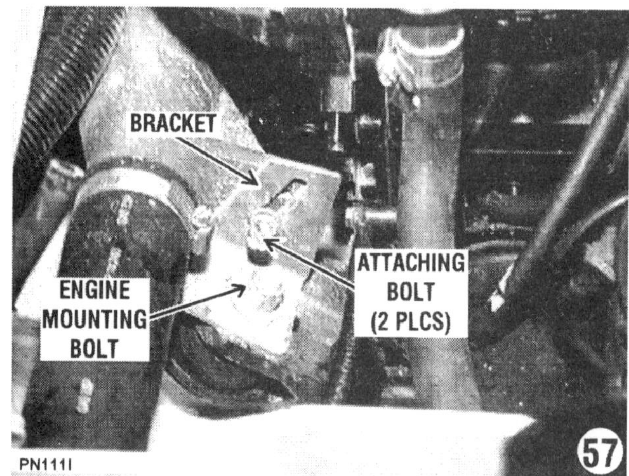

portion of the hull. This fact was discovered during removal. With the bolts properly positioned, tighten them to a torque value of 40 ft lb (54Nm).

56- Secure the stay bracket to the expansion chamber, if it was removed. Install the bolts securing the expansion chamber stay bracket to the flywheel cover. Tighten the bolts securely.

57- Install the aft bracket for the expansion chamber. One of the engine support bracket studs is used to secure the lower part of the

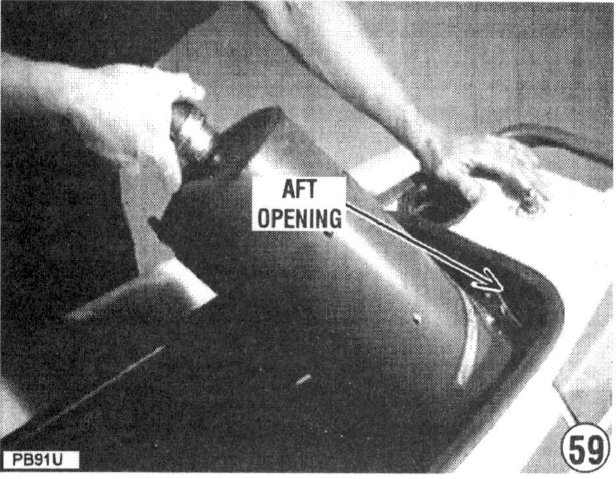

bracket. Remove the temporarily installed nut, install the bracket onto the stud, and then tighten the engine mounting bracket nut to a torque value of 45 ft lbs (61Nm).

58- Connecvt the large cooling hose to the expansion chamber and the small hose to the exhaust manifold. Secure the hoses in place with the clamps.

59- Work the exhaust muffler into the engine compartment and aft into position. The large opening at the aft end faces up and the metal strip on the bottom of the muffler fits into the groove in the bottom of the hull.

60- Check to be sure a couple of hose clamps are on the forward muffler hose. Shift the muffler forward and at the same time slide the forward hose over the expansion chamber exhaust fitting. Secure both ends of the hose with the hose clamps.

61- Verify a hose clamp is on the exhaust hose. Slide the large exhaust hose down over the muffler fitting and secure it in place with the hose clamp.

Tighten the driveshaft coupler on the aft end of the engine to a torque value of 160 ft lbs (217Nm) -- solid coupler; or just hand tight if bonded rubber style coupler.

CLOSING WORDS

TAKE TIME and use **PATIENCE** to connect all fuel, oil, and electrical lines back into their original positions. In many cases , the length of the hose and its particular shape will indicate where it is to be connected.

Be sure to "bleed" the oil injection system, because air entered the lines during the overhaul work. Add oil to the tank, as required. If the engine was not operated for an extended length of time, drain -- by siphoning -- the old fuel from the tank -- it may smell like rotten eggs -- and add **FRESH** fuel.

8-5 CLEANING AND INSPECTING ALL ENGINES

The success of the overhaul work is largely dependent on how well the cleaning and inspecting tasks are completed. If some parts are not thoroughly cleaned, or if an unsatisfactory unit is allowed to be returend to service through negligent inspection, the time and expense involved in the overhaul work will not be justified with peak engine performance and long operating life.

Therefore, the procedures in the following sections should be followed closely and the work perfomed with patience and attention to detail.

Reed Block Service

Disassemble the reed block housing by first removing the screws securing the reed stoppers and reed petals to the housing. After the screws are removed, lift the stoppers and petals from the housing.

Clean the gasket surfaces of the housing. Check the surfaces for deep grooves, cracks or any distortion which could cause leakage.

Replace the reed block housing if it is damaged. The reed petals should fit flush against their seats, and not be preloaded against their seats or bend away from their seats.

Petal Clearance

The maximum petal clearance for all engines covered in this manual is 0.015" (0.38mm)

If the reed petal is distorted, rarely, if ever, can the petal be successfully straightened. **THEREFORE**, it must be replaced.

Do not remove the reed valves unless they, or the stoppers, are to be replaced. The reed **MUST** be replaced in sets.

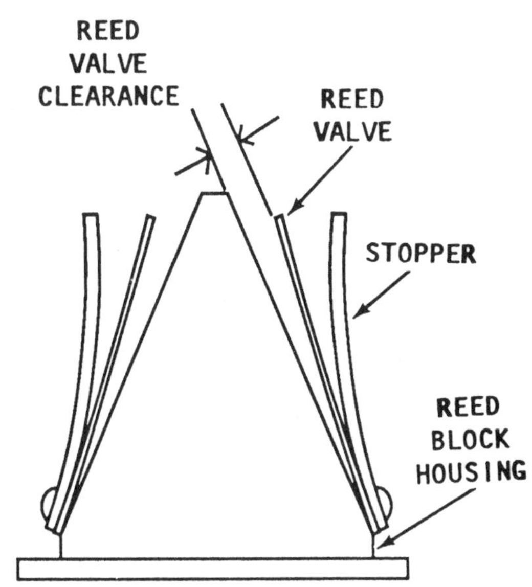

The maximum reed valve clearance is 0.015" (0.38mm), for all engines.

Apply Loctite® to the threads of the reed retaining screws. Tighten each screw gradually, starting in the center and working outward across the reed block. Tighten the screws to a torque value of 3.5 ft lbs (5Nm).

TAKE CARE not to mar the special rubber coating on the reed block housing.

Crankshaft Service

Clean the crankshaft with solvent and wipe the journals dry with a lint free cloth.

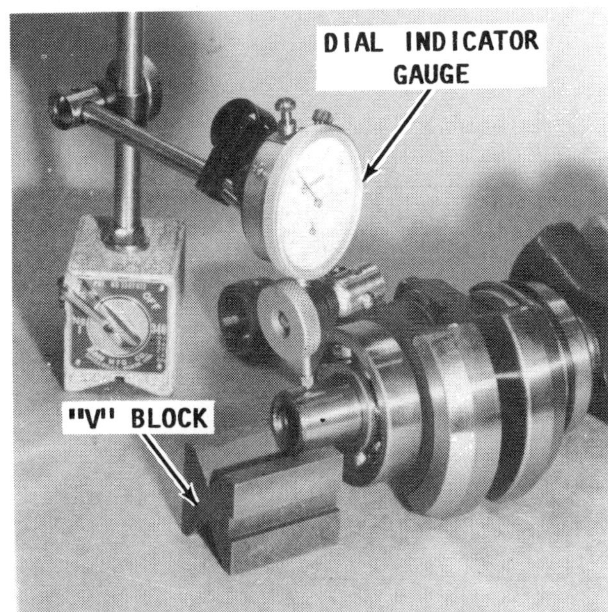

Typical setup of a crankshaft with "V" blocks and dial indicator to measure main bearing journals on an assembled two-stroke crankshaft. The bearings and connecting rod will not affect the reading.

Inspect the main journals and connecting rod journals for cracks, scratches, grooves, or scores. Inspect the crankshaft oil seal surface for nicks, sharp edges or burrs which might damage the oil seal during installation or might cause premature seal wear. **ALWAYS** handle the crankshaft carefully to avoid damaging the highly finished journal surfaces. Blow out all oil passages with compressed air. The oil passageway leads from the rod to the main bearing journal. **TAKE CARE** not to blow dirt into the main bearing journal bore.

Inspect the threads at both ends for signs of abnormal wear. Check the crankshaft for runout by supporting it on two **"V"** blocks at the main bearing surfaces.

Install a dial indicator gauge above the main bearing journals. Rotate the crankshaft and measure the runout (or the out-of-round) and the taper at both ends (and in the center journal on all two-cylinder models).

The following crankshaft runout limits are recommended by the manufacturer.

CRANKSHAFT RUNOUT LIMITS

Series	Standard	Limit
Polaris	N/A	N/A
	N/A	N/A
Fuji	.00015" (.04mm)	.004" (.1mm)

"N/A" indicates figures not available at press time.

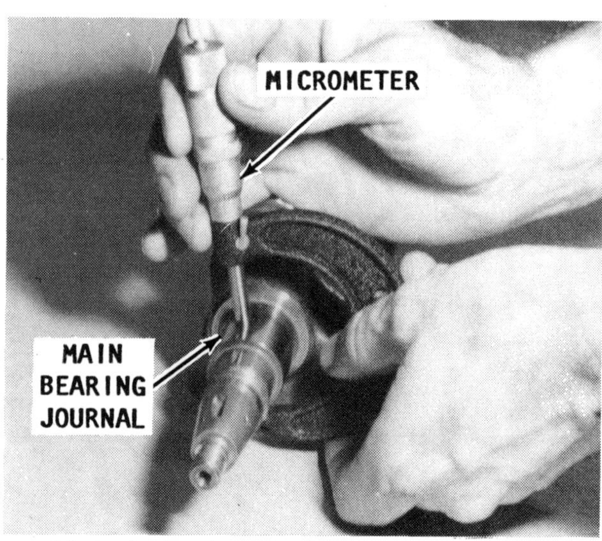

Method of using a micrometer to take the first measurement on a crankshaft main journal for out-of-round and taper.

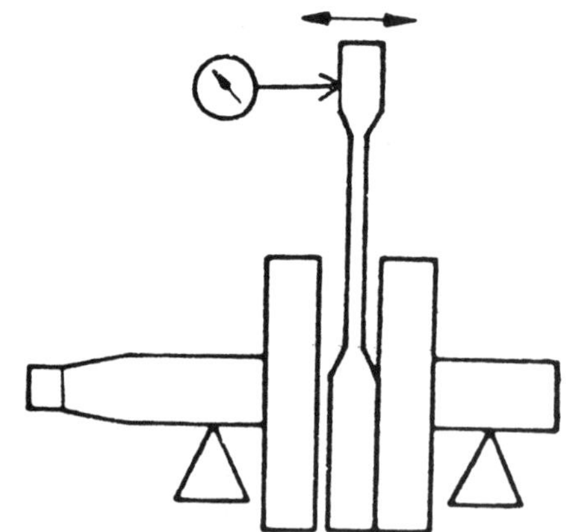

The simplest line drawing possible to depict the setup of the crankshaft and piston rod to check for axial rod "play", as explained in the text.

If **"V"** blocks or a dial indicator are not available, a micrometer may be used to measure the diameter of the journal. Make a second measurement at right angles to the first. Check the difference between the first and second measurement for out-of-round condition. If the journals are tapered, ridged, or out-of-round by more than the specification allows, the journals should be reground, or the crankshaft replaced.

Any out-of-round or taper shortens bearing life.

Crankshaft Repair Words

The manufacturer makes it very clear -- specialized equipment and personnel with a sound knowledge of crankshaft repair and straightening are required to perform crank-

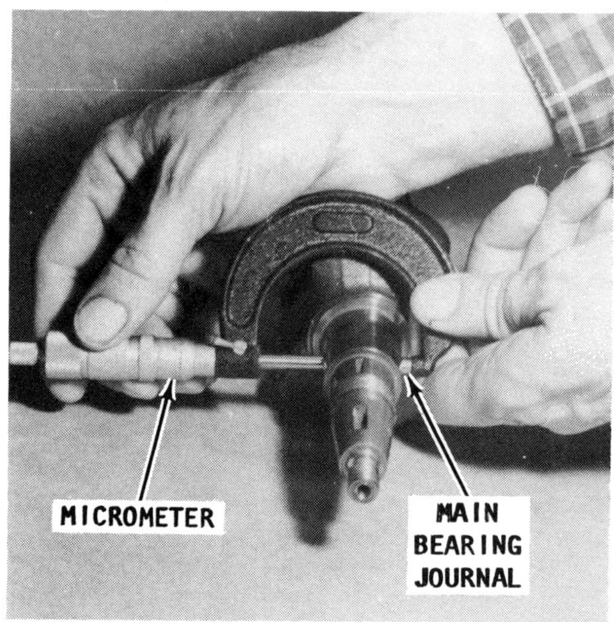

Taking a second measurement on a crankshaft main bearing journal at 90° to the first measurement.

shaft work safely and correctly. Crankshaft repair should only be performed by trained Polaris service technicians or machine shop journeymen with the expertise and proper equipment.

Connecting rod axial "play" is checked at the piston end to determine the amount of wear at the crankshaft end of the connecting rod. Checking the "play" is described in the following paragraphs.

Setup the crankshaft in the "V" blocks. Setup a dial indicator to touch the flat surface of the piston end of the rod. Now, hold the crankshaft steady in the "V" blocks and at the same time, rock the piston end of the rod along the same axis as the crankshaft. If the dial indicator needle moves through more than 0.08" (2mm), for all engines covered in this manual, the "play" is considered excessive.

To check the connecting rod side clearance at the crankshaft, first insert a feeler gauge between the connecting rod and the counterweight of the crankshaft. Acceptable clearances are as follows:

CONNECTING ROD (BIG END) SIDE CLEARANCE

Series	Standard	Service Limit
Polaris	.014" - .020" (.36 - .51mm)	.022" (.56mm)
Fuji	.012" - .024" (.30 - .61mm)	.026" (.66mm)

CONNECTING ROD (BIG END) RADIAL CLEARANCE

Series	Standard	Service Limit
Polaris	.0015" - .0018" (.038 - .046mm)	.0021" (.053mm)
Fuji	N/A	N/A

"N/A" indicates figures not available at press time.

Inspect the crankshaft oil seal surfaces to be sure they are not grooved, pitted, or scratched. Replace the crankshaft if it is severely damaged or worn. Check all crankshaft bearing surfaces for rust, water marks, chatter marks, uneven wear or overheating. Clean the crankshaft surfaces with 320-grit carborundum cloth. **NEVER** spin-dry a crankshaft ball bearing with compressed air.

Clean the crankshaft and crankshaft ball bearing with solvent. Dry the parts, but not the ball bearing, with compressed air. Check the crankshaft surfaces a second time. Replace the crankshaft if the surfaces cannot be cleaned properly for satisfactory service. If the crankshaft is to be installed for service, lubricate the surfaces with light oil. **DO NOT** lubricate the crankshaft ball bearing at this time.

After the crankshaft has been cleaned, grasp the outer race of the crankshaft ball bearing installed on the lower end of the crankshaft, and attempt to work the race back-and-forth. There should not be excessive "play". A very slight amount of side "play" is acceptable because there is only about 0.001" (.025mm) clearance in the bearing.

Lubricate the ball bearing with light oil. Check the action of the bearing by rotating the outer bearing race. The bearing should have a smooth action and no rust stains. If the ball bearing sounds or feels rough or catches, the bearing should be removed and discarded.

Connecting Rod Service

Inspect the connecting rod bearings for rust or signs of bearing failure. **NEVER** intermix new and used bearings. If even one bearing in a set needs to be replaced, all bearings at that location **MUST** be replaced.

Clean the inside diameter of the piston pin end of the connecting rod with crocus cloth.

Clean the connecting rod **ONLY** enough to remove marks. **DO NOT** continue, once the marks have disappeared.

Assemble the piston end of the connecting rod with loose needle bearings, caged needle bearings, or no needle bearing, depending on the model being serviced. Insert the piston pin and check for vertical "play". The piston pin should have **NO** noticeable vertical "play".

If the pin is loose or there is vertical "play" check for and replace the worn part/s.

Inspect the piston pin and matching rod end for signs of heat discoloration. Overheating is identified as a bluish bearing

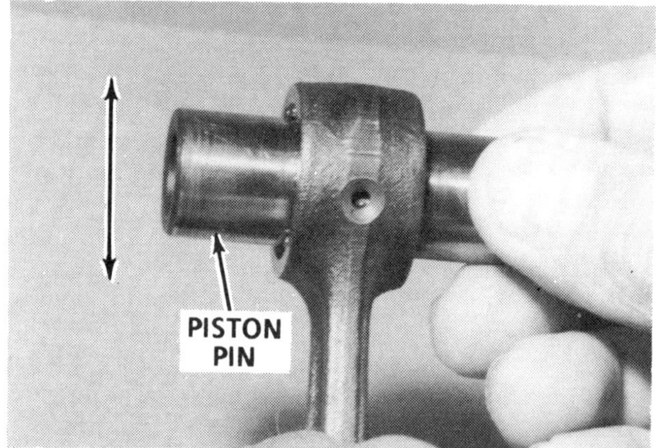

Using crocus cloth to clean the piston end of any connecting rod.

surface color and is caused by inadequate lubrication or by operating the engine at excessive high rpm.

Piston Service

Inspect each piston for evidence of scoring, cracks, metal damage, cracked piston pin boss, or worn pin boss. Be especially critical during inspection if the craft and engine has been submerged. If the piston pin is bent, the piston and pin **MUST** be replaced as a set for two reasons. First, a bent pin will damage the boss when it is removed. Secondly, a piston pin is not sold as a separate item.

Check the piston ring grooves for wear, burns distortion, or loose locating pins. During an overhaul, the rings should be replaced to ensure lasting repair and proper powerhead performance after the work is completed. Clean the piston dome, ring grooves and the piston skirt. Clean carbon deposits from the ring grooves using the recessed end of a broken piston ring.

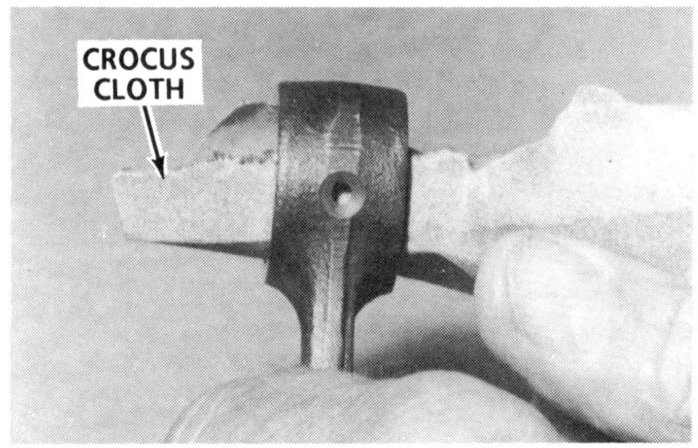

Check the piston end of a connecting rod for vertical free "play" using a piston pin.

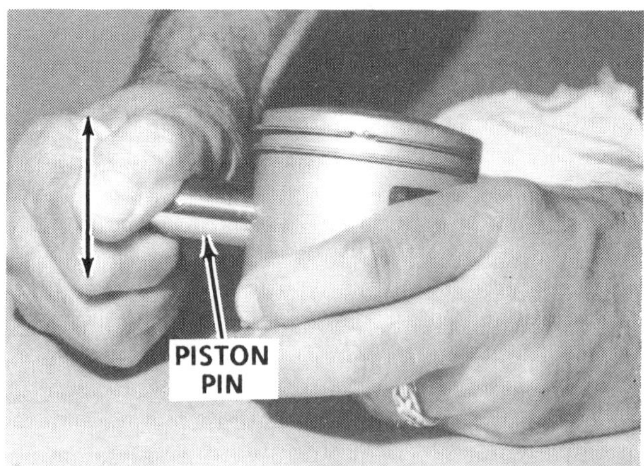

Check free "play" between the piston pin and the piston boss. There should be NO "play".

NEVER use a rectangular ring to clean the groove for a tapered ring, or use a tapered ring to clean the groove for a rectangular ring.

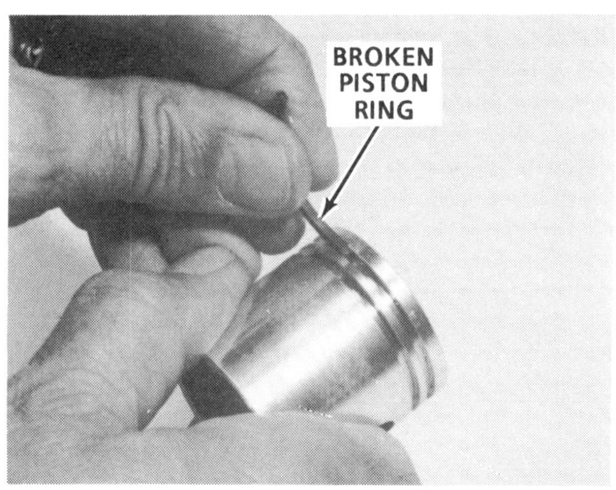

*Method of using part of a broken piston ring to clean the ring groove. Exercise **CARE** not to disturb the locating pin in each groove.*

Locating pins prevent the rings from rotating and catching on a port in the cylinder wall.

It is believed, this crown seized with the cylinder wall when the unit was operated at high rpm and the timing was not adjusted properly. At the same instant, the rod apparently pulled the lower part of the piston downward, severing it from the crown.

NEVER use an automotive-type ring groove cleaner, because such a tool may loosen the piston ring locating pins.

Clean carbon deposits from the top of the piston using a soft wire brush, carbon removal solution or by sand blasting. If a wire brush is used, **TAKE CARE** not to burr or round machined edges. Clean the piston skirt with crocus cloth.

The pitted damage to this piston crown was probably caused by a broken piston ring working its way into the combustion chamber. The little "hills" then became "hot" spots on the crown, contributing to "dieseling" after the engine was shut down.

The rings on this piston became stuck due to lack of adequate lubrication, incorrect timing, or overheating.

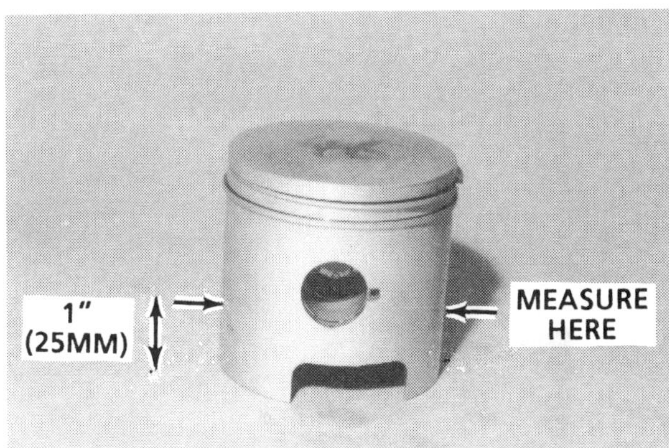

Measure the piston diameter at the specified distance above the bottom of the piston skirt.

The rings on this piston were broken, possibly during installation, and then caused extensive damage to the piston. Ring parts found their way into the combustion chamber and caused damage to the piston crown.

Install the piston pin through the first boss only. Check for vertical free "play". There should be **NO** vertical free "play". The presence of "play" is an indication the piston boss is worn. The piston is manufactured from a softer material than the piston pin. Therefore, the piston boss will wear more quickly than the pin.

Excessive piston skirt wear **CANNOT** be visually detected. Therefore, good shop practice dictates, the piston skirt diameter be measured with a micrometer.

Piston diameters are are measured at a definite distance up from the bottom of the skirt at right angles to the piston pin axis.

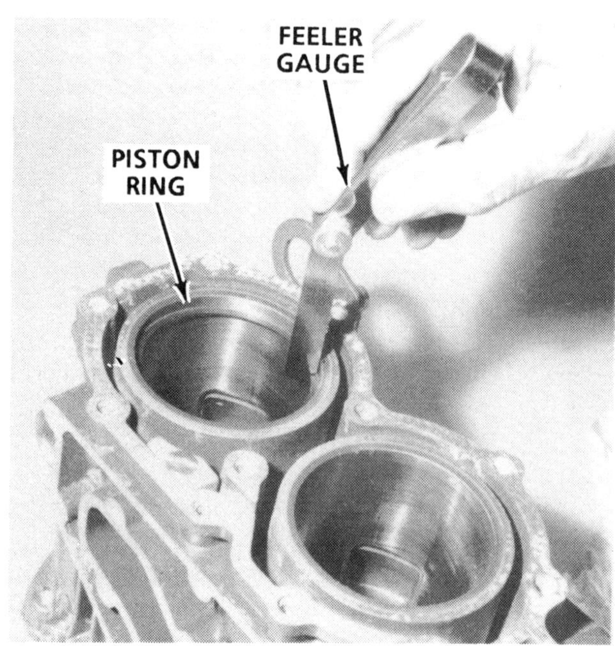

Example of measuring the piston ring end gap with a feeler gauge.

Ring End Gap

Before the piston rings are installed onto the piston, the ring end gap clearance for each ring must be determined. The purpose of the piston rings is to prevent the blowby of gases in the combustion chamber. This cannot be achieved unless the correct oil film thickness is left on the cylinder wall.

This thin coating of oil acts as a seal between the cylinder wall and the face of the piston ring. An excessive end gap will allow blowby and the cylinder will lose compression. An inadequate end gap will scrape too much oil from the cylinder wall and limit lubrication. Lack of adequate lubrication will cause excessive heat and wear.

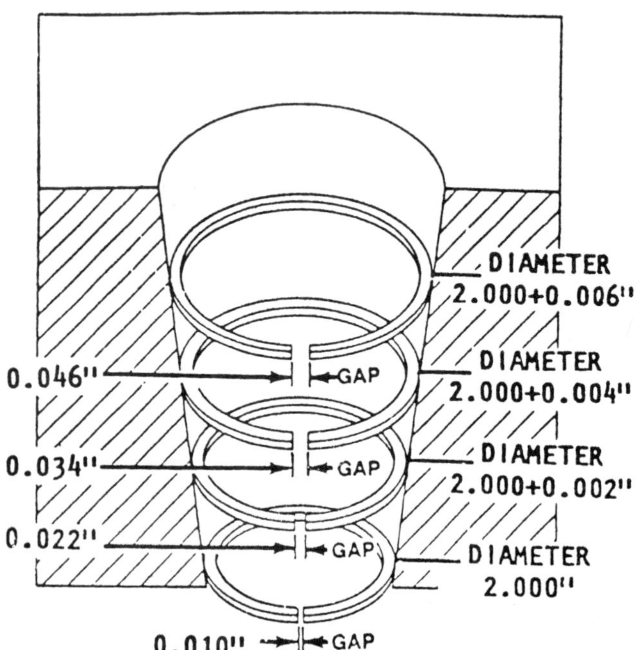

Simple line drawing to illustrate typical cylinder taper. The taper drastically affects ring and end gap, as indicated.

IDEALLY the ring end gap measurement should be taken **AFTER** the cylinder bore has been measured for wear and taper **AND** after any corrective work, such as boring or honing, has been completed.

IF the ring end gap is measured with a taper to the cylinder wall, the diameter at the lower limit of ring travel will be smaller than the diameter at the top of the cylinder.

IF the ring is fitted to the upper part of a cylinder with a taper, the ring end gap will not be great enough at the lower limit of ring travel. Such a condition could result in a broken ring and/or damage to the cylinder wall and/or damage to the piston and/or damage to the cylinder head.

IF the cylinder is to be only honed, not bored, **OR** if only cleaned, not honed, the ring end gap should be measured at the lower limit of ring travel.

The ring should be inserted into the cylinder just above the ports to perform this measurement, using the piston crown. Be sure the ring is parallel to the surface of the block, then measure the ring end gap with a feeler gauge.

If the end gap is greater than the amount listed, replace the entire ring set.

If the end gap is less than 0.008-0.016" (0.20-0.40mm) for both top and bottom rings, carefully file the ends of the ring --

just a little at a time -- until the correct end gap is obtained.

Inspect the piston ring locating pins to be sure they are tight. There is one locating pin in each ring groove. If the locating pins are loose, the piston **MUST** be replaced.

Oversized Pistons and Rings

Scored cylinders can be saved for further service by reboring the cylinder and installing oversize pistons and rings. **HOWEVER**, if the scoring is over 0.0075" (0.13mm) deep, the cylinder cannot be effectively re-bored for continued use.

Check with the local dealer for oversize piston availability.

If oversize pistons are not available, the local marine shop may have the facilities to "knurl" the piston, making it larger.

For detailed information on piston ring end gap, refer to the specifications listed in the Appendix section.

Cylinder Service

Inspect each cylinder and cylinder bore for cracks or other damage. Remove traces of carbon with a fine wire brush on a shaft attached to an electric drill or use a carbon remover solution.

STOP: If a cylinder is to be submerged in a carbon removal solution, the crankcase bleed system **MUST** be removed from the cylinder -- if so equipped -- to prevent damage to hoses and check valves.

Use an inside micrometer or telescopic gauge and micrometer to check the cylinders for wear. Check the bore for an out-of-round and/or oversize bore condition. If the bore is tapered, out-of-round or worn more than the wear limit specified by the manufacturer, the cylinders should be re-bored -- provided oversize pistons and rings are available.

NOTE: Be sure to check with the dealer **PRIOR** to reboring. If oversize pistons and matching rings are not available, the cylinder **MUST** be replaced.

GOOD WORDS

Oversize piston weight is approximately the same as a standard size piston. Therefore, it is **NOT** necessary to re-bore all of the cylinders just because on cylinder requires reboring.

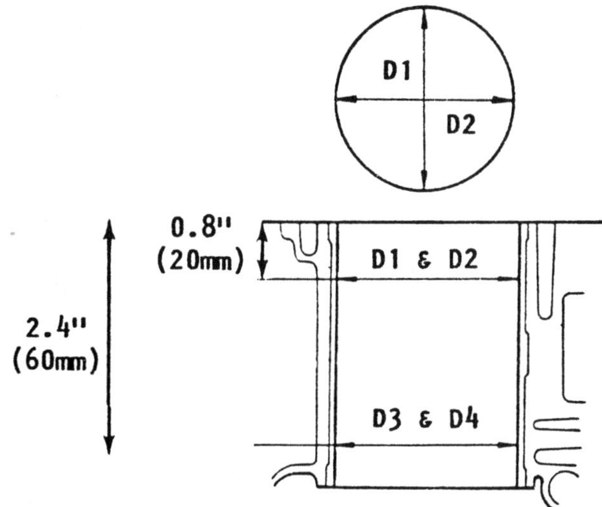

Top view diagram and cross-section of a typical cylinder to indicate where measurements are to be taken for wear limit, taper, and out-of-round limits.

Four inside cylinder bore measurements must be taken for each cylinder to determine an out-of-round condition, the maximum taper, and the maximum bore diameter.

In the accompanying illustration, measurements D1 and D2 are diameters measured at the top of the cylinder at right angles to each other. Measurements D3 and D4 are diameters measured from the top of the cylinder at right angles to each other.

Out-of-Round

Measure the cylinder diameter at D1 and D2. The manufacturer requires the difference between the two measurements should be less than 0.002" (0.050mm) for **ALL** models.

Maximum Taper

Measure the cylinder diameter at D1, D2, D3, and D4. Take the largest of the D1 or D2 measurements and subtract the smallest measurement at D3 or D4. The answer to the subtraction -- the cylinder taper -- should be less than 0.002" (0.05mm), for **ALL** models covered in this manual.

Bore Wear Limit

The maximum cylinder diameter D1, D2, D3, and D4 must not exceed the bore wear limits indicated in the following table **BEFORE** the bore is rebored for the **FIRST** time. These limits are only imposed on original parts because the sealing ability of the rings would be lost, resulting in power loss, increased engine noise, unnecessary vibration, piston slap, and excessive oil consumption.

The limits indicated are usually 0.003 to 0.005" (0.83 to 0.127mm) above the standard bore. Therefore, if the bore is resized, it may sustain another 0.003 to 0.005" (0.83 to 0.127mm) wear before a second reboring is required -- provided the oversize pistons and matching rings are available.

For complete standard bore size specifications, refer to the Appendix.

Piston Clearance

Piston clearance is the difference between a maximum piston diameter and a minimum cylinder bore diameter. If this clearance is excessive, the engine will develop the same symptoms as for excessive cylinder bore wear -- loss of ring sealing ability, loss of power, increased engine noise, unnecessary vibration, and excessive oil consumption.

Maximum piston diameter was described earlier in this section. Minimum cylinder bore diameter is usually determined by measurement D3 or D4 also described earlier in this section.

If the piston clearance exceeds the limits outlined in the following table, either the piston or the cylinder block **MUST** be replaced.

Calculate the piston clearance by subtracting the maximum piston skirt diameter from the maximum cylinder bore measurement and compare the results for the model being serviced.

For detailed piston-to-cylinder clearance specifications, refer to the Appendix.

HONING CYLINDER WALLS

Hone the cylinder walls lightly to seat the new piston rings, as outlined in this section. If the cylinders have been scored, but are not out-of-round or the bore is rough, clean the surface of the cylinder with a cylinder hone as described in the following procedures.

SPECIAL WORDS

If overheating has occurred, check and resurface the spark plug end of the cylinder block, if necessary. This can be accomplished with 240-grit sandpaper and a small flat block of wood.

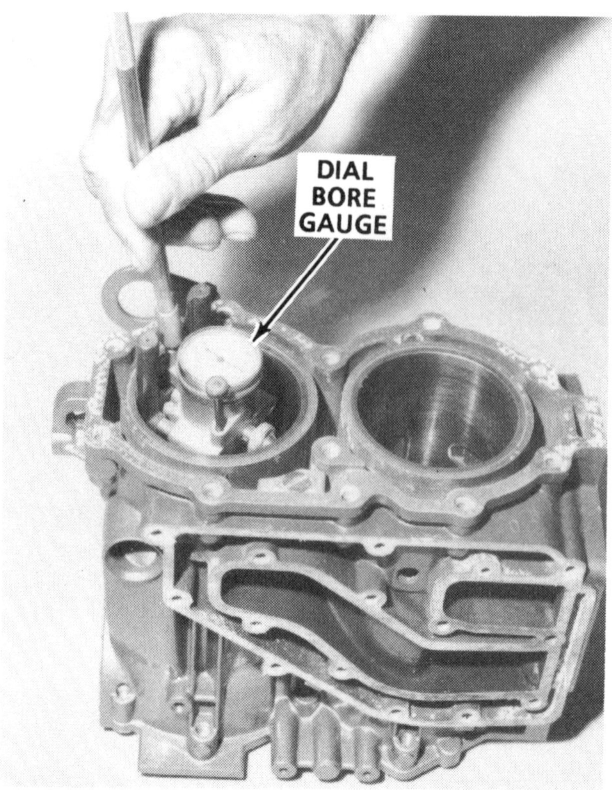

Effective method of checking the cylinder taper of any cylinder using an inside micrometer. One measurement should be taken near the top and another near of the bottom of the cylinder. The difference between the two is the amount of taper.

To ensure satisfactory engine performance and long life following the overhaul work, the honing work should be performed with patience, skill, and in the following sequence:

a- Follow the hone manufacturer's recommendations for use of the hone and for cleaning and lubricating during the honing operation. A "Christmas tree" hone may also be used.

b- Pump a continuous flow of honing oil into the work area. If pumping is not practical, use an oil can. Apply the oil generously and frequently on both the stones and work surface.

c- Begin the stroking at the smallest diameter. Maintain a firm stone pressure against the cylinder wall to assure fast stock removal and accurate results.

d- Expand the stones as necessary to compensate for stock removal and stone wear. The best cross-hatch pattern is obtained using a stroke rate of 30 complete cycles per minute. Again, use the honing oil generously.

e- Hone the cylinder walls **ONLY** enough to de-glaze the walls.

f- After the honing operation has been completed, clean the cylinder bores with hot water and detergent. Scrub the walls with a stiff bristle brush and rinse thoroughly with hot water. The cylinders **MUST** be thoroughly cleaned to prevent any abrasive material from remaining in the cylinder bore. Such material will cause rapid wear of new piston rings, the cylinder bore, and the bearings.

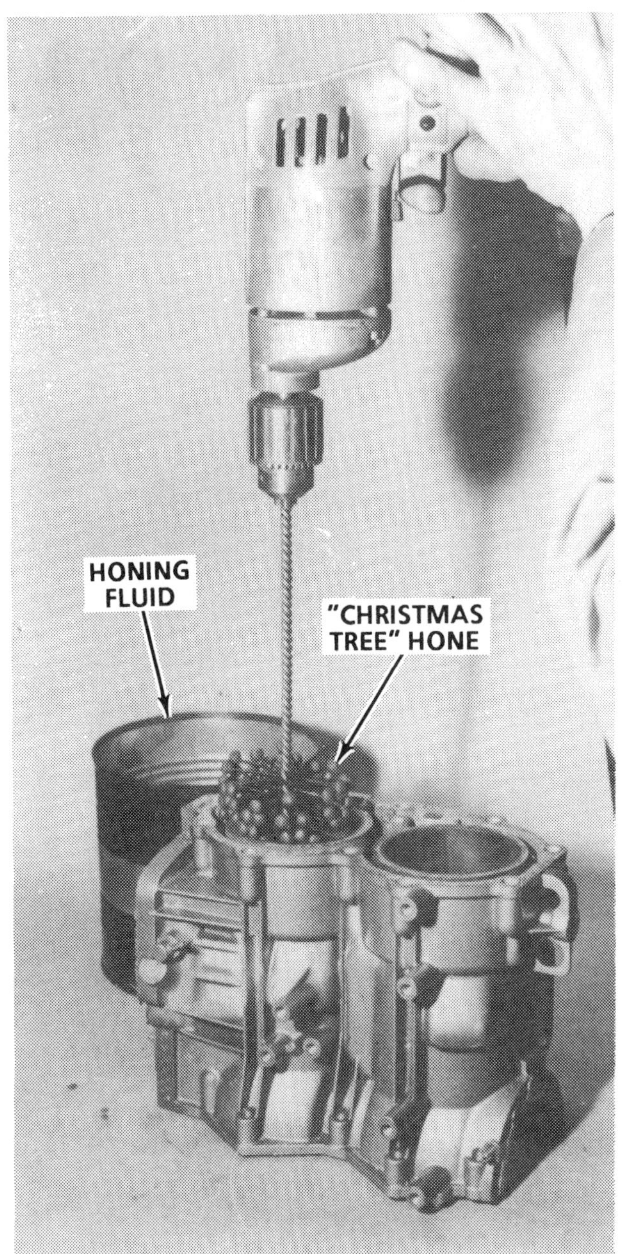

Excellent example of refinishing a cylinder wall using an electric drill and "Christmas Tree" hone. ***ALWAYS*** *keep the tool moving in long even strokes the entire cylinder depth. Use a continuous* ***LIBERAL*** *amount of honing fluid.*

The wall of this cylinder were damaged beyond repair when a piston ring broke and worked its way into the combustion chamber.

g- After cleaning, swab the bores several times with engine oil and a clean cloth, and then wipe them dry with a clean cloth. **NEVER** use kerosene or gasoline to clean the cylinders.

h- Clean the remainder of the cylinder block to remove any excess material spread during the honing operation.

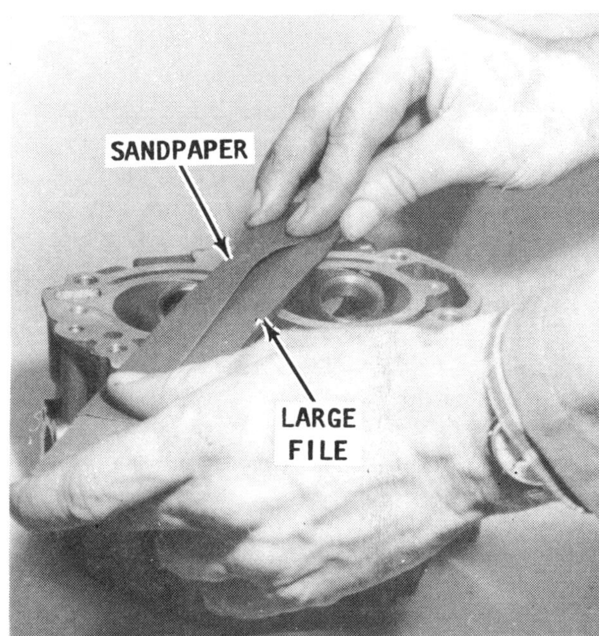

A ***LARGE*** *file and piece of wet sandpaper may be used to resurface a head or cylinder block when a suitable flat surface is not available.*

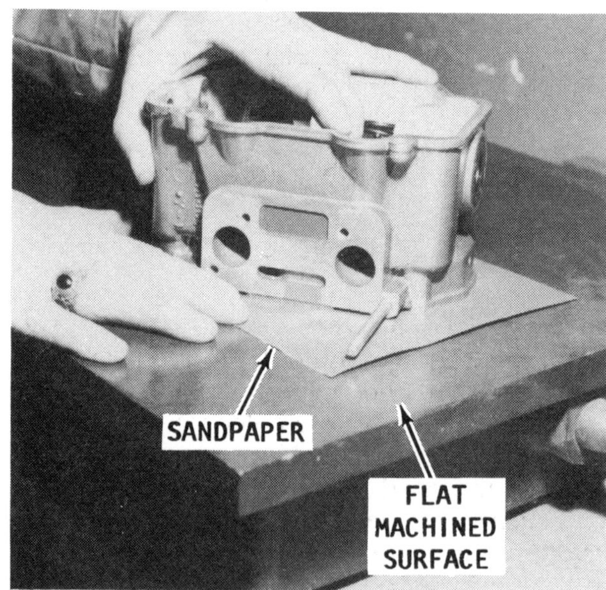

A machined surface and wet sandpaper may be used to resurface a cylinder head, as explained in the text. A piece of glass or mirror may also be used, but do NOT attempt this task on a workbench.

Block and Cylinder Head Warpage

First, check to be sure all old gasket material has been removed from the contact surfaces of the block and the cylinder head. Clean both surfaces down to shiny metal, to ensure a true measurement.

Next, place a straight edge across the gasket surface. Check under the straight edge with a 0.004" (0.1mm) feeler gauge. Move the straight edge to at least eight different locations. If the feeler gauge can pass under the straight edge -- anywhere contact with the other is made -- the surface will have to be resurfaced.

The block or the cylinder head may be resurfaced by placing the warped surface on 400-600 grit **WET** sandpaper, with the sandpaper resting on a **FLAT MACHINED** surface. If a machined surface is not available a large piece of glass or mirror may be used. **DO NOT** attempt to use a workbench or similar surface for this task. A workbench is never perfectly flat and the block or cylinder head will pickup the imperfections of the surface and the warpage will be made worse.

Sand -- work -- the warped surface on the wet sandpaper using large figure **"8"** motions. Rotate the block, or head, through 180° (turn it end-for-end) and spend an equal amount of time in each position to avoid removing too much material from one side.

This cylinder head was removed from the same engine as the pitted piston shown on Page 8-67. In order to return this head to a serviceable condition, the combustion chamber surface must be refinished to a smooth surface using a stone. If left in its present condition, the sharp corners of the gouges will become "hot" spots resulting in pre-ignition and possible "dieseling" after the engine is shut down.

If a suitable flat surface is not available, the next best method is to wrap 400-600 grit wet sandpaper around a **LARGE** file. Draw the file as evenly as possible in one sweep across the surface. Do not file in one place. Draw the file in many, many directions to get as even a finish as possible.

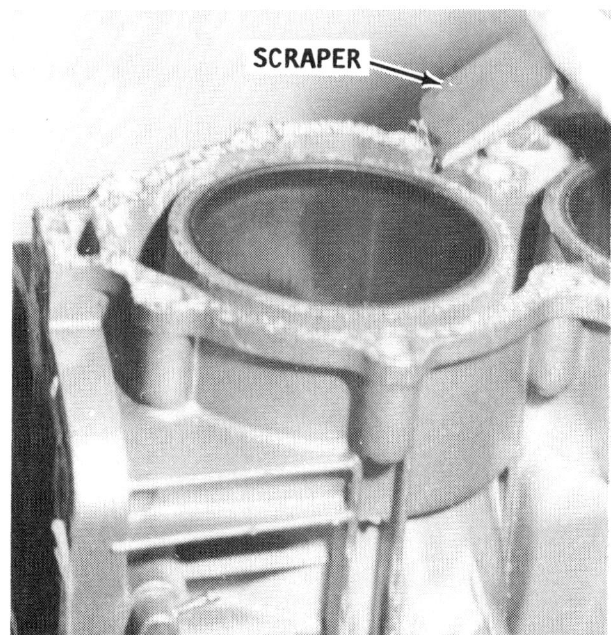

*When servicing any block, all traces of gasket material **MUST** be removed to ensure a good seal with a new gasket. If checking for warpage, even the smallest trace of foreign material would give a false reading.*

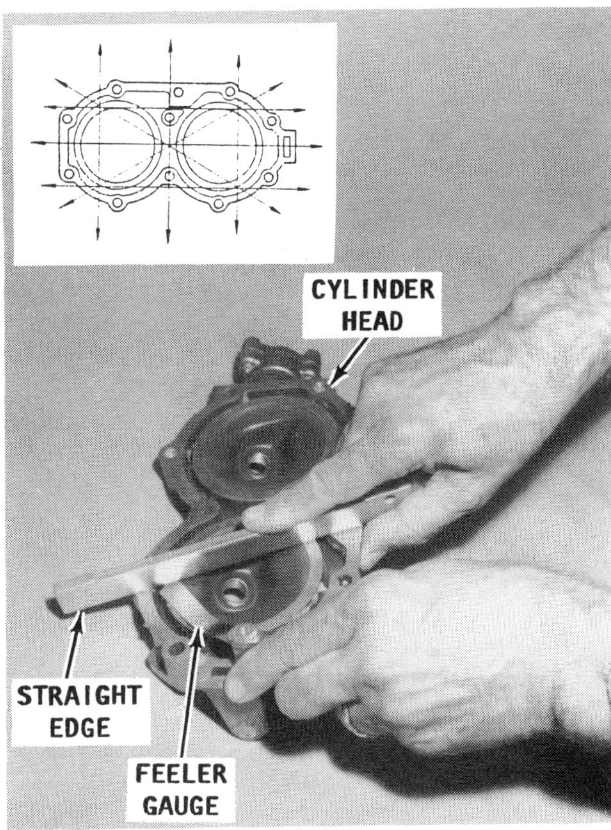

Any cylinder head surface may be checked for warpage using a straight edge and feeler gauge. Move the straight edge to at least eight different positions, as indicated in the superimposed line drawing.

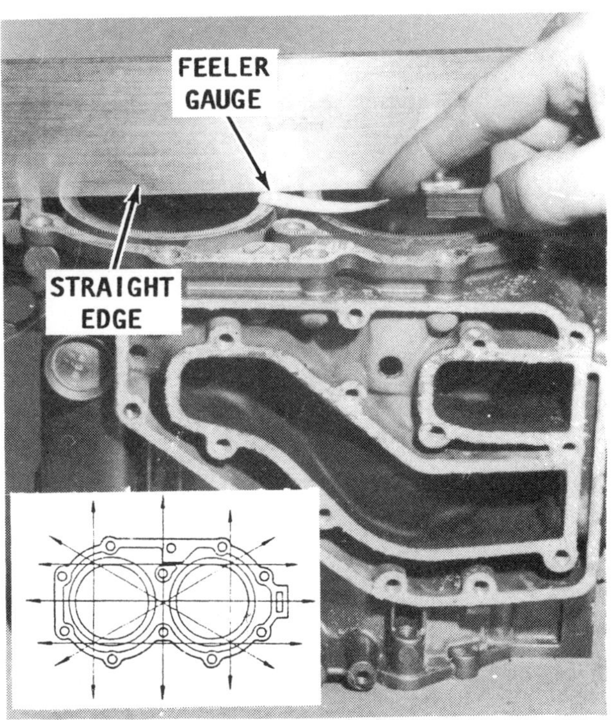

A straight edge and feeler gauge must be used to check engine-to-head mating surface on any block for warpage.

As the work moves along, check the progress with the straight edge and feeler gauge. Once the 0.004" (0.1mm) feeler gauge will no longer slide under the straight edge, consider the work completed.

If the warpage cannot be reduced using one of the described methods, the block or cylinder head should be replaced.

ONE LAST CHANCE

If the warpage cannot be reduced and it is not possible to obtain new items -- and the warped part must be assembled for further use -- there is a strong possibility a water leak will occur at the head gasket. In an effort to prevent such a water leak, the following procedures listed in "Sealing Surface Words" may be performed during assembly.

SEALING SURFACE WORDS CRITICAL READING

Because of the high temperatures and pressures, the sealing surfaces of the cylinder head and the block are the most prone to water leaks. No sealing agent is recommended **BECAUSE** it is almost impossible to apply an even coating, which would be essential to ensure a air/water tight seal.

Some head gaskets are supplied with a "tacky" coating on both surfaces applied at the time of manufacture. This "tacky" substance will provide an even coating all around. Therefore, no further sealing agent is required.

HOWEVER, if a slight water leak should be noticed following completed assembly work and engine start up, **DO NOT** attempt to stop the leak by tightening the head bolts beyond the recommended torque value. Such action will only aggravate the problem and most likely distort the head. **FURTHERMORE**, tightening the bolts, which are case hardened aluminum, may force the bolt beyond its elastic limit and cause the bolt to fracture. A fractured bolt must usually be drilled out and the hole re-tapped to accommodate an oversize bolt, etc. Avoid such a situation.

Probable causes and remedies of a new head gasket leaking are:

a- Sealing surfaces not thoroughly cleaned of old gasket material. Disassemble and remove **ALL** traces of old gasket.

b- Damage to the machined surface of the head or the block. The remedy for this damage is the same as for the next case "c".

c- Permanently distorted head or block. Spray a light **EVEN** coat of any type metallic spray paint on both sides of a new head gasket. Use only metallic paint -- any color will do. Regular spray paint does not have the particle content required to provide the extra sealing properties this procedure requires.

Assemble the block and head with the gasket while the paint is still **TACKY**. Install the head bolts and tighten in the recommended sequence and to the proper torque value and **NO** more!

Allow the paint to set for at least 24 hours before starting the engine. Consider this procedure as a temporary "band aid" type solution until a new head may be purchased or other permanent measures can be performed.

Under normal circumstances, if procedures have been followed to the letter, the head gasket will not leak.

8-6 SEALANTS, ADHESIVES, LUBRICANTS, AND FUEL STABILIZERS

It is common practice for the larger manufacturers of personal watercraft to market their own line of products for use on their craft. Polaris chemical engineers have developed such a line available for use with Polaris personal watercraft. Throughout this manual, the authors recommend the application of the manufacturer's products as a first choice. All products listed are alternatives of equal value, and may be used with confidence if the manufacturer's line is not available.

Sealants and Adhesives

Four sealants are recommended and they are **NOT** interchangeable. Each is designed to perform under a different set of conditions. Follow the directions on the package for cleaning and preparing surfaces. Sealants and adhesives **MUST** be applied **ONLY** to clean and dry parts. Apply sparingly -- excessive amounts may block oil passageways and cause serious damage.

Loctite Lock N' Seal is a non-hardening, non-permanent locking compound. This material is recommended for application to the threads of load bearing fasteners. Loctite helps prevent loosening of the bolt due to vibration, thread wear, and corrosion.

Loctite Stud N' Bearing Mount compound is also a non-hardening, non-permanent compound locking compound. This material is recommended for application to the threads of load bearing fasteners submerged under water.

Loctite Superflex is a water resistant silicone sealer which provides a very effective flexible seal and is able to withstand high temperatures. This sealer is recommended for use where gaskets are required next to a metal surface at high temperature, for example, on the exhaust manifold cover gaskets. Loctite Superflex is blue in color.

3-Bond 1215 - Recommended by Polaris (Polaris P/N 2871557), this sealant is a non-hardening material and is recommended for metal-to-metal joints such as crankcase halves. This substance is resistant to oil and gasoline.

Lubricants

Different lubricants are recommended in the lubrication procedures presented in this manual. Generally, these lubricants are **NOT** interchangeable, each is designed to perform under varying conditions. Following are descriptions of several lubricants recommended.

Polaris All Season grease is a general marine lubricant, recommended for application to bearings, bushings and oil seals.

Shell Alvania EP1 is a general marine lubricant, chemically formulated to resist salt water. This lubricant is also recommended for application to bearings, bushings, and oil seals.

Polaris Premium 2 Cycle lubricant is a two-stroke engine oil. This oil reduces carbon deposits and ensures maximum protection against engine wear. It contains an ashless detergent to keep piston rings "free". Oil additives are usually not recommended by the manufacturer, and in some cases the use of such a substance may invalidate the warranty.

Fuel Stabilizer

"Sta-Bil" Fuel Conditioner and Stabilizer is recommended during engine operation and during the storage period. This fluid absorbs water in the fuel system and protects against corrosion.

If used during operation, this fuel additive will prevent the formation of gum and varnish deposits and greatly extend the period between required carburetor overhauls.

When added to the fuel during storage, the additive will prevent the fuel from "souring" for up to twelve full months.

9 ELECTRICAL

9-1 INTRODUCTION

The battery, cranking system, temperature warning system, rpm limiter, oil sensor system, trim system, and bilge pump are all considered subsystems of the electrical system. Each of these areas will be covered in this chapter beginning with the battery. The charging system is considered a subsystem of ignition and is covered in detail in Chapter 7 -- Ignition.

9-2 BATTERIES

The battery is one of the most important parts of the electrical system. Because of its job and the consequences, (failure to perform when needed), the best advice is to purchase a well-known brand, with an extended warranty period, from a reputable dealer.

Personal Watercraft Batteries

The battery developed for use in a personal watercraft is the same kind used for many years on motorcycles. Personal watercraft batteries are required to perform under the most rigorous conditions.

Personal watercraft batteries have a much heavier exterior case than the usual automobile battery to withstand the violent pounding and shocks imposed on it as the craft moves through rough water. The battery case must also be vibration and seepage resistant. Therefore, a personal watercraft battery should always be the best the owner can afford.

The caps of marine batteries are "spill proof" to prevent acid from spilling into the bilges when the craft is subjected to violent maneuvers. These special caps are designed to vent gases regardless of the battery position, even when upside down.

Because of these features, the personal watercraft battery will recover from a low charge condition and give satisfactory service over a much longer period of time.

Battery Construction

A battery consists of a number of positive and negative plates immersed in a solution of diluted sulfuric acid. The plates contain dissimilar active materials and are kept apart by separators. The plates are grouped into what are termed elements. Plate straps on top of each element connect all of the positive plates and all of the negative plates into groups.

The battery is divided into cells which hold a number of the elements apart from

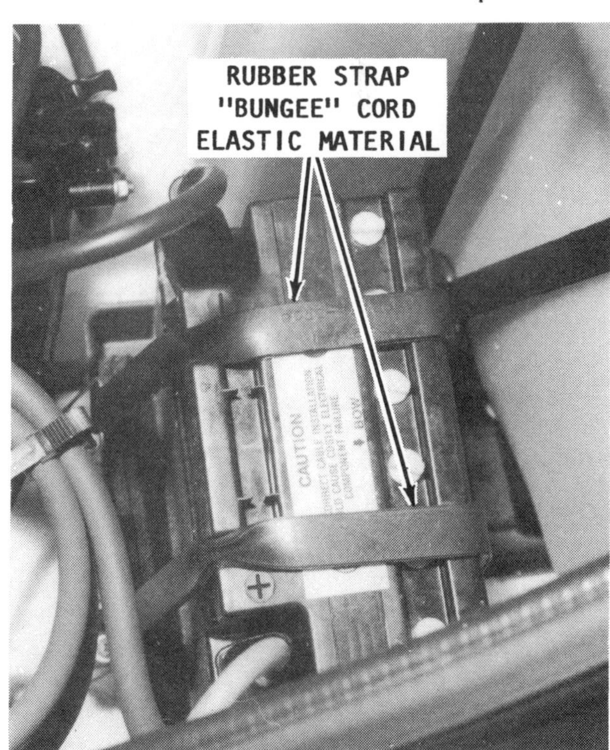

On a Personal Watercraft, two heavy-duty elastic rubber straps, or "bungee" type cords do an excellent job of holding the battery in place, during violent maneuvers.

the others. The entire arrangement is contained within a hard plastic case. The top is a one-piece cover and contains the filler caps for each cell. The terminal posts protrude through the top where the battery connections for the craft are made. Each of the cells is connected to its neighbor in a positive-to-negative manner with a heavy strap called the cell connector.

Battery Ratings

Four different methods are used to measure and indicate battery electrical capacity:

1- Ampere-hour rating
2- Cold cranking performance
3- Reserve capacity
4- Watt-hour rating

Ampere-Hour

The ampere-hour rating of a battery refers to the battery's ability to provide a set amount of amperes for a given amount of time under test conditions at a constant temperature of 80° (27°C). Amperes x hours equals ampere-hour rating. Therefore, if the battery is capable of supplying 3 amperes of current for 5 consecutive hours, the battery is rated as a 15 ampere-hour battery.

The ampere-hour rating is useful for some service operations, such as slow charging or battery testing.

The recommended ampere-hour rating for a personal watercraft battery is 15-19 ampere-hours.

Cold Cranking Performance

Cold cranking performance is measured by first cooling a fully charged battery to 0°F (-18°C), and then testing it for 5 seconds to determine the maximum current flow. In this manner the cold cranking amperes rating is the number of amperes available to be drawn from the battery before the voltage drops below 9.8 volts. A typical cold cranking ampere rating for a personal watercraft battery is 240 amps.

Reserve Capacity

The reserve capacity of a battery is considered the length of time -- in minutes --at 80°F (27°C) a 25 ampere current can be maintained before the voltage drops below 10.5 volts. This test is intended to provide an approximation of how long the engine, including electrical accessories such as bilge pump, could operate satisfactorily if the stator assembly or charging coil did not produce sufficient current. A typical rating is 25 minutes.

Watt-Hour

Watt-hour is a very useful rating of battery power. It is determined by multiplying the number of ampere hours times the voltage. Therefore, a 12-volt battery rated at 16 ampere-hours would be rated at 192 watt-hours (16 x 12 = 192).

If possible, the new battery should have a power rating equal to or higher than the unit it is replacing.

Battery Installation

Every battery installed in a personal watercraft must be secured in a well-protected ventilated area. If the battery area lacks adequate ventilation, hydrogen gas which is given off during charging could become very explosive. This is especially true if the gas is concentrated and confined.

Battery Service

The battery requires periodic servicing and a definite maintenance program will ensure extended life. If the battery should

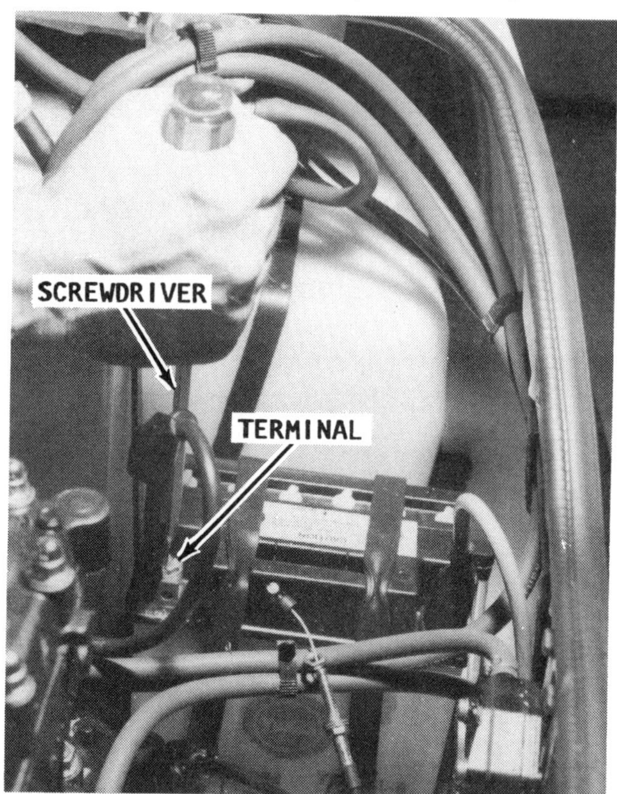

The battery cables can usually be disconnected using a slotted screwdriver or socket wrench.

test satisfactorily, but still fails to perform properly, one of five problems could be the cause.

1- Slow speed engine operation for long periods of time resulting in an undercharged condition.

2- A defect in the charging system. A faulty stator assembly or charging coil, defective rectifier, or high resistance somewhere in the system could cause the battery to become undercharged.

3- Failure to maintain the battery in good order. This might include a low level of electrolyte in the cells; loose or dirty cable connections at the battery terminals; or possibly an excessively dirty battery top.

Remove the battery from the engine compartment for service. Inspect and service the battery, cables and connections. Check for signs of corrosion. Inspect the battery case for cracks or bulges, dirt, acid, and electrolyte leakage.

Make sure the breather pipe is attached to the battery and not pinched by any part of the engine compartment.

Cleaning

Dirt and corrosion should be cleaned from the battery just as soon as it is discovered. Any accumulation of acid film or dirt will permit current to flow between the terminals. Such a current flow will drain the battery over a period of time. In time an acid film on top of the battery will rot the two rubber hold down straps.

Clean the exterior of the battery with a solution of diluted ammonia or a soda solution to neutralize any acid which may be present. Flush the cleaning solution off with clean water. **TAKE CARE** to prevent any of the neutralizing solution from entering the cells, by keeping the caps tight.

A poor contact at the terminals will add resistance to the charging circuit. This resistance will cause the voltage regulator to register a fully charged battery, and thus cut down on the stator assembly or charging coil output adding to the low battery charge problem.

Scrape the battery posts clean with a suitable tool or with a stiff wire brush. Clean the inside of the cable clamps to be sure they do not cause any resistance in the circuit.

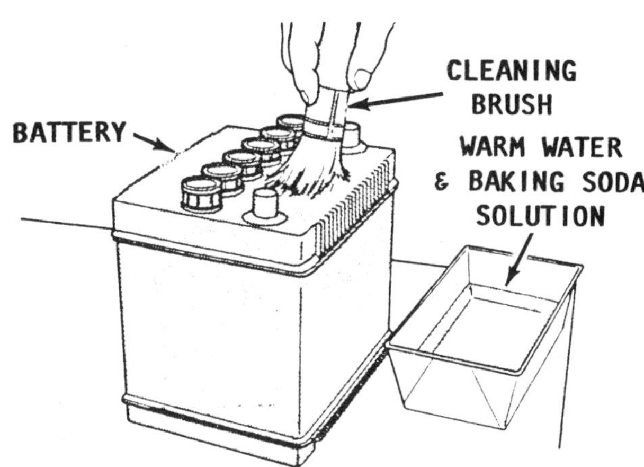

The battery top should be kept clean. A water and baking soda solution applied with a brush will neutralize battery electrolyte and clean the top nicely.

Battery Sulfation

Battery sulfation can be identified by the formation of white crystals forming on top of exposed battery plates. This usually occurs when a battery has been stored in a discharged condition over a period of time and the electrolyte has evaporated leaving the tops of the plates exposed to the air. This condition may also occur if the battery is not serviced at regular intervals during the season and the electrolyte not replenished when low.

Once white crystals form, the plates are permanently damaged and the battery will never hold a charge. It must be replaced.

Filling the Battery

Remove each filler cap using a pair of pliers. Fill each cell to the proper level with distilled water. The level of electro-

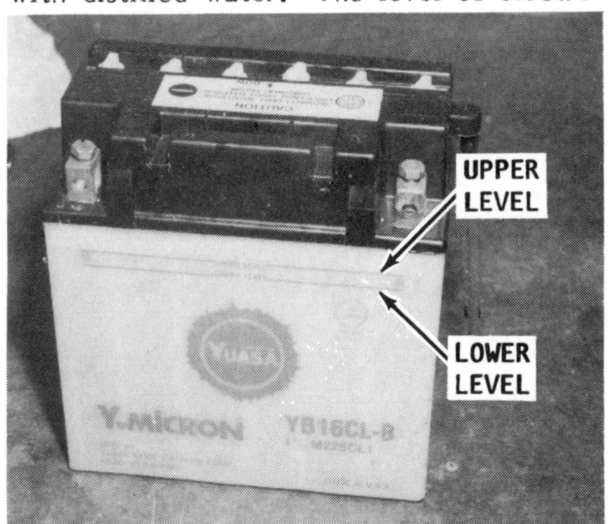

*If the battery is not a "Maintenance Free" type, it **MUST** be filled correctly. The proper procedure for this task is detailed in the text.*

lyte should be between the upper and lower level marks during operation. When filling the battery, **WITHOUT CHARGING,** fill to the upper line and allow the battery to stand for twenty minutes. Check the level again and replenish, as necessary. If filling the battery **PRIOR** to charging, fill to the lower lever only, because the electrolyte level will rise during charging.

During hot weather and periods of heavy use, the electrolyte level should be checked more often than during normal operation. Add distilled water to bring the level of electrolyte in each cell to the proper level. **TAKE CARE** not to overfill, because adding an excessive amount of water will cause loss of electrolyte and any loss will result in poor performance, short battery life, and will contribute quickly to corrosion. **NEVER** add electrolyte from another battery. **NEVER** add sulphuric acid because such acid would make the electrolyte too stong and permanently damage the battery. **NEVER** add ordinary tap water. The minerals in tap water will shorten battery life.

Battery Testing

A hydrometer is a device to measure the percentage of sulphuric acid in the battery electrolyte in terms of specific gravity. When the condition of the battery drops from fully charged to discharged, the acid leaves the solution and enters the plates, causing the specific gravity of the electrolyte to drop.

Use a temperature corrected hydrometer to test the specific gravity of the electrolyte. At 20°C (68°F) the hydrometer reading should be 1.20-1.26 on the scale.

It may not be common knowledge, but hydrometer floats are calibrated for use at 80°F (27°C). If the hydrometer is used at any other temperature, hotter or colder, a correction factor must be applied. Remember, a liquid will expand if it is heated and will contract if cooled. Such expansion and contraction will cause a definite change in the specific gravity of the liquid, in this case the electrolyte.

A quality hydrometer will have a thermometer/temperature correction table in the lower portion, as shown in the accompanying illustration. By knowing the air temperature around the battery and from the table, a correction factor may be applied to the specific gravity reading of the hydrometer float. In this manner, an accurate determination may be made as to the condition of the battery.

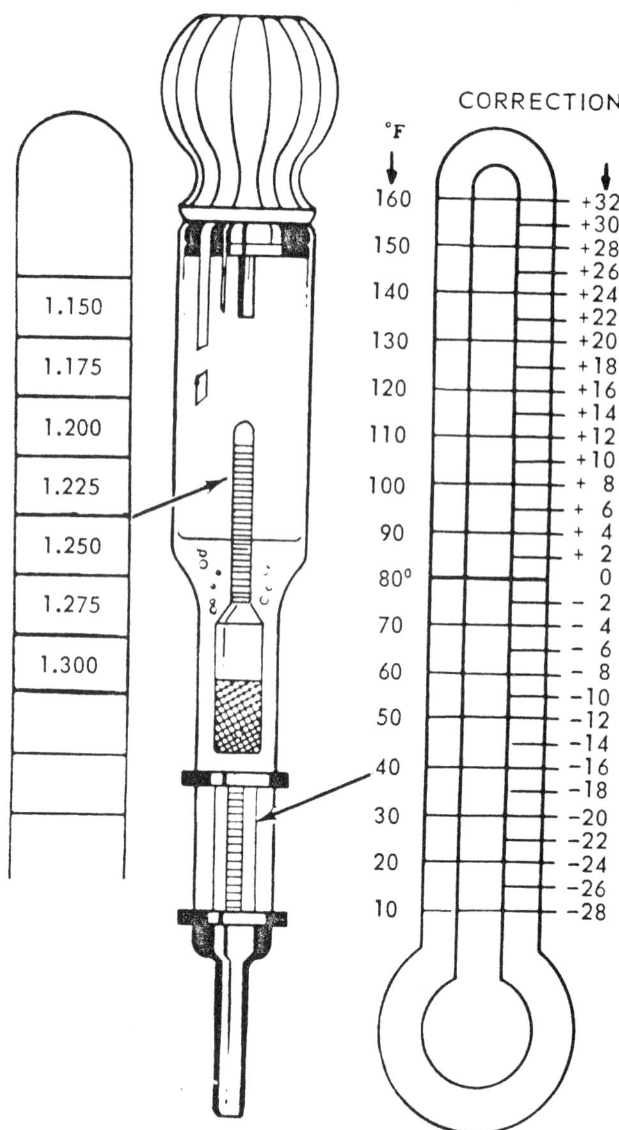

If the battery is not a "Maintenance Free" type, a check of the electrolyte in the battery should be on the maintenance schedule of any craft. A hydrometer reading of 1.300, or in the Green band, indicates the battery is in satisfactory condition. If the reading is 1.150 or in the Red bank, the battery must be charged. Observe the six safety points listed in the text when using a hydrometer.

The following six points should be observed when using a hydrometer.

1- NEVER attempt to take a reading immediately after adding water to the battery. Allow at least 1/4 hour of charging at the recommended rate to thoroughly mix the electrolyte with the new distilled water. This time will also allow for the necessary gases to be created.

2- ALWAYS be sure the hydrometer is clean inside and out as a precaution against contaminating the electrolyte.

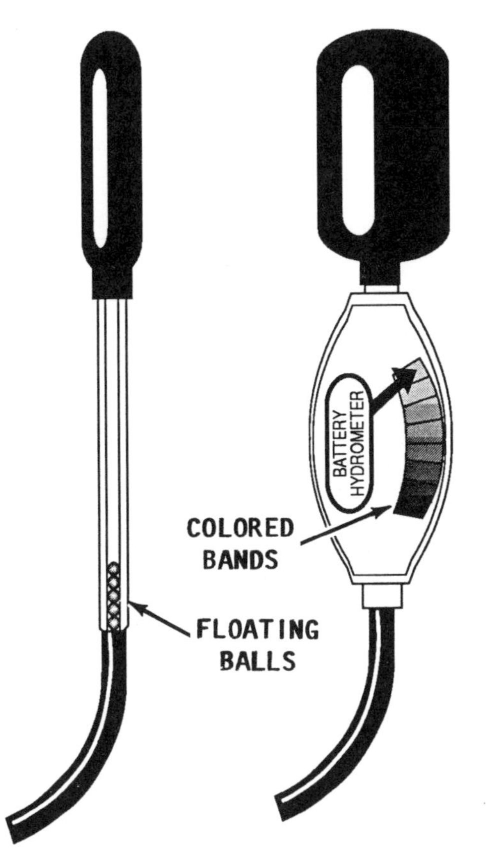

Two types of battery hydrometers, slightly different than the "old style". Their action is explained in the text.

3- If a thermometer is an integral part of the hydrometer, draw liquid into it several times to ensure the correct temperature before taking a reading.

4- **BE SURE** to hold the hydrometer vertically and suck up liquid only until the float is free and floating.

5- **ALWAYS** hold the hydrometer at eye level and take the reading at the surface of the liquid with the float free and floating.

Disregard the light curvature appearing where the liquid rises against the float stem. This phenomenon is due to surface tension.

6- **DO NOT** drop any of the battery fluid on the boat or on your clothing, because it is extremely caustic. Use water and baking soda to neutralize any battery liquid that does accidentally drop.

After withdrawing electrolyte from the battery cell until the float is barely free, note the level of the liquid inside the hydrometer.

Reading the Hydrometer

Different manufacturers of battery hydrometers use different methods of displaying specific gravity readings. Three methods are used:

Red, white, and green bands
Small floating balls
Graduated scale

Colored Band Hydrometer

When using a battery hydrometer with colored bands of red, white, and green, the electrolyte level is interpreted as follows:

Draw electrolyte from the battery cell until the float is barely free, note the level of the liquid inside the hydrometer.

If the level is within the green band range for all cells, the condition of the battery is satisfactory. If the level is within the white band for all cells, the battery is in fair condition.

If the level is within the green or white band for all cells except one, which registers in the red, the cell is shorted internally. No amount of charging will bring the battery back to satisfactory condition.

If the level in all cells is about the same, even if it falls in the red band, the battery may be recharged and returned to service. If the level fails to rise above the red band after charging, the only solution is to replace the battery.

Small Float Ball Hydrometer

If using a battery hydrometer with small floating balls, the electrolyte level is interpreted as follows:

If four or five balls are floating on top of the electrolyte drawn into the hydrometer, the battery is 100% charged. If three balls are floating, the battery is 75% charged. If two balls are floating, the battery is 50% charged. If only one ball is floating the battery is 25% charged. If no balls float, the battery is 0% charged.

If the same number of balls float for each cell, the condition of the battery is satisfactory.

If all the cells but one float balls, the cell is shorted internally. No amount of charging will bring the battery back to satisfactory condition.

If the number of floated balls for all cells is about the same, even if it is only one or two, the battery may be recharged and returned to service. If the number of balls fails to increase after charging, the only solution is to replace the battery.

Graduated Scale Hydrometer

If using a battery hydrometer with a graduated scale, note the level of the liquid inside the hydrometer. If the level is between the 1.280 and 1.265 mark, the battery is 100% charged. If the level is closer to the 1.210 mark, the battery is 75% charged. If the level is closer to the 1.160 mark, the battery is 50% charged. If it is closer to the 1.120 mark the battery is only 25% charged. If the level is below 1.100 the battery is at 0% charge.

If the level is the same and above 1.160 for all cells, the condition of the battery is satisfactory. If the level is the same and above 1.120 for all cells, the battery is in fair condition.

If the level is the same for all cells except one, which registers low on the scale, the cell is shorted internally. No amount of charging will bring the battery back to satisfactory condition.

If the level in all cells is about the same, even if it registers low on the scale, the battery may be recharged and returned to service. If the level fails to rise above 1.210 after charging, the only solution is to replace the battery.

Charging the Battery

If it is necessary to charge the battery, leave the filler caps lightly resting on the cell openings to allow the gases to escape.

The charging current should not exceed 1.9 amps for 10 hours.

NEVER use a fast charger, such as those found at automotive service stations, to charge a personal watercraft battery. These chargers operate at too high an output, ideal for automotive batteries, but very damaging to personal watercraft batteries. If such a battery is fast charged, excessive heat will be generated which will warp the plates and cause internal shorting. The plates will also shed active material which will accumulate at the bottom of the battery and cause internal shorting.

If, under the recommended charging procedures, the temperature of the electrolyte rises above 115°F (45°C) during charging, reduce the charging rate and increase the charging time proportionately. If possible, stop, allow the battery to cool and then resume charging for the recommended length of time.

After charging, push each filler cap into place and wipe the top of the battery clean before installation.

Installing the Battery

Check to be sure the battery is fastened securely in position. The hold-down rubber straps should be tight enough to prevent any movement of the battery in the holder.

Clean the battery posts and battery cable ends with a wire brush to ensure good, clean connections. If the battery posts or cable terminals are corroded, the cables should be cleaned separately with a baking soda solution and a wire brush. Apply a thin coating of Multi-purpose Lubricant to the posts and cable clamps before making the connections. The lubricant will help to prevent corrosion.

Identify the "POS" or "+" and "NEG" or "-" embossed symbols. Correctly connect the battery cables to the battery terminals. If the cables are connected incorrectly, the ignition system **WILL** be destroyed the first time the engine is started.

Check to be sure the breather pipe is attached to the battery and not pinched by any part of the engine compartment.

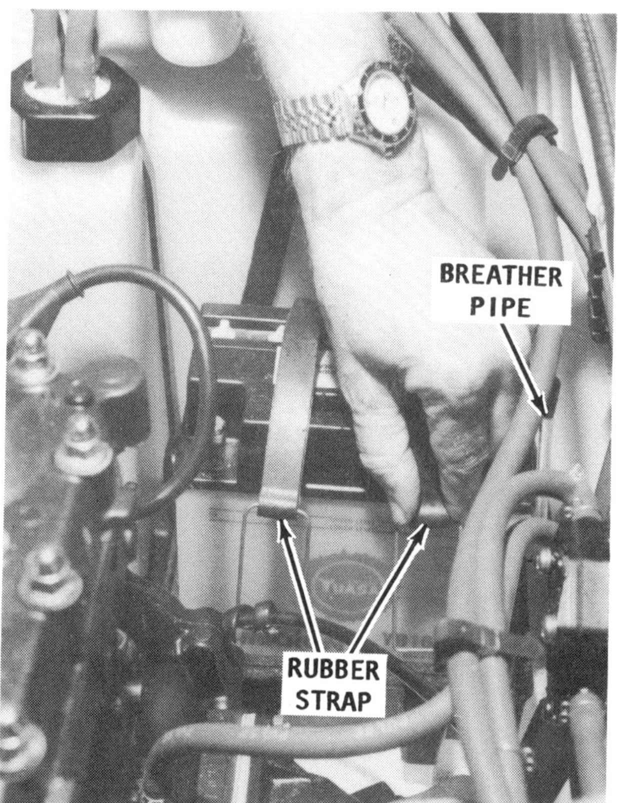

The breather pipe must not be pinched by any item in the engine compartment, after the holding straps are secured.

Jumper Cables

If a booster battery is used for starting an engine, the jumper cables must be connected correctly and in the proper sequence to prevent damage to either battery, or to the diodes in the rectifier/regulator.

ALWAYS connect a cable from the positive terminal of the dead battery to the positive terminal of the good battery **FIRST.** **NEXT**, connect one end of the other cable to the negative terminal of the good battery and the other end to a good ground on the engine. **DO NOT** connect the negative jumper from the good battery to the negative terminal of the low battery. Such action will almost always cause a spark which could ignite gases escaping through the vent holes in the battery filler caps. Igniting the gases may result in an explosion destroying the battery and causing severe personal **INJURY.**

By making the negative (ground) connection on the engine, if an arc is created, it will not be near the battery.

DISCONNECT the battery ground cable before replacing a stator assembly or charging coil, or before connecting any type of meter to the ignition system.

If it is necessary to use a trickle charger on a dead battery, **ALWAYS** disconnect one of the cables from the battery **FIRST**, to prevent burning out the diodes in the regulator/rectifier.

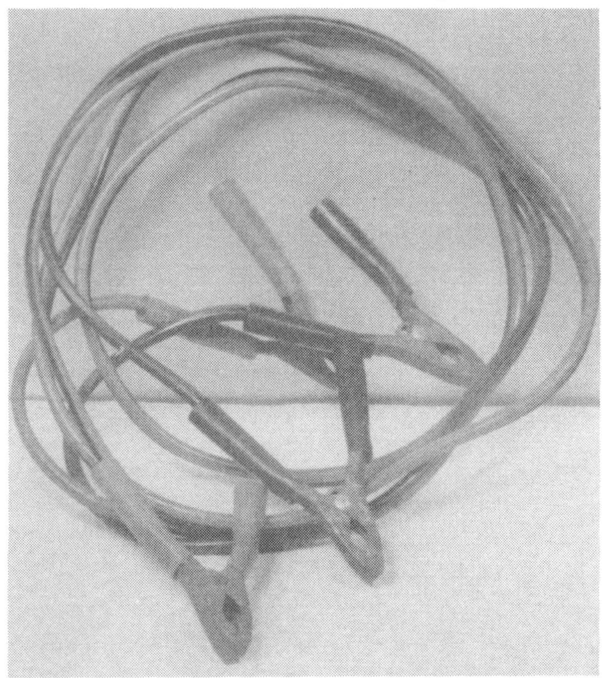

A common set of heavy-duty jumper cables. The booster battery must be connected correctly and in the proper sequence, as outlined in the text. Proper procedure will prevent damage to the battery or the diodes in the rectifier/regulator.

NEVER use a trickle charger as a booster to start the engine because the diodes will be **DAMAGED.**

Storage

If the craft is to be laid up for the winter or for more than a few weeks, special attention must be given to the battery to prevent complete discharge or possible damage to the terminals and wiring. Before putting the craft in storage, disconnect and remove the battery. Clean the battery thoroughly of any dirt or corrosion, and then charge it to full specific gravity reading. After the battery is fully charged, store the battery in a clean cool dry place where it will not be damaged or knocked over, preferably on a couple blocks of wood. Storing the battery up off the deck, will permit air to circulate freely around and under the battery and will help to prevent condensation.

NEVER store the battery with anything on top of it or cover the battery in such a manner as to prevent air from circulating around the filler caps. All batteries, both new and old, will discharge during periods of storage, more so if they are hot than if they remain cool. Therefore, the electrolyte level and the specific gravity should be checked at regular intervals. A drop in the specific gravity reading is cause to charge the battery back to a full reading.

One of the largest personal watercraft battery manufacturers recommends that a battery in storage should be checked every two weeks and charged if necessary. The electrolyte level can also be replenished at this time to prevent sulfation occurring.

In cold climates, care should be exercised in selecting the battery storage area. A fully charged battery will freeze at about 60° below zero. A discharged battery, almost dead, will have ice forming at about 19° above zero.

9-3 TACHOMETER

An accurate tachometer can be connected to any engine. Such an instrument provides an indication of engine speed in revolutions per minute (rpm). This is accomplished by measuring the number of electrical pulses per minute generated in the primary circuit of the ignition system.

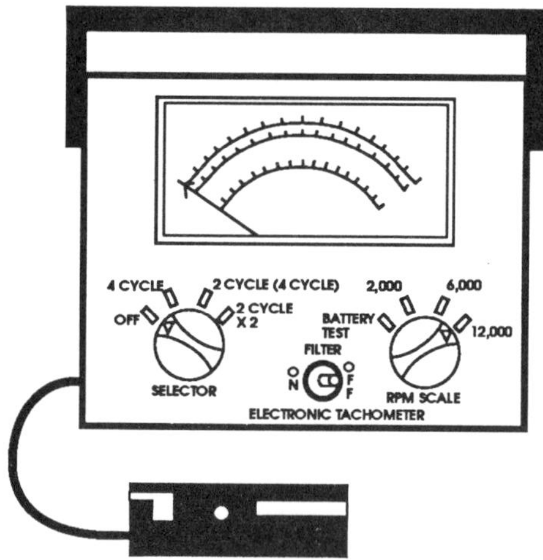

Maximum engine performance can only be obtained through proper tuning using a tachometer.

The manufacturer recommends the use of a digital tachometer, P/N 8712100 with an upper limit of 6,500 rpm or P/N 8712500 with an upper limit of 40,000 rpm. Newer tachometers such as these are more accurate due to solid state electronic circuits which eliminates the need for relays. The indicator window gives an accurate reading at a glance. Additionally, the electronic parts susceptible to moisture are coated to prolong their life.

Refer to the instructions provided with the tester to verify engine rpm. Generally, on a two-stroke engine with a CDI type ignition, the tachometer leads may be connected across the charging coil leads or spark plug high tension leads. If connecting to the charging coil leads, refer to Chapter 7 for the lead color identification.

Components of the ignition circuit and the charging circuit on a Model 650 are housed on the back side of the flywheel, covered and protected by the flywheel.

9-4 ELECTRICAL SYSTEM GENERAL INFORMATION

The electrical system consists of three separate circuits:

Charging circuit
Cranking motor circuit
Ignition circuit

Charging Circuit

Components of the charging circuit include:

Permanent magnets attached to the inside perimeter of the flywheel.
A charging coil installed on the stator plate.
A rectifier/regulator located elsewhere on the engine.
An external battery
Necessary wiring to connect it altogether.

The negative side of the rectifier/regulator is grounded. The positive side of the rectifier/regulator is connected to the battery. The negative side of the battery is connected to a good ground on the engine.

The alternating current generated in the stator windings passes to the regulator/rectifier. The rectifier/regulator changes the alternating current (AC) to direct current (DC) to charge the 12-volt battery.

The stator is located behind and protected by, the flywheel. Therefore, the stator, including the charging coil, seldom causes problems in the charging circuit. Most problems in the charging circuit can usually be traced to the rectifier/regulator or to the battery. If either the stator or the rectifier/regulator fails the troubleshooting tests, the defective unit cannot be repaired, it **MUST** be replaced.

See Chapter 7 to service and test components in the charging system.

Cranking Motor Circuit

The cranking motor circuit consists of a cranking motor and a starter-engaging mechanism. A starter relay is used as a heavy-duty switch to carry the heavy current from the battery to the cranking motor. The starter relay is actuated by depressing the **START** button.

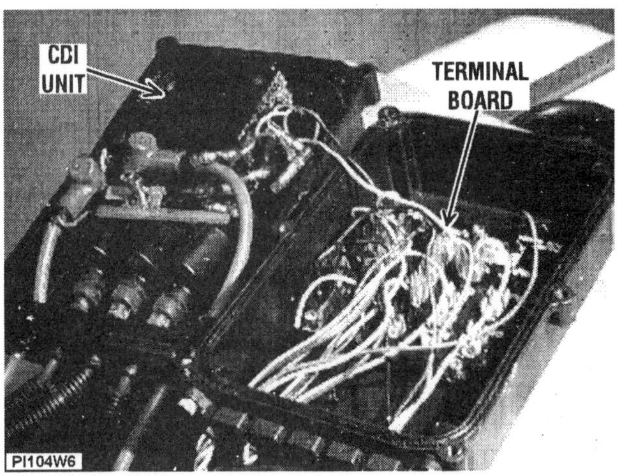

Opening the electrical box reveals cranking, ignition and charging circuit components. The unit should be placed in a convenient location for testing purposes.

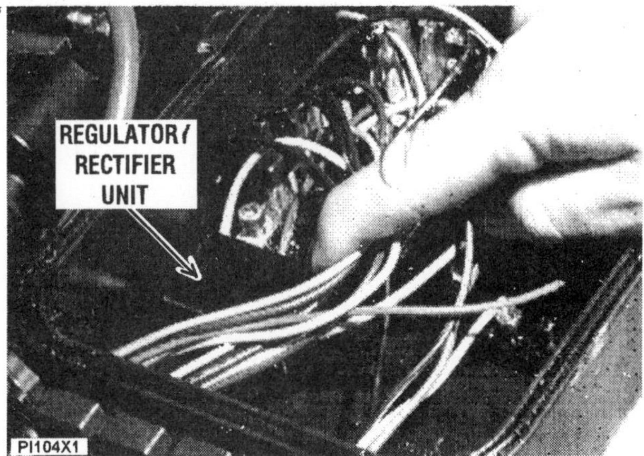

Location of the "LR21" rectifier/regulator unit from an SLT750 Model. Other Models may use an "LR23" module. If servicing this unit, check to be sure the module is properly identified.

Ignition Circuit

The ignition circuit is covered extensively in Chapter 7, Ignition.

9-5 CRANKING MOTOR CIRCUIT SERVICE

DESCRIPTION

As the name implies, the sole purpose of the cranking motor circuit is to control operation of the cranking motor to crank the engine until the air/fuel mixture ignites and the flywheel develops enough torque to continue under its own power. The circuit includes a solenoid or magnetic switch to connect or disconnect the motor from the battery. The operator controls the switch with a push button.

The cranking motor is a series wound electric motor which draws heavy current from the battery. It is designed to be used only for short periods of time to crank the engine for starting. To prevent overheating the motor, cranking should not be continued for more than 30-seconds without allowing the motor time to cool -- say at least three minutes. Actually, this time can be spent in making preliminary checks to determine why the engine fails to start.

Theory of Operation

Some watercraft manufacturers have employed various methods for cranking the engine. On Polaris Personal Watercraft, the same method is used to transmit power from the cranking motor to the engine crankshaft through a combination of meshed gears or clutch mechanisms.

All models have a reduction gear housed in the flywheel cover containing the Bendix drive. The armature shaft has a splined end which meshes inside the reduction gear. When the cranking motor is operated, the pinion gear -- a part of the reduction gear assembly -- moves forward and meshes with the teeth on the flywheel ring gear. The simple line drawing on this page illustrates this arrangement.

When the start button is depressed, the idle gear is driven by the cranking motor. A one-way clutch on the idle gear shaft slides forward, engaging the reduction gear -- at the forward end of the shaft -- with the flywheel ring gear. After the engine starts and the start button is released, engine rpm rises above cranking rpm. The rollers inside the clutch move in a different direction at increased speed and no longer

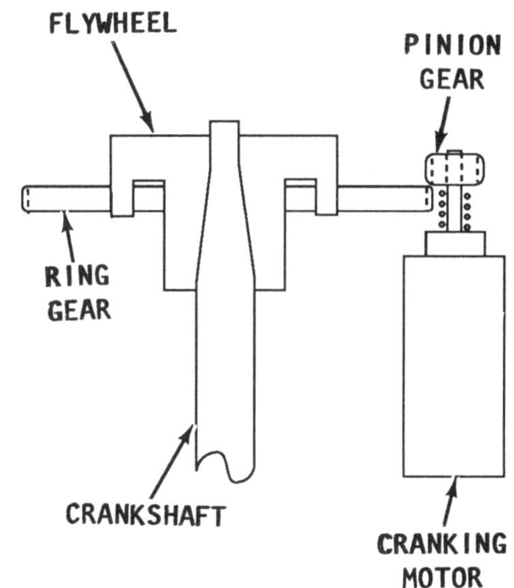

Line drawing to depict the power train method for transferring energy from the cranking motor to the engine flywheel for all Models.

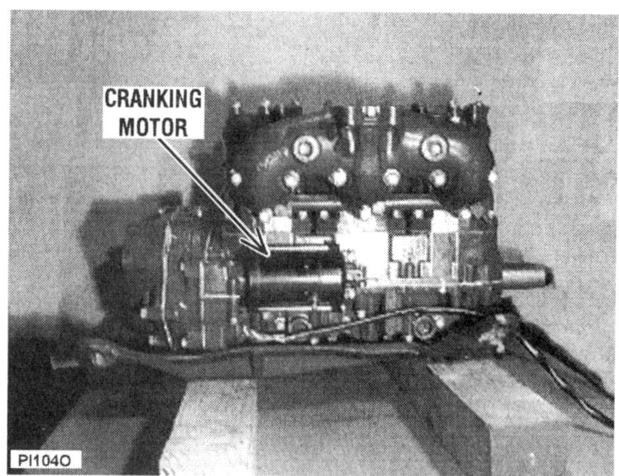

Location of the cranking motor for all engines covered. The installation per Model may differ slightly.

contact the outer race of the one-way clutch. Therefore, the one-way clutch ceases to rotate and slides back on the shaft disengaging the reduction gear from the flywheel gear.

Good Words
Cranking Motor Noises

The sound of the motor during cranking is a good indication of how the cranking motor is operating -- properly or not. Naturally, temperature conditions will affect the speed at which the cranking motor is able to crank the engine. The speed of cranking a cold engine will be much slower than when cranking a warm engine. An experienced operator will learn to recognize the favorable sounds of the engine cranking under various conditions.

Faulty Symptoms

If the cranking motor spins, but fails to crank the engine, the cause is usually a corroded or gummy Bendix drive. The drive should be removed, cleaned, and inspected for signs of corrosion or wear.

View of the reduction gear assembly seated behind the inner magneto cover on a Model SLT750. On early Models, the inner cover must be "pulled" to remove the reduction gear.

If the cranking motor cranks the engine too slowly, the following conditions are possible causes, with corrective actions given which may be taken:

a- Battery charge is low. Charge the battery to full capacity.
b- High resistance connections at the battery, solenoid, or motor. Clean and tighten all connections.
c- Undersize battery cables. Replace cables with sufficient size.

If the cranking motor must be removed for inspection or replacement, most engines covered in this manual require the exhaust elbow, manifold and chamber be removed first.

All engine cranking problems fall into one of three problem areas:

1- The cranking motor fails to rotate.
2- The cranking motor spins rapidly, but fails to crank the engine.
3- The cranking motor cranks the engine, but too slowly.

The cranking motor does not require periodic maintenance or lubrication. If the motor fails to perform properly, the checks outlined in the previous paragraph should be performed. Naturally, the motor must be removed, if the listed corrective actions outlined in this section do not restore the motor to satisfactory operation.

CRANKING MOTOR TROUBLESHOOTING

Before wasting too much time troubleshooting the cranking motor circuit, the following checks should be made. Many times, the problem will be corrected.

a- Battery fully charged.
b- Main fuse is "good" (not blown).
c- All electrical connections clean and tight.
d- Wiring in good condition -- insulation not worn or frayed.

The following troubleshooting procedures are presented in a logical sequence.

Do not operate the cranking motor for more than 15 seconds. Prolonged cranking motor operation will cause overheating and damage the motor.

After each test, allow the cranking motor to cool for a minute or so.

Never depress the **START** button to activate the cranking motor while the engine is operating. Such action will damage the pinion and/or flywheel gears.

Cranking Circuit Tests

The following procedures require the cranking motor leads, starter solenoid, and battery. The cranking motor solenoid is housed inside the electrical box. The battery of course, is underneath the electrical box, or mounted to a bulkhead. The cranking motor and solenoid need not be removed. This series of tests will involve dynamic testing of the cranking motor system -- a digital multimeter must be used.

During the dynamic tests, the leads must make contact **WHILE** the engine is being cranked. **ALSO**, both meter leads must be removed from the solenoid terminals **WHILE** the engine is being cranked and **BEFORE** the engine has stopped cranking. If these precautions are not followed, the voltmeter may be damaged. Ask an assistant to press the start switch.

a- Remove all spark plug leads from the plugs and ground the ends to the engine to prevent accidental engine start.

b- Obtain a voltmeter and select the VDC scale. Observe the starter solenoid. Both large terminals have Red leads connected. One terminal has the battery cable connected and the other terminal has the lead from the cranking motor connected.

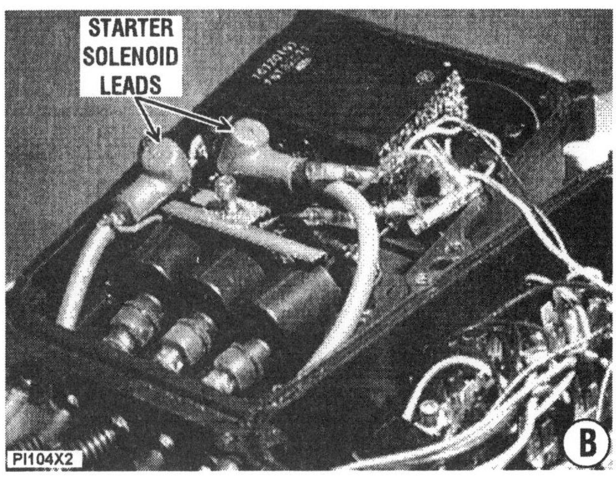

Disconnect the Red engagement coil wire from the starter solenoid. Connect the Black tester wire to an appropriate ground and the Red lead to the Red harness wire at the solenoid. Press the start button. The meter should read battery voltage. If so, proceed directly to step **c**. Otherwise, place the Black tester lead on ground and check for voltage at the large relay in terminal, the circuit breaker in and out terminals, and across both sides of the start switch, while the start button is in the **START** position. Repair or replace any defective parts.

c- Reconnect the starter solenoid. Connect the Black tester lead to the positive battery terminal and the Red tester lead to the solenoid end of the battery-to-solenoid cable. Press the start button. If the reading is less than 0.1V DC, proceed with step **d**. Otherwise, clean the battery-to-solenoid cable ends or replace the cable.

d- Connect the Black tester lead to the solenoid end of the battery-to-solenoid cable. Connect the Red tester lead to the solenoid end of the solenoid-to-starter cable. Press the start button. If the reading is less than 0.1V DC, proceed to step **e**. Otherwise, the starter solenoid must be replaced.

e- Connect the Black tester lead to the solenoid end of the solenoid-to-starter cable. Connect the red tester lead to the starter end of this cable as well. Press the start button. If the reading is less than 0.1V DC, proceed to step **f**. Otherwise, clean the ends of the solenoid-to-starter cable or if necessary, replace the cable.

f- Connect the Black tester lead to the cranking motor frame. Connect the Red tester lead to the negative battery terminal. Press the start button. If the reading is less than 0.1V DC, continue to step **g**. Otherwise, clean the ends of the negative battery cable at the battery and engine block (or engine mounting plate). Check to be sure the bolts are secure, not loose. The ground path from the cranking motor to the negative battery terminal must be complete.

g- If all these tests indicate a good condition, yet the cranking motor still fails to adequately rotate, the motor must be replaced.

For detailed instructions on removing and installing the cranking motor, refer to Section 8-3; Complete Service of Three-Cylinder Engines or Section 8-4; Complete Service of Two-Cylinder Engines -- Chapter 8.

Once the work is complete, close up the electrical box, after checking to be sure all connections are properly secured. Install the electrical box and secure it in place with the attaching hardware or restraining straps. Insert and tighten the spark plugs to a torque value of 18ft lbs (25Nm). Install the spark plug high tension leads.

9-6 TESTING OTHER ELECTRICAL COMPONENTS

This section provides testing procedures for other electrical parts and subsystems. If a unit fails the testing, the faulty part **MUST** be replaced. In most cases, removal and installation is through attaching hardware.

Temperature Warning Circuitry Test

A temperature sensor located on the water outlet manifold or exhaust manifold detects engine coolant water temperature. If the water reaches 160° or 180° F (71° or 82° C) depending on model, a warning buzzer or light warns the operator of the overheat condition. The electronic warning switch is normally open. Once activated, the switch closes, completing the current flow from the wire harness through the warning buzzer or light and to ground.

On 1992 - 1993 Models, remove the lead from the temperature sensor -- switch -- located on the water outlet manifold. Connect the lead securely to the engine ground. Start the engine -- the warning buzzer should "sound off". Shut down the engine immediately. If the warning buzzer did not activate, inspect leads for an open circuit or short, or replace the warning buzzer assembly.

1994 - 1995 SL650 Models only -- Remove the electrical box cover to access the terminal board. Disconnect the tan lead from the temperature **BUZZER** and connect it securely to the engine ground. Start the engine. If the temperature warning buzzer does **NOT** activate, the unit should be replaced, if the wiring is not at fault.

MFD equipped Models -- During an overheat condition, the temperature switch closes, illuminating a warning light on the Multi Function Display (MFD). On Fuji manufactured engines, the temperature probe is located on the water outlet manifold. On Polaris manufactured engines, the temperature sensor is located on the exhaust manifold. To test the circuit, ground the sensor wire from the MFD while the engine is running -- the light should illuminate, the LCD (Liquid Crystal Display) should indicate "HOT".

Location of the temperature warning sensor on an SLT750 Model. Other models may have a sensor located on the exhaust manifold as explained in the text.

Temperature Sensor Test

Obtain a digital ohmmeter and set the dial to the Ohms position. Disconnect the sensor leads and make contact with one test probe to the switch terminal, make contact with the other probe to engine ground. The meter should indicate an open circuit -- the temperature switch is normally open. The sensor may be heated in water -- **NEVER** an open flame. Heat a pan of water on the stove and have a thermometer handy to measure the water temperature. Submerge only the sensor end in the water as shown, without the sensor touching the sides or the bottom of the pan. Observe the meter reading as the water temperature rises. Stir the water with the thermometer to evenly distribute the heat.

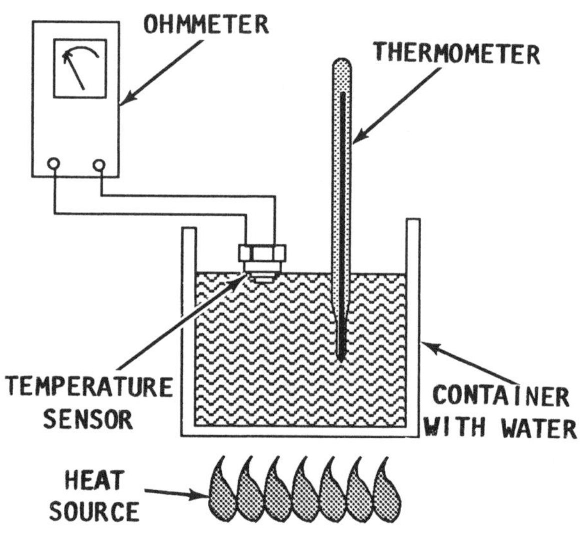

Line drawing to depict the setup for testing the operation of the temperature sensor, as outlined in the text.

Again, expose **ONLY** the sensor, not the leads, to the water bath. Once the temperature reaches the activation level -- 160° F (71°C) or 180°F (82°C), depending on Model -- measuring the switch resistance should indicate 0.4 ohms.

RPM Limiter Operation

The rpm limiter circuitry -- including the CDI module -- protects the engine by limiting rpm in the event of an overheat condition. If the thermal switch detects engine overheat, the warning light will illuminate; the MFD will indicate "HOT"; and the rpm should be limited to approximately 4,200 or 4,500rpm depending on Model.

Some Models also have a circuit to detect a low fuel condition, coupled to the rpm limiter. During a low fuel condition, the engine is limited to 4,200rpm; the warning light illuminates; and the MFD will indicate "FUEL". This protects the engine from running out of fuel during high rpm settings and gives the operator more time to reach shore.

Important Words

The problem causing the rpm limitation when one or both of these conditions exist must be corrected before the limiter deactivates. **Both** of the following conditions must be met before the rpm limitation ceases:

a -- The condition causing the fault must be corrected. Fill the tank or cool the engine.

b-- The throttle must be released and allow the rpm to drop below the threshold limit.

Rpm Limiter Troubleshooting

The following procedures are listed in logical order to troubleshoot the rpm limiting circuitry per Model. Refer to the wiring diagrams in the Appendix for the Model being serviced.

1995 SLX780
1996 SL780, SLT780, and SLX780

The listed Models limit rpm to 4,200rpm due to an overheat and/or low fuel condition.

Additionally, the **1996 SL900** and **SLTX** Models may be tested using the same following procedures, however, the ignition timing is retarded to 8° BTDC.

A-- Check to be sure the fuel supply is sufficient -- at least half full. If the fuel level was low, fill the tank and re-test.

B-- Remove the electrical box from its secured position. Carefully open the box ensuring the large **O**-ring remains intact. Place the electrical box in a convenient location for testing components. Check to be sure all Tan leads in the terminal board area are isolated from a grounding source. If a short exists, repair or replace the faulty part.

C-- Disconnect the CDI module Grey lead from the terminal board. If the system **CONTINUES** to limit rpm or retard timing, continue following the procedures listed below. If the system does **NOT** limit rpm or retard timing, proceed directly to Step **D**.

The Gray lead may be shorted to ground. Verify the lead condition and terminal board connection. Repair if necessary and re-test.

The fault may be caused by improper ignition timing. For detailed procedures to verify timing, see Chapter 7 -- Ignition.

The CDI may be faulty. The CDI module is **NOT** serviceable and must be replaced if necessary.

D-- Connect the CDI module Gray lead to the terminal board. Disconnect the Pink lead from the fuel sender. Ground the Pink lead from the "LR54" module -- similar in appearance to the rectifier/regulator module. If the system does **NOT** continue to limit rpm or retard timing, proceed with the following steps. If the system **CONTINUES** to limit or retard timing, proceed directly to Step **E**.

The fuel level system is probably at fault. Inspect leads and terminal board, repair if necessary and retest.

If the fault does not appear to be within the wiring, remove the fuel sender from the tank and test the sender unit as described on Page 9-14.

E-- Connect the Pink leads to the terminal board. Disconnect the temperature Tan lead at the terminal board. If the system does **NOT** limit rpm or retard timing, proceed with the following steps. If the system **CONTINUES** to limit rpm or retard timing, go directly to Step **F**.

Inspect the cooling system for leaks, restrictions or other visible problems which may cause the overheating condition.

Inspect the temperature sensor Tan lead by the sending unit for signs of shorting to ground. Repair or replace as necessary.

If the lead is intact, test the temperature sensor as outlined on Page 7-12.

F-- At this point, it is likely the CDI module must be replaced. Consult a local dealer for availability.

1997 SL780, SLT780, SL1050 and SLTX

A-- Remove the electrical box from its secured position. Carefully open the box ensuring the large **O**-ring remains intact. Place the electrical box in a convenient location for testing components. Check to be sure all Tan leads in the terminal board area are isolated from a grounding source. If a short exists, repair or replace the faulty part.

B-- Disconnect the temperature sensor Tan lead from the terminal board. If the system does **NOT** limit rpm or retard timing, proceed with the following steps. If the system **CONTINUES** to limit rpm, the CDI module must be replaced.

Inspect the cooling system for leaks, restrictions or other visible problems which may cause the overheating condition.

Inspect the temperature sensor Tan lead by the sending unit for signs of shorting to ground. Repair or replace as necessary.

If the lead is intact, test the temperature sensor as outlined on Page 7-12.

Fuel Sender Unit Testing

The sending unit "measures" fuel level by varying the resistance value as the height of the float arm changes in the fuel tank. The resistance can be tested with an ohmmeter by disconnecting the sender leads at the electrical box. For all Models so equipped, the values are as follows:

Full tank (float up) -- 35 Ohms
Empty tank (float down) -- 240 Ohms

For the most accurate results, the test should be performed at room temperature. The test results should be within ± 10%.

Oil Level Sender Operation

An oil level sensor monitors available engine oil. If the level drops below a given threshold, an audible buzzer is activated on Models not equipped with an MFD. The MFD on Models so equipped will activate a bright flashing light.

Audible Buzzer Test

A-- Refer to the wiring diagram in the Appendix for the Model being serviced. Place the electrical box in a convenient location on the craft to perform the following procedures.

B-- Carefully open the box ensuring the large **O**-ring remains intact. Identify the Orange lead terminal or buzzer harness connection. Start the engine and check for battery voltage at the Orange lead or harness connection. Shut down the engine. If battery voltage is **NOT** present, the problem is probably in the rectifier/regulator module, wiring harness, or stator assembly. Inspect these components and repair or replace as necessary.

C-- If battery voltage is present, connect the buzzer Tan lead to ground. Start the engine and listen for the buzzer to activate within a few seconds. Shut down the engine. If the buzzer did not work, inspect the leads between the buzzer and electrical box. Repair any damaged leads. If leads are in good condition, replace the buzzer.

If the buzzer activated after engine start-up, the oil level sensor should be replaced.

Oil Level Sensor Test Models with MFD Only

If the oil level is not represented or the low oil warning is not indicated on the MFD, perform the following procedures.

Move the electrical box to a convenient location for testing internal components. Obtain a digital ohmmeter. Remember, the tests should be performed at room temperature, readings may vary -- depending on tester -- by ±10%. Refer to the appropriate wiring diagram in the Appendix for specific values and lead colors.

Open the box and disconnect the oil sender wires. Measure the resistance of the leads and compare results with the schematic specifications.

Inspect each of the electrical connections -- including the electrical harness -- at the back of the MFD. All connections should be clean and tightened securely. Oil sender resistance values are as follows:

Full (float up) -- 35 Ohms
Empty (float down) -- 240 Ohms

Replace the oil sender unit if necessary.

Trim System Motor Test

The electric trim adjustment system allows the operator to manipulate the vertical movement of the steering nozzle. This adjustment permits compensating for varying center of gravity adjustments, which minimizes the "porpoising" of the craft. Non-MFD equipped Models indicate vertical positioning of the steering nozzle via a mechanical link.

Remove the four attaching bolts securing the trim motor assembly to the inside of the hull at the aft end of the craft. Lift the assembly and push in against the plastic trim arm to release the retaining pin. Remove the trim arm and spring.

Remove the outer cover free of the trim base plate. Apply 12VDC to the motor leads in one direction and verify motor operation. Reverse the leads and apply 12VDC -- the motor should rotate in the opposite direction. Check to be sure the motor rotates freely. If the motor does not function properly, the assembly must be replaced.

Move the trim motor assembly into place and secure it with the four attaching bolts. Now, outside the hull, push in on the plastic trim arm and install the retaining pin.

Trim System Indicator Cable Replacement

Although the pitch indicator functions mechanically, procedures to repair the gauge is included in this section.

Remove the four attaching bolts securing the trim motor assembly to the inside of the hull at the aft end of the craft. Lift the assembly and push in against the plastic trim arm to release the retaining pin.

Remove the retaining pin, trim arm, and spring. Loosen, then remove the cable retaining screw from the trim arm.

Remove the position indicator cable from the housing. Lubricate the cable **O**-ring and push a new cable into the housing. Attach the cable end with the retaining screw and fasten it securely.

Before assembling the unit, check to be sure the motor shaft oil seal is in good condition. Verify the motor operates properly.

Begin assembling by installing the spring, trim arm, and the retaining pin. Position the trim motor in place in the hull and secure it with the four attaching bolts tightened securely.

Multi Function Display Words

A complete list of all the functions available to the operator using an MFD is not given. However, if the unit does not operate properly, the following helpful corrective suggestions are presented.

Check to be sure the battery connections are secured and free of corrosion. If disconnecting the battery, be sure to disconnect the negative lead first. When connecting the battery leads, connect the negative cable last. Be sure to contact the lead to the terminal just one time, because repeated attempts at contacting the terminal may affect the microprocessor of the MFD.

Check the integrity of the fuse located inside the electrical box.

Check to be sure the MFD connector pins are secure and free of corrosion. Verify the power and ground leads are tight.

Check to be sure all accessory inputs to the MFD are functioning properly. For example, if the speedometer appears to be indicating false readings, verify the condition of the speedometer pitot tube -- it should be clear of obstructions. Fuel and oil senders should be checked -- the leads **MUST** be securely fastened. A loose lead may cause short to ground.

If the MFD continues to malfunction, disconnect the battery for approximately two hours. This may help to "clear" the microprocessor. Permanent data such as engine operating time will remain intact. Connect the battery negative cable first. If the MFD continues to display inaccurate information the unit should be replaced.

Electric Bilge Pump

The electric bilge unit receives battery power from the "LR23" or "LR52" module any time the engine is operating. To test the operation of the pump, disconnect the bilge leads and apply an alternate power source and ground. If the bilge pump does not operate from an alternate power source, the unit should be replaced because it is not serviceable.

If the bilge pump does operate from an alternate power source, obtain an ohmmeter and check to be sure the Orange bilge pump lead terminal -- in the electrical box -- supplies battery voltage. If battery voltage is not indicated, check to be sure the connection is intact, or the specific "LR" module should be replaced. If battery power is indicated, check to be sure the pump is grounded properly.

ELECTRICAL BASICS

See wiring diagrams in the appendix for more information

MODEL	ALTERNATOR OUTPUT	SPARK PLUG NGK	CHAMP	PLUG GAP
1992-1995 SL650/STD SL750 SLT750	120 Watts @ 4500 RPM	BPR7ES	RN3C	.024-.028 (.6-.7 MM)
1995 SLX780	120 Watts @ 4500 RPM	BPR8ES	RN2C	.024-.028 (.6-.7 MM)
1996 SL\SLT 700	60 Watts @ 4500 RPM	BPR7ES	RN3C	.024-.028 (.6-.7 MM)
1996 Hurricane	60 Watts @ 4500 RPM	BPR7ES	RN3C	.024-.028 (.6-.7 MM)
1996 SL780 SLT 780 SLX 780	120 Watts @ 4500 RPM	BPR8ES	RN2C	.024-.028 (.6-.7 MM)
1996 SL 900 STTX	60 Watts @ 4500 RPM	BPR8ES	RN2C	.024-.028 (.6-.7 MM)
1997 SL 700 Deluxe SLT 700	60 Watts @ 4500 RPM	BPR7ES	RN3C	.024-.028 (.6-.7 MM)
1997 SL 700 Hurricane	60 Watts @ 4500 RPM	BPR8ES	RN2C	.024-.028 (.6-.7 MM)
1997 SL 780 SLT 780 SLX 780	120 Watts @ 4500 RPM	BPR8ES	RN2C	.024-.028 (.6-.7 MM)
1997 SL 900	60 Watts @ 4500 RPM	BPR8ES	RN2C	.024-.028 (.6-.7 MM)
1997 SL 1050	60 Watts @ 4500 RPM	BPR8ES	RN2C	.024-.028 (.6-.7 MM)
1997 SLTX	60 Watts @ 4500 RPM	BPR8ES	RN2C	.024-.028 (.6-.7 MM)

Alternate Spark Plugs

For NGK BPR8ES -- Use NGK BR9ES
For NGK BPR7ES -- Use NGK BR8ES

Manufacturer Suggests: Use only Resistor type spark plugs in all Polaris products.

10
JET PUMP

10-1 INTRODUCTION

Chapter Coverage

Polaris personal watercraft use single stage axial flow jet pumps. The axial pump used on all models covered in this manual, have a two-piece impeller shaft -- actually a long drive shaft and a short impeller shaft. Be sure to note the number of bearings and proper positioning of oil seals during disassembly.

A general description of jet pumps follows in this Section. Both mixed and axial flow jet pumps are presented for review.

The drive train of a typical watercraft is comprised of a coupler at the aft end of the engine, a long driveshaft, and the jet pump assembly. Procedures for removing, cleaning and inspecting, and then installation of each subsystem of the drive train will be discussed.

A description of impellers and impeller-to-case measurements are given in Section 10-2.

Procedures to service the jet pump are presented in Section 10-3.

Bearing housing and driveshaft inspection procedures are provided in Section 10-4.

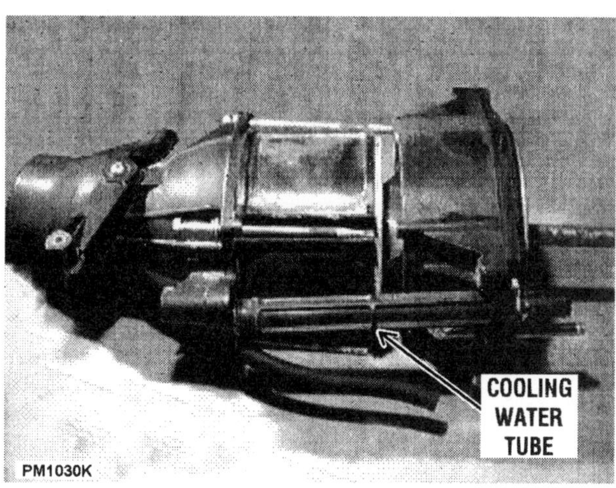

Jet pump removed from a Hurricane Model, with the cooling water tube clearly visible.

SPECIAL WORDS ON JET PUMPS

A very small gap must be maintained between the impeller blade and the pump case circumference or wear ring. Due to cavitation, ventilation, and other types of wear, the gap will increase between these components. If the impeller-to-pump case clearance becomes excessive, or the pump case is deeply grooved, the case **MUST** be replaced.

An integral wear ring lines the inner circumference of the pump housing on early model jet pumps manufactured from 1992-1993. These wear rings are replaceable, however, it is likely the ring inner surface must be machined after installation to achieve the proper impeller-to-case (wear ring) clearance. As the gap between the impeller and case increases, pump efficiency is lost. A loss of thrust will occur.

Keep a record of all shim material removed from the craft. Check each piece for signs of wear or damage. If any material should need replacement, be sure to use the appropriate

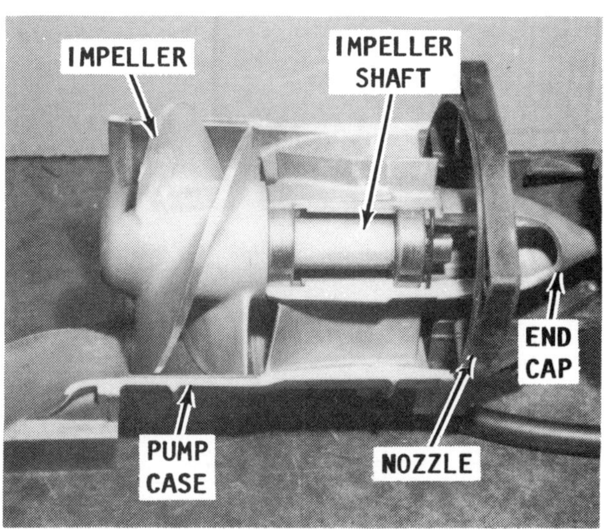

Cutaway of a typical demonstration-type axial flow jet pump. Polaris watercraft use a similar design. Major parts are identified.

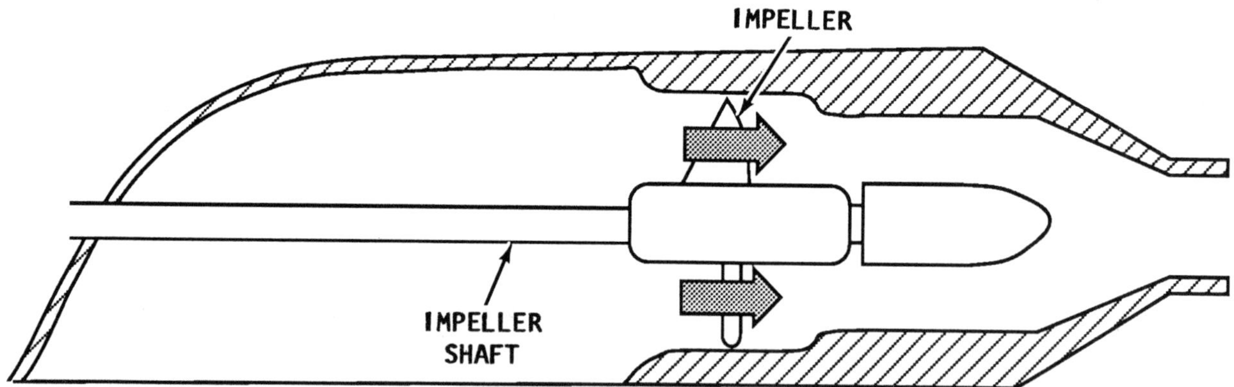

Simple cross-section line drawing to depict water flow through an axial flow jet pump. The water passes through parallel to the axis of the impeller rotation.

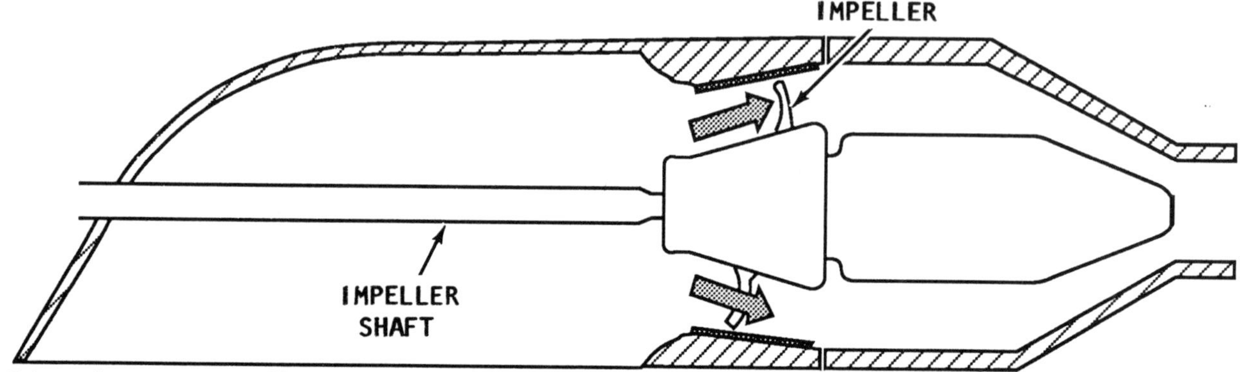

Line drawing to illustrate water flow through a mixed flow jet pump. Notice how the water is directed upward at an angle to the axis of impeller rotation.

measuring device to obtain the exact thickness of shim material to be purchased. Improper installation of shim material may lead to excessive vibration and possible failure of watercraft components.

JET PUMP DESCRIPTION

Axial Flow

Water in an axial pump moves on a single axis, as depicted in the adjacent illustration, thus the word "axial". In simple terms, water is ingested and discharged parallel to the axis of impeller rotation. The inside of the pump case is parallel to the water flow.

Mixed Flow

Many experts consider the mixed flow system to be the most efficient for personal watercraft application. The water flow is discharged at an inclined angle to the axis of impeller rotation, as indicated by the arrows in the accompanying illustration. The impeller blades are inclined in such a manner to cause the water to be thrown off at a 45 degree angle to the impeller shaft within the pump case.

From the pump case, the water then flows through the guide vane casting and forced out through the outlet nozzle and steering nozzle. The vanes in the guide vane casting help straighten the water flow through the nozzle. The inside of the pump case is inclined at an angle to the water flow. Therefore, this design pump incorporates shim material to space the impeller blades at a specified distance from the pump case. Mixed flow pump efficiency may decrease as much as 30% if the impeller-to-pump case clearance increases by only 0.04" (1mm).

Axial or Mixed?

Axial and mixed flow pumps are intended for different applications. One is not necessarily better than the other. The axial pump can be operated at higher engine rpm and is more suitable for racing. A mixed flow pump adds thrust at low engine rpm due to the pump design. Therefore, the mixed flow pump performs more efficiently at low engine rpm. This type pump is more suitable for a craft equipped to carry extra weight or intended to pull a water skier.

A jet pump seldom requires service except for the following:

Impeller-to-pump case clearance becomes excessive.
Impeller is damaged.
The oil seal around the impeller shaft fails -- allowing water to enter hull.
Silicone seal around the perimeter of the pump case allows the pump to ingest air.

10-2 IMPELLERS

Impellers provide thrust using a combination of water flow and water pressure in a closed environment.

The blades of an impeller overlap, thus water is trapped and forced through the impeller into the pump case and outlet nozzle. The water moving past the impeller resists cavitation because it is under pressure. Blades in the pump case are angled exactly opposite to those of the impeller. These blades redirect the water to establish a concentrated flow through the nozzle.

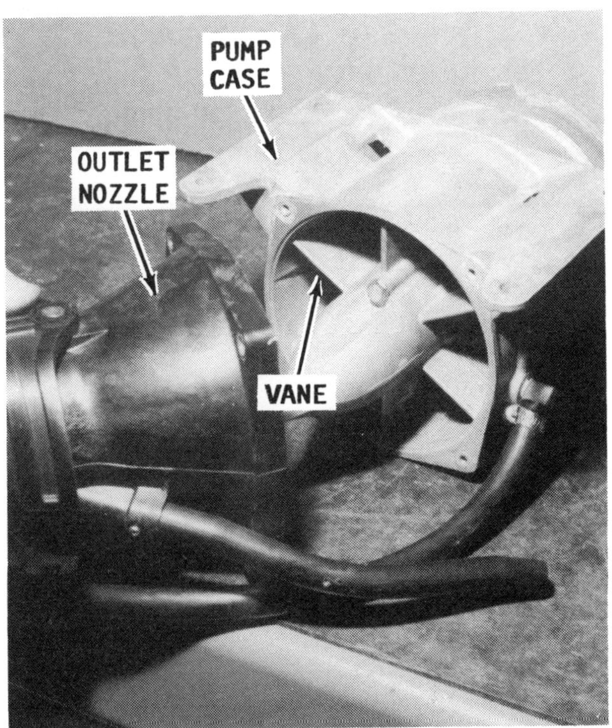

The vanes in a typical pump case straighten water flow and the conical shape of the outlet nozzle increases flow velocity.

The manufacturer recommends a radius of about 1/64" (0.3-0.5mm) for the leading edge of the impeller blade. A sharper radius will result in cavitation and a greater radius will reduce pump efficiency.

Cavitation Burns

Cavitation burns are the worst enemy of an impeller. These burns literally "eat away" material from the impeller, leaving holes and weakening the impeller structure. In extreme cases entire blades have been known to "depart" from the impeller hub because the cavitation burns at the base of the blade were so severe.

Cavitation burns are the result of imperfections (damage) on the impeller blades or air mixed with the water flow.

Cavitation burns may be caused by:

Wave jumping - air is sucked into the pump as the craft leaves the surface of the water.

Worn seals between the impeller and intake housings -- allowing air to enter.

Leading edge of the impeller is damaged.

To explain exactly what causes cavitation burns, first, let us examine the last cause. If the impeller leading edge is damaged and "mushrooms" over the blade, a low pressure area will form under the mushroomed lip.

Under atmospheric pressure -- at sea level -- water boils at 212°F. Anytime there is an imperfection on the surface of an impeller blade and water passes over that imperfection, the water pressure is lowered. When the pressure is lowered, water will boil at a much lower temperature. Therefore, air bubbles will form in the boiling water under the mushroomed lip. The bubbles will then creep down the surface of the blade and accumulate at the impeller hub. A high pressure area is formed at the base of the blade. Here, the air bubbles will collapse and reform back into water with a release of energy. This energy is absorbed by the impeller and the end result is impeller material is "eaten away".

The impeller should be dressed to a slightly rounded edge. It is impossible to

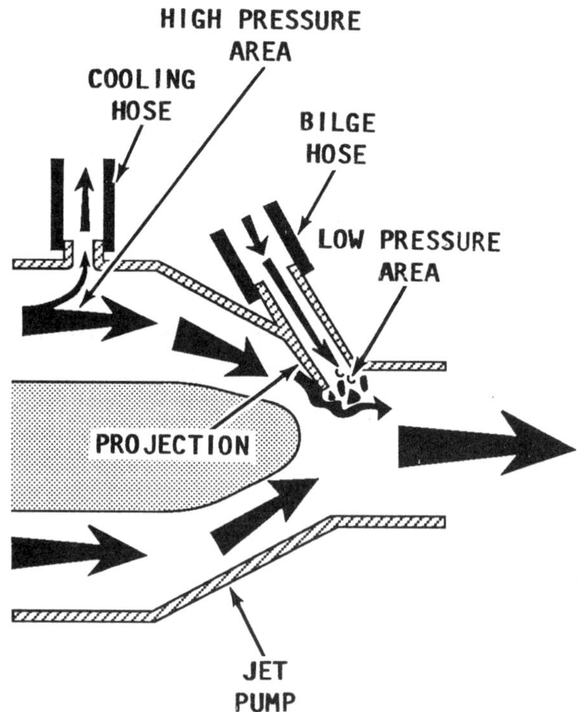

The direction of water flow through the cooling hose and the bilge hose is opposite. The cooling hose routes water from the pump to the engine. Water is "vacuumed" from the bilge through the bilge hose to the pump where it is forced out through the nozzle.

achieve a perfectly straight sharp edge, as one side will always be convex and the other side concave. The concave side will form an area of low pressure and encourage cavitation as described above.

The other two causes of cavitation -- defective seals and wave jumping -- occurs when air is sucked in with the water flow. The same principles apply, as just described, in the previous paragraphs. Air bubbles will creep down the surface of the blade and accumulate at the impeller hub. A high pressure area is formed at the base of the blade. Here, the air bubbles will collapse and reform back into water with a release of energy. As explained earlier, this energy is absorbed by the impeller and the end result -- impeller material is "eaten away".

Cooling Water and Bilge Hoses

On most pumps, two hoses are attached to the jet pump housing. One hose channels some of the water flowing through the pump to the exhaust manifold -- the hottest part of the engine -- to cool the block during operation. The other hose siphons water from the bottom of the hull, and is referred to as the bilge system. The cooling hose has water flowing **FROM** the pump. The bilge hose has water flowing **TO** the pump. Both hoses attach to the pump housing in a similar manner.

What determines the direction of water flow?

The answer to this question is in a simple explanation of high and low pressure areas along the inside surface of the housing.

Close inspection of the area inside the housing at the cooling water hose fitting reveals a smooth rounded shoulder with no obstruction or obstacle to impede the flow of water. A high pressure area develops here and draws the water down the hose attached to the fitting.

Further inspection of the area inside the housing at the bilge hose fitting reveals a small protrusion around the opening. This protrusion causes disturbance to the water flow and creates an area of low pressure. This low pressure area will have the effect of emptying air and water from the hose aft with the impeller water flow -- similar to the action of a vacuum cleaner, as depicted in the accompanying illustration. If the other end of the hose is submerged in water inside the hull, the water will be "vacuumed" out. In this manner, the bilge system drains the bilge.

Many an owner -- with good intentions -- thinking his action would smooth water flow and increase pump performance -- has filed the protrusion described in the previous paragraph smooth with the surrounding area of the housing. During operation of the water craft, the bilge system worked in **REVERSE.** Water was pumped from the impeller hous-

The bilge fitting protrusion extends into the water flow area, causing a low pressure area and a "vacuuming" effect.

ing into the bilge area, quickly filling the bilge and possibly sinking the craft before the rider had time to realize what was happening.

Impeller-to-Pump Case Clearance

SPECIAL WORDS

In order to determine impeller clearance, the ride plate and rock grate must be removed to gain access to the inlet scoop and impeller.

The following procedure may be performed with the watercraft elevated on saw horses enabling a person to work underneath, or with the watercraft on its port side, allowing access to the rock grate and impeller. If available, a trailer may be used for ease of component accessibility, if the beams do not impair service.

If the impeller-to-pump case clearance is to be determined with the watercraft raised on saw horses or trailer, the battery may be left secured in the craft.

If the watercraft is to be positioned on its side, first disconnect the electrical leads at both battery terminals, and then remove the battery from the watercraft.

Remove the attaching bolts securing the ride plate to the hull. The ride plate is secured to the hull with a sealant, take time and carefully pry the plate away from the hull. Remove the bolt and washer securing the rock grate to the inlet scoop. Insert a feeler gauge between each impeller blade and the pump case and determine the clearance. Make a note of all three clearances, one for each blade. Calculate the average of the clearances.

On a new craft, the clearance will be what the manufacturer refers to as the "standard" clearance. The maximum allowable clearance value is known as the "service limit".

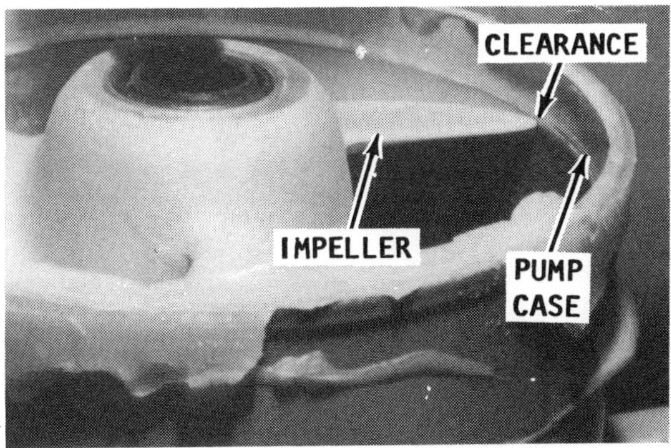

A feeler gauge is inserted between the impeller blade and the pump case to measure clearance.

Impeller-To-Case Clearance

Standard: .002 - .008" (.05 - .20mm)
Service Limit: .020" (.5mm)

If the clearance is less than the maximum value, no action is required.

If the clearance is more than the service limit, inspect the condition of the pump case. If the pump case has scratches deeper than .04" (1mm), replace the pump case. If the pump case is satisfactory, the impeller must be the problem. Visually inspect the impeller for nicks, scratches, pitting or a "mushroomed" edge. If the cause of the excessive clearance cannot be determined visually, the pump must be disassembled and the pump case measured and compared with specifications.

The jet pump must be removed from the craft to perform this work.

10-3 JET PUMP SERVICE

SPECIAL WORDS ON SPECIAL TOOLS

The following list of special tools are required in order to perform some of the necessary work on the jet pump. Some tasks become extremely difficult, if not impossible, without the special tools or some type of equivalent. Manufacturers' part numbers are given. In specific steps of the procedures -- whenever possible -- a special effort has been made to give some type of equivalent or homemade tool which may be substituted. However, the cost of a special tool may be offset by money saved if an expensive part is not damaged.

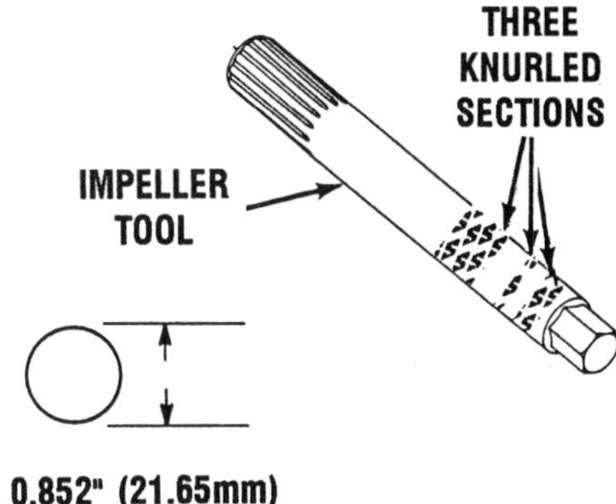

The impeller removal tool is used to remove the impeller from the impeller shaft of all models covered in this manual. The driveshaft coupler removal tool looks very similar, but only has 2 knurled sections and a diameter of 0.787" (21.0mm).

SPECIAL TOOLS FOR JET PUMP SERVICE

Part No.	Special Tool
P/N 2871036	-- Impeller Removal Tool
P/N 2871037	-- Coupler Removal Tool
P/N 2871343	-- Pump Alignment Tool

Removal

Disconnect both battery cables from the battery and remove the battery from the craft.

CAUTION

Non-sealed batteries **MUST** be removed to avoid spilling electrolyte into the bilge through the battery vent holes when the craft is tilted to one side.

Rotate the fuel control valve to the **"OFF"** position. Drain or remove the fuel tank. Drain the oil from the oil tank.

Place an old blanket or equivalent on the workshop floor to cushion the edges of the craft and prevent the craft from sliding while the work progresses. Rotate the craft -- lower the port side; left -- on the blanket to expose the rock grate, and then block the hull to hold the desired position. If a trailer or dolly is available, these may serve as a steady platform during pump removal if there is ample height above ground to perform the work.

GOOD WORDS

Before starting service work, the impeller-to-pump case clearance should be checked. The rock grate must be removed to check the clearance. If the rock grate cannot be removed first, the ride plate must be removed also.

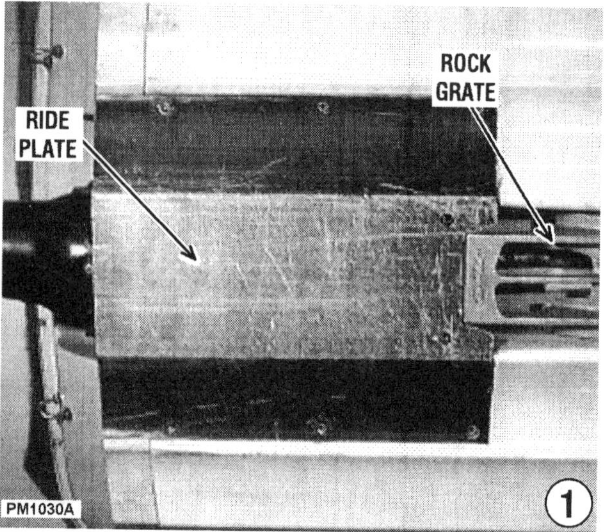

Removal and Disassembling

1- Remove the eight attaching bolts securing the ride plate over the jet pump. Using a wide-blade screwdriver or blunt chisel, carefully pry the ride plate from the hull. Work back and forth on each side, a little at a time to loosen the plate. NOTE: This is **NOT** an easy nor quick task! The plate is securely fastened with a silicone sealant.

Carefully pull the ride plate away from the hull.

2- Remove the bolt securing the rock guard -- or intake grate -- to the hull. Pull the rock grate free of the hull.

3- Remove the nut securing the steering rod bolt to the pump nozzle. Remove the connecting bolt from the nozzle arm. Disconnect the reverse gate and trim connectors, if equipped. Note -- if equipped with a reverse gate, move the lever to the reverse position -- the link is easily removed in this manner.

4- Loosen the hose clamps around the cooling hose and siphon hose through-hull fittings inside the aft end of the engine compartment. Pull the cooling and siphon hoses free of the fittings. The jet pump cannot be separated from the watercraft if these hoses are attached.

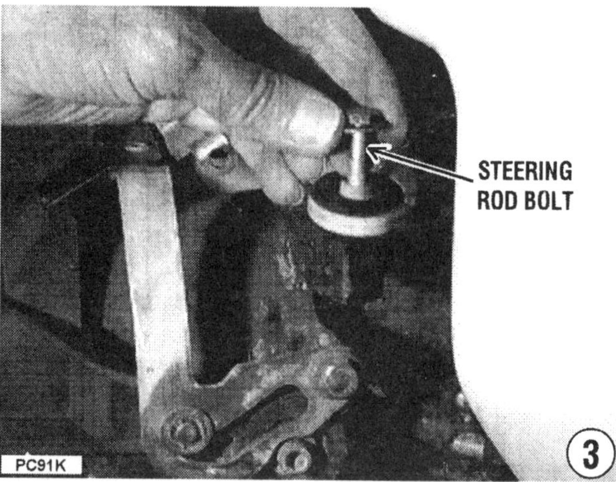

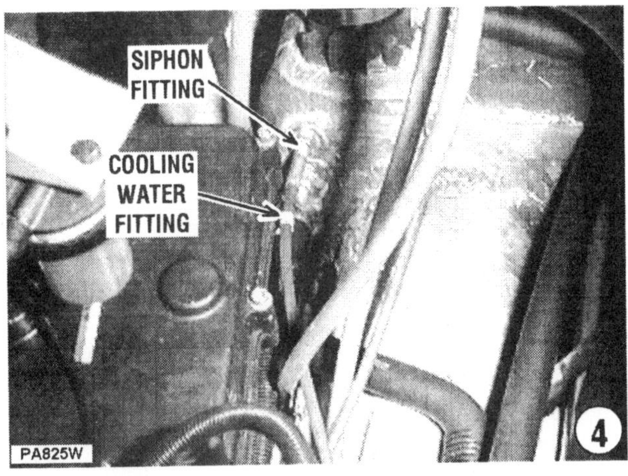

5- Remove the four mounting nuts securing the jet pump to the hull. The reverse gate and steering nozzle have been removed from the jet pump for photographic clarity in the illustration.

6- Gently pry the jet pump away from the hull, using a blunt chisel or wide blade screwdriver. Work evenly around the perimeter of the through-hull fitting and pump mating surfaces.

7- Pull the jet pump away from the hull of the boat. Move the pump straight aft, to avoid damaging the splines of the driveshaft.

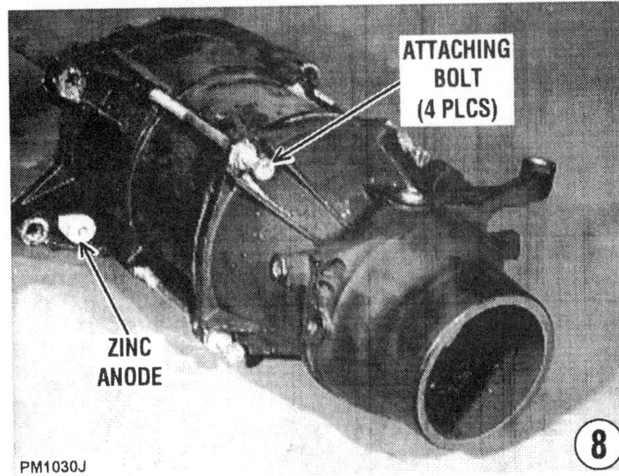

8- Place the jet pump assembly on a suitable work surface. Remove the four long bolts securing the pump sections together. Observe the zinc anode, located on the jet pump. This anode must be tightened securely against the pump casing, to aid in the prevention of corrosion by electrolysis.

9- Tap the case flanges with a soft-head mallet to assist in separating the pump sections. Use a screwdriver or blunt chisel to carefully pry the sections apart. Pull the aft case section away from the pump body and over the driveshaft.

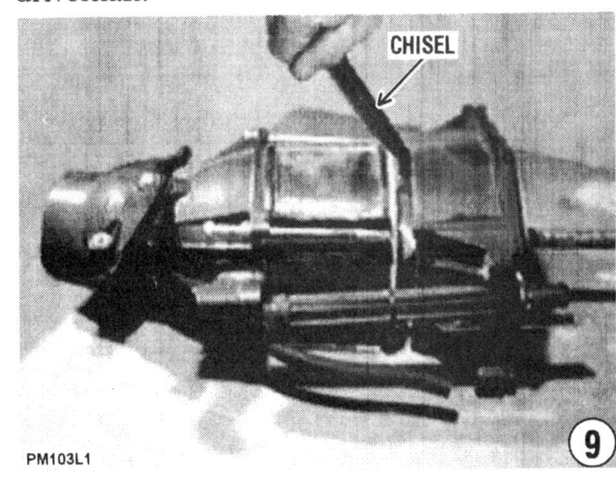

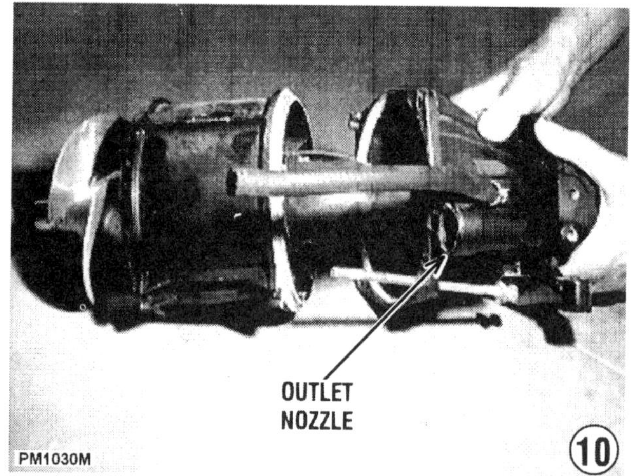

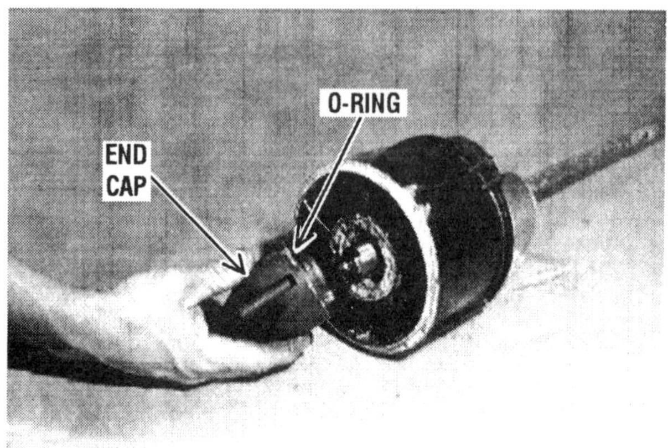

10- Pull the stationary jet nozzle from the guide vane case housing.

11- Remove the three allen screws securing the "cone" or end cap to the guide vane case. Tap the end cap with a soft-head mallet to jar it loose, then pull the cap from the pump case.

Impeller Removal

12- Obtain special impeller removal tool P/N 2871036. Clean the slotted end of the impeller shaft of any lubricant remaining after the end cap was removed. Clamp this end of the driveshaft into a vice equipped with soft jaws. Insert the special impeller tool into the internal splines inside the impeller. Seat the tool in place firmly with a soft-head mallet.

13- With the correct size wrench on the tool, loosen the impeller from the threads -- right hand threads -- on the impeller shaft. Lift the impeller from the pump case. Be careful to save the rubber damper located at the end of the impeller shaft -- if so equipped.

14 - Remove the bearing from the front end of the guide vane case; inspect for damage.

15 - Pull the impeller shaft from the case.

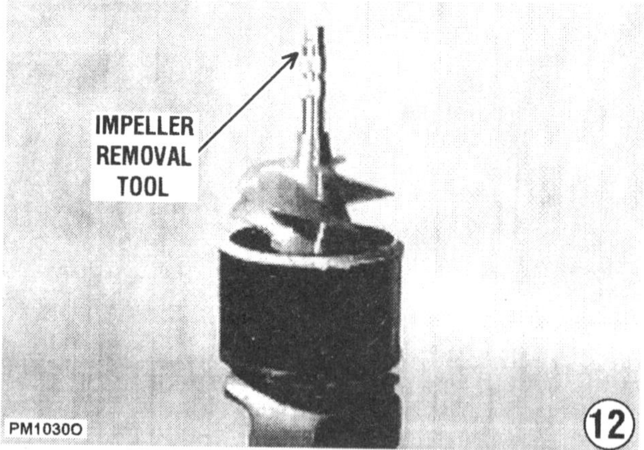

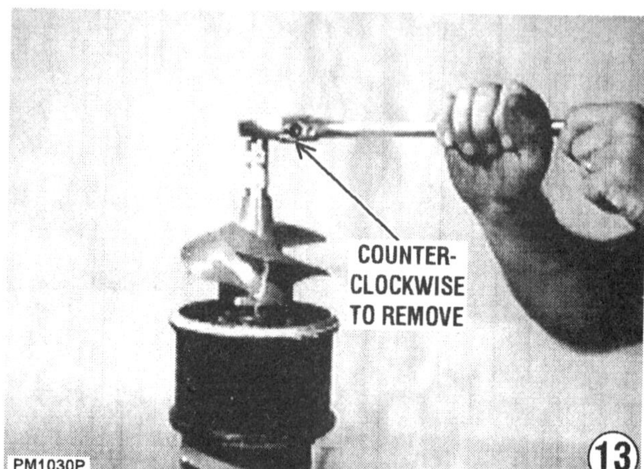

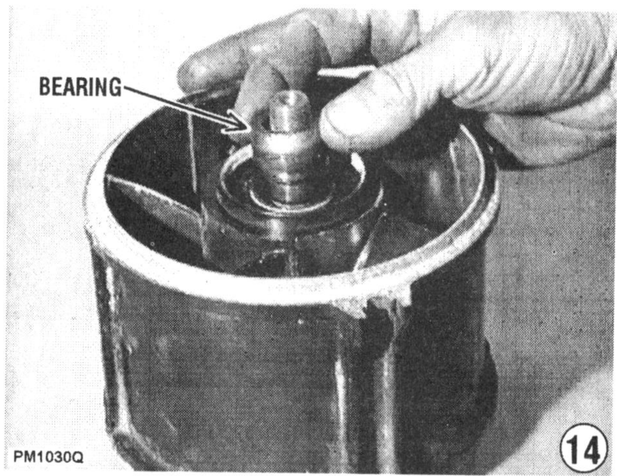

16 - Carefully remove the **O**-ring. If the bearings are to be removed for inspection or replacement, proceed to Step **17**, otherwise proceed directly to the next section, Cleaning and Inspecting, following Step **17**.

17- Remove the seal spacer. Use a slim long punch and push the spacer between the bearings aside, allowing access to the rear bearing. Tap out the aft bearing first. Next, tap the forward bearing spacer aside to expose the inner bearing races. Drive out the aft end bearing and remove the spacer. Drive out the other bearing and seals.

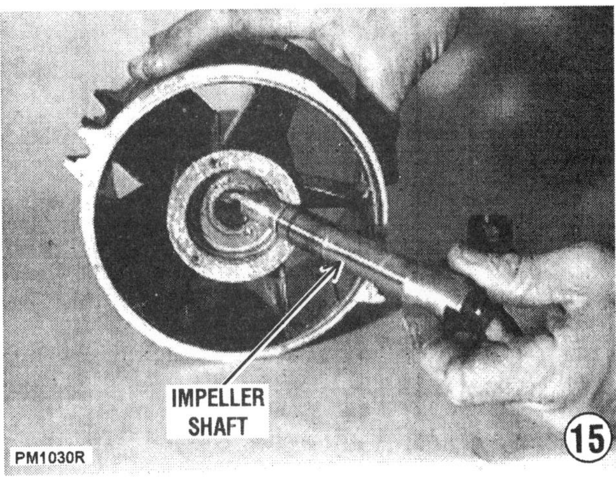

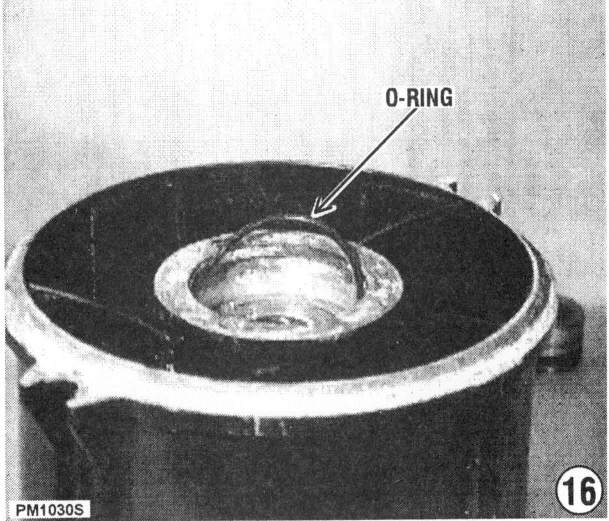

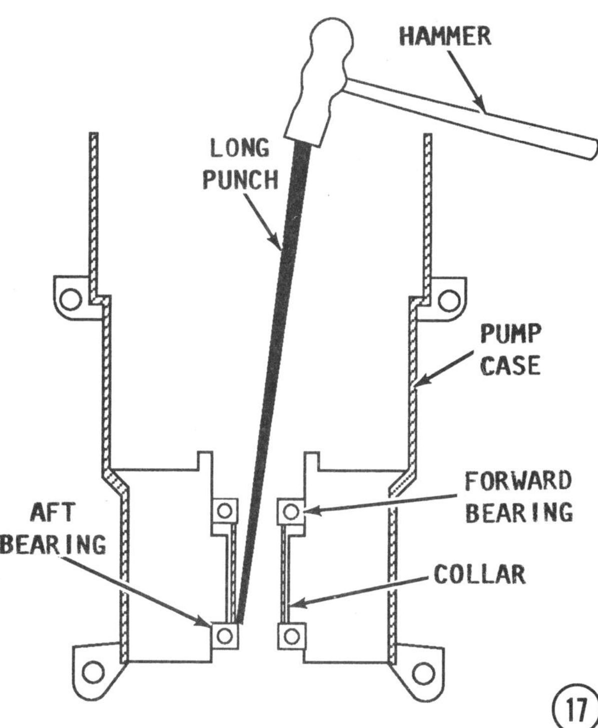

CLEANING AND INSPECTING

Inspect the leading edge of each impeller blade. Ideally, the edge should be rounded to a 1/32" radius, **NOT** a knife-edge! The cross-section shape of an impeller blade is similar to that of an airplane wing.

If the craft is consistently operated in shallow water with a sandy or "pebbly" bottom, the leading edge of the impeller blades will become rounded or will become chipped with missing pieces.

Very little material may be removed from the impeller blades during reconditioning work. Therefore, if the edge is rounded off or corroded, a file can be carefully used to "dress" the edge.

If an impeller blade is chipped more than 1/4", the impeller should be replaced. Any chip smaller than 1/4" can be evenly filed away along the length of the blade. Whatever work is done to one blade must be done to all blades in order to maintain impeller balance. If the impeller becomes excessively out of balance, a definite vibration condition will be quite noticeable as the craft moves through the water. Such vibration will also cause premature bearing and shaft wear.

Measure the outside diameter of the impeller. If the impeller is worn close to **OR** beyond the service limit, it must be replaced.

ASSEMBLING

The following procedures cover assembly of the complete jet pump -- all parts. The following procedures outline detailed instructions to assemble virtually all parts of jet pumps installed on the personal watercraft covered in this manual. Minor changes occurred in the design of the jet pump between 1993 and 1994. From 1994 to present, the jet pump design is relatively similar.

Where differences occur, they will be clearly identified. Refer to the following exploded drawings for clarity of jet drive components and design changes.

As the assembly work progresses, if a particular item was not removed or disassembled -- simply skip the steps involved for the part and proceed directly to the work required.

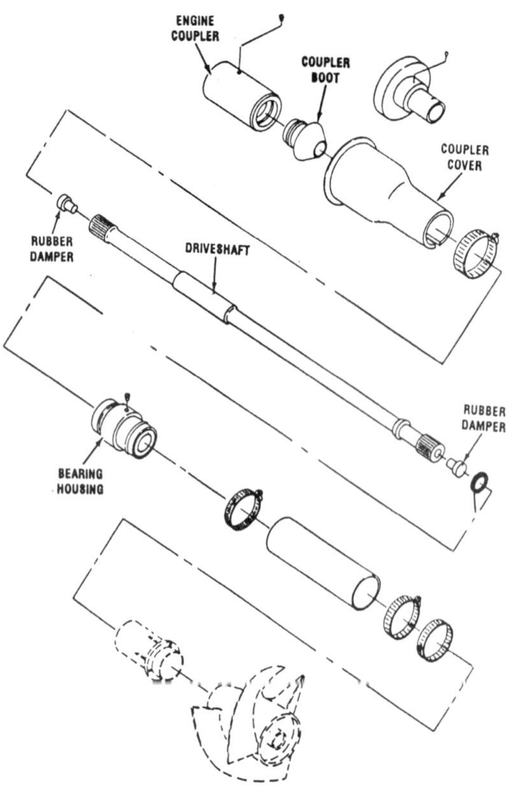

Drive shaft and coupler system for all Models 1992 - 1995. The engine coupler equipped with a harmonic balancer is used only on early build 1992 engines. June 1992 through 1995 are equipped with a coupler and rubber boot as shown.

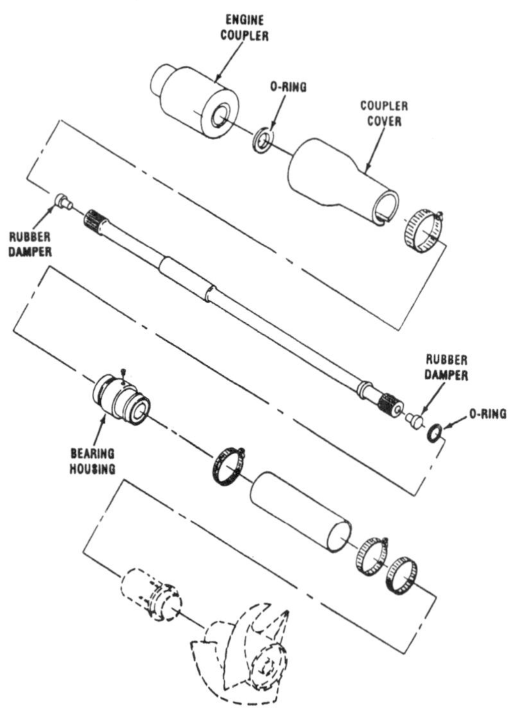

Exploded diagram of the driveshaft for all Models, 1996 to current. If servicing such a system, check to be sure the rubber dampers are fitted properly during installation

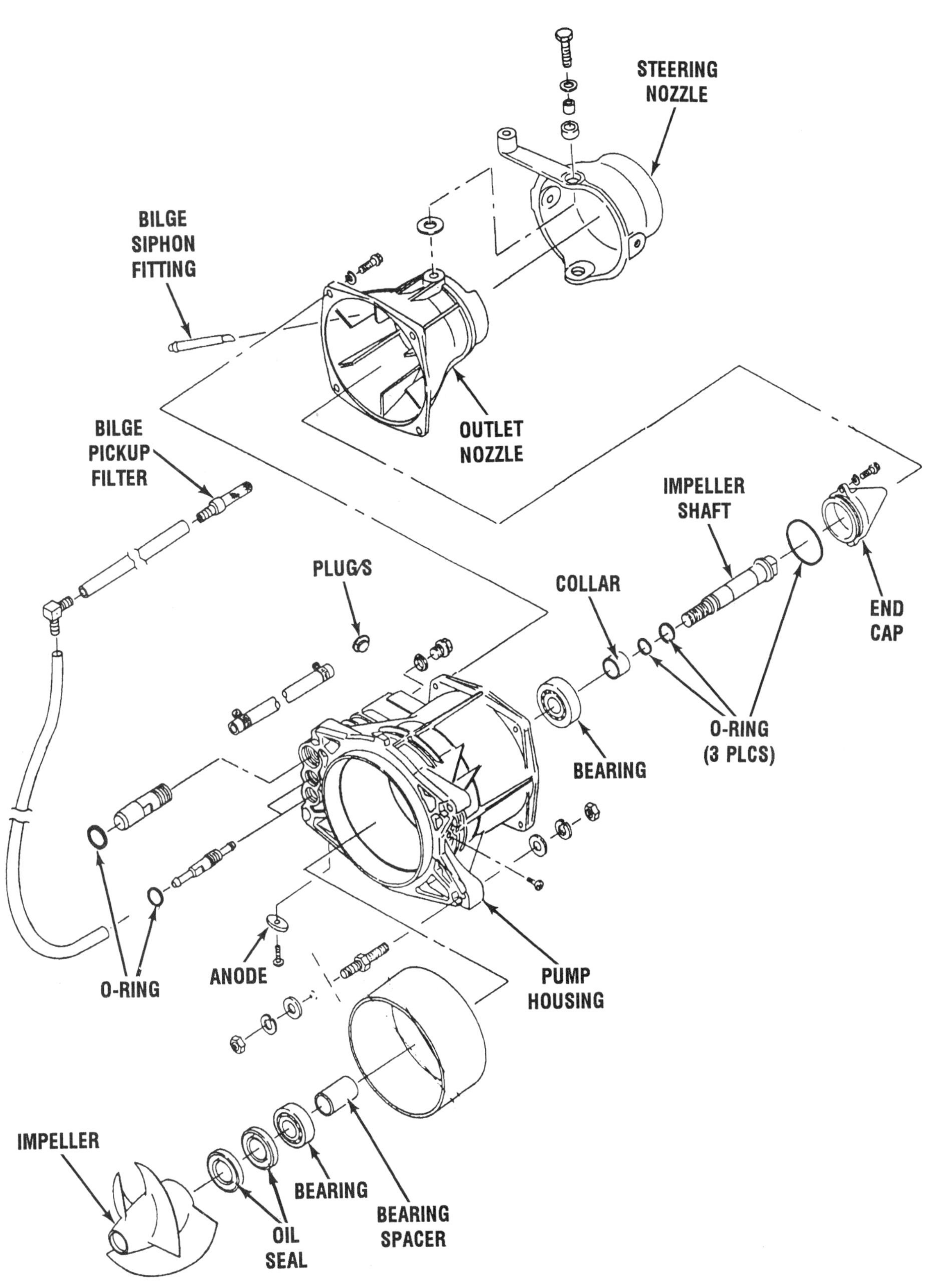

Exploded drawing of the jet pump assembly installed on all 1992 - 1993 Models. If installing such a jet pump, check to be sure the pump housing is flush against the hull to avoid leakage or cavitation.

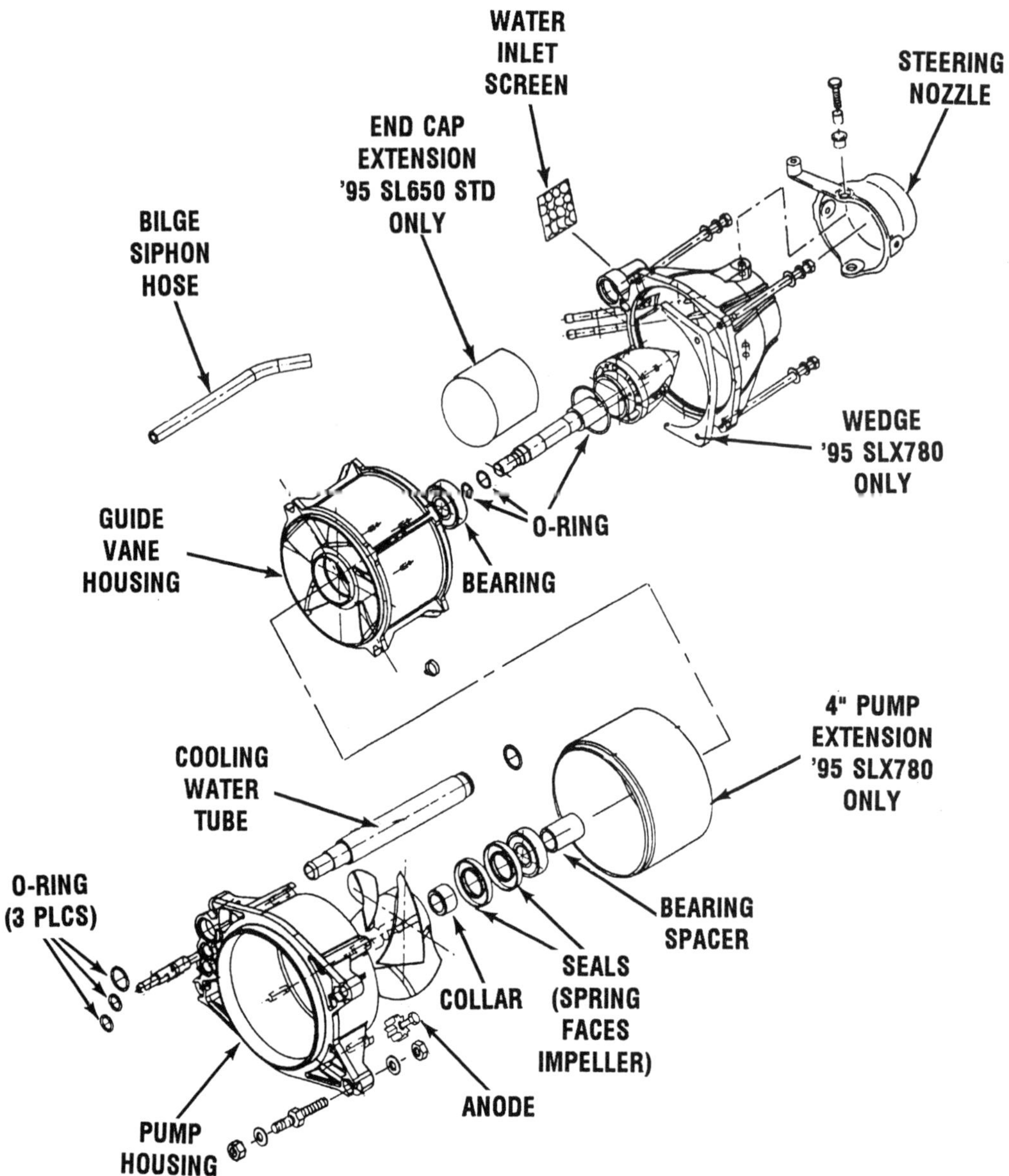

Exploded drawing of the jet pump installed on all Models manufactured from 1994 - 1995, including 1996 Models SL700, SLT700, and Hurricane.

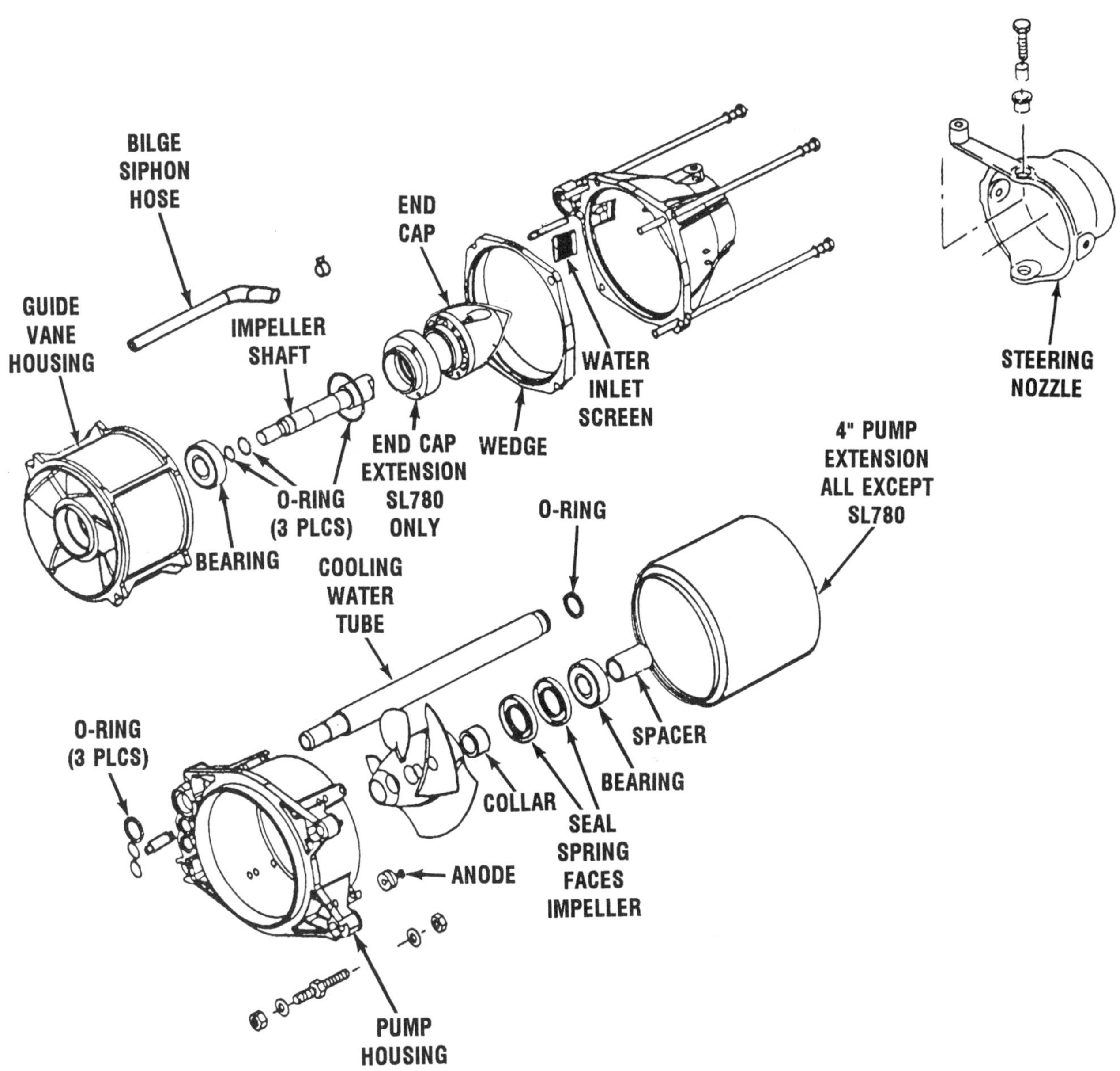

Exploded drawing of the jet pump installed on all 1996 to current SLX, SLT780, SL780 Models as indicated, including SL900, SL1050 and SLTX Models manufactured from 1996 to current.

Pump Housing
Assembly

If the bearings were removed, continue the procedures in the following non-numbered paragraphs. Otherwise, proceed to Step **1**.

Pack both bearings with Shell Alvania EP1 or equivalent water resistant multipurpose lubricant. Install the aft pump bearing into the pump case hub. The bearing is correctly installed when the outer bearing race seats against a shoulder in the hub. Insert the spacer from the forward end of the pump case - if so equipped. Install the forward bearing. The bearing is correctly positioned inside the hub when the inner bearing race contacts the spacer.

Pack the lips of the two oil seals with the same lubricant and install them into the forward face of the pump case. **BOTH** oil seals are installed with the lip facing **UP** -- forward when installed. The first seal is correctly installed when the back side seats against the bearing. The back side of the second seal must seat against the lip of the first oil seal.

Insert the bushing into the forward end of the pump case hub.

1- Lower the impeller shaft into the guide vane casting. Apply grease to the threads of the impeller shaft.

2- Slide the **O**-ring over the shaft -- stretching it carefully -- and position it in the impeller shaft groove.

3- Position the impeller in place on the shaft. Rotate the impeller clockwise to secure it hand tight. Lower the guide vane housing assembly onto a vise with soft jaws or wood holding the impeller shaft in place. Tighten the impeller onto the shaft to a torque value of 100 ft lbs (136Nm).

4- Install the pump housing section on the guide vane case, encasing the impeller. Rotate the forward case housing to be sure the impeller does not bind with the inner surface of the casing.

5- Pack the end cap with marine grease. Slide a new **O**-ring onto the cap. Position the cap in place on the guide vane case. Install the attaching bolts and tighten them to a torque value of 72 in lbs (8Nm).

6- Rotate the pump case assembly around so the aft section is pointing up. Lower the outlet and steering nozzle onto the pump case. Thread the long bolts just "finger tight" at this time. Remove the assembly from the vise.

7- Install new **O**-rings on the cooling water tube and bilge siphon fittings, if applicable. As

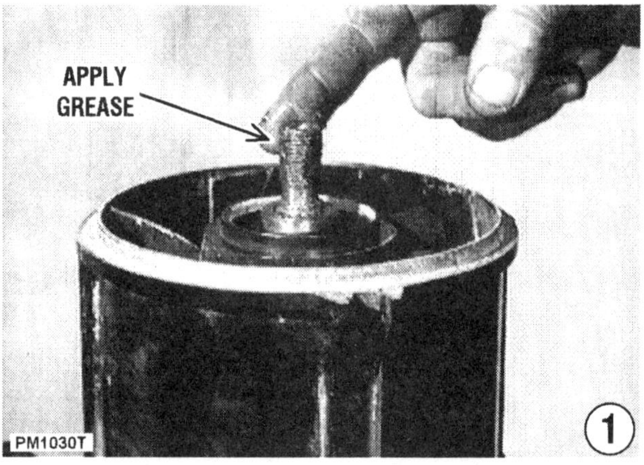

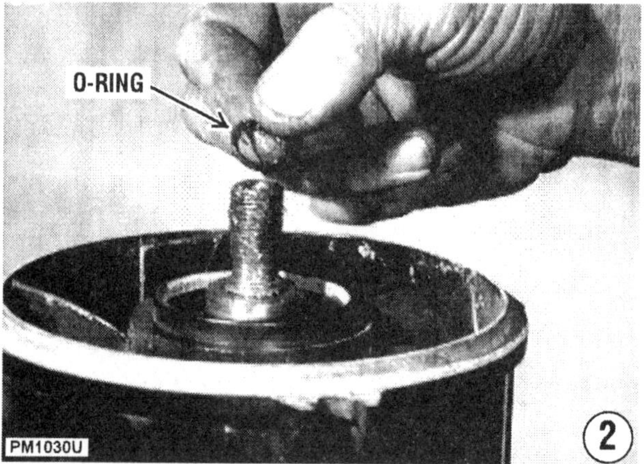

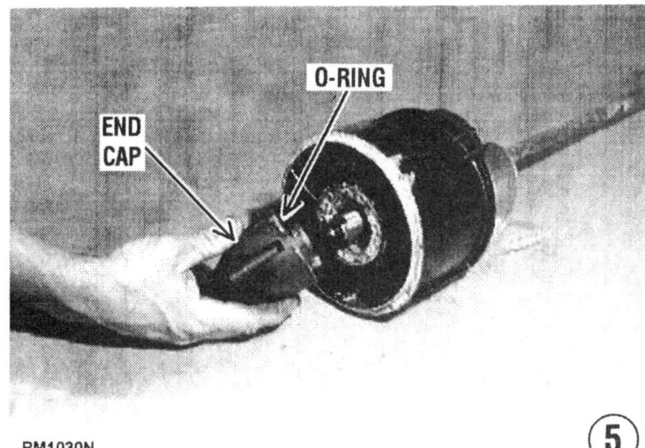

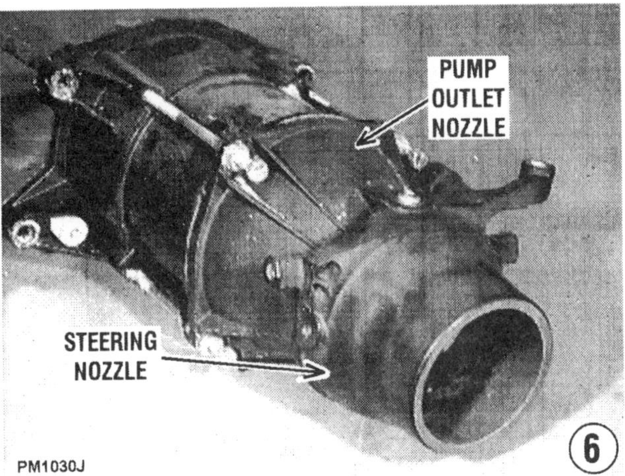

discussed previously, most watercraft models are equipped with one or two bilge siphon fittings. Some models may not have these siphon fittings, only an electric bilge pump is used -- models manufactured 1997 and on.

8- Position a new **O**-ring onto the aft end of the driveshaft. Check to be sure the rubber vibration damper is in good condition and installed properly. Insert the driveshaft into the impeller. Install the other small rubber damper onto the forward end of the driveshaft. Apply a liberal amount of grease on the large pump-

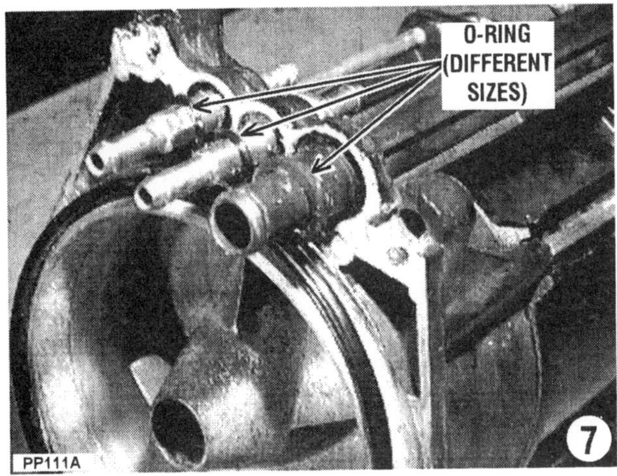

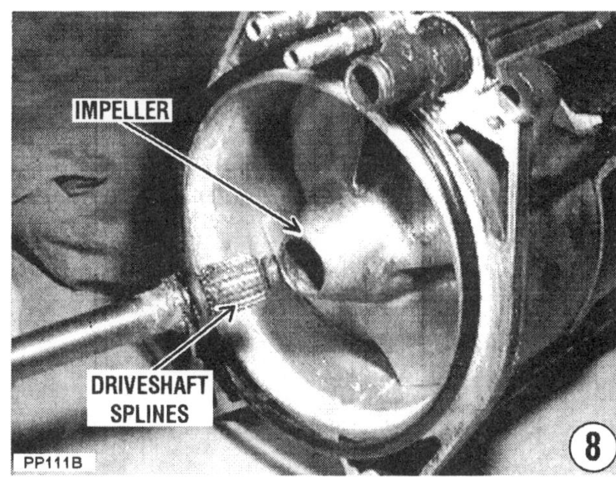

to-hull **O**-ring. Slide the large **O**-ring into place on the pump housing.

9- Obtain the service of an assistant. Apply a liberal amount of silicone sealant around the intake grate bolt hole inside the engine compartment. Apply sealant to the attaching bolt. Position the intake grate -- sometimes referred to as a rock guard -- in place on the bottom of the hull. Insert the attaching bolt from underneath, the assistant places a washer and the attaching nut on the bolt in the engine compartment. Tighten the nut securely. The engine was removed for photographic clarity of the attaching bolt hole.

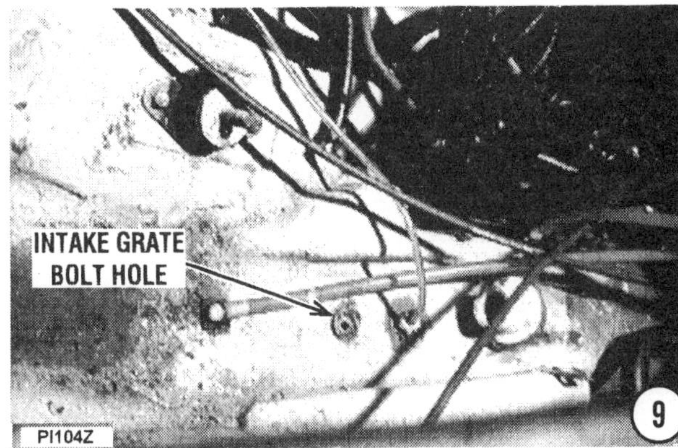

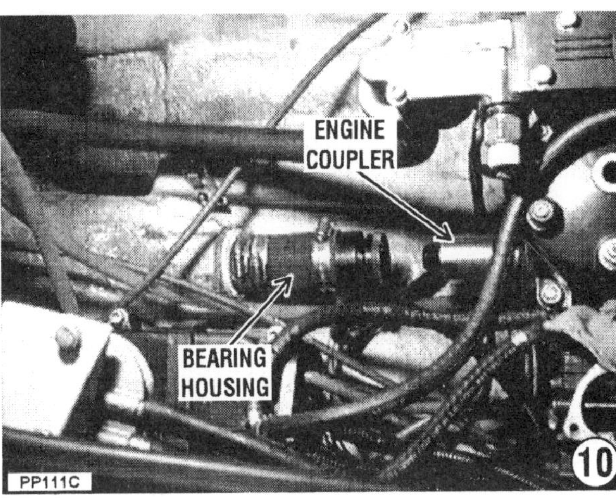

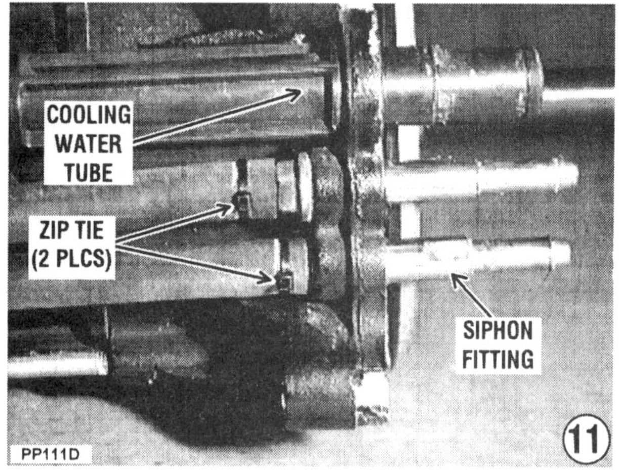

10- Check to be sure the engine coupler and forward bearing housing are properly positioned and secured in place.

11- Attach the short bilge siphon hose/s to the forward fittings on the pump housing. Secure the hoses in place with zip ties.

12- Apply an even bead of silicone sealant around the perimeter of the forward cooling hose and bilge fittings.

GOOD WORDS

Driveshaft/Pump Installation

Wrap electrical tape around the splines of the forward driveshaft prior to installation. Apply grease to the tape. This will prevent the splines from damaging the bearing housing seals during assembly. Install new **O**-rings on the driveshaft. Apply a coating of grease to these **O**-rings.

Once the driveshaft is installed past the bearing housing, remove the tape and apply grease to the splines.

13- Obtain the services of an assistant. Position the jet pump assembly in place on the watercraft, carefully guiding the driveshaft

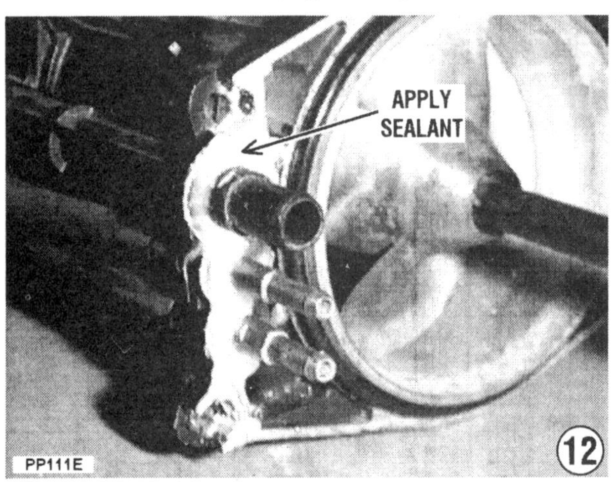

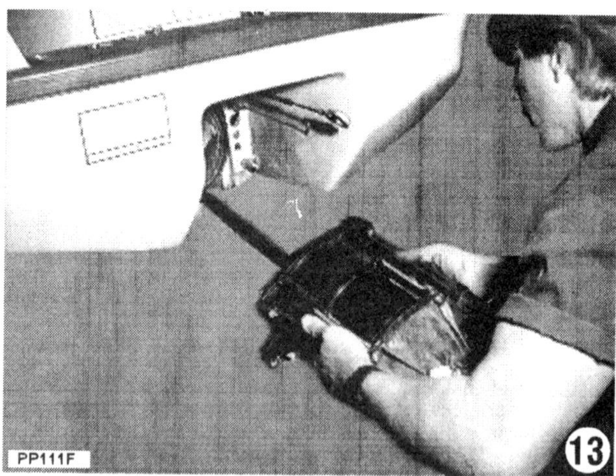

through the bearing housing. Push the pump forward evenly, over the four mounting studs.

The assistant ensures the rubber vibration damper **REMAINS** on the forward end of the driveshaft **AND** continues to guide the driveshaft forward into the boot of the coupler. The coupler boot was equipped on craft manufactured before 1996. Without the aid of an assistant, guiding the driveshaft into the coupler and keeping the damper in place is not an easy task. The alignment tool specified earlier in this chapter, will **INDICATE** the proper alignment of the bearing housing to the driveshaft coupler, however, the shaft must still be inserted properly into the boot by hand.

Once the driveshaft is in the coupler, rotate the shaft slightly, while carefully pushing the jet pump forward. The splines of the shaft will mesh with the grooves in the coupler. Install the four pump housing attaching nuts and tighten them in a criss-cross pattern, to a torque value of 28 ft lbs (38Nm). A socket wrench with a long extension is required for this task. Air tools should not be used to tighten these bolts.

Now, tighten the outlet nozzle attaching bolts to a torque value of 18 ft lbs (24Nm). Check to be sure the mating surface of the pump housing and thru-hull fitting seat properly together. An improper fit will lead to leakage or cavitation. Check to be sure the drive shaft is properly aligned.

14- Position the steering control rod on the nozzle post. Insert the attaching hardware and tighten securely. Move the steering nozzle to hard port and starboard, checking to be sure the linkage does not bind and control movement is smooth.

Move the directional control lever into the full reverse position. Install the reverse gate control rod, if so equipped. Attach the nozzle trim linkage, if applicable.

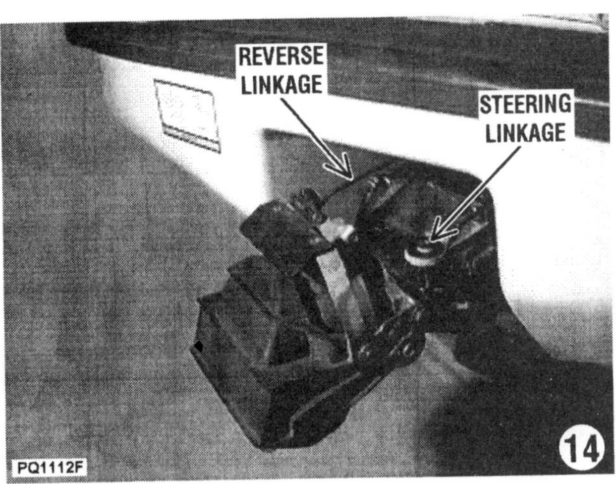

15- Position the ride plate on the hull, ensuring the rock grate indexes properly with the plate. Install the attaching screws and tighten them to a torque value of 8 ft lbs (11Nm).

FINISHING TASKS

If the craft was turned on its side, rotate it right side up. Install the battery and connect the cables to the proper terminals. Install or fill the oil tank. The oil injection system **MUST** be bled before running the engine -- refer to Chapter 6 -- **FUEL AND OIL** for detailed procedures.

Move the craft to a body of water or connect a garden hose to the engine flush fitting on the water outlet manifold.

Start the engine and allow the rpm's to stabilize at idle speed **FOR JUST A FEW SECONDS,** and then turn the water on.

Adjust the water flow until a small trickle is discharged from the bypass outlet thru-hull fitting.

When the engine is to be shut down, turn the water off **FIRST** -- raise the aft portion of the hull -- **WHILE THE ENGINE IS OPERATING AT IDLE** -- "rev" the engine just a **COUPLE** of times to clear water from the exhaust system -- and them shut it down.

NEVER allow the engine to operate without cooling water for more than 15 seconds.

CAUTION

Water must circulate through the jet pump -- to and from the engine, anytime the engine is operating. Circulating water will prevent overheating -- which could cause damage to moving engine parts and possible engine seizure.

NEVER, AGAIN NEVER operate the engine at high speed with a flush device attached. The engine, operating at high speed with such a device attached, would **RUNAWAY** from lack of a load on the impeller, causing extensive damage.

10-4 BEARING HOUSING AND DRIVESHAFT SERVICE

The driveshaft bearing housing serves as a support for the driveshaft and prevents water from entering the engine compartment from the pump intake.

Inspection of the bearing housing requires removal of the driveshaft and pump, the procedures are given in the previous section. Loosen the hose clamps securing the housing assembly to the thru-hull fitting. Remove the housing.

Inspect the bearings and seals for signs of wear or damage. This assembly is not serviceable, the unit must be replaced.

1- Clean the entire surface of the drive shaft. Inspect the driveshaft bearing surfaces for signs of wear. Place the driveshaft on a set of Vee blocks at location "**B**" and position a dial indicator over the center of the shaft at location "**A**". Slowly rotate the shaft and observe the dial indicator. A shaft in satisfactory condition will have a runout less than 0.005" (0.13mm). If the runout exceeds this value, the driveshaft must be replaced.

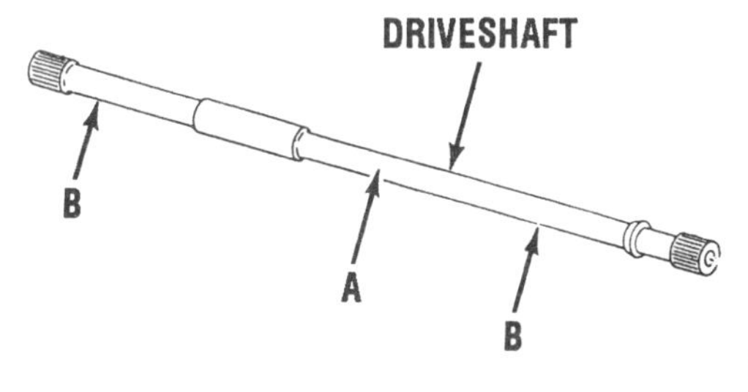

The bottom line is definitely "the jet pump moves the craft". All other components and systems, including the engine, contribute to the efficiency of the jet pump. Therefore, adequate maintenance and adjustment of the pump is critical to ensure a "fun day" on the water for the owner and her/his guest/s.

11
CONTROL ADJUSTMENTS

11-1 INTRODUCTION

Control adjustments on Polaris Personal Watercraft includes changing the length of the steering cable, and trim system, speedometer operation and inspection -- if equipped. Normally, adjustments are not necessary for extended periods of time, unless the craft has experienced an accident or the control has been disconnected to service other components.

Detailed procedures to adjust the steering, reverse, and trim cables and speedometer inspection are presented in this short chapter. Adjustment of the throttle cable and the choke cable is thoroughly covered in Chapter 6 -- **FUEL**.

11-2 STEERING CABLE ADJUSTMENT ALL MODELS

If the steering nozzle is not positioned exactly in center of the transom opening, with an equal clearance on both sides, the steering cable **MUST** be adjusted to prevent the craft from veering to port or starboard when the handle bar is positioned for "dead ahead".

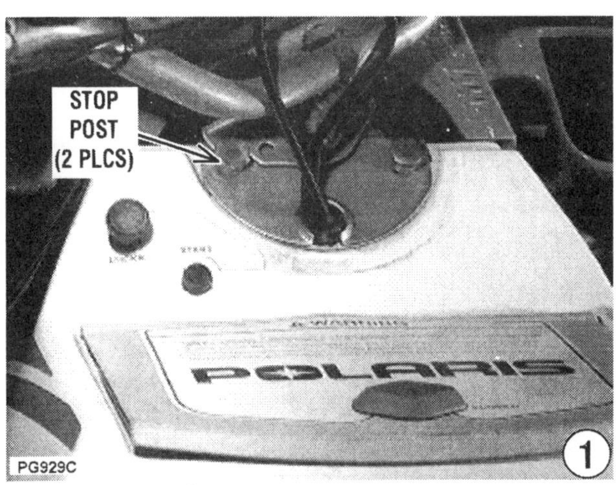

Inspection

1- Remove the handlebar pad. Looking down at the handlebar column and stop bracket, visually center the handlebar by measuring the distance from the stop bracket to the stop post on each side. The gaps should be equidistant. Move the handlebar left or right until the distance on each stop post and bracket is equal.

2- At the stern, place a straight edge across the horizontal center line -- across the **CENTER** -- of the steering nozzle or trim nozzle. The trim nozzle mountings bolts can be used as a reference for the center line for models equipped with the manufacturer's trim system feature.

Measure the distance from the straight edge to the hull on the left and right side. Measure at right angles to the straight edge and at equal distances from the nozzle -- as shown in the illustration -- for the most accurate measurement readings. These two readings should be equal when the handlebar is straight.

Cable Adjustment

1- Minor adjustments can be made by manipulating the cable rod end/s. Loosen the jam nut on the cable rod end. Remove the end

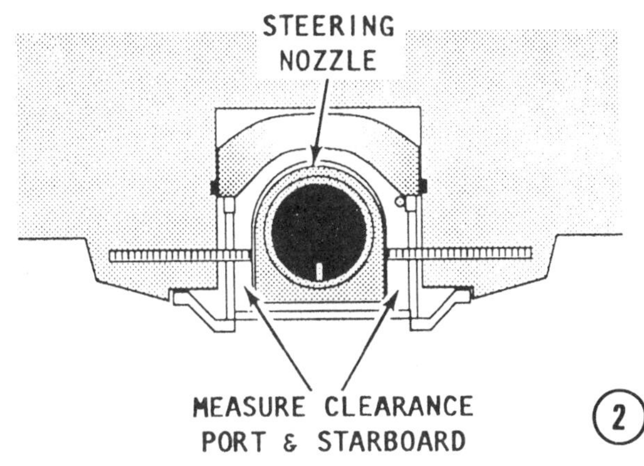

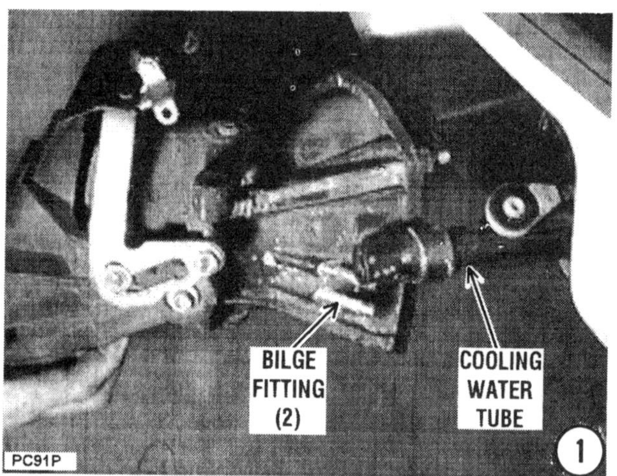

from the top of the steering nozzle - or steering arm on 1992 - 1994 models. Keep the handlebar in a "dead ahead" position. Turn the rod end in or out to adjust, without moving the handlebar. Attach the cable rod end and tighten the bolt. If attaching the steering arm cable rod, tighten the bolt to a torque value of 8 ft lbs (11Nm).

GOOD WORDS

The cable **MUST** be threaded a minimum of 3/8" (10mm) into the rod end. Do not attempt major adjustments by manipulating the rod ends.

Apply Loctite 242 to the threads of the steering cable and tighten the jamnut securely. Check to be sure the rod end is positioned parallel to the mounting surface.

2- If major adjustments must be made, use only the main cable adjuster nuts. These adjuster nuts are positioned at the forward end of the cable near the steering mechanism. Open the front compartment and remove the storage container for access to the bottom nut. On SLT models, the steering cable is mounted on a bracket. The cable and hull must be re-sealed with RTV silicone if adjustments are made at the through-hull end on all 1992-1994 models. On 1995 models, the cable is **NOT** adjustable at the through-hull end fitting.

Obtain two 11/16" open end wrenches -- or the second a crescent if necessary -- hold one main adjuster nut and loosen the other. Adjust the length of the cable to center the steering nozzle. Tighten the cable end nuts securely upon satisfactory adjustment.

Turn the handlebar to port and starboard, through the entire steering range. The bar should move smoothly -- without binding. If

any resistance is noticed, check the cable conditioning, rod end positioning, and routing to locate and correct the cause.

11-3 REVERSE CABLE ADJUSTMENT

On models equipped with reverse capability, periodic adjustment of the cable is not necessary. If the reverse gate is not operating smoothly, inspect the cable, linkage, fasteners and pivots for damage. If a new cable must be installed, adjust it using the following method.

1- Place the shift lever in the full **REVERSE** position.

2- Loosen the cable adjuster nuts located on the aft end of the craft -- both inside the engine compartment and outside by the jet pump.

3- Adjust the cable until the reverse gate collars engage fully with a "locking" motion into the reverse actuator locating slot when slight upward pressure is applied to the reverse gate. The gate must travel freely into this "locking" notch, it should not have to be forced into place. Once the cable is properly positioned, tighten the adjuster nuts securely.

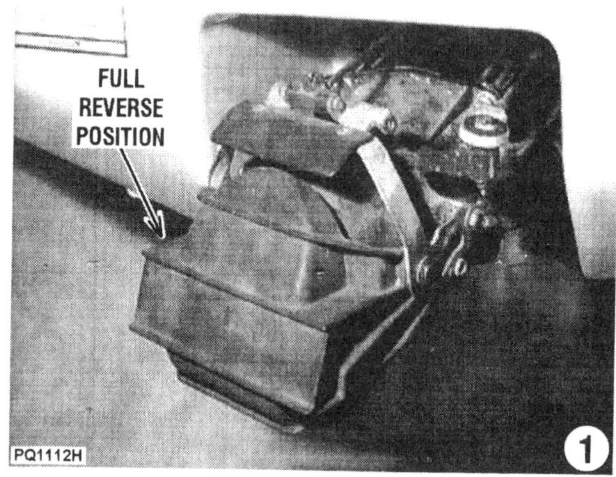

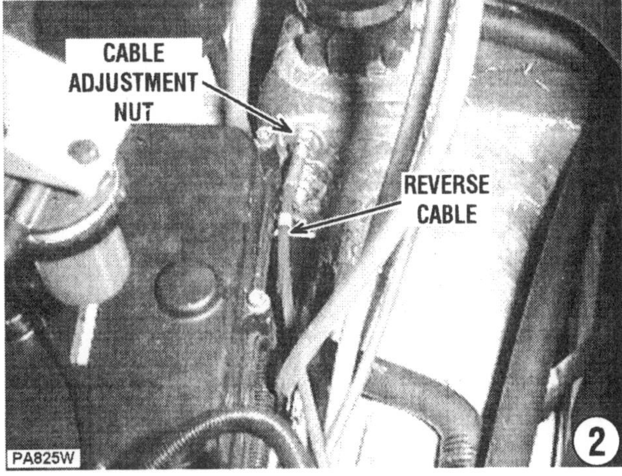

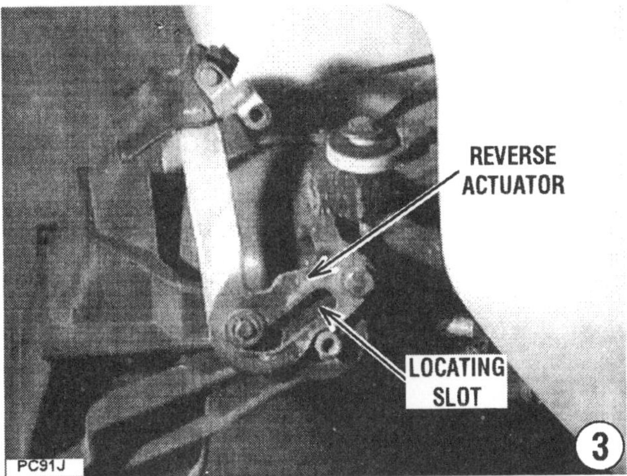

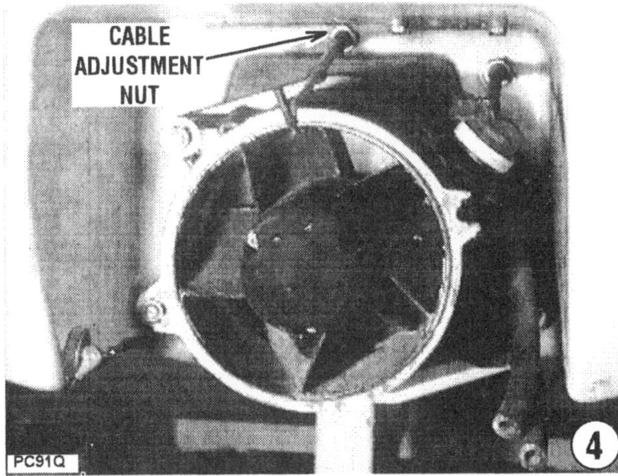

4- Move the reverse lever from **REVERSE** to the **FORWARD** position. Check to be sure the reverse gate is clear of -- up away from -- the steering nozzle. Readjust the reverse cable if necessary. Apply marine grade silicone sealant to the cable threads and adjuster nuts. The accompanying illustration shows the adjuster nut in place. The jet pump has been removed for photographic clarity.

11-4 TRIM NOZZLE

The electric attitude trim system is operated by a rocker switch located on the left handlebar. Activation of the switch causes an electric motor -- located in the aft section of the craft -- to move the steering nozzle on its horizontal axis via a control link rod, moving the nozzle up and down. This in turn changes the nose attitude of the watercraft.

On models equipped with a Multi Function Display (MFD), the trim angle indicator is integral to the display panel. On models not featuring an MFD, a cable connected links to an indicator mounted in the dash to inform the operator of the trim system positioning.

ADJUSTMENT

This section applies to all 1993 - 1994 models equipped with an electric trim adjustment system and mechanical trim indicator. 1995 to current Models with an electric indicator on the MFD will be covered after this section.

Mechanical Indicator

a- In a test tank or with the craft in a body of water with restraining lines to prevent the craft from moving. If a body of water is not available, install a garden hose to the flush fitting. Start the engine and move the trim indicator to the neutral position.

CAUTION

Water must circulate through the jet pump -- to and from the engine, anytime the engine is operating. Circulating water will prevent overheating -- which could cause damage to moving engine parts and possible engine seizure.

If a test tank or body or water is not available, a garden hose should be attached to the flush fitting.

NEVER, AGAIN NEVER, operate the engine at high speed with a flush device attached. An engine operating at high speed with such a device attached, would **RUNAWAY** from lack of a load on the impeller shaft, causing extensive damage.

b- Once the trim indicator is centered, shut down the engine.

c- Place a straight edge on the **VERTICAL** centerline of the trim nozzle.

d- Measure the distance from the straight-edge to the stationary nozzle on the jet pump -- both top and bottom. The distance between the straightedge and pump nozzle should be equal in both locations.

e- If adjustment is necessary, remove the cotter pin from the nozzle end of the trim rod cable.

f- Loosen the locknut at the trim motor end of the rod cable. Rotate the trim rod to shorten or lengthen it to obtain an equal distance between the trim and stationary nozzles at top and bottom.

g- Start the engine and move the trim indicator to the "full up" position. Shut down the engine.

h- Inspect the clearance between the top steering nozzle bolt and the trim nozzle. This clearance gap must be at least 0.015" (0.4mm). Readjust the trim rod slightly if it is necessary to provide the proper clearance.

Electric Trim Indicator Adjustment

Models manufactured from 1995 to current may have an electric trim indicator, located within the Multi Function Display. The adjustment process is less time consuming than the mechanical method.

Never operate the engine for more than 15 seconds while the watercraft is out of the water. Serious damage to the engine will occur.

a- Start the engine. Activate the trim switch in the full up or full down position until a "ratcheting" (metallic clicking sound) is heard at the trim motor -- aft portion of the watercraft. The MFD will automatically readjust the indicator!

b- Shut down the engine.

11-5 SPEEDOMETER INSPECTION

The speedometer is operated by water pressure on models so equipped. A pick-up fitting is located on the jet pump, by the rock grate -- intake grate. The fitting supplies positive pressure to the speedometer, which translates the pressure into a speed through the water indication, in mph.

a- If the speedometer is inoperative, check the hose for any restrictions which may cause a loss in pressure. Check to be sure it is properly connected to the pick-up fitting aft, and at the speedometer. The fitting **MUST** be aligned with the pump intake grate to operate accurately.

b- Check to be sure the pick-up fitting is clear of foreign debris or obstructions -- seaweed, grass, etc. To clear the hose and fitting of such debris, disconnect the hose from the speedometer and apply a little pressure into the hose.

c- To test the speedometer accurately, it is best to obtain a Mity Vac or similar portable vacuum unit. Connect the device to the speedometer fitting. Apply pressure and compare the reading to the table listed below.

SPEEDOMETER PRESSURE TABLE

5psi	21mph
10psi	30mph
15psi	37mph
20psi	42mph
25psi	47mph
30psi	51mph

Upon completion of the test, re-connect the hose and check for proper operation.

APPENDIX

METRIC CONVERSION CHART

LINEAR

inches	X 25.4	= millimetres (mm)
feet	X 0.3048	= metres (m)
yards	X 0.9144	= metres (m)
miles	X 1.6093	= kilometres (km)
inches	X 2.54	= centimetres (cm)

AREA

inches^2	X 645.16	= millimetres^2 (mm^2)
inches^2	X 6.452	= centimetres^2 (cm^2)
feet^2	X 0.0929	= metres^2 (m^2)
yards^2	X 0.8361	= metres^2 (m^2)
acres	X 0.4047	= hectares (10^4 m^2) (ha)
miles^2	X 2.590	= kilometres^2 (km^2)

VOLUME

inches^3	X 16387	= millimetres^3 (mm^3)
inches^3	X 16.387	= centimetres^3 (cm^3)
inches^3	X 0.01639	= litres (l)
quarts	X 0.94635	= litres (l)
gallons	X 3.7854	= litres (l)
feet^3	X 28.317	= litres (l)
feet^3	X 0.02832	= metres^3 (m^3)
fluid oz	X 29.60	= millilitres (ml)
yards^3	X 0.7646	= metres^3 (m^3)

MASS

ounces (av)	X 28.35	= grams (g)
pounds (av)	X 0.4536	= kilograms (kg)
tons (2000 lb)	X 907.18	= kilograms (kg)
tons (2000 lb)	X 0.90718	= metric tons (t)

FORCE

ounces - f (av)	X 0.278	= newtons (N)
pounds - f (av)	X 4.448	= newtons (N)
kilograms - f	X 9.807	= newtons (N)

ACCELERATION

feet/sec^2	X 0.3048	= metres/sec^2 (m/S^2)
inches/sec^2	X 0.0254	= metres/sec^2 (m/s^2)

ENERGY OR WORK (watt-second - joule - newton-metre)

foot-pounds	X 1.3558	= joules (j)
calories	X 4.187	= joules (j)
Btu	X 1055	= joules (j)
watt-hours	X 3500	= joules (j)
kilowatt - hrs	X 3.600	= megajoules (MJ)

FUEL ECONOMY AND FUEL CONSUMPTION

miles/gal	X 0.42514	= kilometres/litre (km/l)

Note:
235.2/(mi/gal) = litres/100km
235.2/(litres/100 km) = mi/gal

LIGHT

footcandles	X 10.76	= lumens/metre^2 (lm/m^2)

PRESSURE OR STRESS (newton/sq metre - pascal)

inches HG (60° F)	X 3.377	= kilopascals (kPa)
pounds/sq in	X 6.895	= kilopascals (kPa)
inches H_2O (60° F)	X 0.2488	= kilopascals (kPa)
bars	X 100	= kilopascals (kPa)
pounds/sq ft	X 47.88	= pascals (Pa)

POWER

horsepower	X 0.746	= kilowatts (kW)
ft-lbf/min	X 0.0226	= watts (W)

TORQUE

pound-inches	X 0.11299	= newton-metres (N·m)
pound-feet	X 1.3558	= newton-metres (N·m)

VELOCITY

miles/hour	X 1.6093	= kilometres/hour (km/h)
feet/sec	X 0.3048	= metres/sec (m/s)
kilometres/hr	X 0.27778	= metres/sec (m/s)
miles/hour	X 0.4470	= metres/sec (m/s)

TEMPERATURE

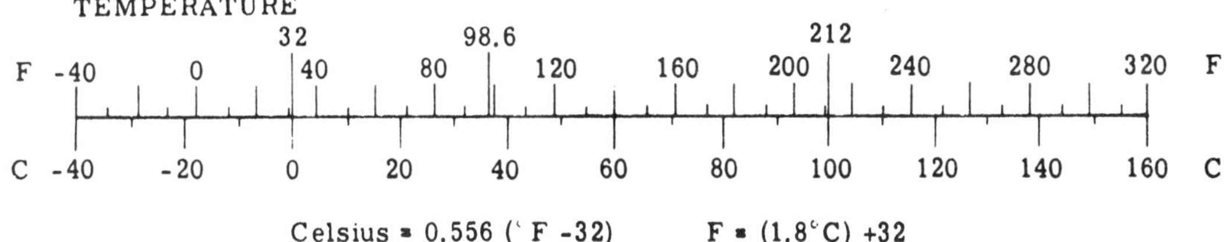

Celsius = 0.556 (°F -32)   F = (1.8°C) +32

SAE TAP DRILL SIZES

Thread Size	Drill Size	Thread Size	Drill Size
#0-80	3/64	1/2-13	27/64
#1-64	53	1/2-20	29/64
#1-72	53	9/16-12	31/64
#2-56	51	9/16-18	33/64
#2-64	50	5/8-11	17/32
#3-48	5/64	5/8-18	37/64
#3-56	45	3/4-10	21/32
#4-40	43	3/4-16	11/16
#4-48	42	7/8-9	49/64
#5-40	38	7/8-14	13/16
#5-44	37	1-8	7/8
#6-32	36	1-12	59/64
#6-40	33	1 1/8-7	63/64
#8-32	29	1 1/8-12	1 3/64
#8-36	29	1 1/4-7	1 7/64
#10-24	24	1 1/4-12	1 11/64
#10-32	21	1 1/2-6	1 11/32
#12-24	17	1 1/2-12	1 27/64
#12-28	4.6mm	1 3/4-5	1 9/16
1/4-20	7	1 3/4-12	1 43/64
1/4-28	3	2-4 1/2	1 25/32
5/16-18	F	2-12	1 59/64
5/16-24	I	2 1/4-4 1/2	2 1/32
3/8-16	O	2 1/2-4	2 1/4
3/8-24	Q	2 3/4-4	2 1/2
7/16-14	U	3-4	2 3/4
7/16-20	25/64		

METRIC TAP DRILL SIZES

Tap Size	Drill Size	Decimal Equivalent	Nearest Fraction
3 x .50	#39	0.0995	3/32
3 x .60	3/32	0.0937	3/32
4 x .70	#30	0.1285	1/8
4 x .75	1/8	0.125	1/8
5 x .80	#19	0.166	11/64
5 x .90	#20	0.161	5/32
6 x 1.00	#9	0.196	13/64
7 x 1.00	16/64	0.234	15/64
8 x 1.00	J	0.277	9/32
8 x 1.25	17/64	0.265	17/64
9 x 1.00	5/16	0.3125	5/16
9 x 1.25	5/16	0.3125	5/16
10 x 1.25	11/32	0.3437	11/32
10 x 1.50	R	0.339	11/32
11 x 1.50	3/8	0.375	3/8
12 x 1.50	13/32	0.406	13/32
12 x 1.75	13/32	0.406	13/32

ENGINE TORQUE VALUES

ENGINE FASTENERS	FUJI Engine	POLARIS Engine	FUJI Engine	POLARIS Engine
Air Intake Cover Center Bolt (8mm) Air Intake Cover Screws	REFER TO CHAPTER 8 - ENGINE OR CHAPTER 6 - FUEL			
Crankcase Bolts (8mm)	16 ft lb	22 ft lb	22 Nm	30 Nm
Crankcase Bolts (10mm)	26 ft lb	28 ft lb	35 Nm	38 Nm
Cylinder Head Nuts (8mm)	18 ft lb	-	24 Nm	-
Cylinder Head Bolts (8mm)	-	20 ft lb	-	27 Nm
Cylinder Head Cover Bolts (8mm)	-	22 ft lb	-	30 Nm
Cylinder Base Nuts (10mm)	28 ft lb	40 ft lb	38 NM	54 Nm
Carburetor Mounting Nuts (Mikuni) (8mm)	16 ft lb	-	22 Nm	-
Carburetor Mounting Bolts (Keihin) (6mm)	-	108 in lb	-	12 Nm
Engine Mount Plate (To Engine) Bolts	45 ft lb	50 ft lb	61 Nm	68 Nm
Engine Mount Nuts	45 ft lb	45 ft lb	61 Nm	61 Nm
Engine Mount to Hull (5/16"-18)	14 ft lb	14 ft lb	19 Nm	19 Nm
Exhaust Manifold Bolts (8mm)	16 ft lb	22 ft lb	22 Nm	30 Nm
Exhaust Manifold to Pipe Bolts (12mm)	45ft lb	45 ft lb	61 Nm	61 Nm
Flywheel Nut 14mm	-	90 ft lb	-	122 Nm
Flywheel Nut 16mm 18mm	55 ft lb 65 ft lb	-	75 Nm 88Nm	-
Flywheel Housing Bolt (6mm)	78 in lb	108 in lb	9 Nm	12 Nm
Flywheel Housing Bolt (8mm)	16 ft lb	22 ft lb	22 Nm	30 Nm
Intake Manifold Nuts/Bolts (6mm)	78 in lb	108 in lb	9 Nm	12 Nm
Oil Pump Bolts/Screws (5mm)	48 in lb	60 in lb	5 Nm	7 Nm
Starter Motor Mounting Bolts	108 in lb	108 in lb	12 Nm	12 Nm
Spark Plug	18 ft lb	18 ft lb	24 Nm	24 Nm
Water Outlet Manifold Bolts (6mm)	78 in lb	108 in lb	9 Nm	12 Nm

ENGINE SPECIFICATIONS

SERIES	Piston-to-Cylinder Clearance		Piston Ring End Gap	BORE & STROKE INCHES (mm)
	Standard	Limit		
	1992		1992	
SL650	0.004 - 0.006" (0.1 - 0.15mm)	0.008" (0.2mm)	0.008 - 0.016" (0.2 - 0.41mm)	2.559 x 2.559" (65 x 65mm)
	1993		1993	
SL650	0.004 - 0.006" (0.1 - 0.15mm)	0.008" (0.2mm)	0.008 - 0.016" (0.2 - 0.41mm)	2.559 x 2.559" (65 x 65mm)
SL750	0.004 - 0.006" (0.1 - 0.15mm)	0.008" (0.2mm)	0.008 - 0.018" (0.2 - 0.46mm)	2.774 x 2.559" (69.72 x 65mm)
	1994		1994	
SL650	0.005- 0.007" (0.13 - 0.18mm)	0.008" (0.2mm)	0.008 - 0.016" (0.2 - 0.41mm)	2.559 x 2.559" (65 x 65mm)
SL750	0.004 - 0.006" (0.1 - 0.16mm)	0.008" (0.2mm)	0.008 - 0.018" (0.2 - 0.46mm)	2.774 x 2.559" (69.72 x 65mm)
SLT750	0.004 - 0.006" (0.1 - 0.16mm)	0.008" (0.2mm)	0.008 - 0.018" (0.2 - 0.46mm)	2.774 x 2.559" (69.72 x 65mm)
	1995		1995	
SL650	0.005 - 0.007" (0.13 - 0.18mm)	0.009" (0.23mm)	0.008 - 0.016" (0.2 - 0.41mm)	2.559 x 2.559" (65 x 65mm)
SL650 STD	0.005 - 0.007" (0.13 - 0.18mm)	0.009" (0.23mm)	0.008 - 0.016" (0.2 - 0.41mm)	2.559 x 2.559" (65 x 65mm)
SL750	0.004 - 0.006" (0.1 - 0.16mm)	0.008" (0.2mm)	0.008 - 0.018" (0.2 - 0.46mm)	2.774 x 2.559" (69.72 x 65mm)
SLT750	0.004 - 0.006" (0.1 - 0.16mm)	0.008" (0.2mm)	0.008 - 0.018" (0.2 - 0.46mm)	2.774 x 2.559" (69.72 x 65mm)
SLX780	0.0035 - 0.005" (0.09 - .13mm)	0.008" (0.2mm)	0.012 - 0.02" (0.3 - 0.5mm)	2.807 x 2.559" (71.3 x 65mm)
	1996		1996	
SL700	0.004 - 0.005" (0.10 - .13mm)	0.010" (0.25mm)	0.01 - 0.018" (0.25 - 0.45mm)	2.189 x 2.667" (81 x 68mm)
SLT700	0.004 - 0.005" (0.10 - .13mm)	0.010" (0.25mm)	0.01 - 0.018" (0.25 - 0.45mm)	2.559 x 2.559" (65 x 65mm)

ENGINE SPECIFICATIONS

SERIES	Piston-to-Cylinder Clearance		Piston Ring End Gap	BORE & STROKE Inches (mm)
	Standard	Limit		
	1996		1996	
Hurricane	0.004 - 0.005" (0.10 - .13mm)	0.010" (0.25mm)	0.01 - 0.018" (0.25 - 0.45mm)	3.189 x 2.677" (81.3 x 68mm)
SL780	0.0035 - 0.005" (0.09 - .13mm)	0.008" (0.2mm)	0.012 - 0.018" (0.3 - 0.45mm)	2.807 x 2.559" (71.3 x 65mm)
SLT780	0.0035 - 0.005" (0.09 - .13mm)	0.008" (0.2mm)	0.012 - 0.018" (0.3 - 0.45mm)	2.807 x 2.559" (71.3 x 65mm)
SLX780	0.0035 - 0.005" (0.09 - .13mm)	0.008" (0.2mm)	0.012 - 0.018" (0.3 - 0.45mm)	2.807 x 2.559" (71.3 x 65mm)
SL900	0.004 - 0.005" (0.1 - .13mm)	0.010" (0.25mm)	0.01 - 0.018" (0.25 - 0.45mm)	2.933 x 2.677" (74.5 x 68mm)
SLTX	0.004 - 0.005" (0.1 - .13mm)	0.010" (0.25mm)	0.01 - 0.018" (0.25 - 0.45mm)	3.189 x 2.677" (81 x 68mm)
	1997		1997	
SL700	0.004 - 0.005" (0.10 - .13mm)	0.010" (0.25mm)	0.01 - 0.018" (0.25 - 0.45mm)	3.189 x 2.677" (81 x 68mm)
SL700 Deluxe	0.004 - 0.005" (0.10 - .13mm)	0.010" (0.25mm)	0.01 - 0.018" (0.25 - 0.45mm)	3.189 x 2.677" (81 x 68mm)
SLT700	0.004 - 0.005" (0.10 - .13mm)	0.010" (0.25mm)	0.01 - 0.018" (0.25 - 0.45mm)	3.189 x 2.677" (81 x 68mm)
Hurricane	0.004 - 0.005" (0.10 - .13mm)	0.010" (0.25mm)	0.01 - 0.018" (0.25 - 0.45mm)	3.189 x 2.677" (81 x 68mm)
SL780	0.003 - 0.005" (0.08 - .13mm)	0.008" (0.2mm)	0.012 - 0.02" (0.3 - 0.5mm)	2.807 x 2.559" (71.3 x 65mm)
SLT780	0.003 - 0.005" (0.08 - .13mm)	0.008" (0.2mm)	0.012 - 0.02" (0.3 - 0.5mm)	2.807 x 2.559" (71.3 x 65mm)
SL900	0.004 - 0.005" (0.10 - .13mm)	0.010" (0.25mm)	0.01 - 0.018" (0.25 - 0.45mm)	2.933 x 2.677" (74.5 x 68mm)
SL1050	0.0055 - 0.0065" (0.14 - .16mm)	0.010" (0.25mm)	0.01 - 0.02" (0.25 - 0.55mm)	3.189 x 2.677" (81 x 68mm)
SLTX	0.0055 - 0.0065" (0.14 - .16mm)	0.010" (0.25mm)	0.01 - 0.02" (0.25 - 0.55mm)	3.189 x 2.677" (81 x 68mm)

ENGINE SPECIFICATIONS

SERIES	ENGINE MANUF.	N0. OF CYL.'s	ENGINE DISP. INCHES (cc)	BORE & STROKE INCHES (mm)	H.P. rpm	NUMBER OF CARB's & SIZE
		1992			1992	
SL650	FUJI	3	39.482" (647cc)	2.559 x 2.559" (65 x 65mm)	68 @ 6,500	(3) 38mm Mikuni
		1993			1993	
SL650	FUJI	3	39.482" (647cc)	2.559 x 2.559" (65 x 65mm)	68 @ 6,500	(3) 38mm Mikuni
SL750	FUJI	3	45.401" (744cc)	2.774 x 2.559" (69.72 x 65mm)	78 @ 6,500	(3) 38mm Mikuni
		1994			1994	
SL650	FUJI	3	39.482" (647cc)	2.559 x 2.559" (65 x 65mm)	68 @ 6,350	(3) 38mm Mikuni
SL750	FUJI	3	45.401" (744cc)	2.774 x 2.559" (69.72 x 65mm)	80 @ 6,150	(3) 38mm Mikuni
SLT750	FUJI	3	45.401" (744cc)	2.774 x 2.559" (69.72 x 65mm)	80 @ 6,250	(3) 38mm Mikuni
		1995			1995	
SL650	FUJI	3	39.482" (647cc)	2.559 x 2.559" (65 x 65mm)	68 @ 6,500	(3) 38mm Mikuni
SL650 STD	FUJI	3	39.482" (647cc)	2.559 x 2.559" (65 x 65mm)	68 @ 6,500	(3) 38mm Mikuni
SL750	FUJI	3	45.401" (744cc)	2.774 x 2.559" (69.72 x 65mm)	80 @ 6,000	(3) 38mm Mikuni
SLT750	FUJI	3	45.401" (744cc)	2.774 x 2.559" (69.72 x 65mm)	80 @ 6,000	(3) 38mm Mikuni
SLX780	FUJI	3	47.537" (779cc)	2.807 x 2.559" (71.3 x 65mm)	90+ @ 6,250	(3) 38mm Mikuni
		1996			1996	
SL700	POLARIS	2	42.716" (700cc)	2.189 x 2.667" (81 x 68mm)	80 @ 6,250	(2) CDK II 38mm
SLT700	POLARIS	2	42.716" (700cc)	2.559 x 2.559" (65 x 65mm)	68 @ 6,500	(2) CDK II 38mm

ENGINE SPECIFICATIONS

SERIES	ENGINE MANUF.	N0. OF CYL.'s	ENGINE DISP. INCHES (cc)	BORE & STROKE INCHES (cc)	H.P. rpm	NUMBER OF CARB's & SIZE
		1996			1996	
Hurricane	FUJI	2	42.716" (700cc)	3.189 x 2.677" (81.3 x 68mm)	90+ @ 6,700	(2) CDK II 38mm
SL780	FUJI	3	47.537" (779cc)	2.807 x 2.559" (71.3 x 65mm)	90+ @ 6,250	(3) 38mm Mikuni
SLT780	FUJI	3	47.537" (779cc)	2.807 x 2.559" (71.3 x 65mm)	90+ @ 6,250	(3) 38mm Mikuni
SLX780	FUJI	3	47.537" (779cc)	2.807 x 2.559" (71.3 x 65mm)	90+ @ 6,250	(3) 38mm Mikuni
SL900	POLARIS	3	54.250" (889cc)	2.933 x 2.677" (74.5 x 68mm)	100+ @ 6250	(3) CDK II 38mm
SLTX	POLARIS	3	64.074" (1050cc)	3.189 x 2.677" (81 x 68mm)	115+	(3) CDK II 38mm
		1997			1997	
SL700	FUJI	2	42.716" (700cc)	3.189 x 2.677" (81 x 68mm)	82.5 @ 6,200	(1) CDK II 40mm
SL700 Deluxe	FUJI	2	42.716" (700cc)	3.189 x 2.677" (81 x 68mm)	85 @ 6,350	(2 CDK II 38mm
SLT700	FUJI	2	42.716" (700cc)	3.189 x 2.677" (81 x 68mm)	85 @ 6,350	(2) CDK II 38mm
Hurricane	FUJI	2	42.716" (700cc)	3.189 x 2.677" (81 x 68mm)	90 @ 6,700	(2) CDK II 38mm
SL780	FUJI	3	47.537" (779cc)	2.807 x 2.559" (71.3 x 65mm)	93 @ 6,250	(3) 38mm Mikuni
SLT780	FUJI	3	47.537" (779cc)	2.807 x 2.559" (71.3 x 65mm)	93 @ 6,350	(3) 38mm Mikuni
SL900	POLARIS	3	54.921" (900cc)	2.933 x 2.677" (74.5 x 68mm)	107 @ 6,350	(3) CDK II 38mm
SL1050	POLARIS	3	64.074" (1050cc)	3.189 x 2.677" (81 x 68mm)	119 @ 6,500	(3) CDK II 40mm
SLTX	POLARIS	3	64.074" (1050cc)	3.189 x 2.677" (81 x 68mm)	119 @ 6,500	(3) CDK II 38mm

TUNE-UP ADJUSTMENTS

SERIES	IGNITION SYSTEM	TIMING DEGREES BTDC	SPARK PLUG TYPE	SPARK PLUG GAP	IDLE RPM (IN WATER)
	1992			**1992**	
SL650	CDI Flywheel Magneto	24° at 3,000 rpm	NGK BPR7ES Champion RN-3C	.028" (.7mm)	1300 RPM
	1993			**1993**	
SL650	CDI Flywheel Magneto	18° at 3,000 rpm	NGK BPR7ES Champion RN-3C	.028" (.7mm)	1300 RPM
SL750	CDI Flywheel Magneto	16° at 3,000 rpm	NGK BPR7ES Champion RN-3C	.028" (.7mm)	1250±50 RPM
	1994			**1994**	
SL650	CDI Flywheel Magneto	18° at 3,000 rpm	NGK BPR7ES Champion RN-3C	.028" (.7mm)	1300±50 RPM
SL750	Digital CDI Flywheel Magneto	25° at 3,000 rpm	NGK BPR7ES Champion RN-3C	.028" (.7mm)	1200-1300 RPM
SLT750	Digital CDI Flywheel Magneto	25° at 3,000 rpm	NGK BPR7ES Champion RN-3C	.028" (.7mm)	1200-1300 RPM
	1995			**1995**	
SL650	CDI Flywheel Magneto	18° at 3,000 rpm	NGK BPR7ES Champion RN-3C	.028" (.7mm)	1200-1300 RPM
SL650 STD	CDI Flywheel Magneto	18° at 3,000 rpm	NGK BPR7ES Champion RN-3C	.028" (.7mm)	1200-1300 RPM
SL750	Digital CDI Flywheel Magneto	24° at 3,000 rpm	NGK BPR7ES Champion RN-3C	.028" (.7mm)	1200-1300 RPM
SLT750	Digital CDI Flywheel Magneto	24° at 3,000 rpm	NGK BPR7ES Champion RN-3C	.028" (.7mm)	1200-1300 RPM
SLX780	Digital CDI Flywheel Magneto	28° at 3,000 rpm	NGK BPR8ES Champion RN-2C	.028" (.7mm)	1250±50 RPM
	1996			**1996**	
SL700	Digital CDI	18° at 3,000 rpm	NGK BPR7ES Champion RN-3C	.028" (.7mm)	1250-1300 RPM
SLT700	Digital CDI	18° at 3,000 rpm	NGK BPR7ES Champion RN-3C	.028" (.7mm)	1250-1300 RPM

TUNE-UP ADJUSTMENTS

SERIES	IGNITION SYSTEM	TIMING DEGREES BTDC	SPARK PLUG TYPE	SPARK PLUG GAP	IDLE RPM (IN WATER)
	1996			**1996**	
Hurricane	Digital CDI	20° at 3,000 rpm	NGK BPR7ES Champion RN-3C	.028" (.7mm)	1300-1400 RPM
SL780	Digital	** 28° at 3,000 rpm	NGK BPR8ES Champion RN-2C	.028" (.7mm)	1200-1300 RPM
SLX780	Digital	28° at 3,000 rpm	NGK BR8ES Champion RN-3C	.028" (.7mm)	1200-1300 RPM
SLT780	Digital	28° at 3,000 rpm	NGK BR8ES Champion RN-2C	.028" (.7mm)	1200-1300 RPM
SL900	Digital	18° at 3,000 rpm	NGK BR8ES Champion RN-3C	.028" (.7mm)	1250-1300 RPM
SLTX	Digital	18° at 3,000 rpm	NGK BR8ES Champion RN-2C	.028" (.7mm)	1250-1300 RPM
	1997			**1997**	
SL700	Digital CDI	18° at 3,000 rpm	NGK BR8ES Champion RN-2C	.028" (.7mm)	1200-1300 RPM
SL700 Deluxe	Digital CDI	18° at 3,000 rpm	NGK BPR7ES Champion RN-3C	.028" (.7mm)	1200-1300 RPM
SLT700	Digital CDI	18° at 3,000 rpm	NGK BPR7ES Champion RN-3C	.028" (.7mm)	1200-1300 RPM
Hurricane	Digital CDI	20° at 3,000 rpm	NGK BR8ES Champion RN-2C	.028" (.7mm)	1300-1400 RPM
SL780	Digital CDI	24° at 3,000 rpm	NGK BR8ES Champion RN-2C	.028" (.7mm)	1200-1300 RPM
SLT780	Digital CDI	24° at 3,000 rpm	NGK BR8ES Champion RN-2C	.028" (.7mm)	1200-1300 RPM
SL900	Digital/ Sequential Ignition	18° at 3,000 rpm	NGK BR8ES Champion RN-2C	.028" (.7mm)	1200-1300 RPM
SL1050	Digital/ Sequential Ignition	18° at 3,000 rpm	NGK BR8ES Champion RN-2C	.028" (.7mm)	1200-1300 RPM
SLTX	Digital/ Sequential Ignition	18° at 3,000 rpm	NGK BR8ES Champion RN-2C	.028" (.7mm)	1200-1300 RPM

CARBURETOR INSTALLATION AND BENCH ADJUSTMENTS

Year	Model	No. Of Carbs.	Carburetor Identification	High Speed Screw Turns Out	Low Speed Screw Turns Out
1992	SL 650	3	Mikuni Super BN	Fwd 3/8, Ctr 3/8, Aft 1/4	All Three 1-3/8 Note 1
1993	SL650	3	Mikuni Super BN	Fwd 7/8, Ctr 3/8 Aft 5/8	All Three 1/4
	SL 750	3	Mikuni Super BN	Fwd 7/8, Ctr 1/2, Aft 5/8	All Three 1/2
1994	SL 650 Eng. S/N 94-0001 - 94-02010	3	Mikuni Super BN	Fwd 3/4, Ctr 1/4, Aft 1/2	All Three 1-1/4
	SL 650 Eng. S/N 94-02011 & On	3	Mikuni Super BN	Fwd 7/8, Ctr 1/2 Sft 3/4	All Three 1 full turn
	SL 750	3	Mikuni Super BN	Fwd 1-1/4, Ctr 3/8, Aft 7/8	All Three 1/2
	SLT 750	3	Mikuni Super BN	Fwd 1-1/4, Ctr 3/8, Aft 7/8	All Three 1/2
1995	SL 650	3	Mikuni Super BN	Fwd 1-1/8, Ctr 1/4, Aft 7/8	All three 1 full turn
	SL650 STD	3	Mikuni Super BN	Fwd 1-1/8, Ctr 1/4, Aft 7/8	All three 1 full turn
	SL 750	3	Mikuni Super BN	Fwd 1, Ctr 1/2, Aft 3/4	All Three 1/2
	SLT 750	3	MikuniSuper BN	Fwd 1, Ctr 1/2, Aft 3/4	All Three 1/2
	SLX 780	3	Mikuni Super BN	Fwd 7/8, Ctr 3/4, Aft 1-1/8	All Three 1/2
1996	SL 700	2	Keihin CDK II	Both 1-1/2	Both 5/8
	SLT 700	2	Keihin CDK II	Both 1-1/2	Both 5/8
	Hurricane	2	Keihin CDK II	No H/P Screw	Both 5/8

CARBURETOR INSTALLATION AND BENCH ADJUSTMENTS

Year	Model	No. Of Carbs.	Carburetor Identification	High Speed Screw Turns Out	Low Speed Screw Turns Out
1996 Cont.	SL780 EC78PWE-02/05	3	Mikuni Super BN	Fwd 3/4, Ctr 3/4 Aft 1 full turn	All Three 5/8
	SLT 780 EC78PWE-03/04	3	Mikuni Super BN	Fwd 1-1/8, Ctr 7/8, Aft 1-1/4	All three 1-3/8
	SLX 780 EC78PWE-04	3	Mikuni Super BN	Fwd 1-1/8, Ctr 7/8, Aft 1-1/4	All three 1-3/8
	SL 900	3	Keihin CDKII	No H/P Screw	All Three 5/8
	SLTX	3	Keihin CDKII	No H/P Screw	1 Full Turn
1997	SL 700	1	Keihin CDK II	No H/P Screw	Single 7/8
	SL 700 Deluxe	2	Keihin CDK II w/accel. Pump	Both 1-5/8 to 1-3/4	Both 5/8
	SLT 700	2	Keihin CDK II w/Accel. Pump	Both 1-5/8 to 1-3/4	Both 5/8
	Hurricane	2	Keihin CDK II	No H/P Screw	Both 5/8
1997	SL 780	3	Mikuni Super BN	All Three 1/8 ± 1/8	All Three 1-1/4
	SLT 780	3	Mikuni Super BN	All Three 1/8 ± 1/8	All Three 1-1/4
	SL 900	3	Keihin CDKII w/Accel. Pump	No H/P Screw	All Three 5/8
	SL 1050	3	Keihin CDKII	No H/P Screw	All Three 7/8
	SLTX	3	Keihin CDKII w/Accel. Pump	No H/P Screw	1 Full Turn

Mekuni carburetors have separate fuel pump.
Keihin carburetors have integral fuel pump.

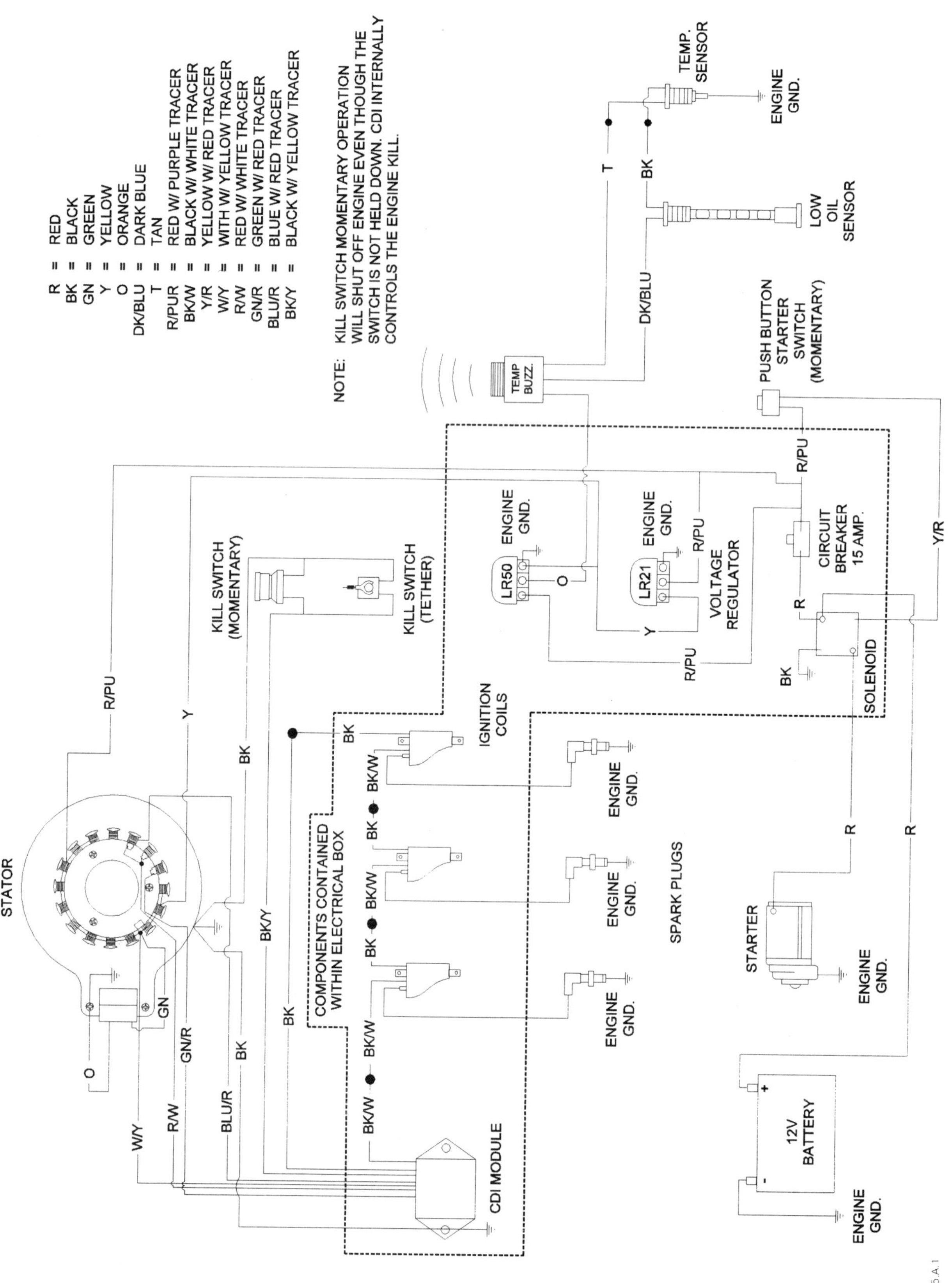

***SL650 -- 1992** Wiring diagram and color code identification. Major items are identified. Components within the dashed border are housed inside the electrical box.*

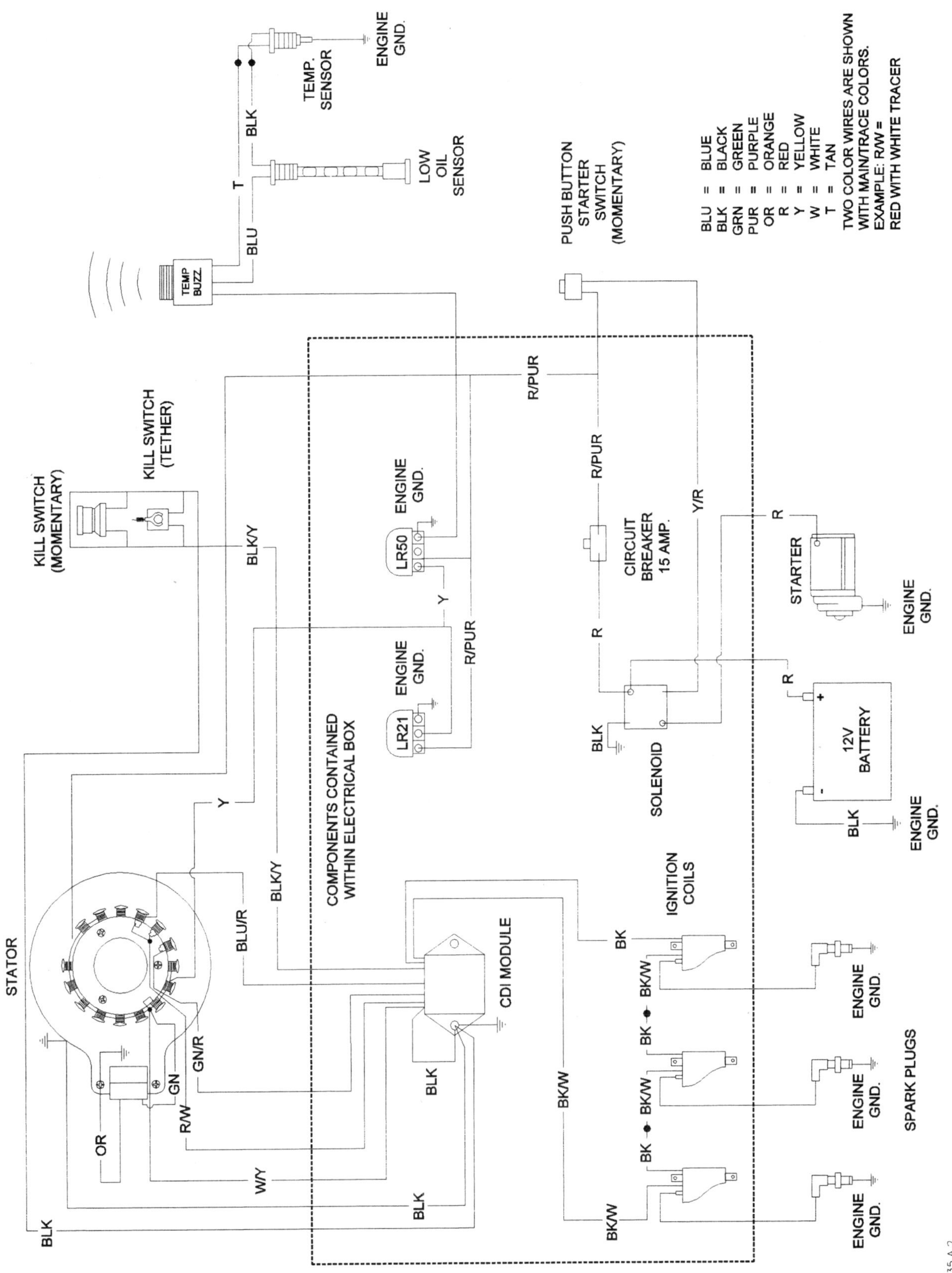

SL650 -- 1993 *Wiring diagram and color code identification. Major items are identified. Components within the dashed border are housed inside the electrical box.*

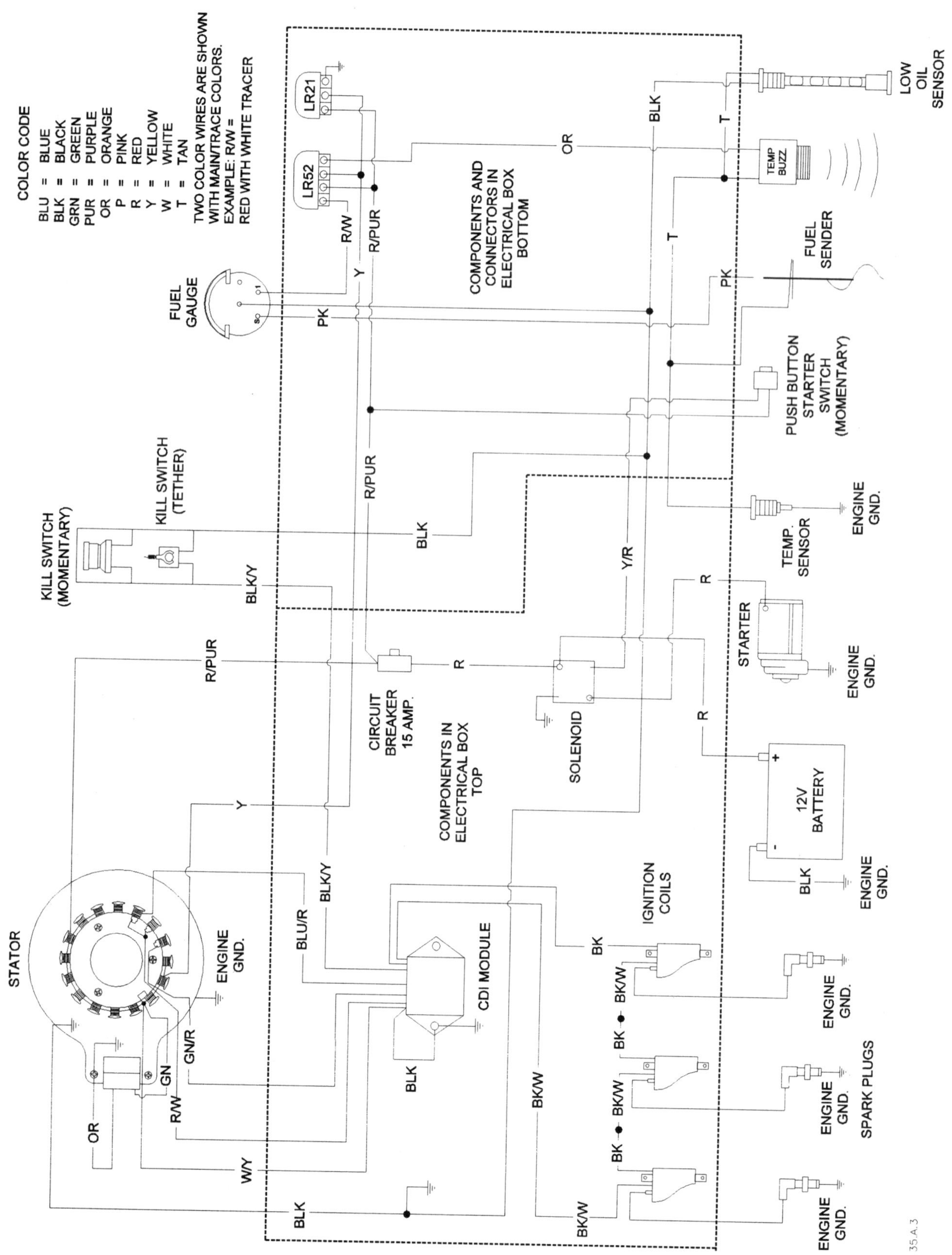

SL650 -- 1994 *Wiring diagram and color code identification. Major items are identified. Components within the dashed border are housed inside the electrical box.*

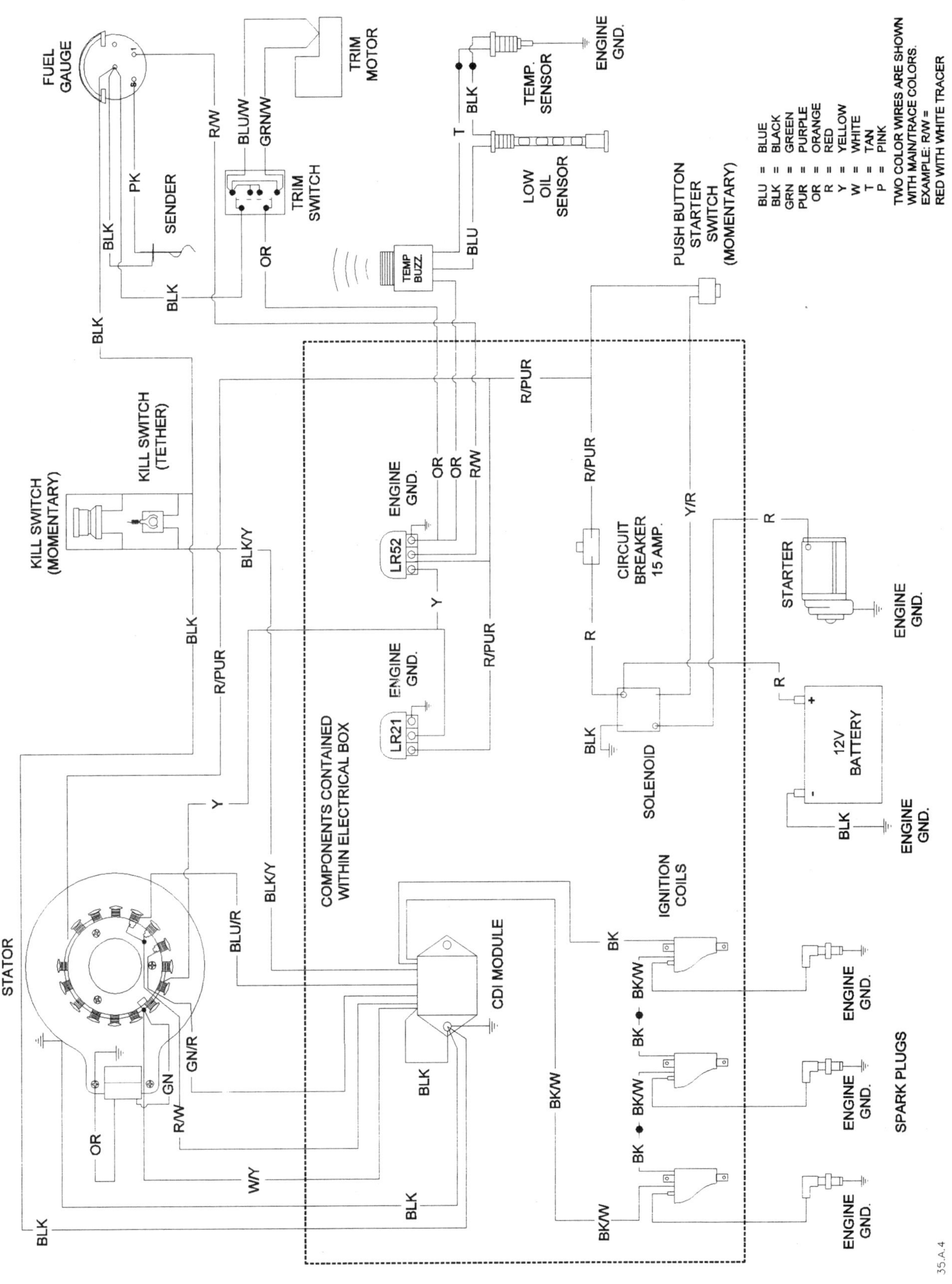

SL750 -- 1993 *Wiring diagram and color code identification. Major items are identified. Components within the dashed border are housed inside the electrical box.*

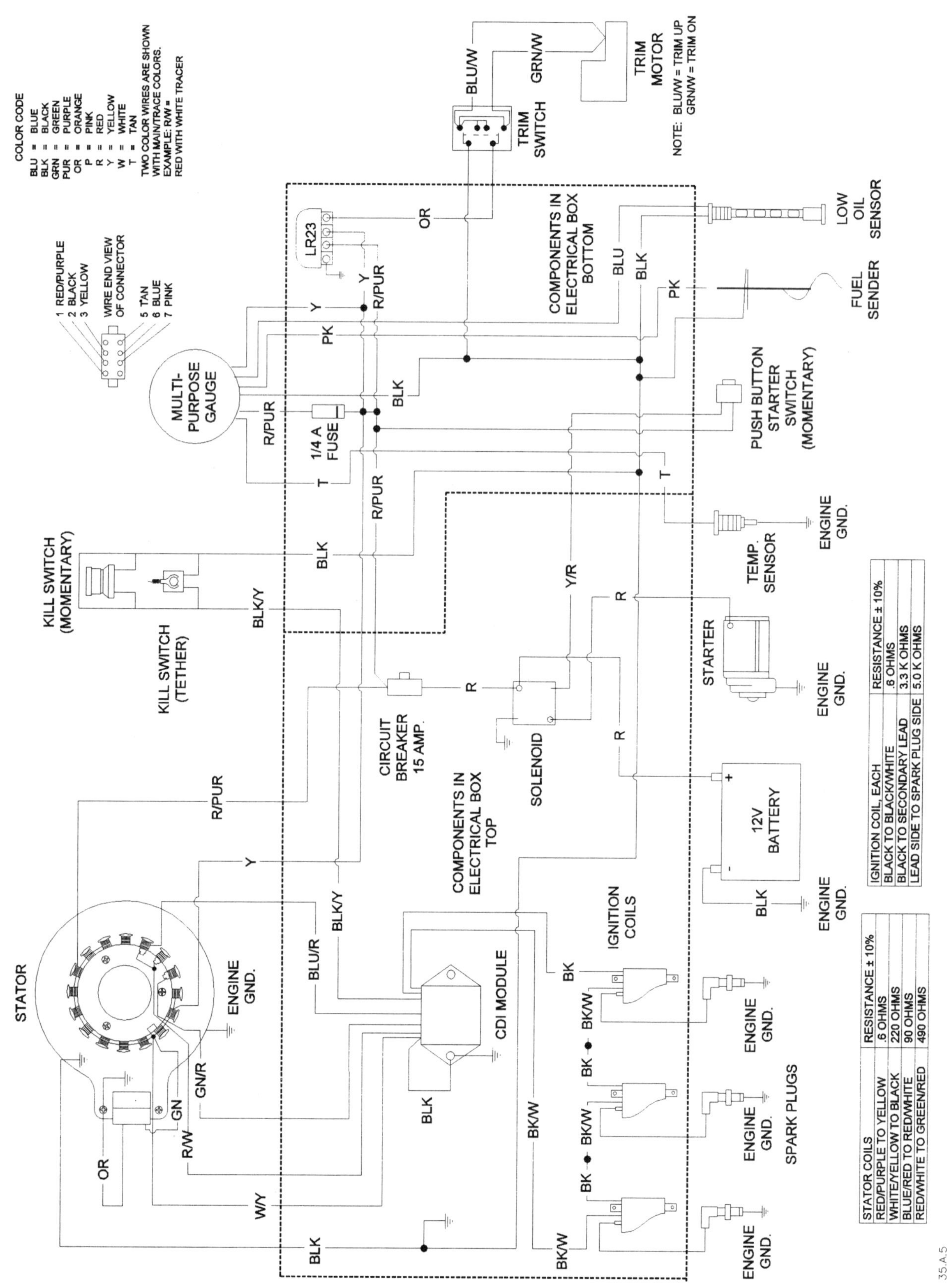

IGNITION COIL, EACH	RESISTANCE ± 10%
BLACK TO BLACK/WHITE	.6 OHMS
BLACK TO SECONDARY LEAD	3.3 K OHMS
LEAD SIDE TO SPARK PLUG SIDE	5.0 K OHMS

STATOR COILS	RESISTANCE ± 10%
RED/PURPLE TO YELLOW	.6 OHMS
WHITE/YELLOW TO BLACK	220 OHMS
BLUE/RED TO RED/WHITE	90 OHMS
RED/WHITE TO GREEN/RED	490 OHMS

SL750 -- 1994 *Wiring diagram and color code identification. Major items are identified. Components within the dashed border are housed inside the electrical box.*

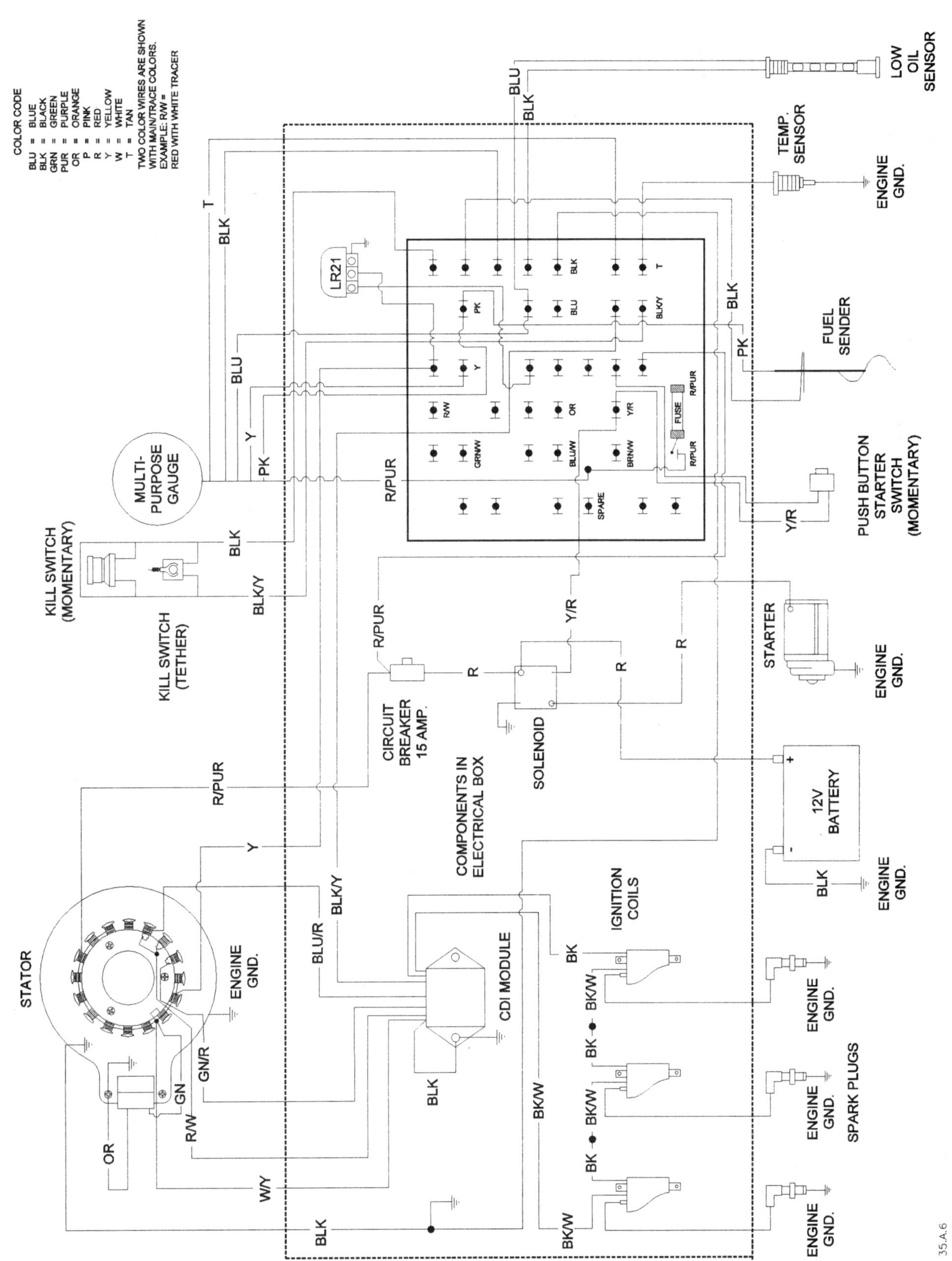

SLT750 -- 1994 *Wiring diagram and color code identification. Major items are identified. Components within the dashed border are housed inside the electrical box.*

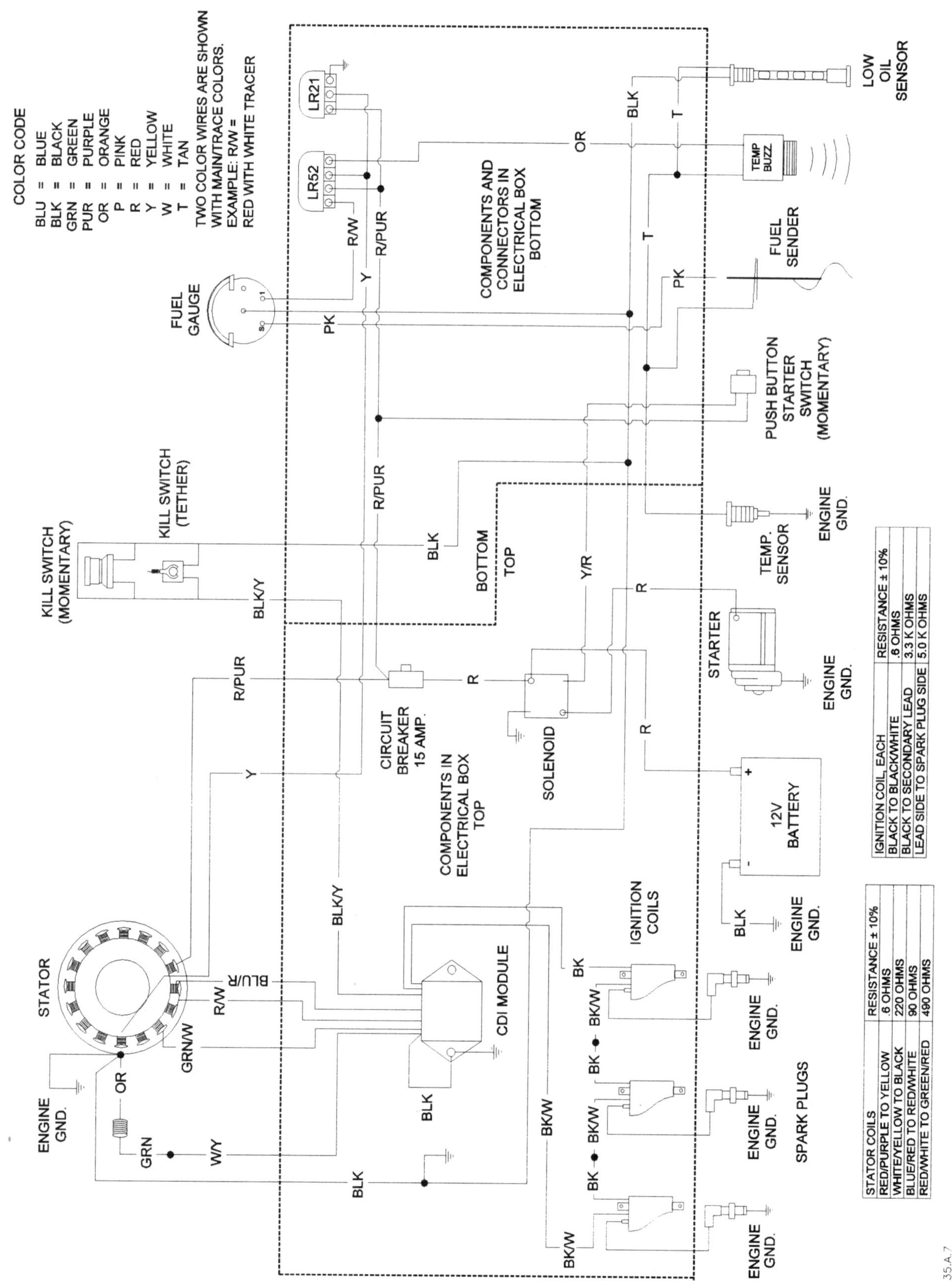

IGNITION COIL, EACH	RESISTANCE ± 10%
BLACK TO BLACK/WHITE	.6 OHMS
BLACK TO SECONDARY LEAD	3.3 K OHMS
LEAD SIDE TO SPARK PLUG SIDE	5.0 K OHMS

STATOR COILS	RESISTANCE ± 10%
RED/PURPLE TO YELLOW	.6 OHMS
WHITE/YELLOW TO BLACK	220 OHMS
BLUE/RED TO RED/WHITE	90 OHMS
RED/WHITE TO GREEN/RED	490 OHMS

35.A.7

SL650 -- 1995 *Wiring diagram and color code identification. Major items are identified. Components within the dashed border are housed inside the electrical box.*

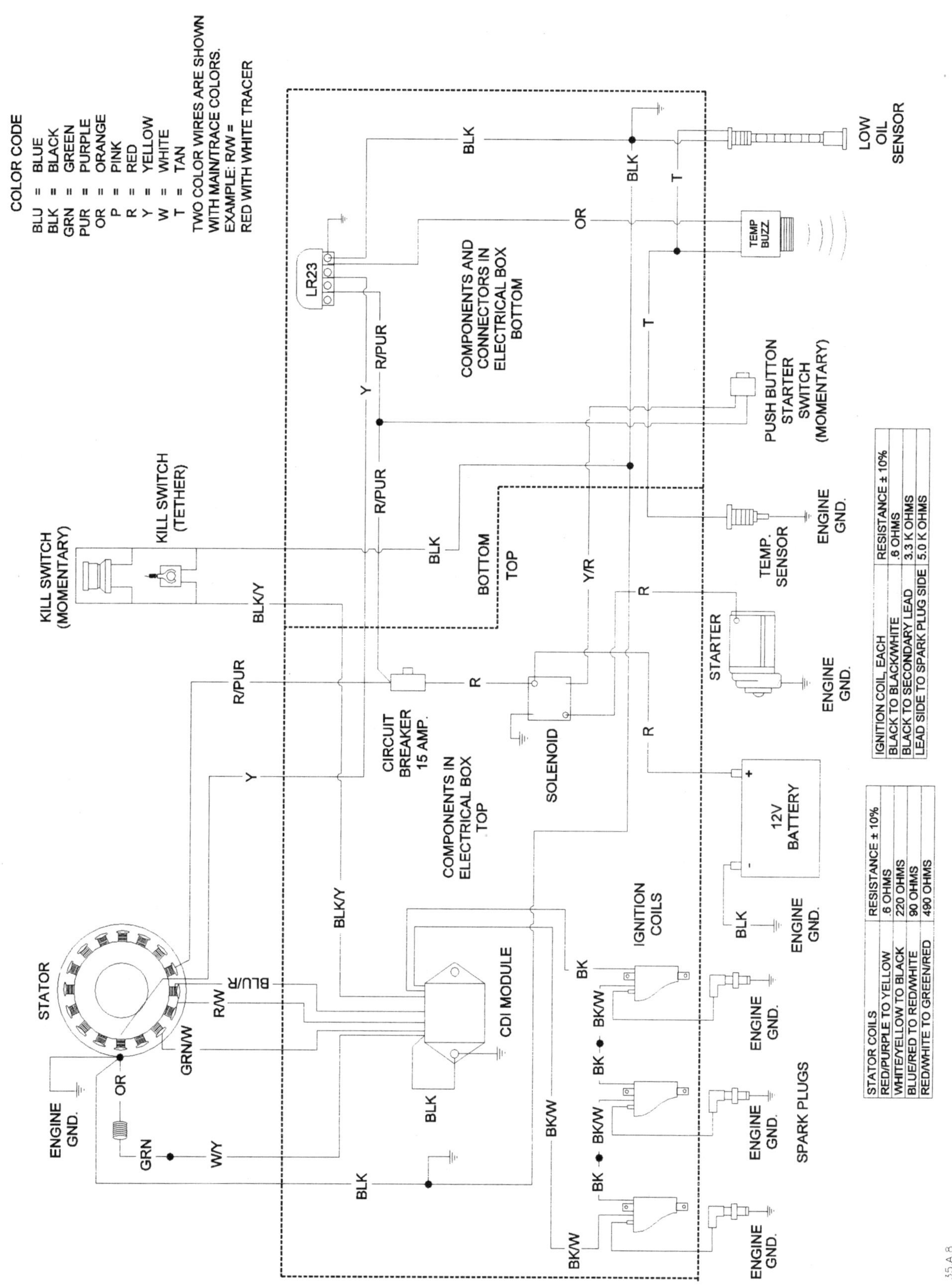

IGNITION COIL, EACH	RESISTANCE ± 10%
BLACK TO BLACK/WHITE	.6 OHMS
BLACK TO SECONDARY LEAD	3.3 K OHMS
LEAD SIDE TO SPARK PLUG SIDE	5.0 K OHMS

STATOR COILS	RESISTANCE ± 10%
RED/PURPLE TO YELLOW	.6 OHMS
WHITE/YELLOW TO BLACK	220 OHMS
BLUE/RED TO RED/WHITE	90 OHMS
RED/WHITE TO GREEN/RED	490 OHMS

SL650 STD -- 1995 *Wiring diagram and color code identification. Major items are identified. Components within the dashed border are housed inside the electrical box.*

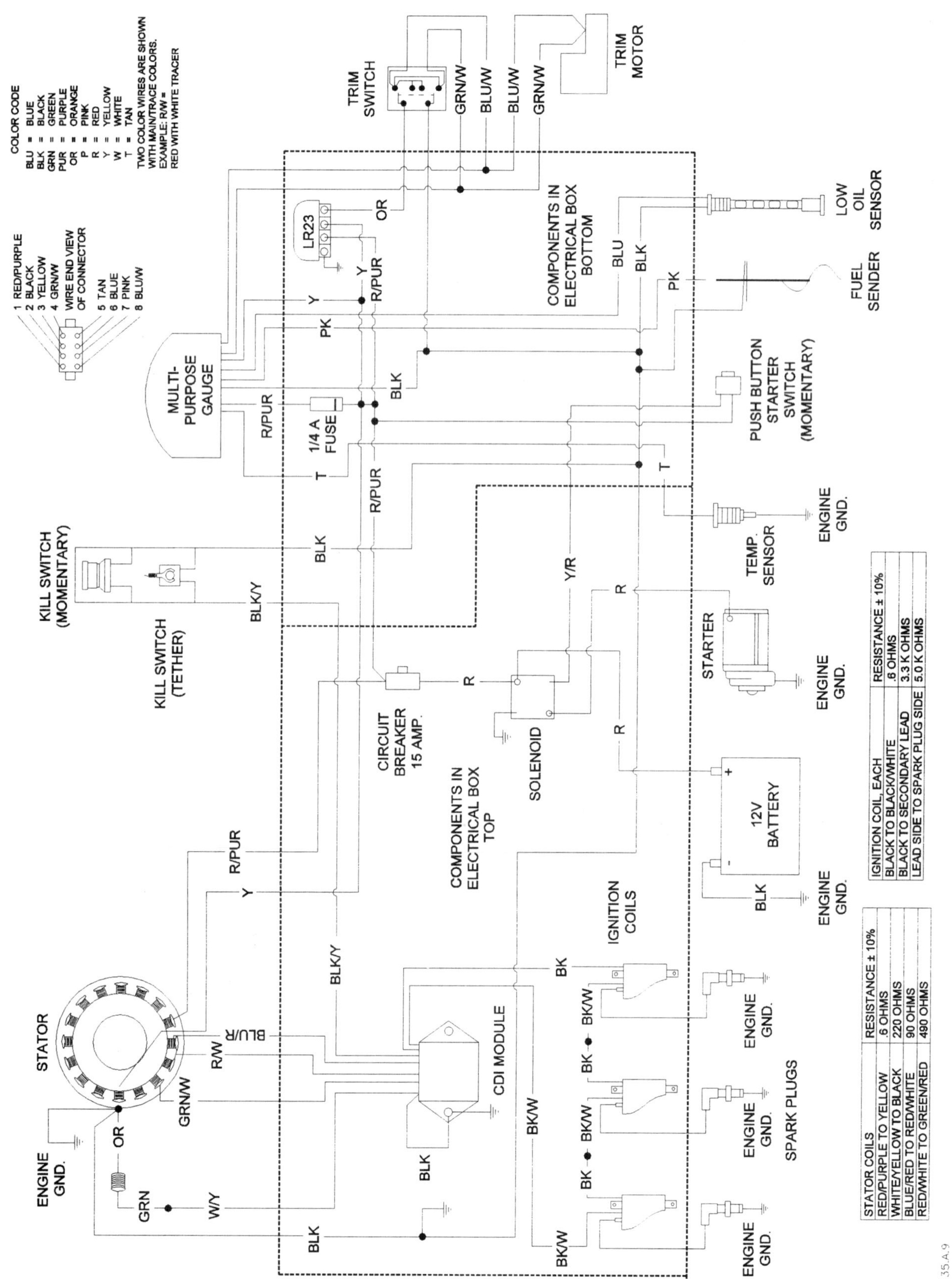

STATOR COILS	RESISTANCE ± 10%
RED/PURPLE TO YELLOW	.6 OHMS
WHITE/YELLOW TO BLACK	220 OHMS
BLUE/RED TO RED/WHITE	90 OHMS
RED/WHITE TO GREEN/RED	490 OHMS

IGNITION COIL, EACH	RESISTANCE ± 10%
BLACK TO BLACK/WHITE	.6 OHMS
BLACK TO SECONDARY LEAD	3.3 K OHMS
LEAD SIDE TO SPARK PLUG SIDE	5.0 K OHMS

***SL750 -- 1995** Wiring diagram and color code identification. Major items are identified. Components within the dashed border are housed inside the electrical box.*

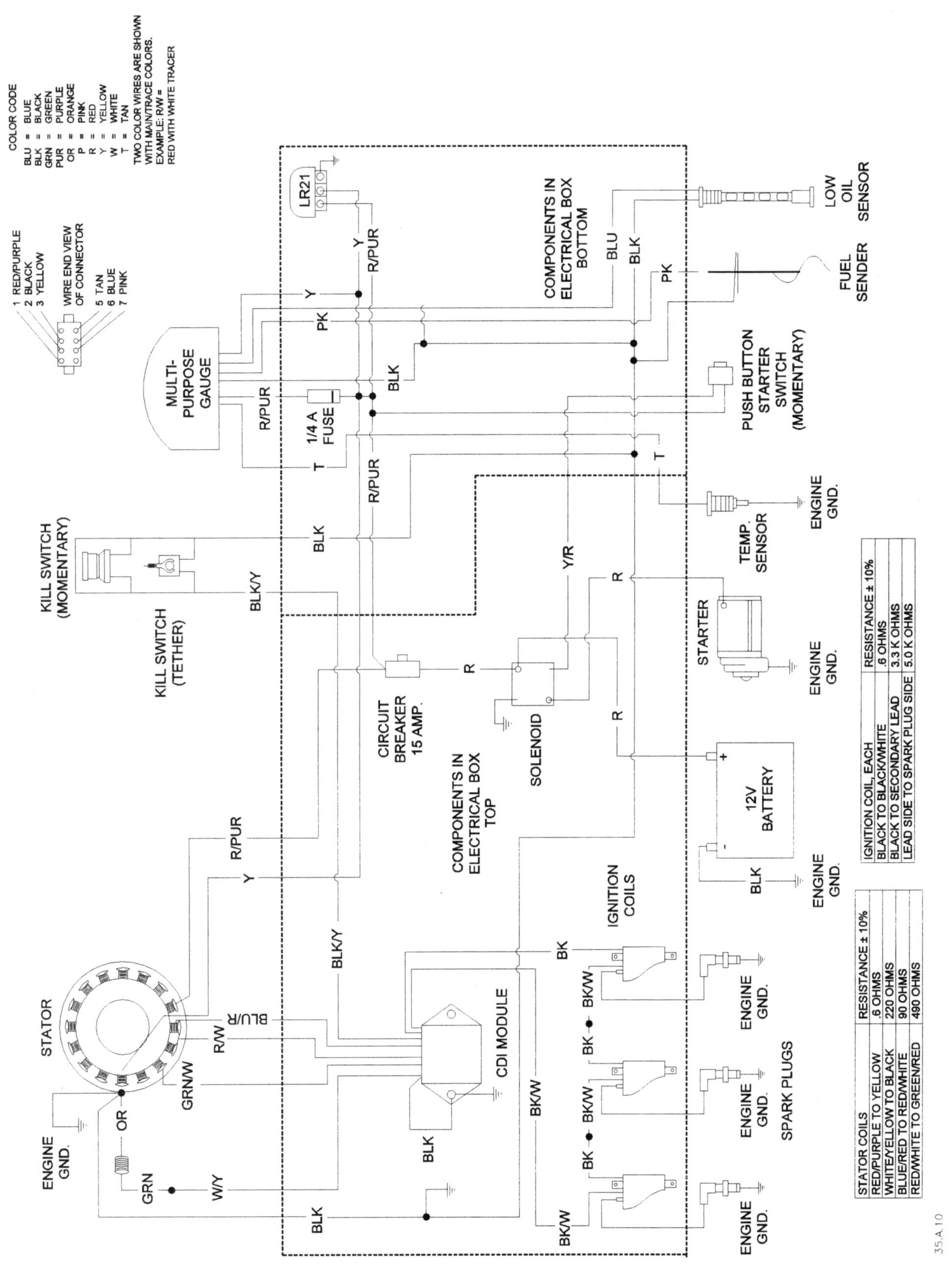

IGNITION COIL, EACH	RESISTANCE ± 10%
BLACK TO BLACK/WHITE	.6 OHMS
BLACK TO SECONDARY LEAD	3.3 K OHMS
LEAD SIDE TO SPARK PLUG SIDE	5.0 K OHMS

STATOR COILS	RESISTANCE ± 10%
RED/PURPLE TO YELLOW	.6 OHMS
WHITE/YELLOW TO BLACK	220 OHMS
BLUE/RED TO RED/WHITE	90 OHMS
RED/WHITE TO GREEN/RED	490 OHMS

***SLT750 -- 1995** Wiring diagram and color code identification. Major items are identified. Components within the dashed border are housed inside the electrical box.*

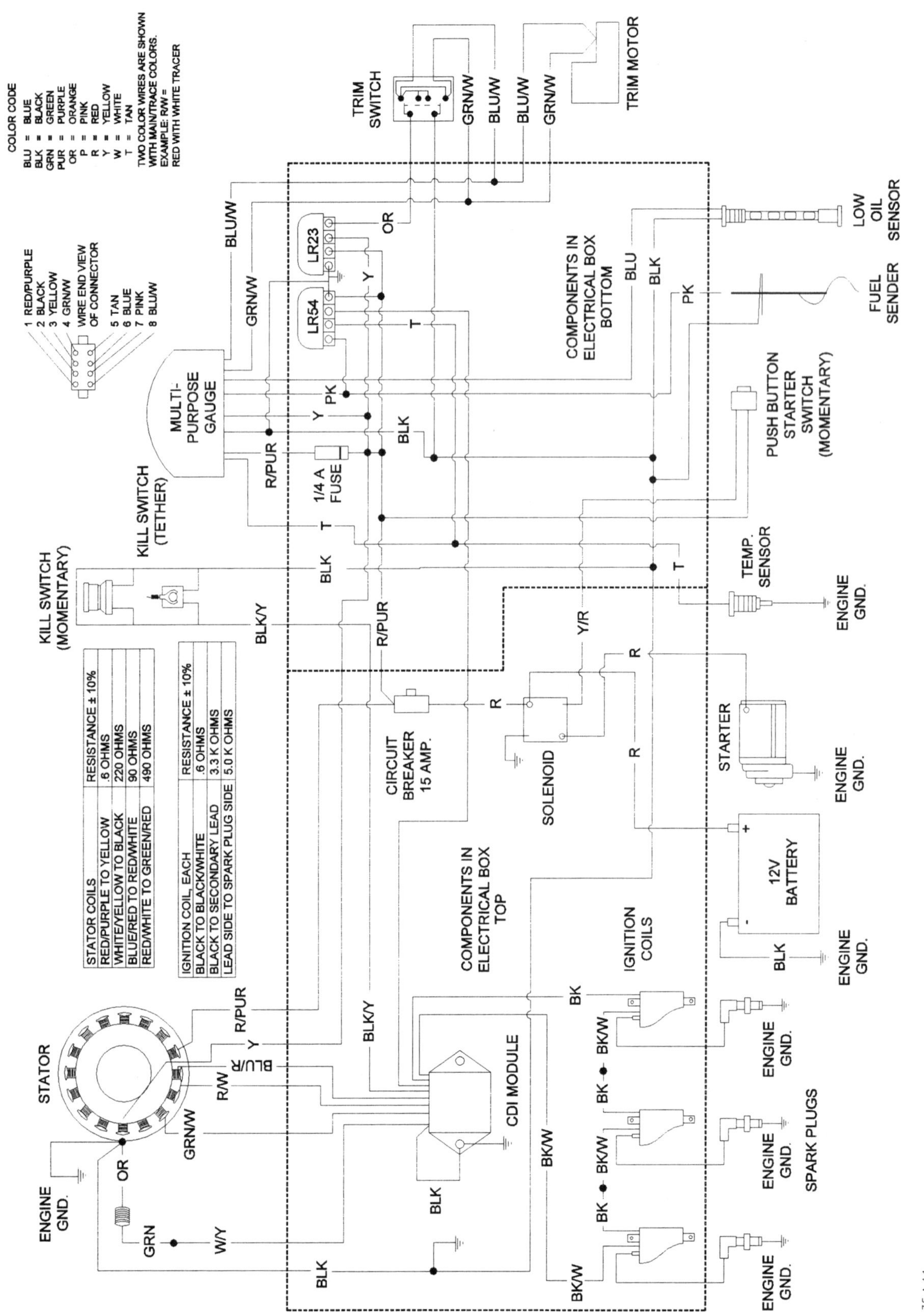

STATOR COILS	RESISTANCE ± 10%
RED/PURPLE TO YELLOW	.6 OHMS
WHITE/YELLOW TO BLACK	220 OHMS
BLUE/RED TO RED/WHITE	90 OHMS
RED/WHITE TO GREEN/RED	490 OHMS

IGNITION COIL, EACH	RESISTANCE ± 10%
BLACK TO BLACK/WHITE	.6 OHMS
BLACK TO SECONDARY LEAD	3.3 K OHMS
LEAD SIDE TO SPARK PLUG SIDE	5.0 K OHMS

SLX780 -- 1995 *Wiring diagram and color code identification. Major items are identified. Components within the dashed border are housed inside the electrical box.*

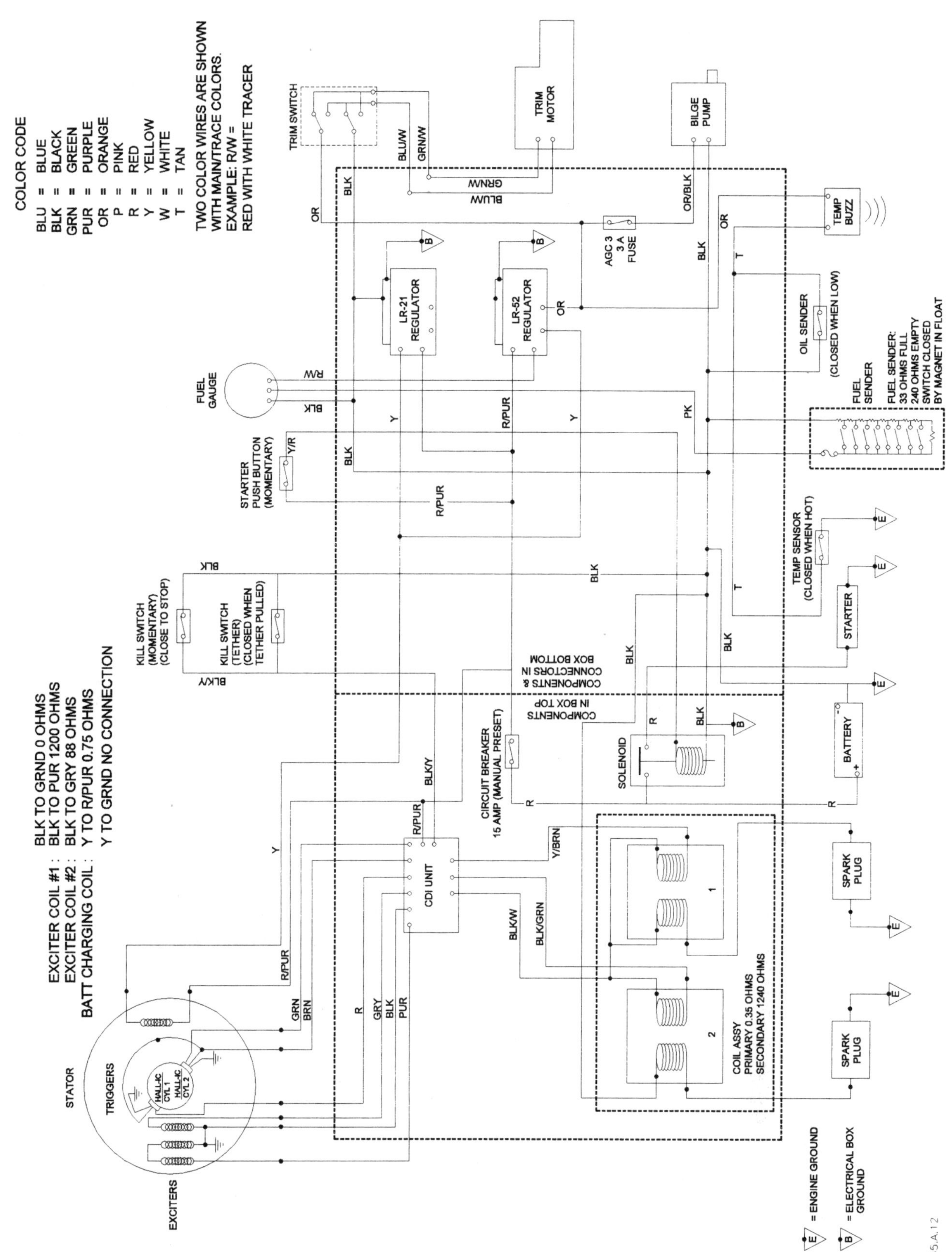

***SL700 -- 1996** Wiring diagram and color code identification. Major items are identified. Components within the dashed border are housed inside the electrical box.*

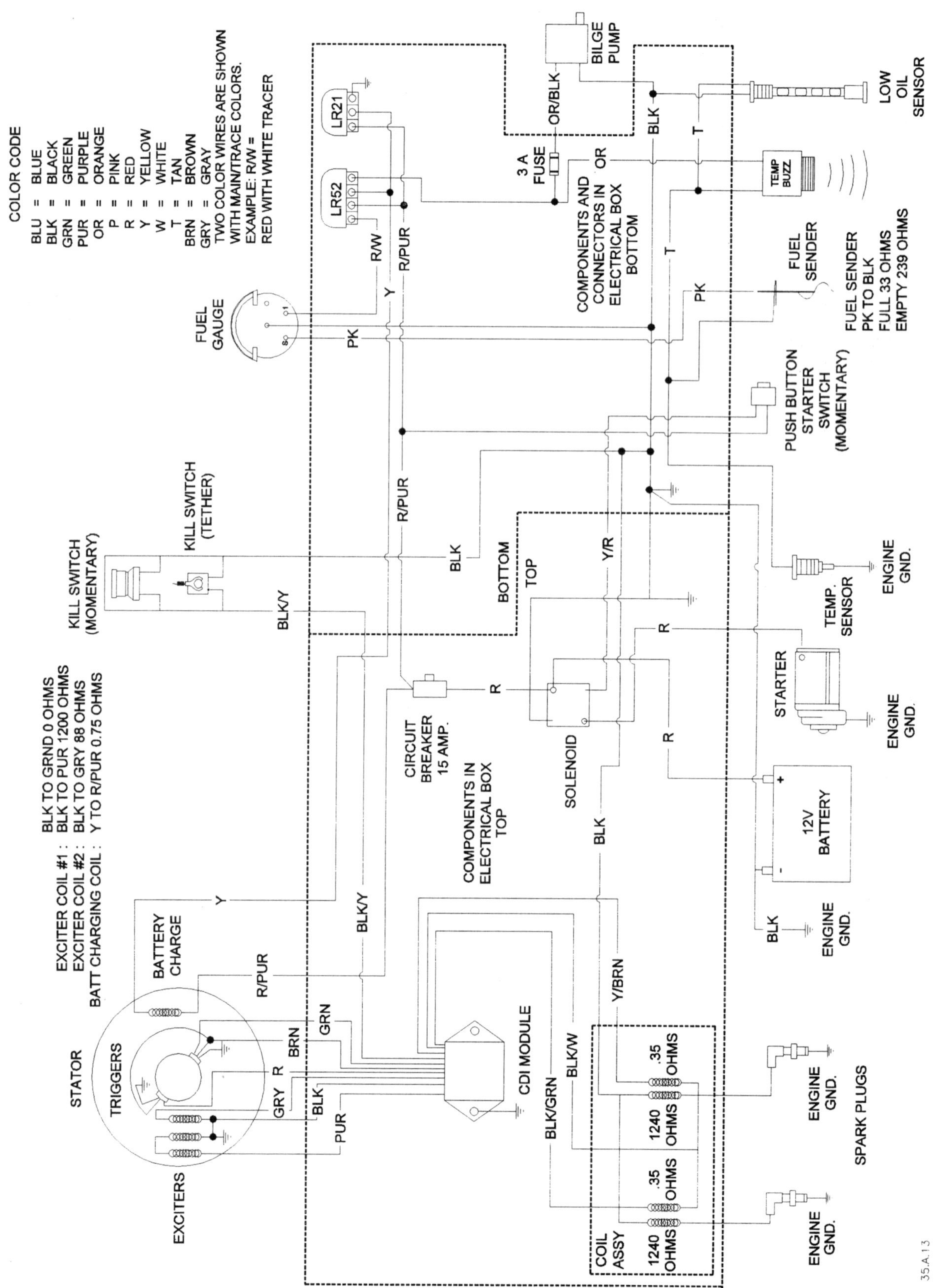

SLT700 -- 1996 *Wiring diagram and color code identification. Major items are identified. Components within the dashed border are housed inside the electrical box.*

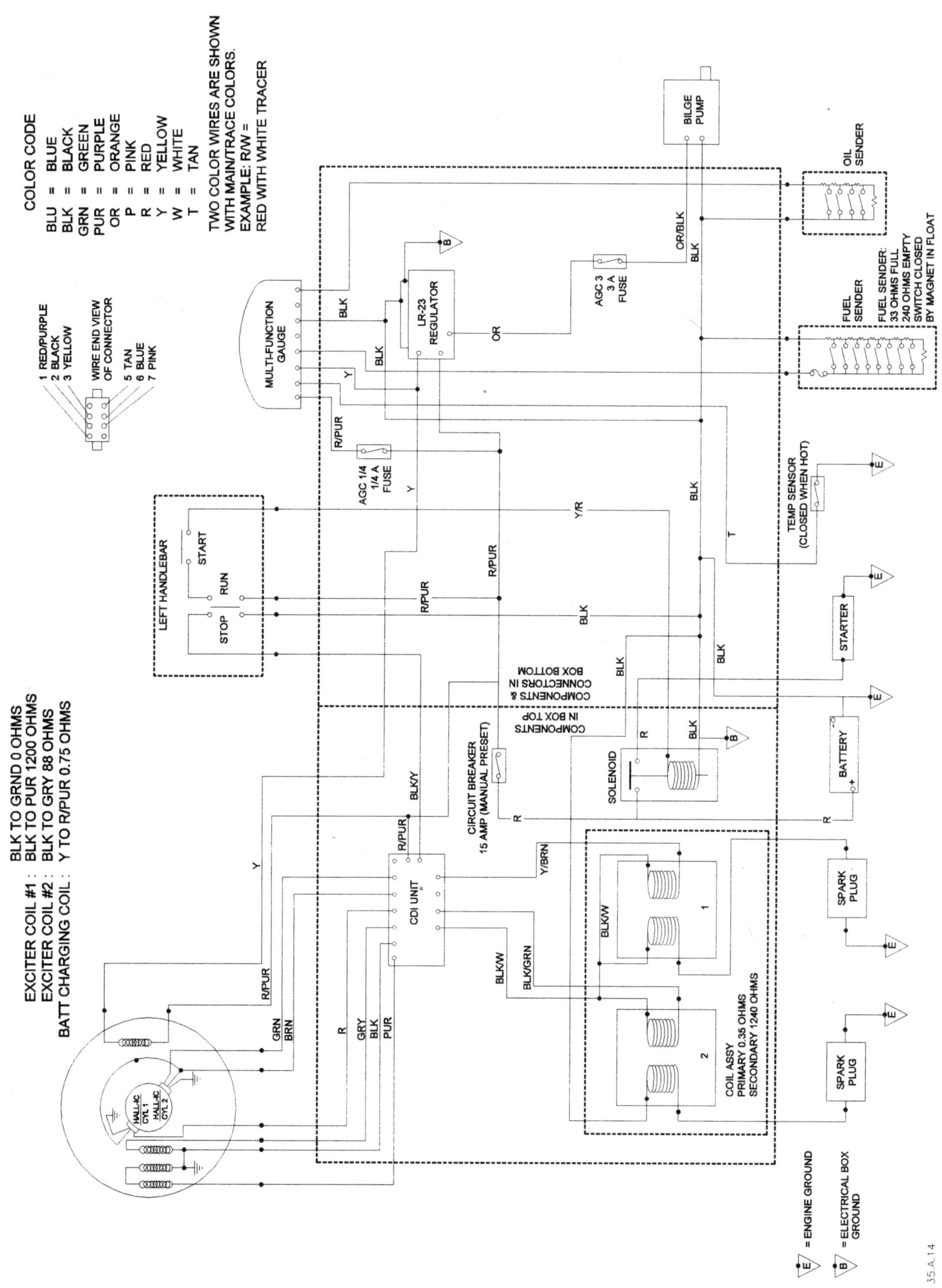

***Hurricane* -- 1996** *Wiring diagram and color code identification. Major items are identified. Components within the dashed border are housed inside the electrical box.*

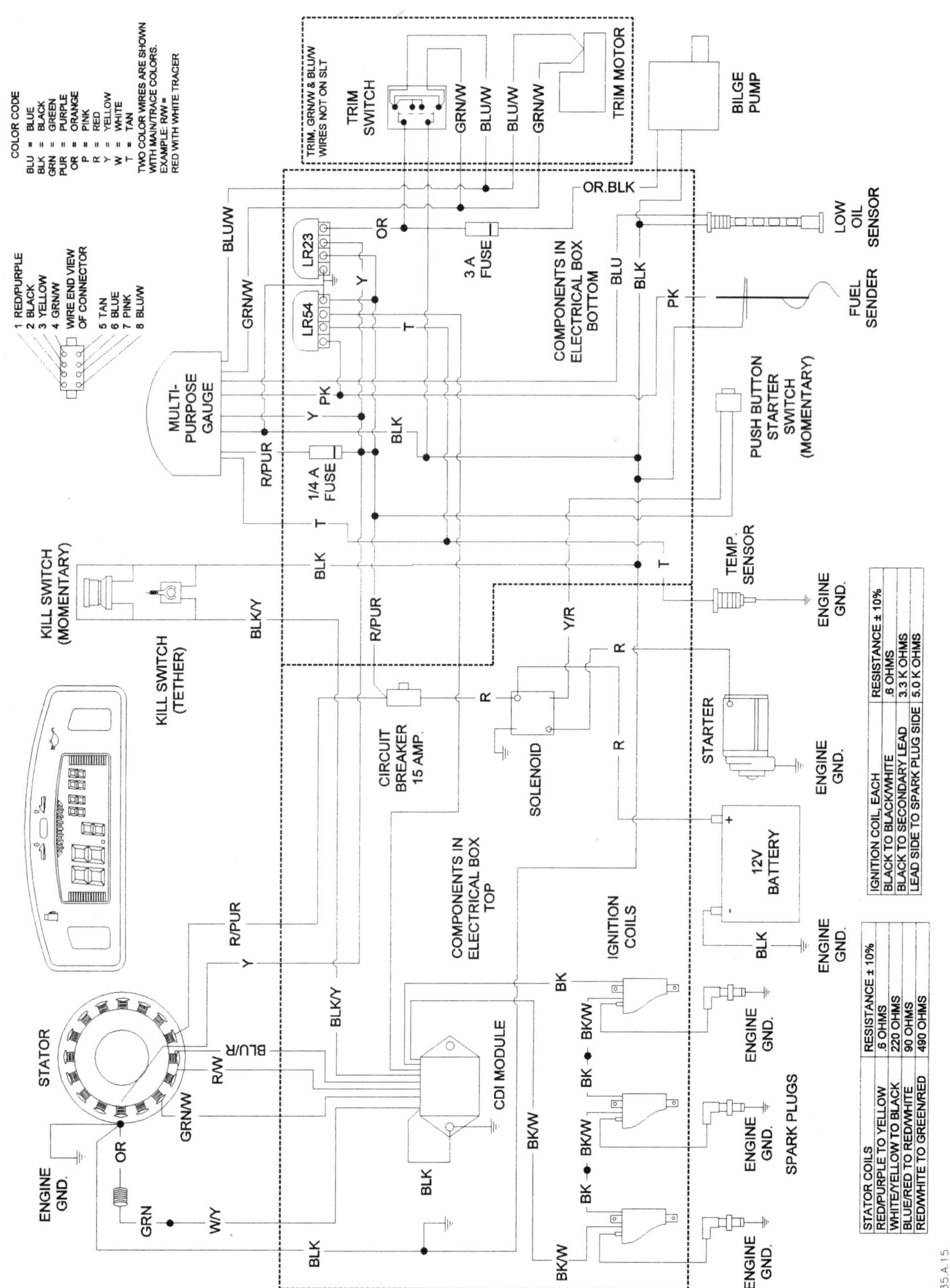

IGNITION COIL, EACH	RESISTANCE ± 10%
BLACK TO BLACK/WHITE	.6 OHMS
BLACK TO SECONDARY LEAD	3.3 K OHMS
LEAD SIDE TO SPARK PLUG SIDE	5.0 K OHMS

STATOR COILS	RESISTANCE ± 10%
RED/PURPLE TO YELLOW	.6 OHMS
WHITE/YELLOW TO BLACK	220 OHMS
BLUE/RED TO RED/WHITE	90 OHMS
RED/WHITE TO GREEN/RED	490 OHMS

SL780, SLT780, SLX780 -- 1996. *Wiring diagram and color code identification. Major items are identified. Components within the dashed border are housed inside the electrical box.*

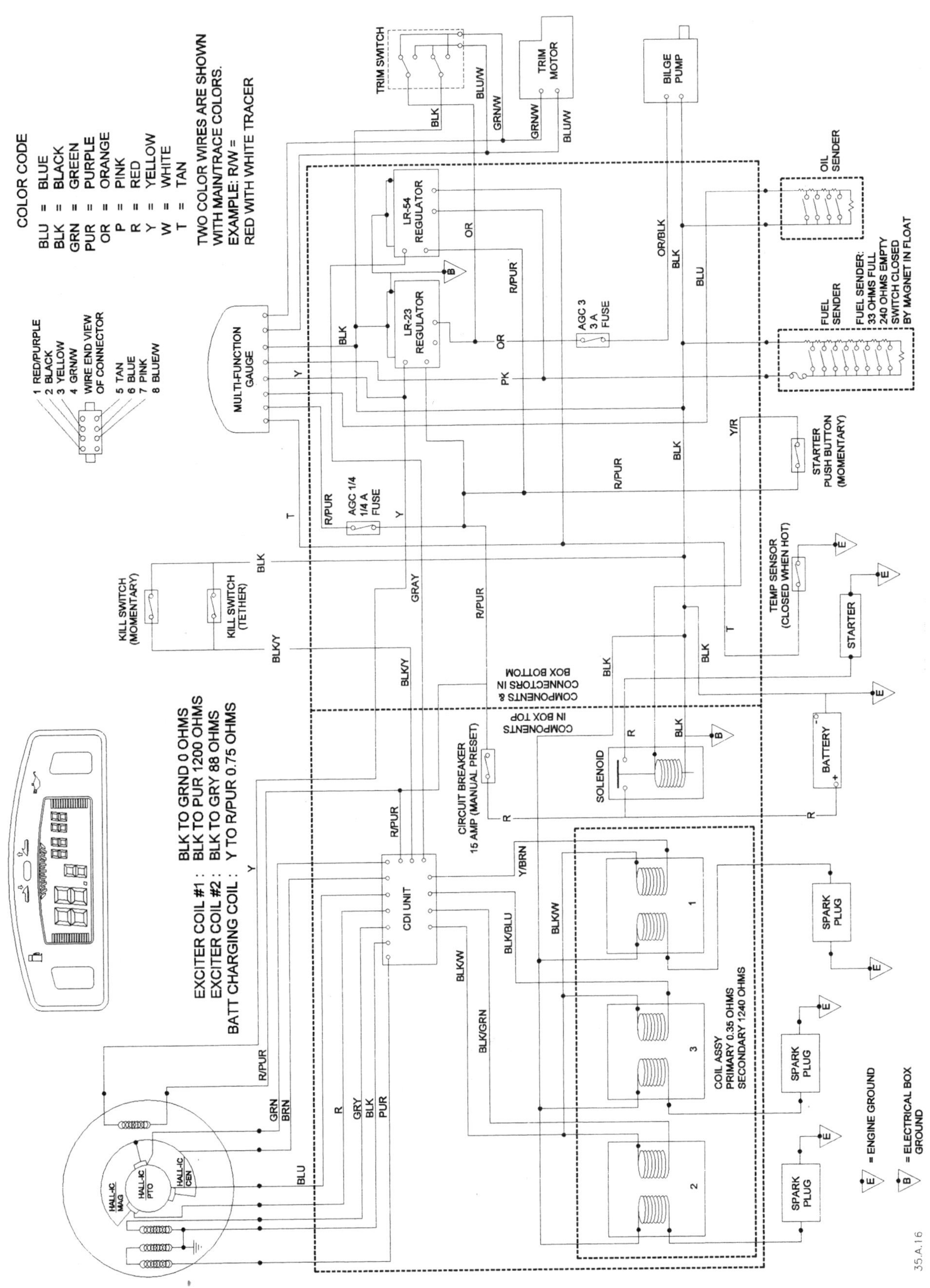

SL900 -- 1996 *Wiring diagram and color code identification. Major items are identified. Components within the dashed border are housed inside the electrical box.*

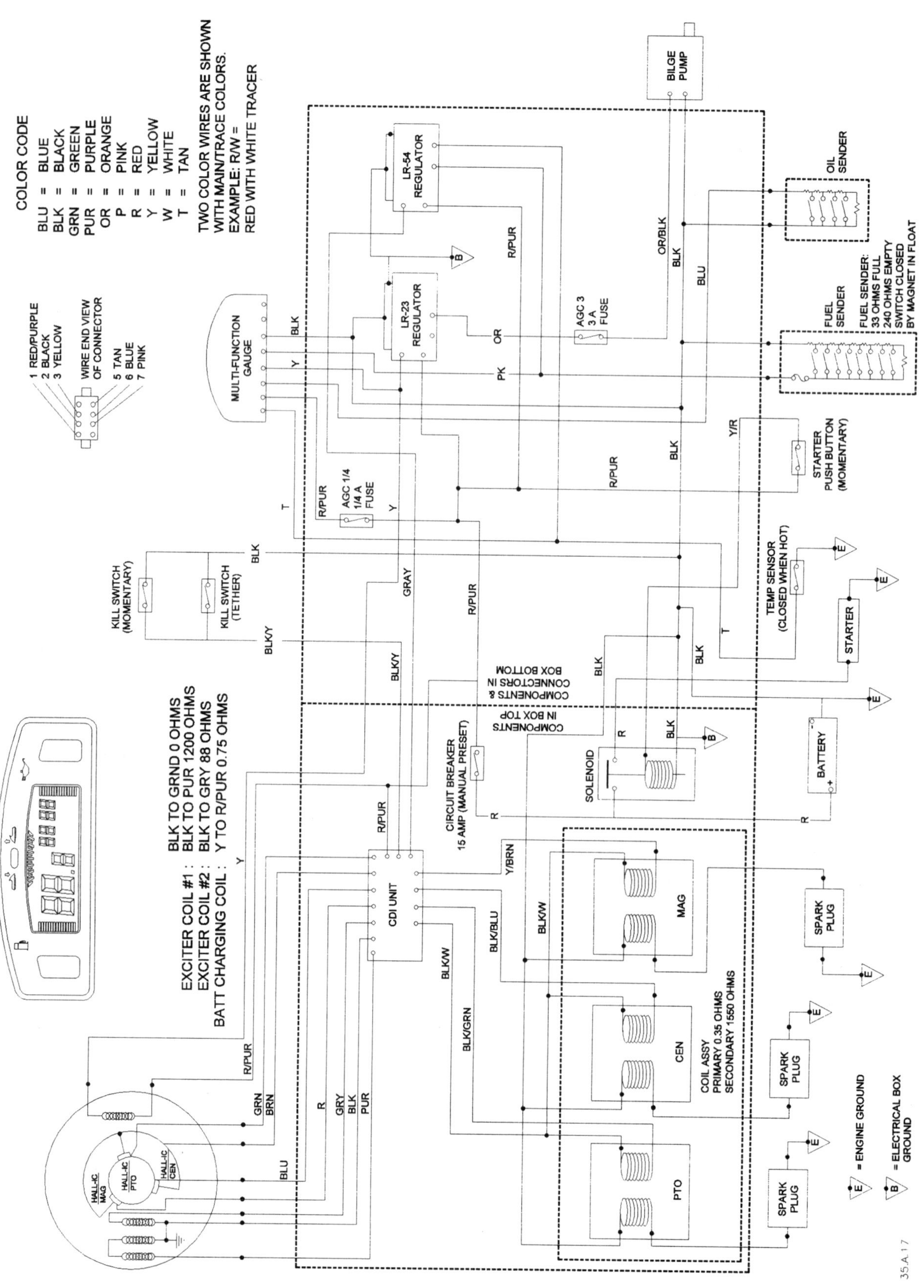

SLTX -- 1996 *Wiring diagram and color code identification. Major items are identified. Components within the dashed border are housed inside the electrical box.*

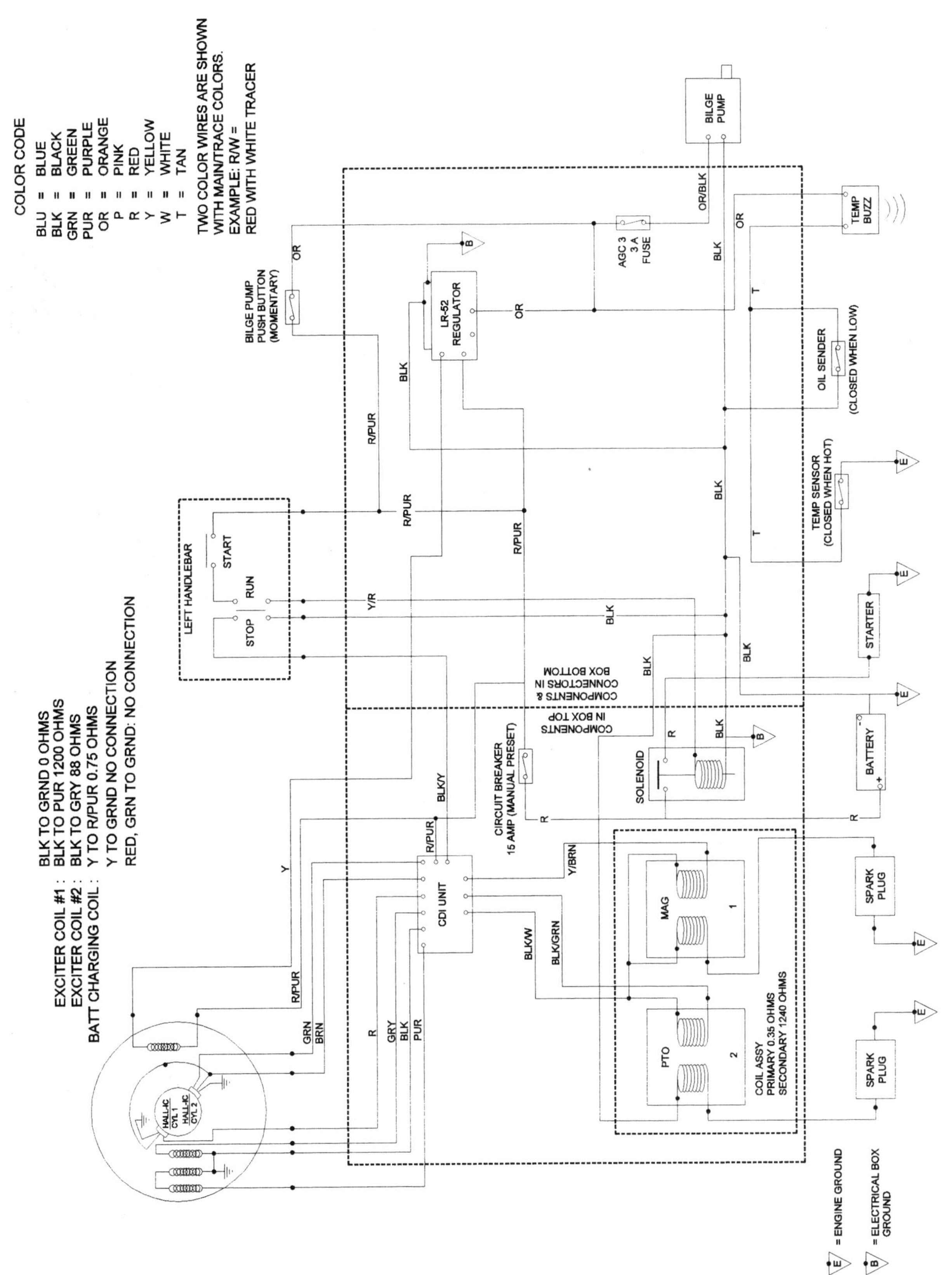

SL700 -- 1997 *Wiring diagram and color code identification. Major items are identified. Components within the dashed border are housed inside the electrical box.*

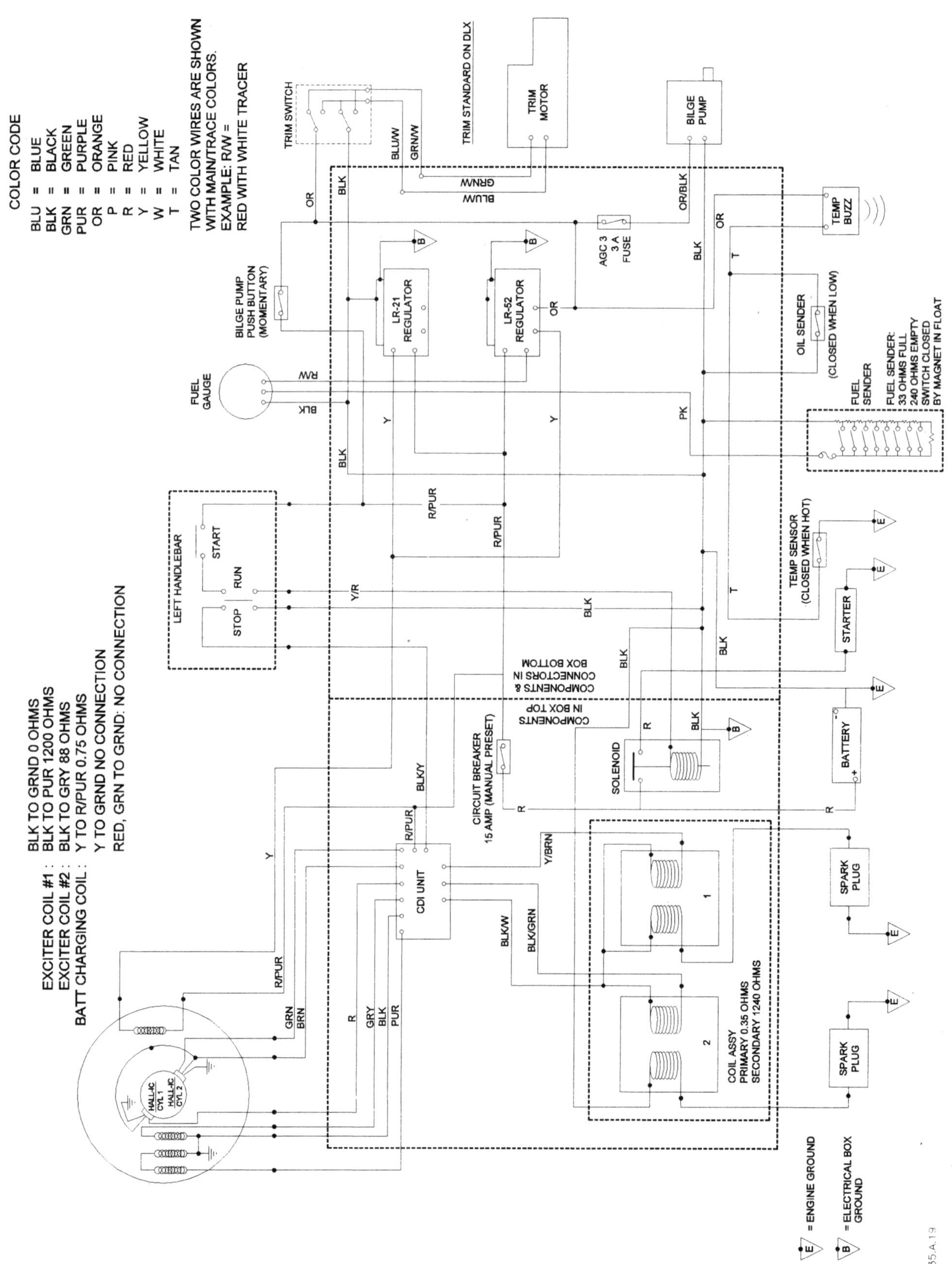

SLT700, SL700 DELUXE *-- Wiring diagram and color code identification. Major items are identified. Components within the dashed border are housed inside the electrical box.*

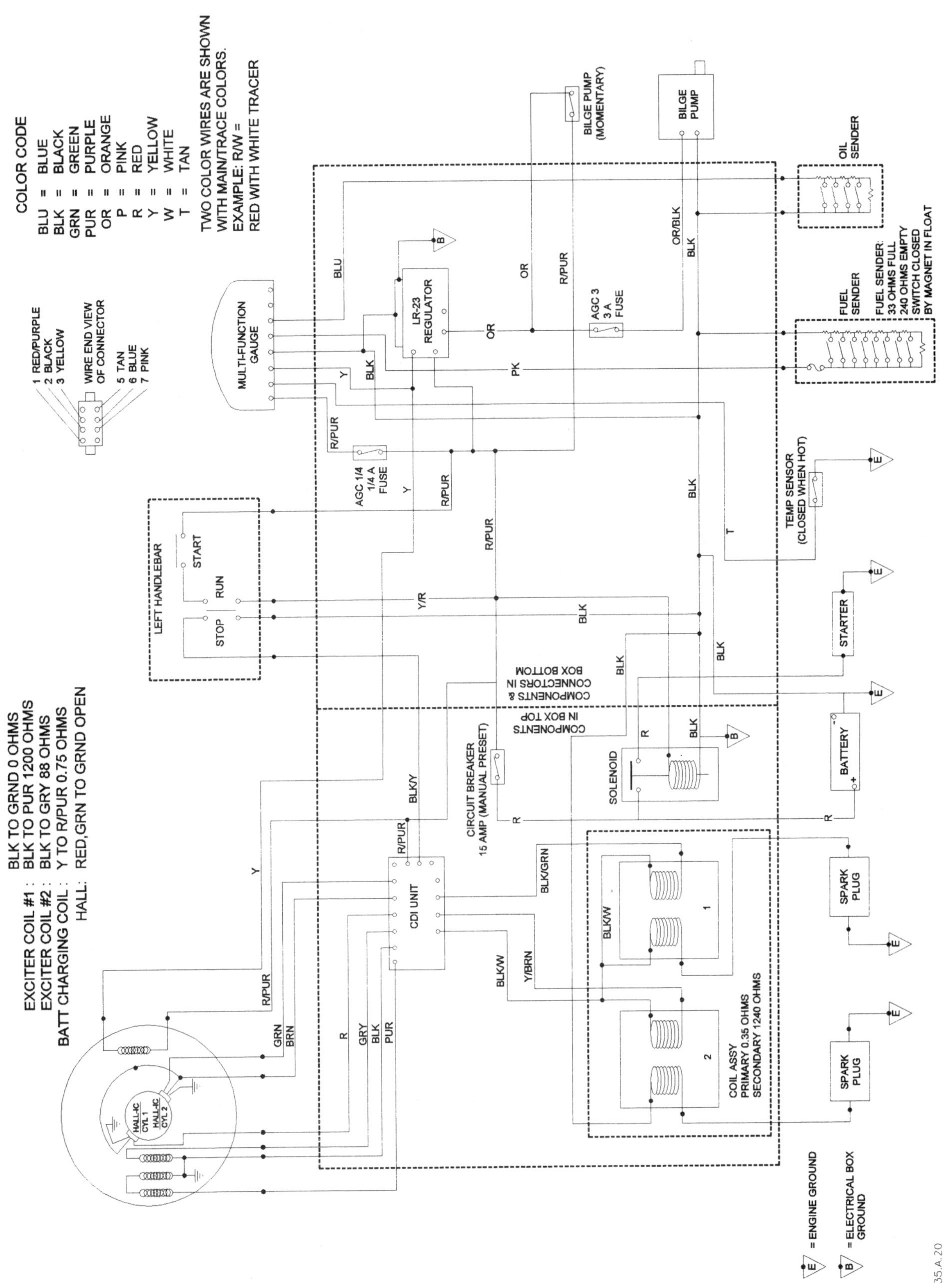

HURRICANE -- 1997 *Wiring diagram and color code identification. Major items are identified. Components within the dashed border are housed inside the electrical box.*

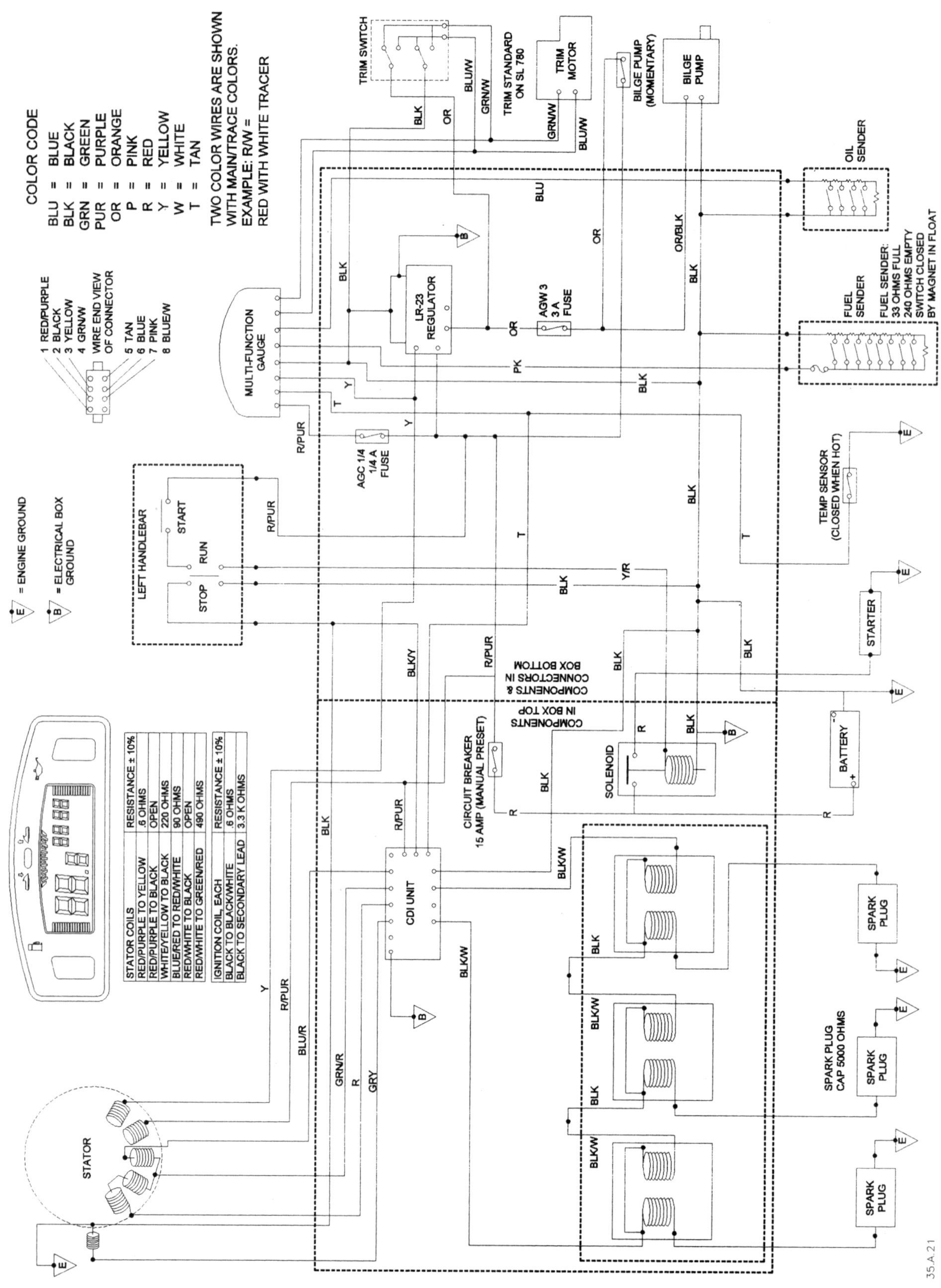

STATOR COILS	RESISTANCE ± 10%
RED/PURPLE TO YELLOW	.6 OHMS
RED/PURPLE TO BLACK	OPEN
WHITE/YELLOW TO BLACK	220 OHMS
BLUE/RED TO RED/WHITE	90 OHMS
RED/WHITE TO BLACK	OPEN
RED/WHITE TO GREEN/RED	490 OHMS

IGNITION COIL, EACH	RESISTANCE ± 10%
BLACK TO BLACK/WHITE	.6 OHMS
BLACK TO SECONDARY LEAD	3.3 K OHMS

SL780, SLT 780 -- 1997 *Wiring diagram and color code identification. Major items are identified. Components within the dashed border are housed inside the electrical box.*

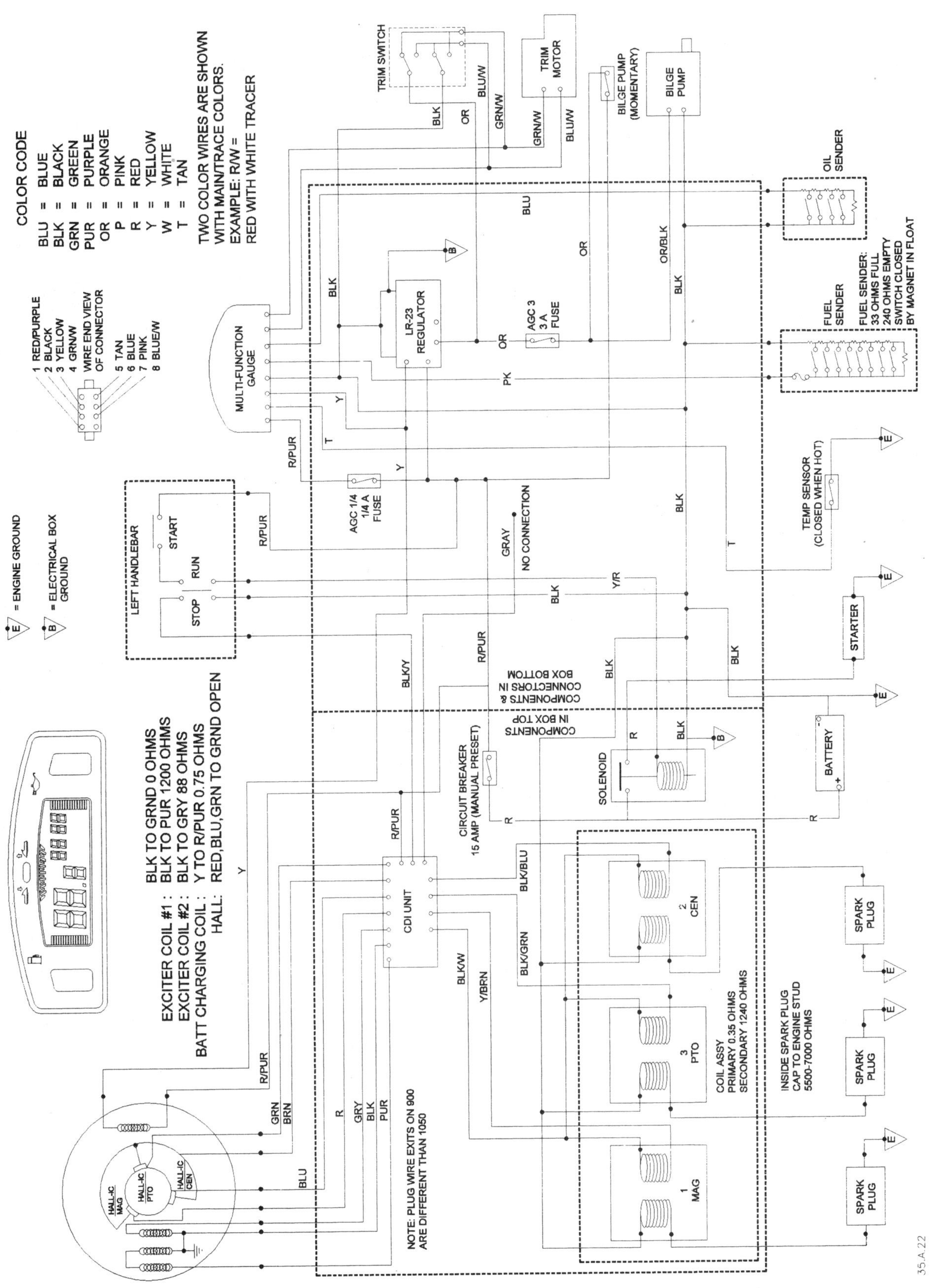

SL900 -- 1997 *Wiring diagram and color code identification. Major items are identified. Components within the dashed border are housed inside the electrical box.*

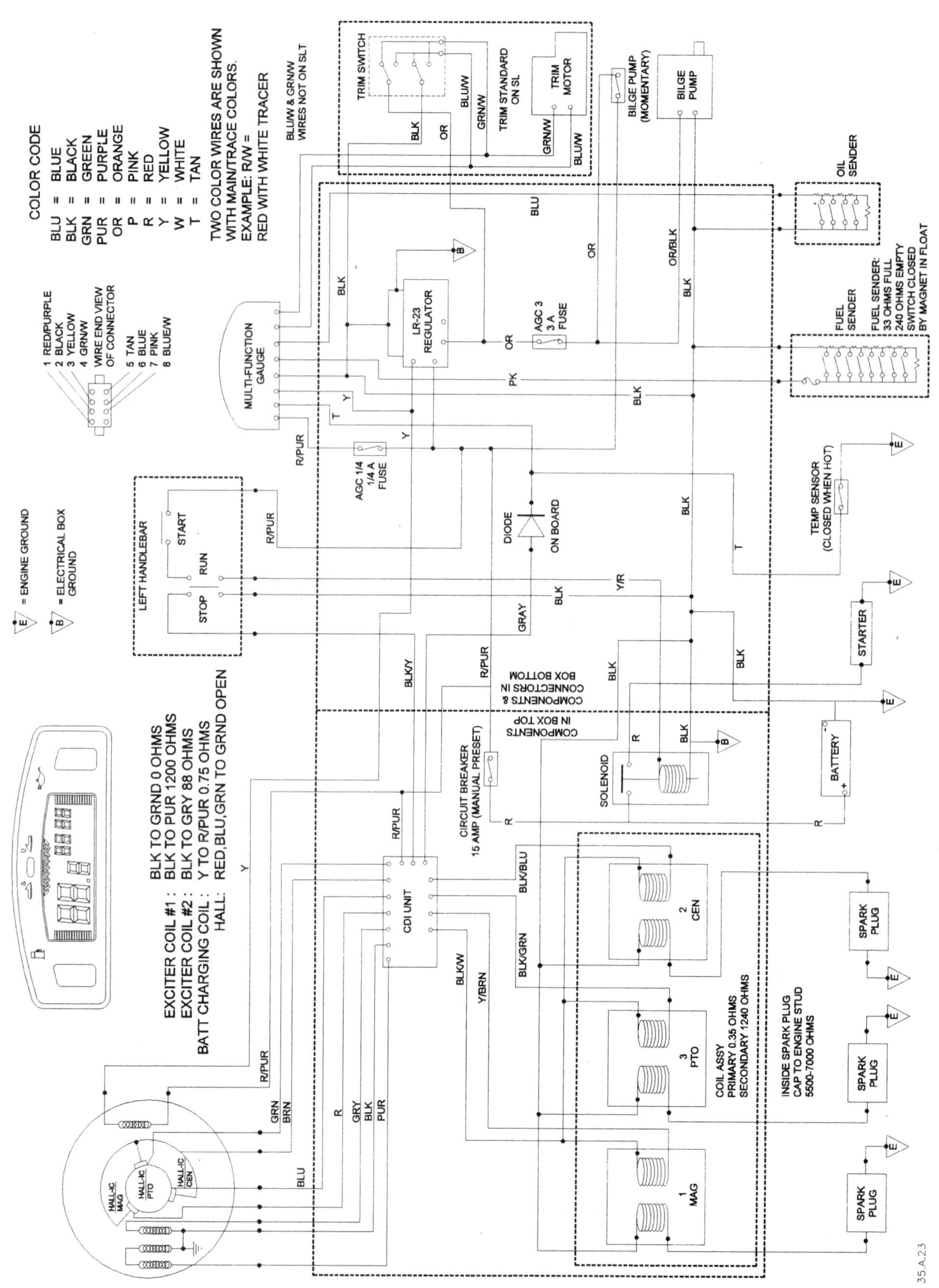

***SL1050, SLTX -- 1997** Wiring diagram and color code identification. Major items are identified. Components within the dashed border are housed inside the electrical box.*

NOTES & NUMBERS

NOTES & NUMBERS